Springer Complexity

Springer Complexity is an interdisciplinary program publishing the best research and academic-level teaching on both fundamental and applied aspects of complex systems – cutting across all traditional disciplines of the natural and life sciences, engineering, economics, medicine, neuroscience, social and computer science.

Complex Systems are systems that comprise many interacting parts with the ability to generate a new quality of macroscopic collective behavior the manifestations of which are the spontaneous formation of distinctive temporal, spatial or functional structures. Models of such systems can be successfully mapped onto quite diverse "real-life" situations like the climate, the coherent emission of light from lasers, chemical reaction-diffusion systems, biological cellular networks, the dynamics of stock markets and of the internet, earthquake statistics and prediction, freeway traffic, the human brain, or the formation of opinions in social systems, to name just some of the popular applications.

Although their scope and methodologies overlap somewhat, one can distinguish the following main concepts and tools: self-organization, nonlinear dynamics, synergetics, turbulence, dynamical systems, catastrophes, instabilities, stochastic processes, chaos, graphs and networks, cellular automata, adaptive systems, genetic algorithms and computational intelligence.

The two major book publication platforms of the Springer Complexity program are the monograph series "Understanding Complex Systems" focusing on the various applications of complexity, and the "Springer Series in Synergetics", which is devoted to the quantitative theoretical and methodological foundations. In addition to the books in these two core series, the program also incorporates individual titles ranging from textbooks to major reference works.

Editorial and Programme Advisory Board

Understanding Complex Systems

Founding Editor: J.A. Scott Kelso

Future scientific and technological developments in many fields will necessarily depend upon coming to grips with complex systems. Such systems are complex in both their composition – typically many different kinds of components interacting simultaneously and nonlinearly with each other and their environments on multiple levels – and in the rich diversity of behavior of which they are capable.

The Springer Series in Understanding Complex Systems series (UCS) promotes new strategies and paradigms for understanding and realizing applications of complex systems research in a wide variety of fields and endeavors. UCS is explicitly transdisciplinary. It has three main goals: First, to elaborate the concepts, methods and tools of complex systems at all levels of description and in all scientific fields, especially newly emerging areas within the life, social, behavioral, economic, neuro- and cognitive sciences (and derivatives thereof); second, to encourage novel applications of these ideas in various fields of engineering and computation such as robotics, nano-technology and informatics; third, to provide a single forum within which commonalities and differences in the workings of complex systems may be discerned, hence leading to deeper insight and understanding.

UCS will publish monographs, lecture notes and selected edited contributions aimed at communicating new findings to a large multidisciplinary audience.

Applications of Nonlinear Dynamics

Model and Design of Complex Systems

Edited by

Visarath In
SPAWAR, San Diego

Patrick Longhini
SPAWAR, San Diego

and

Antonio Palacios
SDSU

Springer

Visarath In
Space and Naval Warfare Systems Center
Code 2373
53560 Hull Street
San Diego, CA 92152-5001
USA
Visarath@spawar.navy.mil
Visarath@euler.sdsu.edu

Patrick Longhini
Space and Naval Warfare Systems Center
Code 2373
53560 Hull Street
San Diego, CA 92152-5001
USA

Antonio Palacios
Department of Mathematics & Statistics
San Diego State University
San Diego, CA 92182-7720
USA
palacios@euler.sdsu.edu

ISBN: 978-3-540-85631-3 e-ISBN: 978-3-540-85632-0

DOI 10.1007/978-3-540-85632-0

Understanding Complex Systems ISSN: 1860-0832

Library of Congress Control Number: 2008936465

Cover design: WMXDesign GmbH

Printed on acid-free paper

9 8 7 6 5 4 3 2 1

springer.com

Organizers

Bruno Ando, University of Catania
Adi Bulsara, SPAWAR, San Diego
Salvatore Baglio, University of Catania
Visarath In, SPAWAR, San Diego
Ljupco Kocarev, University of California, San Diego
Patrick Longhini, SPAWAR, San Diego
Joseph Neff, SPAWAR, San Diego
Antonio Palacios, San Diego State University
Toshimichi Saito, Hosei University
Michael F. Shlesinger, Office of Naval Research
Hiroyuki Torikai, Hosei University

SPONSOR:
Office of Naval Research (ONR)
875 N. Randolph Street, Suite 1475
Arlington, VA 22217

Preface

The field of applied nonlinear dynamics has attracted scientists and engineers across many different disciplines to develop innovative ideas and methods to study complex behavior exhibited by relatively simple systems. Examples include: population dynamics, fluidization processes, applied optics, stochastic resonance, flocking and flight formations, lasers, and mechanical and electrical oscillators. A common theme among these and many other examples is the underlying universal laws of nonlinear science that govern the behavior, in space and time, of a given system. These laws are universal in the sense that they transcend the model-specific features of a system and so they can be readily applied to explain and predict the behavior of a wide ranging phenomena, natural and artificial ones. Thus the emphasis in the past decades has been in explaining nonlinear phenomena with significantly less attention paid to exploiting the rich behavior of nonlinear systems to design and fabricate new *devices* that can operate more efficiently.

Recently, there has been a series of meetings on topics such as Experimental Chaos, Neural Coding, and Stochastic Resonance, which have brought together many researchers in the field of nonlinear dynamics to discuss, mainly, theoretical ideas that may have the potential for further implementation. In contrast, the goal of the 2007 ICAND (International Conference on Applied Nonlinear Dynamics) was focused more sharply on the implementation of theoretical ideas into actual devices and systems. Thus the meeting brought together scientists and engineers from all over the globe to exchange research ideas and methods that can bridge the gap between the fundamental principles of nonlinear science and the actual development of new technologies. Examples of some of these new and emerging technologies include: (magnetic and electric field) sensors, reconfigurable electronic circuits, nanomechanical oscillators, chaos-based computer chips, nonlinear nano-detectors, nonlinear signal processing and filters, and signal coding.

The 2007 ICAND meeting was held in Hawaii, at Poipu Beach, Kauai on September 24–27, 2007. The waters off Poipu Beach are crystal clear and provided a truly beautiful atmosphere to hold a meeting of this kind. The invited speakers at this seminal meeting on applied nonlinear dynamics were drawn from a rarefied mix. They included a few well-established researchers in the field of nonlinear dynamics as well as a "new breed" of pioneers (applied physicists, applied mathematicians, engineers, and biologists) who are attempting to apply these ideas in laboratory and, in some cases, industrial applications. The discussions in the meeting cover broad topics ranging from the effects of noise on dynamical systems to symmetry mathematics in the analyses of coupled nonlinear systems to microcircuit designs in implementation of these nonlinear systems. The meeting also featured, as already stated, some novel theoretical ideas that have not yet made it to the drawing board, but show great promise for the future. The organizers also attempted to give some exposure to much younger researchers, such as advanced graduate students and postdocs, in the form of posters. The meeting set aside singificant amount of time and provided many opportunities outside of presentation setting to promote the discussions and foster collaborations amongs the participants.

The organizers extend their sincerest thanks to the principal sponsors of the meeting: Office of Naval Research (Washington, DC), Office of Naval Research-Global (London), San Diego State University (College of Sciences), and SPAWAR Systems Center San Diego. In particular, we wish to acknowledge Dr. Michael Shlesinger from the Office of Naval Research (Washington DC) for his support and encouragement. In addition, we extend our grateful thanks, in specific, to Professor Antonio Palacios and Dan Reifer at SDSU for their hardwork in making the financial transactions as smoothly as possible despite many obstacles thrown in their way. We also want to thank our colleagues who chaired the sessions and to the numerous individuals who donated long hours of labor to the success of this meeting. Finally, we thank Spinger-Verlag for their production of an elegant proceedings.

San Diego, USA
May 2008

V. In
P. Longhini
A. Palacios

Contents

Invited Speakers

Construction of a Chaotic Computer Chip

William L. Ditto, K. Murali and Sudeshna Sinha

Chaotic systems are great pattern generators and their defining feature, sensitivity to initial conditions, allows them to switch between patterns exponentially fast. We exploit such pattern generation by "tuning" representative continuous and discrete chaotic systems to generate all logic gate functions. We then exploit exponential sensitivity to initial conditions to achieve rapid switching between all the logic gates generated by each representative chaotic element. With this as a starting point we will present our progress on the construction of a chaotic computer chip consisting of large numbers of individual chaotic elements that can be individually and rapidly morphed to become all logic gates. Such a chip of arrays of morphing chaotic logic gates can then be programmed to perform higher order functions (such as memory, arithmetic logic, input/output operations, ...) and to rapidly switch between such functions. Thus we hope that our reconfigurable chaotic computer chips will enable us to achieve the flexibility of field programmable gate arrays (FPGA), the optimization and speed of application specific integrated circuits (ASIC) and the general utility of a central processing unit (CPU) within the same computer chip architecture. Results on the construction and commercialization of the ChaoLogix™ chaotic computer chip will also be presented to demonstrate progress being made towards the commercialization of this technology (http://www.chaologix.com).

1 Introduction

It was proposed in 1998 that chaotic systems may be utilized to design computing devices [1]. In the early years the focus was on proof-of-principle schemes that demonstrated the capability of chaotic elements to do universal computing. The

W.L. Ditto (✉)
J. Crayton Pruitt Family Department of Biomedical Engineering, University of Florida, Gainesville, FL 32611-6131, USA; ChaoLogix, Inc. 101 S.E. 2nd Place, Suite 201 A, Gainesville, FL 32601, USA, e-mail: william.ditto@bme.ufl.edu

V. In et al. (eds.), *Applications of Nonlinear Dynamics*, Understanding Complex Systems, DOI 10.1007/978-3-540-85632-0_1, © Springer-Verlag Berlin Heidelberg 2009

distinctive feature of this alternate computing paradigm was that they exploited the sensitivity and pattern formation features of chaotic systems.

In subsequent years, it was realized that one of the most promising direction of this computing paradigm is its ability to exploit a single chaotic element to reconfigure into different logic gates through a threshold based morphing mechanism [2–5]. In contrast to a conventional field programmable gate array element, where reconfiguration is achieved through switching between multiple single purpose gates, reconfigurable chaotic logic gates (RCLGs) are comprised of chaotic elements that morph (or reconfigure) logic gates through the control of the pattern inherent in their nonlinear element. Two input RCLGs have recently been realized and shown to be capable of reconfiguring between all logic gates in discrete circuits [6–9]. Additionally such RCLGs have been realized in prototype VLSI circuits (0.13 μm CMOS, 30 Mhz clock cycles) that employ two input reconfigurable chaotic logic gates arrays (RCGA) to morph between higher order functions such as those found in a typical arithmetic logic unit (ALU) [10].

In this article we first recall the theoretical scheme for flexible implementation of all these fundamental logical operations utilizing low dimensional chaos [2, 3], and the specific realisation of the theory in a discrete-time and a continuous-time chaotic circuit. Then we will present new results on the design of reconfigurable multiple input gates. Note that multiple input logic gates are preferred mainly for reasons of space in circuits and also many combinational and sequential logic operations can be realized with these logic gates, in which one can minimize the propagation delay. Such a multiple input CGA would make RCLGs more power efficient, increase their performance and widen their range of applications. Here we specifically demonstrate a three input RCLG by implementing representative fundamental NOR and NAND gates with a continuous-time chaotic system.

2 Concept

In order to use the rich temporal patterns embedded in a nonlinear time series efficiently one needs a mechanism to extract different responses from the system, in a controlled manner, without much run-time effort. Here we employ a threshold based scheme to achieve this [11–13].

Consider the discrete-time chaotic map, with its state represented by a variable x, as our *chaotic chip* or *chaotic processor*. In our scheme all the basic logic gate operations (AND, OR, XOR, NAND, NOR, NOT) involve the following simple steps:

1. Inputs:
   $x \to x_0 + I_1 + I_2$ for 2-input gates such as the AND, OR, XOR, NAND and NOR operations, and
   $x \to x_0 + I$ for the 1-input gate such as the NOT operation.

Here x_0 is the initial state of the system, and the input value $I = 0$ when logic input is 0 and $I = V_{in}$ when logic input is 1 (where V_{in} is a positive constant).

2. Dynamical update, i.e. $x \to f(x)$

   where $f(x)$ is a strongly nonlinear function.

3. Threshold mechanism to obtain output V_0 :

   $V_0 = 0$ if $f(x) \leq E$, and

   $V_0 = f(x) - E$ if $f(x) > E$

   where E is the threshold.

   This is interpretated as logic output 0 if $V_0 = 0$ and Logic Ouput 1 if $V_0 \sim V_{in}$.

Since the system is chaotic, in order to specify the inital x_0 accurately one needs a controlling mechanism. Here we will employ a threshold controller to set the inital x_0. So in this example we use the clipping action of the threshold controller to achieve the initialization, and subsequently to obtain the output as well.

Note that in our implementation we demand that the input and output have equivalent definitions (i.e. 1 unit is the same quantity for input and output), as well as among various logical operations. This requires that constant V_{in} assumes the same value throughout a network, and this will allow the output of one gate element to

Table 1 Necessary and sufficient conditions, derived from the logic truth tables, to be satisfied simultaneously by the nonlinear dynamical element, in order to have the capacity to implement the logical operations AND, OR, XOR, NAND, NOR and NOT with the same computing module

Logic Operation	Input Set (I_1, I_2)	Output	Necessary and Sufficient Condition
AND	(0, 0)	0	$f(x_0) < E$
	(0,1)/(1,0)	0	$f(x_0 + V_{in}) < E$
	(1,1)	1	$f(x_0 + 2V_{in}) - E = V_{in}$
OR	(0, 0)	0	$f(x_0) < E$
	(0,1)/(1,0)	1	$f(x_0 + V_{in}) - E = V_{in}$
	(1,1)	1	$f(x_0 + 2V_{in}) - E = V_{in}$
XOR	(0, 0)	0	$f(x_0) < E$
	(0,1)/(1,0)	1	$f(x_0 + V_{in}) - E = V_{in}$
	(1,1)	0	$f(x_0 + 2V_{in}) < E$
NOR	(0, 0)	1	$f(x_0) - E = V_{in}$
	(0,1)/(1,0)	0	$f(x_0 + V_{in}) < E$
	(1,1)	0	$f(x_0 + 2V_{in}) < E$
NAND	(0, 0)	1	$f(x_0) - E = V_{in}$
	(0,1)/(1,0)	1	$f(x_0 + V_{in}) - E = v_{in}$
	(1,1)	0	$f(x_0 + 2V_{in}) < E$
NOT	0	1	$f(x_0) - E = V_{in}$
	1	0	$f(x_0 + V_{in}) < E$

Table 2 One specific solution of the conditions in Table 1 which yields the logical operations AND, OR, XOR, NAND and NOT, with $V_{in} = \frac{1}{4}$. Note that these theoretical solutions have been fully verified in a discrete electrical circuit emulating a logistic map [6]

Operation	AND	OR	XOR	NAND	NOT
x_0	0	1/8	1/4	3/8	1/2
E	3/4	11/16	3/4	11/16	3/4

easily couple to another gate element as input, so that gates can be "wired" directly into gate arrays implementing compounded logic operations.

In order to obtain all the desired input-output responses of the different gates, we need to satisfy the conditions enumerated in Table 1 simultaneously. So given a dynamics $f(x)$ corresponding to the physical device in actual implementation, one must find values of threshold and initial state satisfying the conditions derived from the Truth Tables to be implemented.

For instance, Table 2 shows the exact solutions of the initial x_0 and threshold E which satisfy the conditions in Table 1 when

$$f(x) = 4x(1-x)$$

The constant $V_{in} = \frac{1}{4}$ is common to both input and output and to all logical gates.

3 Continuous-Time Nonlinear System

We now present a somewhat different scheme for obtaining logic responses from a continuous-time nonlinear system. Our processor is now a continuous time system described by the evolution equation $d\,\mathbf{x}\,/dt = \mathbf{F}\,(\mathbf{x},t)$, where $\mathbf{x} = (x_1, x_2, \ldots x_N)$ are the state variables and $\mathbf{F}$ is a nonlinear function. In this system we choose a variable, say x_1, to be thresholded. Whenever the value of this variable exceeds a threshold E it resets to E, i.e. when $x_1 > E$ then (and only then) $x_1 = E$.

Now the basic 2-input 1-output logic operation on a pair of inputs I_1, I_2 in this scheme simply involves the setting of an inputs-dependent threshold, namely the threshold voltage

$$E = V_C + I_1 + I_2$$

where V_C is the dynamic control signal determining the functionality of the processor. By switching the value of V_C one can switch the logic operation being performed.

Again I_1/I_2 has value 0 when logic input is 0 and has value V_{in} when logic input is 1. So the theshold E is equal to V_C when logic inputs are $(0,0)$, $V_C + V_{in}$ when logic inputs are $(0,1)$ or $(1,0)$, and $V_C + 2V_{in}$ when logic inputs are $(1,1)$.

The output is interpreted as logic output 0 if $x_1 < E$, i.e. the excess above threshold $V_0 = 0$. The logic output is 1 if $x_1 > E$, and the excess above threshold $V_0 = (x_1 - E) \sim V_{in}$. The schematic diagram of this method is displayed in Fig. 1.

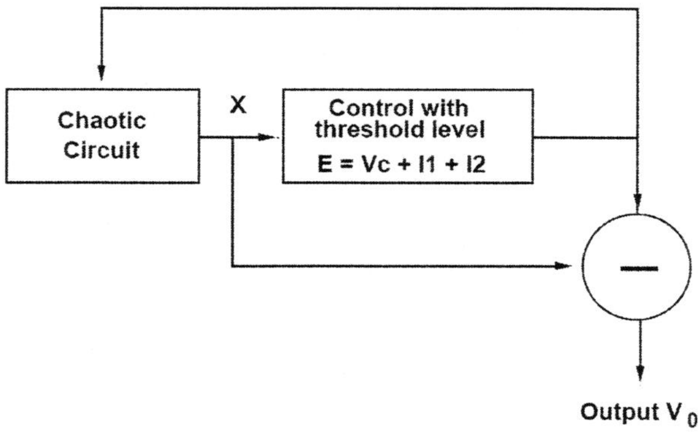

Fig. 1 Schematic diagram for implementing a morphing 2 input logic cell with a continuous time dynamical system. Here V_C determines the nature of the logic response, and the 2 inputs are $I1, I2$

Now for a NOR gate implementation ($V_C = V_{NOR}$) the following must hold true:

(i) when input set is $(0,0)$, output is 1, which implies that for threshold $E = V_{NOR}$, output $V_0 = (x_1 - E) \sim V_{in}$
(ii) when input set is $(0,1)$ or $(1,0)$, output is 0, which implies that for threshold $E = V_{NOR} + V_{in}$, $x_1 < E$ so that output $V_0 = 0$.
(iii) when input set is $(1,1)$, output is 0, which implies that for threshold $E = V_{NOR} + 2V_{in}$, $x_1 < E$ so that output $V_0 = 0$.

For a NAND gate ($V_C = V_{NAND}$) the following must hold true:

(i) when input set is $(0,0)$, output is 1, which implies that for threshold $E = V_{NAND}$, output $V_0 = (x_1 - E) \sim V_{in}$
(ii) when input set is $(0,1)$ or $(1,0)$, output is 1, which implies that for threshold $E = V_{in} + V_{NAND}$, output $V_0 = (x_1 - E) \sim V_{in}$
(iii) when input set is $(1,1)$, output is 0, which implies that for threshold $E = V_{NAND} + 2V_{in}$, $x_1 < E$ so that output $V_0 = 0$.

In order to design a dynamic NOR/NAND gate one has to find values of V_C that will satisfy all the above input-output associations in a robust and consistent manner.

A proof-of-principle experiment of the scheme was realized with the double scroll chaotic Chua's circuit given by the following set of (rescaled) 3 coupled ODEs [14]

$$\dot{x_1} = \alpha(x_2 - x_1 - g(x_1)) \tag{1}$$

$$\dot{x_2} = x_1 - x_2 + x_3 \tag{2}$$

$$\dot{x_3} = -\beta x_2 \tag{3}$$

where $\alpha = 10.$ and $\beta = 14.87$ and the piecewise linear function $g(x) = bx + \frac{1}{2}(a - b)(|x+1| - |x-1|)$ with $a = -1.27$ and $b = -0.68$. We used the ring structure configuration of the classic Chua's circuit [14].

In the experiment we implemented minimal thresholding on variable x_1 (this is the part in the "control" box in the schematic figure). We clipped x_1 to E, if it exceeded E, only in Eq. (2). This has very easy implementation, as it avoids modifying the value of x_1 in the nonlinear element $g(x_1)$, which is harder to do. So then all we need to do is to implement $\dot{x_2} = E - x_2 + x_3$ instead of Eq. (2), when $x_1 > E$, and there is no controlling action if $x_1 \leq E$.

A representative example of a dynamic NOR/NAND gate can be obtained in this circuit implementation with parameters: $V_{in} = 2V$. The NOR gate is realized around $V_C = 0V$. At this value of control signal, we have the following: for input (0,0) the threshold level is at 0, which yields $V_0 \sim 2V$; for inputs (1,0) or (0,1) the threshold level is at 0, which yields $V_0 \sim 0V$; and for input (1,1) the threshold level is at $2V$, which yields $V_0 = 0$ as the threshold is beyond the bounds of the chaotic attractor. The NAND gate is realized around $V_C = -2V$. The control signal yields the following: for input (0,0) the threshold level is at $-2V$, which yields $V_0 \sim 2V$; for inputs (1,0) or (0,1) the threshold level is at $2V$, which yields $V_0 \sim 2V$; and for input (1,1) the threshold level is at $4V$, which yields $V_0 = 0$ [7,8].

So the knowledge of the dynamics allowed us to design a control signal that can select out the temporal patterns emulating the NOR and NAND gates [9]. For instance in the example above, as the dynamic control signal V_C switches between $0V$ to $-2V$, the module first yields the NOR and then a NAND logic response. Thus one has obtained a dynamic logic gate capable of switching between two fundamental logic reponses, namely the NOR and NAND.

4 Design and Construction of a Three-Input Reconfigurable Chaotic Logic Gate

As in Sect. 3, consider a single chaotic element (for inclusion into a RCLG) to be a continuous time system described by the evolution equation: $d\,\mathbf{x}/dt = \mathbf{F}\,(\mathbf{x};t)$ where $\mathbf{x} = (x_1, x_2, \ldots, x_N)$ are the state variables, and $\mathbf{F}$ is a strongly nonlinear function. Again in this system we choose a variable, say x_1, to be thresholded. So whenever the value of this variable exceeds a critical threshold E (i.e. when $x_1 > E$), it re-sets to E.

In accordance to our basic scheme, the logic operation on a set of inputs I_1, I_2 and I_3 simply involves the setting of an inputs-dependent threshold, namely the threshold voltage $E = V_C + I_1 + I_2 + I_3$, where V_C is the dynamic control signal determining the functionality of the processor. By switching the value of V_C, one can switch the logic operation being performed.

$I_{1,2,3}$ has value $\sim 0V$ when logic input is zero, and $I_{1,2,3}$ has value V_{in} when logic input is one. So for input (0,0,0) the threshold level is at V_C; for inputs (0,0,1) or

Table 3 Truth table for NOR gate implementation ($V_{in} = 1.84V$, $V_{NOR} = 0V$)

Input Set (I_1, I_2, I_3)	Threshold E	Output	Logic Output
(0,0,0)	V_{NOR}	$V_0 = (x_1 - E) \sim V_{in}$	1
(0,0,1) or (1,0,0) or (0,1,0)	$V_{NOR} + V_{in}$	$V_0 \sim 0V$ as $x_1 < E$	0
(0,1,1) or (1,1,0) or (1,0,1)	$V_{NOR} + 2V_{in}$	$V_0 \sim 0V$ as $x_1 < E$	0
(0,0,0)	$V_{NOR} + 3V_{in}$	$V_0 \sim 0V$ as $x_1 < E$	0

Table 4 Truth table for NAND gate implementation ($V_{in} = 1.84V$, $V_{NOR} = -3.68V$)

Input Set (I_1, I_2, I_3)	Threshold E	Output	Logic Output
(0,0,0)	V_{NAND}	$V_0 = (x_1 - E) \sim V_{in}$	1
(0,0,1) or (1,0,0) or (0,1,0)	$V_{NAND} + V_{in}$	$V_0 = (x_1 - E) \sim V_{in}$	1
(0,1,1) or (1,1,0) or (1,0,1)	$V_{NAND} + 2V_{in}$	$V_0 = (x_1 - E) \sim V_{in}$	1
(0,0,0)	$V_{NAND} + 3V_{in}$	$V_0 \sim 0V$ as $x_1 < E$	0

(0,1,0) or (1,0,0) the threshold level is at $V_C + V_{in}$; for input (0,1,1) or (1,1,0) or (1,1,0) the threshold level is at $V_C + 2V_{in}$ and for input (1,1,1) the threshold level is $V_C + 3V_{in}$.

As before, the output is interpreted as logic output 0 if $x_1 < E$, and the excess above threshold $V_0 \sim 0$. The logic output is 1 if $x_1 > E$, and $V_0 = (x_1 - E) \sim V_{in}$. Now for the 3-inputs NOR and the NAND gate implementations the input-output relations given in Tables 3 and 4 must hold true. Again, in order to design the NOR or NAND gates, one has to use the knowledge of the dynamics of the nonlinear system to find the values of V_C and V_0 that will satisfy all the input-output associations in a consistent and robust manner.

Consider again the simple realization of the double-scroll chaotic Chua's attractor represented by the set of (rescaled) 3-coupled ODEs given in Eqs. (1), (2), (3). This system was implemented by the circuit shown in Fig. 2, with circuit component values: [$L = 18$ mH, $R = 1710\ \Omega$, $C_1 = 10$ nF, $C_2 = 100$ nF, $R_1 = 220\ \Omega$, $R_2 = 220\ \Omega$, $R_3 = 2.2\ k\ \Omega$, $R_4 = 22\ k\ \Omega$, $R_5 = 22\ k\ \Omega$, $R_3 = 3.3\ k\ \Omega$, $D =$ IN4148, $B_1, B_2 =$ Buffers, OA1 – OA3 : opamp μA741]. The x_1 dynamical variable (corresponding to the voltage V1 across the capacitor C1) is thresholded by a control circuit shown in the dotted box in Fig. 2, with voltage E setting varying thresholds. In the circuit, VT corresponds to the output signal from the threshold controller. Note that, as in the implementation of 2-input gates, we are only replacing $dx_2/dt = x_1 - x_2 + x_3$ by $dx_2/dt = E - x_2 + x_3$ in Eq. (2), when $x_1 > E$, and there is no controlling action if $x_1 \leq E$.

The schematic diagram for the NAND/NOR gate implementation is depicted in Fig. 3. In the representative example shown here, $V_{in} = 1.84$ V. The NOR gate is realized around $V_C = V_{NOR} = 0$ V and the NAND gate is realized with $V_C = V_{NAND} = -3.68$ V (See Tables 3 and 4).

Thus the nonlinear evolution of the element has allowed us to obtain a control signal that selects out temporal patterns corresponding to NOR and NAND gates.

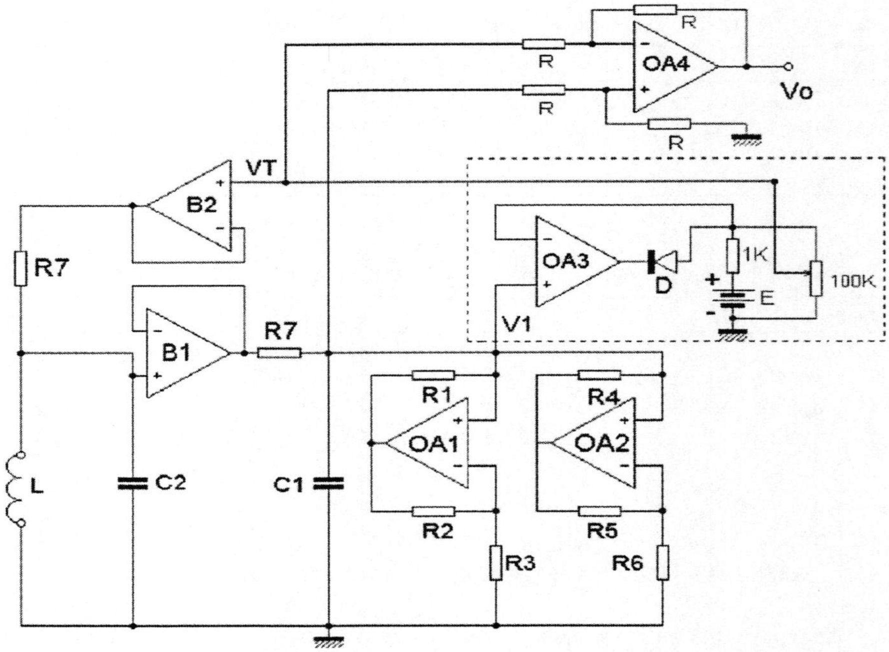

Fig. 2 Circuit module implementing a RCLG that morphs between NAND and NOR logic gates. The diagram represented in the dotted region is the threshold controller. Here $E = V_C + I_1 + I_2 + I_3$ is the dynamically varying threshold voltage. VT is the output signal from the threshold controller and V_0 is the difference voltage signal

For instance in Fig. 4, as the dynamic control signal V_C switches between -3.68 V to 0 V, the element yields first a NAND gate and then morphs into a NOR gate. The fundamental period of oscillation of the Chua's circuit is 0.33 ms. The average latency of morphing between logic gates is 48% of this period.

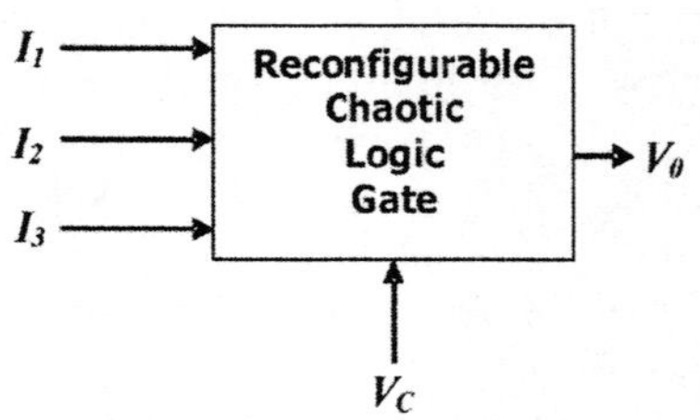

Fig. 3 Symbolic diagram for dynamic 3-Input NOR/NAND logic cell. Dynamic control signal V_C determines the logic operation. In our example, V_C can switch between V_{NAND} giving a NAND gate, and V_{NOR} giving a NOR gate

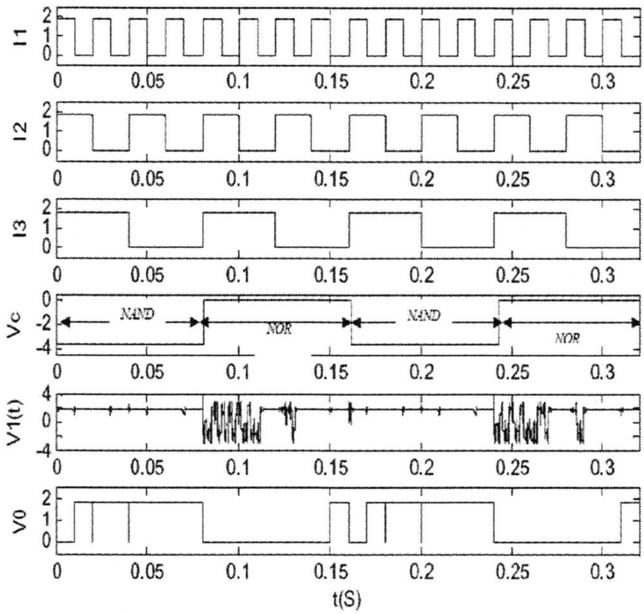

Fig. 4 Voltage timing sequences from top to bottom (PSPICE simulation): (**a**) First input $I1$, (**b**) Second input $I2$, (**c**) Third input $I3$, (**d**) Dynamic control signal V_C, where V_C switches between $V_{NAND} = -3.68$ V and $V_{NOR} = 0$ V (**e**) Output signal $V1$ (corresponding to $x_1(t)$) from the Chua's circuit, (**f**) Recovered logic output signal from $V0$. The fundamental period of oscillation of this circuit is 0.33 mS

5 VLSI Implementation of Chaotic Computing Architectures – Proof of Concept

Recently ChaoLogix Inc. designed and fabricated a proof of concept chip that demonstrates the feasibility of constructing reconfigurable chaotic logic gates, henceforth ChaoGates, in standard CMOS based VLSI (0.18 μm TSMC process operating at 30 Mhz with a 3.1 × 3.1 mm die size and a 1.8 V digital core voltage). The basic building block ChaoGate is shown schematically in Fig. 5. ChaoGates were then incorporated into a ChaoGate Array in the VLSI chip to demonstrate higher order morphing functionality including:

1 A small Arithmetic Logic Unit (ALU) that morphs between higher order arithmetic functions (multiplier and adder/accumulator) in *less than one clock cycle*. An ALU is a basic building block of computer architectures.
2 A Communications Protocols (CP) Unit that morphs between two different complex communications protocols *in less than one clock cycle*: Serial Peripheral Interface (SPI, a synchronous serial data link) and an Inter Integrated Circuit Control bus implementation (I2C, a multi-master serial computer bus).

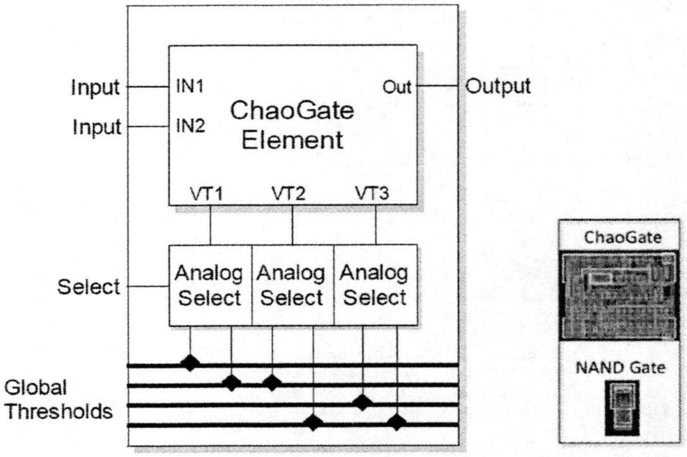

Fig. 5 (*Left*) Schematic of a two-input, one output morphable ChaoGate. The gate logic functionality (NOR, NAND, XOR,) is controlled (morphed), in the current VLSI design, by global thresholds connected to VT1, VT2 and VT3 through analog multiplexing circuitry and (*Right*) a size comparison between the current ChaoGate circuitry implemented in the ChaoLogix VLSI chaotic comuting chip and a typical NAND gate circuit (Courtesy of ChaoLogix Inc.)

While the design of the ChaoGates and ChaoGate Arrays in this proof of concept VLSI chip was not optimized for performance, it clearly demonstrates that ChaoGates can be constructed and organized into reconfigurable chaotic logic gate arrays capable of morphing between higher order computational building blocks. Current efforts are focused upon optimizing the design of a single ChaoGate to levels where they are comparable or smaller to a single NAND gate in terms of power and size yet are capable of morphing between all gate functions in under a single computer clock cycle. Preliminary designs indicate that this goal is achievable and that all gates currently used to design computers may be replaced with ChaoGates to provide added flexibility and performance.

Acknowledgments We acknowledge the support of the Office of Naval Research [N000140211019].

References

1. Sinha, S. and Ditto, W.L. *Phys. Rev. Lett.* **81** (1998) 2156.
2. Sinha, S., Munakata, T. and Ditto, W.L, *Phys. Rev. E* **65** (2002) 036214.
3. Munakata, T., Sinha, S. and Ditto, W.L, IEEE *Trans. Circ. and Systems* **49** (2002) 1629.
4. Sinha, S. and Ditto, W.L. *Phys. Rev. E* **59** (1999) 363.
5. Sinha, S., Munakata, T. and Ditto, W.L *Phys. Rev. E* **65** 036216.
6. Murali, K., Sinha, S. and Ditto, W.L., Proceedings of the STATPHYS-22 Satellite conference Perspectives in Nonlinear Dynamics Special Issue of Pramana 64 (2005) 433.
7. Murali, K., Sinha, S. and Ditto, W.L., *Int. J. Bif. and Chaos (Letts)* **13** (2003) 2669.

8. Murali, K., Sinha S., and I. Raja Mohamed, I.R., *Phys. Letts. A* **339** (2005) 39.
9. Murali, K., Sinha, S., Ditto, W.L., Proceedings of *Experimental Chaos Conference* (ECC9), Brazil (2006) published in *Philosophical Transactions of the Royal Society of London* (Series A) (2007).
10. W. Ditto, S. Sinha and K. Murali, US Patent Number 07096347 (August 22, 2006).
11. Sinha, S., *Nonlinear Systems*, Eds. R. Sahadevan and M.L. Lakshmanan, (Narosa, 2002) 309–328.
12. Murali, K. and Sinha, S., *Phys. Rev. E* **68** (2003) 016210.
13. Ditto, W.L. and Sinha, S., *Philosophical Transactions of the Royal Society of London* (Series A) **364** (2006) 2483–2494.
14. Dimitriev, A.S. et al., J. Comm. Tech. Electronics, **43** (1998) 1038.

Activated Switching in a Parametrically Driven Micromechanical Torsional Oscillator

H.B. Chan and C. Stambaugh

Parametric resonance and parametric amplification are important phenomena that are relevant to many fields of science. For mechanical systems, parametric driving typically involve modulating the spring constant [1, 2] or the moment of inertia near twice the natural frequency of the system. Parametric amplification has proved useful in improving the signal to noise before transduction of the mechanical displacement into an electrical signal [1]. Apart from amplifying a signal, parametric pumping can also reduce the linewidth of the resonance response, opening up new opportunities for biochemical detection using micro- and nano-mechanical devices in viscous environments [3]. Recently the sharp jump in the parametric response of micromechanical oscillators at subcritical bifurcation was used for accurate determination of the natural frequency to deduce device parameters [4].

Depending on the amplitude and frequency of the drive, a parametric oscillator possesses one, two or three stable states. Fluctuations induce transitions between coexisting attractors. The transition rate depends exponentially on the ratio of an activation barrier to the fluctuation intensity [5, 6]. Such dependence bears resemblance to equilibrium systems where the transition rate can be obtained from the height of the free energy barrier. However, the multistability in a parametric oscillator develops only when the system is under strong periodic drive. The system is far from thermal equilibrium and cannot be characterized by free energy [5]. Consequently, the transition rate needs to be calculated from system dynamics. Such fluctuation induced switching has been observed in a number of driven systems, including parametrically driven electrons in a Penning trap [7], micro- and nano-mechanical devices [8–11], radio frequency driven Josephson junctions [12–14] and atoms in magneto-optical traps [15, 16].

Here we describe our investigation of noise-activated switching in a parametrically driven micromechanical torsional oscillator. The electrostatic contribution to the spring constant is modulated near twice the natural frequency of the oscillator. When the parametric modulation is sufficiently strong, oscillations are induced at

H.B. Chan (✉)
Department of Physics, University of Florida, Gainesville, FL 32608, USA

V. In et al. (eds.), *Applications of Nonlinear Dynamics*, Understanding Complex Systems, DOI 10.1007/978-3-540-85632-0_2,

half the modulation frequency. The phase of the oscillation can take on either one of two values that differ from each other by π. When noise is injected into the excitation voltage, the system can occasionally overcome the activation barrier and switch between the two states. The transition rates out of the two states are identical, yielding a dynamical bistable system that is driven out of equilibrium. As the parametric driving frequency approaches a bifurcation value, both the amplitude of oscillation and the activation barrier decrease and eventually become zero at the bifurcation frequency. Near the bifurcation frequency, we find that the activation barrier depends on frequency detuning with a critical exponent of 2, consistent with predicted universal scaling in parametrically driven systems [6]. Away from the vicinity of the bifurcation point, the dependence of the activation barrier on frequency detuning crosses over from quadratic to 3/2th power dependence that is specific to our device.

In our experiment, the micromechanical oscillator is fabricated using a surface micromachining process on a silicon substrate. As shown in Fig. 1a, the oscillator consists of a polysilicon plate (500 μm by 500 μm by 3.5 μm) supported by two torsional rods. By etching away a 2-μm-thick sacrificial silicon oxide layer beneath the top plate, the top plate becomes free to rotate about the torsional rods. The other ends of the torsional rods are anchored to the substrate (Fig. 1b). Two fixed polysilicon electrodes are located below the top plate. One of the electrodes is used to modulate the spring constant electrostatically and the other electrode is used to capacitively detect motions of the top plate.

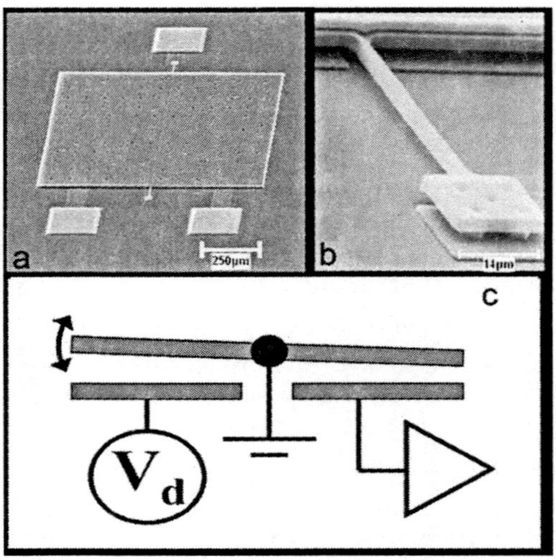

Fig. 1 (**a**) Scanning electron micrograph of the torsional oscillator. The large square in the middle is a movable polysilicon plate. The small squares are wire-bond pads that provide electrical connections to the top plate and the two fixed electrodes. (**b**) Close up on one of the torsional springs. (**c**) Cross sectional schematic of the oscillator with measurement circuitry

Figure 1c shows a cross-sectional schematic of the oscillator with electrical connections. The application of a periodic voltage with dc bias to the left electrode exerts an electrostatic torque on the grounded top plate. Modulations of the restoring torque are generated by the periodic component of the applied voltage. As the plate oscillates, the capacitance between the plate and the detection electrode changes. The detection electrode is connected to a charge sensitive preamplifier followed by a lock-in amplifier that measures the signal at the half the modulation frequency. Measurements were performed at liquid nitrogen temperature and at pressure of less than 1×10^{-6} *torr*. The quality factor Q of the oscillator exceeds 7,500.

The modulations of the spring constant in our torsional oscillator originate mainly from the strongly distance-dependent electrostatic interaction between the top plate and the excitation electrode. The equation of motion of the oscillator is given by:

$$\ddot{\theta} + 2\gamma\dot{\theta} + \omega_0^2\theta = \tau/I \tag{1}$$

where θ is the angular rotation of the top plate, γ is the damping coefficient and ω_0 is the natural frequency of the oscillator. The driving torque τ arises from the electrostatic interaction between the top plate and the driving electrode:

$$\tau = \frac{1}{2}\frac{dC}{d\theta}V_d^2 \tag{2}$$

where C is the capacitance between the top plate and the driving electrode. A Taylor expansion of τ about the equilibrium angular position θ_0 yields:

$$\tau = \frac{1}{2}\left(C'(\theta_o) + C''(\theta_o)\theta + \frac{1}{2}C'''(\theta_o)\theta^2 + \frac{1}{6}C''''(\theta_o)\theta^3\right)V_d^2 \tag{3}$$

where C', C'', C''' and C'''' denote the first, second, third and fourth angular derivative of C respectively. The excitation voltage V_d is a sum of three components:

$$V_d = V_{dc} + V_{ac}\sin(\omega t) + V_{noise}(t) \tag{4}$$

The three terms on the right side of Eq. (4) represent the dc voltage, periodic ac voltage and random noise voltage respectively. V_{dc} is chosen to be much larger than V_{ac} and V_{noise} to partially linearize the dependence of τ on V_{ac} and V_{noise}. The strong angular dependence of the electrostatic torque leads to nonlinear contributions to the restoring torque. Substituting V_d and τin Eqs. (1) and (2) leads to:

$$\ddot{\theta} + 2\gamma\dot{\theta} + \left[\omega_0^2 + \frac{k_o}{I} + \frac{K_e}{I}\cos(\omega t)\right]\theta + \alpha\theta^2 + \beta\theta^3 + D = N(t) + F\cos(\omega t) \tag{5}$$

where $\alpha = -C'''(\theta_o)V_{dc}^2/4I$, $\beta = -c''''(\theta_o)V_{dc}^2/12I$, $D = -C'(\theta_o)V_{dc}^2/2I$, $F = C'(\theta_o)V_{dc}V_{ac}(t)/I$, $k_o = -C''(\theta_o)V_{dc}^2/2$, $N(t) = -C'(\theta_o)V_{dc}V_{noise}(t)/I$ is the effective noise in the excitation and $k_e + -C''(\theta_o)V_{dc}V_{ac}$ is the effective amplitude of the modulation of spring constant. Coupling of the electrostatic torque

$F\cos(\omega t)$ and the nonlinear terms generates an effective modulation of spring constant. However, this contribution is much smaller than the direct electrostatic modulation of the spring (k_e/I). As a result, the $F\cos(\omega t)$ term can be neglected. After redefining the angle to be measured from the equilibrium angular position θ_0, the equation of motion reduces to [17]:

$$\ddot{\theta} + 2\gamma\dot{\theta} + \left[\omega_1^2 + \frac{k_e}{I}\cos(\omega t)\right]\theta + \alpha\theta^2 + \beta\theta^3 = N(t) \tag{6}$$

where $\omega_1^2 = \omega_o^2 + k_o/I$.

We first focus on the response of the oscillator with no injected noise in the excitation. When the electrostatic spring modulation is small, there are no induced oscillations at $\omega/2$. It is necessary for the amplitude of spring modulation to reach a threshold value of $k_T = 4\omega_o\gamma I$ before oscillations are induced at half the modulation frequency in a range close to ω_o. As shown in Fig. 2, there are three ranges of frequencies with different numbers of stable attractors, separated by a supercritical bifurcation point $\omega_{b1} = 2\omega_o + \omega_p$ and a subcritical bifurcation point $\omega_{b1} = 2\omega_o - \omega_p$, where $\omega_p = \sqrt{k_e^2 - k_T^2}/2I\omega_o$. In the first region ($\omega > \omega_{b1} \sim 41174$ rad s^{-1}), no oscillations take place, as the only stable attractor is a zero-amplitude state. At ω_{b1}, there emerge two stable states of oscillations at frequency $\omega/2$ that differ in phase by π but are otherwise identical. This symmetry is illustrated in Fig. 3a, where the period of the induced oscillation is twice that of the parametric driving. In Fig. 3b, both the drive and response are shifted in time by $2\pi/\omega$. While the drive remains unchanged, the response has picked up an extra phase of π. Both oscillation states are valid solutions of Eq. (6). Their amplitudes are the same but their phase differs by π. These two stable states are separated in phase space by an unstable state with zero oscillation amplitude (dotted line in Fig. 2). At frequencies below ω_{b2}($\sim$ 41150 rad s^{-1}), the zero-amplitude state becomes stable, resulting in the coexistence

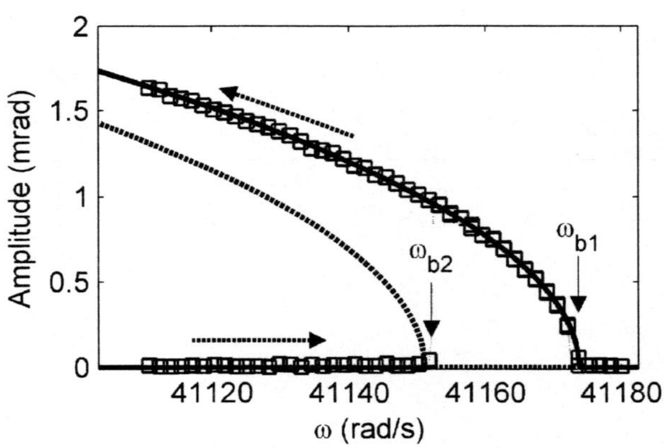

Fig. 2 Oscillation amplitude at $\omega/2$ vs frequency of parametric modulation ω. The solid and dotted lines represent the stable and unstable oscillation states respectively

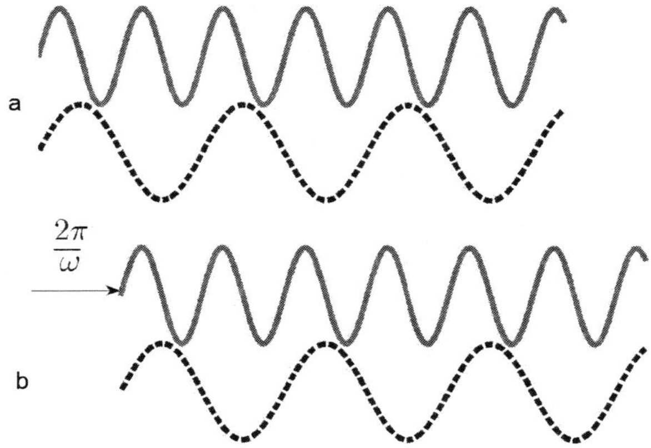

Fig. 3 Two coexisting oscillation states of the parametric oscillator. The red lines represent the parametric modulation. The two blue lines in (**a**) and (**b**) shows the two stable states with the same oscillation frequency but opposite phase

of three stable attractors. These stable states are separated in phase space by two unstable states indicated by the dotted line in Fig. 2.

In the presence of noise in the excitation, the oscillator could be induced to switch between coexisting attractors. Since the parametric oscillator is a driven system that is far from thermal equilibrium and cannot be characterized by free energy, calculation of the escape rate is a non-trivial problem. Theoretical analysis suggests that the rate of escape Γat a particular driving frequency depends exponentially on the ratio of an activation barrier R to the noise intensity I_N [5]:

$$\Gamma = \Gamma_0 \exp(-R/\mathrm{I}_N) \tag{7}$$

In general, R depends on various device parameters including the natural frequency, the parametric driving frequency, the damping constant and the nonlinear coefficients. Near the bifurcation points, the system dynamics is characterized by an overdamped soft mode and R decreases to zero according to $k|\omega - \omega_b|^\xi$ with a critical exponent ξ that is system independent [5]. While the prefactor k might be different for each system, ξ is universal for all systems and depends only on the type of bifurcation. For instance, a Duffing oscillator resonantly driven into bistability undergoes spinodal bifurcations at the boundaries of the bistable region. One stable state merges with the unstable state while the other stable state remains far away in phase space. Recent experiments in micromechanical oscillators [9] and rf-driven Josephson junctions [18] have confirmed the theoretical prediction [5, 19] that the activation barrier scales with critical exponent 3/2 near spinodal bifurcations in driven systems. On the other hand, in a parametric oscillator, the supercritical and subcritical bifurcations involve the merging of two stable oscillation states and an unstable zero-amplitude state (at ω_{b1}) and the merging of two unstable states and a zero-amplitude stable state (at ω_{b2}) respectively. When three states merge together in such pitchfork bifurcations, the critical exponent is predicted to be 2. Away from the bifurcation points, the scaling relationship no longer holds and different exponents were obtained depending on the nonlinearity and damping of the system.

In our experiment, we inject $V_{noise}(t)$ with a bandwidth of ~300 rad s^{-1}centered at ω_o to induce the transitions between stable states in our parametric oscillator. The bandwidth of the noise is much larger than the resonance linewidth. Figure 4a and 4c show respectively the oscillation amplitude and phase at a driving frequency between ω_{b2} and ω_{b1}, in the range of two coexisting attractors. The oscillator resides in the one of the oscillation state for various durations before escaping to the other state. Transitions take place when the phase slips by π. The two oscillation states have the same amplitude. These two attractors can also be clearly identified in the occupation histograms in Fig. 5a and 5b. Figure 4b and 4d show switching events at a driving frequency lower than ω_{b2}, with three attractors. The zero-amplitude state has also become stable. Unlike Fig. 4a, the oscillator switches between two distinct amplitudes. At high amplitude, the phase takes on either one of two values that differ by π. When the oscillator is in the zero-amplitude state, there are large fluctuations of the phase as a function of time as the oscillator moves about the origin. The coexistence of three attractors in phase space is also illustrated in Fig. 5c and 5d for two other driving frequencies.

Figure 6a shows a histogram of the residence time in one of the oscillation states before a transition occurs. The exponential dependence on the residence time indicates that the transitions are random and follow Poisson statistics as expected. An exponential fit to the histogram yields the transition rate. The transition rates out of the two oscillation states are measured to be identical to within experimental

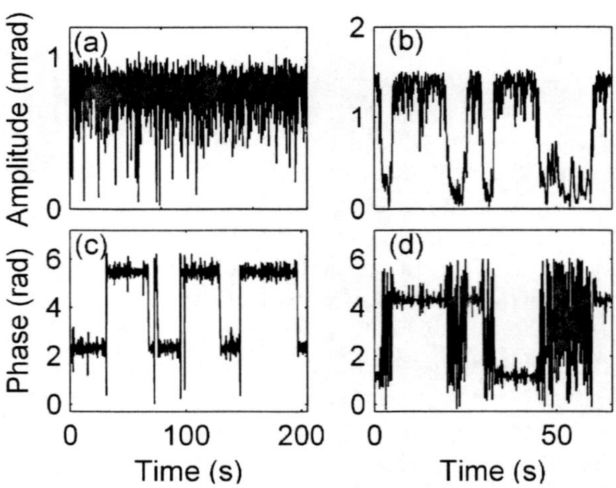

Fig. 4 Oscillation amplitude (**a**) and phase (**c**) for a driving frequency of 41159.366 rad s^{-1}. In the range $\omega_{b2} < \omega < \omega_{b1}$ transitions take place when the phase slips by π (**b**) When ω (41124.705 rad s^{-1}) is lower than ω_{b2} , transitions occur with jumps in the amplitude. (**d**) The phase differs by π for the two high amplitude states. When the oscillator is in the zero-amplitude state, there are large fluctuations in the phase

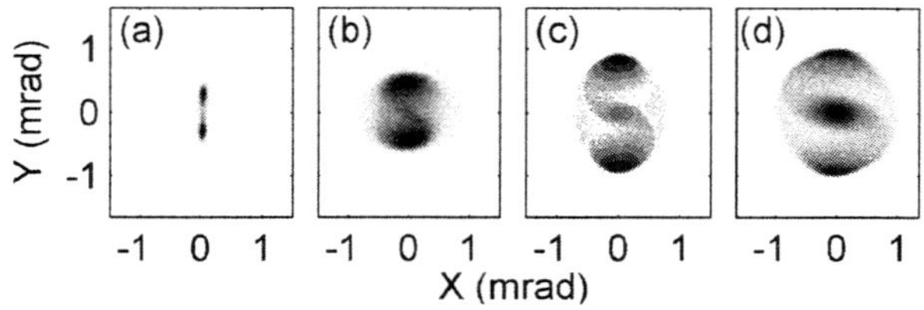

Fig. 5 Phase space occupation for four different modulation frequencies. X and Y represent the amplitudes of the two quadratures of oscillation that are out of phase with each other. Darker grey scale corresponds to higher occupation. (**a**) $\omega = 41171.6$ rad s^{-1}. Two oscillation states coexist near ω_{b1}. (**b**) $\omega= 41163.0$ rad s^{-1}. The two states move further apart as ω decreases. (**c**) $\omega= 41139.8$ rad s^{-1}. The zero-amplitude states becomes stable. (**d**) $\omega= 41124.8$ rad s^{-1}. The occupation of the zero-amplitude state increases as ω decreases

uncertainty at all noise intensities. Figure 6b plots the logarithm of the transition rate as a function of inverse noise intensity. The transition rate depends exponentially on inverse noise intensity, demonstrating that switching between the states is activated in nature. According to Eq. (7), the slope in Fig. 6b yields the activation barrier at the particular detuning frequency. Transitions were also measured for switching out of the zero-amplitude state for $\omega < \omega_{b2}$. These switches were also found to be activated and follow Poisson statistics in a similar manner.

We repeat the above procedure to determine the activation barriers at other driving frequencies. Figure 7a plots the activation barriers R_1 for switching out of the oscillation states and R_2 for switching out of the zero-amplitude state as a function of the driving frequency. When the driving frequency is high, only the zero-amplitude state is stable. With decreasing frequency, two stable oscillation states (separated by an unstable state) emerge at ω_{b1}. As the detuning $\Delta\omega_1 = \omega_{b1} - \omega$ increases, the pair of oscillation states move further apart in phase space and R_1 increases. At ω_{b2}, the zero-amplitude state becomes stable. The appearance of the stable zero-amplitude state is accompanied by the creation of two unstable states separating it in phase space from the two stable oscillation states. Initially, R_2 increases with frequency detuning $\Delta\omega_2 = \omega_{b2} - \omega$ in a fashion similar to R_1. Close to ω_{b2}, R_1 is larger than

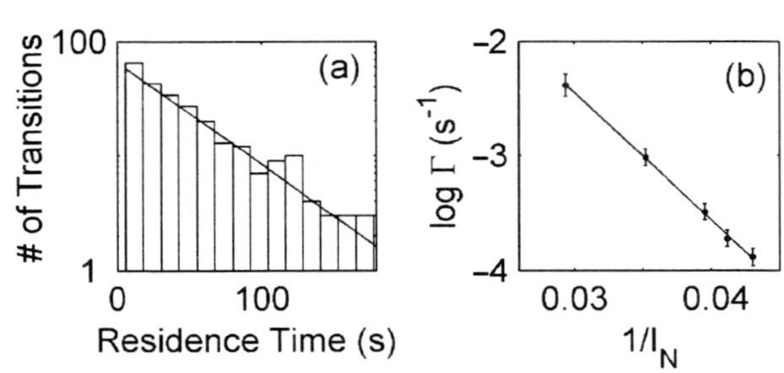

Fig. 6 (**a**) Histogram of the residence time in one of the oscillation states at $\omega = 41130.49$ rad s^{-1} on semi-logarithmic scale, fitted by an exponential decay (solid line). (**b**) Dependence of the logarithm of the transition rate on the inverse noise intensity

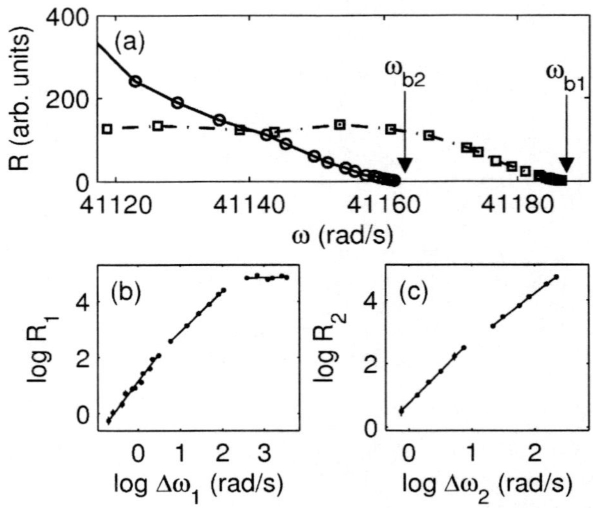

Fig. 7 (**a**) The activation barriers R_1 and R_2 vs. the parametric modulation frequency. (**b**) log R_1 vs. log $\Delta\omega_1$. (**c**) log R_2 vs. log $\Delta\omega_2$. The lines are power law fits to different ranges of $\Delta\omega$

R_2 and the occupation of the oscillation states is higher than the zero-amplitude state. As the frequency decreases, R_2 continues to increase monotonically while R_1 remains approximately constant. As a result, R_1 and R_2 cross each other at ~ 41140 rad s^{-1}, beyond which the occupation of the zero-amplitude state becomes higher than the oscillation states. The dependence of the occupation on frequency detuning was also observed in parametrically driven atoms in magneto-optical traps [15].

Figure 7b and 7c show the dependence of the activation barriers $R_{1,2}$on $\Delta\omega_{1,2}$ on logarithmic scales. When the detuning is small, both R_1 and R_2 show power law dependence on $\Delta\omega$. From the linear fits, the exponents are measured to be 2.0 ± 0.1 and 2.00 ± 0.03 for R_1 and R_2 respectively. Such quadratic dependence of the activation barrier on detuning near the bifurcation points is predicted to be system-independent [5,6] and is expected to occur in other parametrically-driven, nonequilibrium systems such as electrons in Penning traps [7] and atoms in magneto-optical traps [15,16]. Away from the vicinity of the bifurcation point, however, the variation of the activation barrier with frequency detuning becomes device-specific. Figure 7b and 7c show crossovers from the quadratic dependence to different power law dependence with exponents 1.43 ± 0.02 and 1.53 ± 0.02 for R_1 and R_2 respectively. These values obtained in our experiment are distinct from the exponents obtained in parametrically driven electrons in Penning traps [7] because the nonlinearity and damping are different for the two systems.

Recent theoretical predictions indicate that the symmetry in the occupation of the two oscillation states in a parametrically driven oscillator will be lifted when an additional small drive close to frequency $\omega/2$ is applied [20]. A number of phenomena, including strong dependence of the state populations on the amplitude of the small drive and fluctuation enhanced frequency mixing, are expected to occur. Experiments are underway to test such predictions and to reveal other fluctuation phenomena in parametrically driven oscillators.

We are grateful to M. I. Dykman and D. Ryvkine for useful discussions. This work was supported by NSF DMR-0645448.

References

1. D. Rugar and P. Grutter, Phys. Rev. Lett. **67**, 699 (1991).
2. D. W. Carr, S. Evoy, L. Sekaric, et al., Appl. Phys. Lett. **77**, 1545 (2000).
3. L. Sekaric, M. Zalalutdinov, R. B. Bhiladvala, et al., Appl. Phys. Lett. **81**, 2641 (2002).
4. W. H. Zhang, R. Baskaran, and K. L. Turner, Sensors and Actuators a-Physical **102**, 139 (2002).
5. M. I. Dykman and M. A. Krivoglaz, Zh. Eksper. Teor. Fiz. **77**, 60 (1979).
6. M. I. Dykman, C. M. Maloney, V. N. Smelyanskiy, et al., Phys. Rev. E **57**, 5202 (1998).
7. L. J. Lapidus, D. Enzer, and G. Gabrielse, Phys. Rev. Lett. **83**, 899 (1999).
8. J. S. Aldridge and A. N. Cleland, Phys. Rev. Lett. **94**, 156403 (2005).
9. C. Stambaugh and H. B. Chan, Phys. Rev. B **73**, 172302 (2006).
10. R. L. Badzey and P. Mohanty, Nature **437**, 995 (2005).
11. R. Almog, S. Zaitsev, O. Shtempluck, et al., Appl. Phys. Lett. **90**, 013508 (2007).
12. E. V. Sukhorukov and A. N. Jordan, Phys. Rev. Lett. **98**, 136803 (2007).
13. I. Siddiqi, R. Vijay, F. Pierre, et al., Phys. Rev. Lett. **93**, 207002 (2004).
14. I. Siddiqi, R. Vijay, F. Pierre, C. M. Wilson, L. Frunzio, M. Metcalfe, C. Rigetti, R. J. Schoelkopf, M. H. Devoret, D. Vion, and D.Esteve, Phys. Rev. Lett. **94**, 027005 (2005).
15. K. Kim, M. S. Heo, K. H. Lee, et al., Phys. Rev. A **72**, 053402 (2005).
16. K. Kim, M. S. Heo, K. H. Lee, et al., Phys. Rev. Lett. **96**, 150601 (2006).
17. L. D. Landau and E. M. Lifshitz, *Mechanics, Course of theoretical physics vol.1*, (1969).
18. I. Siddiqi, R. Vijay, F. Pierre, et al., cond-mat/0507248 (2005).
19. M. I. Dykman, I. B. Schwartz, and M. Shapiro, Phys. Rev. E **72**, 021102 (2005).
20. D. Ryvkine and M. I. Dykman, Phys. Rev. E **74**, 061118 (2006).

Quantum Nanomechanics

Pritiraj Mohanty

Abstract Quantum Nanomechanics is the emerging field which pertains to the mechanical behavior of nanoscale systems in the quantum domain. Unlike the conventional studies of vibration of molecules and phonons in solids, quantum nanomechanics is defined as the quantum behavior of the entire mechanical structure, including all of its constituents—the atoms, the molecules, the ions, the electrons as well as other excitations. The relevant degrees of freedom of the system are described by macroscopic variables and quantum mechanics in these variables is the essential aspect of quantum nanomechanics. In spite of its obvious importance, however, quantum nanomechanics still awaits proper and complete physical realization. In this article, I provide a conceptual framework for defining quantum nanomechanical systems and their characteristic behaviors, and chart out possible avenues for the experimental realization of *bona fide* quantum nanomechanical systems.

1 Why Quantum Nanomechanics

A Quantum Nano-Mechanical (QnM) system is defined as a structure which demonstrates quantum effects in its mechanical motion. This mechanical degrees of freedom involve physical movement of the entire structure. In its current physical realizations, a typical nanomechanical system may consist of 100 million–100 billion atoms. The mechanical degrees of freedom are therefore described by macroscopic variables.

Experimental access to the quantum realm is *crudely* defined as the regime in which the quantum of energy hf in a resonant mode with frequency f is larger than the thermal energy k_BT. The motivation behind this crude definition of the quantum regime is simple. The motion of a QnM system can be described by a harmonic

P. Mohanty (✉)
Department of Physics, Boston University, 590 Commonwealth Avenue, Boston, MA 02215, USA
e-mail: mohanty@physics.bu.edu

V. In et al. (eds.), *Applications of Nonlinear Dynamics*, Understanding Complex Systems, 25
DOI 10.1007/978-3-540-85632-0_3, © Springer-Verlag Berlin Heidelberg 2009

oscillator potential. In the quantum regime, the harmonic oscillator potential energy levels are discrete. In order to observe the effects of discrete energy levels, smearing by thermal energy—due to finite temperature of the QnM system—must be small compared to the energy level spacing, hf. However, a formal definition of the quantum regime must involve a proper definition of the QnM system itself, which may include a much more general potential. In any case, the condition $hf \geq k_B T$ gives physically relevant parameters: a nanomechanical structure with a normal mode resonance frequency at 1 GHz will enter the quantum regime below a temperature $T \equiv (h/k_B)f = 48$ mK. Since typical dilution cryostats have a base temperature of 10 mK, nanomechanical structures with frequencies above 1 GHz can enable experimental access to the quantum regime [1, 2]. The experimental challenge is then to fabricate structures capable of high gigahertz-range resonance frequencies, and to measure their motion at low millikelvin-range temperatures. Because the resonance frequency increases with decreasing system size, one or many of the critical dimensions of the gigahertz-frequency oscillators will be in the sub-micron or nano scale.

What is the fundamental reason behind a new intiative to physically realize QnM systems in new experiments? Quantum mechanical oscillators have never been realized in engineered structures [3]; our physical understanding of quantum harmonic oscillators come from experiments in molecular systems. Furthermore, the obvious extension may also include applications in quantum computing—any quantum system with discrete energy levels and coherence can be construed as quantum bits. Therefore, imagining QnM systems as potential nanomechanical qubits is not farfetched. From a foundational perspective, study of coherence and tunneling effects in any quantum system, somewhat macroscopic in size, lends itself to relevant questions in quantum measurement—usually in a system-environment coupling framework.

Beyond these obvious interests, I argue that a plethora of new and fundamentally important physical problems can be experimentally studied with QnM systems. These problems range from dissipative quantum systems [4] and quantum decoherence in the measurement problem [5, 6] to phase transition models in condensed matter physics. Furthermore, the structure size of a typical QnM system lies in a regime where the continuum approximation of the elasticity theory is bound to fail [7]. The atomistic molecular dynamics approach also becomes severely limited due to the large number of atoms. The size of 100 million to 100 billion atoms requires multi-scale modeling of the elastic properties of QnM systems, which may require novel approaches to computational modeling of large systems. Currently, the state-of-the-art large-scale computing power of a large cluster can handle a size of 100–200 million atoms. Fundamentally, QnM systems may enable a new formalism that marries quantum descriptions of molecules, usually studied in chemistry, with physicist's approach to mechanical systems, quantum or classical.

This is a list, see Fig. 1, of some of the obvious and not-so-obvious potential applications of QnM systems. Although this list is primarily utilitarian, I argue that uncharted territories bring about unknown concepts. Therefore, it is quite conceivable that—once the experimental activities in QnM systems take off—some yet unknown concept will completely dominate this short list of studies.

Fig. 1 Nanomechanical devices important to the foundation of quantum nanomechanics. **(a)** A silicon nanomechanical beam which can work as a nanomechanical memory element by its controlled transition between two nonlinear states [8,9]. **(b)** A nanomechanical spin-transport device is used to detect and control spins through a nanowire by the associated spin-transfer torque [10]. The device contains a hybrid half-metallic half-ferromagnetic nanowire, which sits on top of a suspended silicon torsion oscillator. **(c)** A nanomechanical beam with electrostatic gate coupling may allow tunable nanomechanical qubit, which will be robust against environmentally induced decoherence due to its macrosopic structure. **(d)** A novel multi-element oscillator structure which allows very high frequency oscillation without compromising detectability of small displacements arising due to high spring constant of a straight-beam oscillator [1]

2 Quantum Nanomechanical Systems: Definitions and Requirements

Nanomechanical systems can be defined as mechanical structures free to move in three dimensions with one or many of the critical dimensions under 100 nm. Quantum nanomechanical systems are structures which under certain conditions demonstrate quantum mechanical behavior in their motion.

2.1 Dimensionality

A formal definition of quantum nanomechanics involves quantum mechanics in the acoustic modes of the structure, which include flexural (bending), torsional, and longitudinal modes. These modes represent a geometric change in the shape of the structure [11, 12]. Therefore an appropriate choice for dimensionality involves how these modes are generated and how they scale as a function of length, width or thickness. In Table 1, we define four distinct dimensions, corresponding to the relative

Table 1 Dimensionality of nanomechanical systems in terms of geometrical parameters, length L, width w and thickness t for rectangular geometry

Geometrical Parameters	Dimensionality	Description
$w,t \ll L$	1D	String limit
$t \ll w$; $w,t \ll L$	Quasi-1D	Thin-beam limit
$t \ll L,w$	2D	Membrane limit
$t \sim w \sim L$	3D	Solid limit

geometric length scales of a rectangular structure. A fundamental distinction among nanomechanical systems with different dimension is the scaling of natural resonant mode frequencies with the natural length scales. For example, in the string limit, resonance frequency varies as $1/L$, whereas in the quasi-1D thin-beam limit, resonance frequency of the natural flexural modes varies as t/L^2 according to the elastic theory of continuous media. The relationship between resonance-mode frequencies and geometric parameters also includes a number of relevant material parameters such as material density ρ, Young's modulus Y, sound velocity v_s, and thermal conductivity κ.

2.2 *Classical and Quantum Regimes*

As listed in Table 2, a nanomechanical structure is described by a number of characteristic length scales, important for describing its mechanical motion in either quantum or classical regime. In addition to the scales corresponding to geometry and acoustic phonon wavelength, thermal length defines how far a phonon, the mechanical mode of vibration, extends within the thermal time $\tau_\beta = \hbar/k_BT$, where τ_β represents the timescale for the system to reach equilibrium with the thermal bath at temperature T. The condition for the entire nanomechanical system to be in the quantum regime, the phonon or the mechanical excitation has to extend over the length of the system or $hv_s/k_BT \geq L$. For example, in a silicon nanomechanical beam, the thermal length is ~ 2 micron at a temperature of ~ 100 mK.

Table 2 Length scales of nanomechanical systems in both classical and quantum regime

Characteristic length scale	Notation	Description
geometrical length	L,w,t	Rectangular Structure
lattice constant	a	Crystal Structure
thermal phonon wavelength	λ_{th}	$\lambda_{th} = hv_s/k_BT$
acoustic phonon wavelength	λ_k	$\lambda_k = 2\pi/k_n$
dissipation length	L_d	$L_{d(n)} = v_s\tau_{d(n)} = v_s(2\pi Q_n/\omega_n)$
decoherence length	L_ϕ	$L_\phi(n) = v_s\tau_\phi(n)$
de Broglie wavelength	λ_B	$h/\sqrt{2\pi mk_BT}$
oscillator length	$L_O(n)$	$\sqrt{\hbar/m\omega_n}$

Therefore, a *fully* QnM system of silicon at a temperature above 100 mK should have critical dimensions less than 2 microns in length, irrespective of the quality factor Q.

Thermal correlation time is crucial in distinguishing a quantum mechanical system from a classical one, particularly in presence of dissipation (characterized by quality factor Q). Consider, for instance, a temperature at which $\tau_\beta < T_n$, where $T_n = 2\pi/\omega_n$ is the period of oscillation for a given mode of vibration. In this case, correlation between the system and the thermal bath is lost before the end of one cycle of oscillation. Even though, the energy in the classical description is lost in Q cycles, the quantum dynamics is independent from cycle to cycle in this regime. Therefore, the first condition to be in the quantum regime is $\tau_\beta > T_n = 2\pi/\omega_n$ or $\hbar\omega > k_B T$. For simple harmonic oscillator motion energy level spacing is $\hbar\omega$, so this condition also takes care of the requirement that thermal smearing must be smaller than the energy level spacing.

Dissipation length is defined through dissipation time $\tau_d = 2\pi Q_n/\omega_n$, where ω_n and Q_n are the resonance angular frequency and the quality factor for a given mode at a specific temperature. This is the characteristic timescale for loss of energy in the system. In the language of phonons, $1/\tau_d$ is then the inelastic scattering rate of the phonon due to its coupling to the intrinsic or extrinsic environmental degrees of freedom. It is important to compare this to the decoherence of the system at the rate $1/\tau_\phi$. Typically, decoherence of the system can occur much faster than the dissipation of energy, $1/\tau_\phi > 1/\tau_D$ or $\tau_\phi < \tau_D$.

2.3 Requirement for the Nanomechanical Structures to be Quantum Mechanical

There are two characterizing aspects of quantum mechanics: coherence and spin. These defining characters manifest in interference effects and statistics, which are not observed in a classical system. Following the spirit of large quantum mechanical systems, one can list two primary requirements for the nanomechanical system to be quantum mechanical. First, quantum coherence of the system must involve proper definition of the relevant physical quantity. Second, macroscopic character of the system must be integrated into the relevant physical quantities to differentiate a quantum nanomechanical system from an idealized point particle in a harmonic oscillator potential.

In a classical description, the motion of a beam or a cantilever can be completely described by its transverse displacement at a single point along its length, in particular for flexural or bending motion. Other physical quantities such as velocity and acceleration can be obtained from the transverse displacement $u(x,t)$ for $x\varepsilon[0,L]$, where x is the coordinate along the beam axis. Instead of $u(x,t)$, one can define the integrated transverse displacement ψ to describe the beam's motion with a single parameter. ψ can be obtained by integrating the appropriate displacement field $u(x,t)$ along the length with proper boundary conditions. Special cases will involve further constraints on ψ to describe special physical situations (for example, for incompressible beams). ψ can be thought of as an order parameter, representing the motion of a macroscopic structure in both linear and nonlinear regimes. Such a definition can also enable a simple formalism for studying phase transition in the Euler instability region: beyond a critical force a straight beam demonstrates transition to two separate phases of broken symmetry, characterized by mean square displacement [8]. The second advantage of the definition of an order parameter is the natural connection to the Bose-Einstein Condensation (BEC) description, which contains the essential physics of a large mechanical system, including classical phase transition, macroscopic quantum coherence and multi-stable potential dynamics.

The quantum mechanics of a nanomechanical system can be described by the order parameter with an amplitude and a phase: $\psi = |\psi|e^{i\phi}$. In the quantum regime, the nanomechanical system becomes a phase coherent system with a "macroscopic" quantum wave function. Since, the quantum motion of nanomechanical systems involve matter waves, the macroscopic nature of the structure can simply be defined in terms of its mass m through a new quantity, M-factor, which is defined by $M = log(m/m_e)$, where m_e represents the mass of an electron. The concept behind this simple definition is two-fold. First, if the M-factor is larger than 10, then it can be considered truly macroscopic from the perspective of our experience in the "everyday world." Figure 2 displays the M-factor for a number of macroscopic

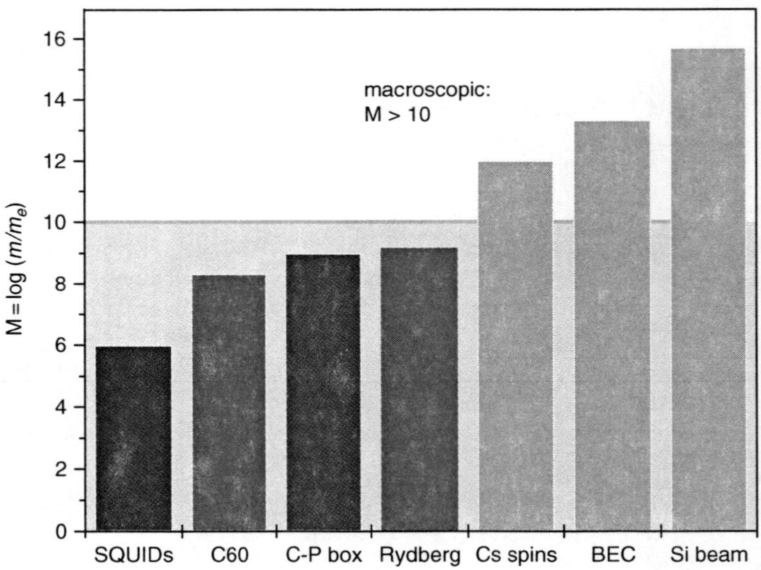

Fig. 2 Macroscopic nature of the quantum nanomechanical systems shown in comparison with other macroscopic quantum systems by the M-factor, which characterizes the mass of the system relative to the electron mass. With this new definition, structures with an M-factor of 10 or higher will have true macroscopic realism. Quantum nanomechanical systems, about $\sim$ 10 microns in size, can in fact be seen by naked eye

quantum systems. Nanomechanical structures in the quantum regime can contain about a billion atoms, and their size in the range of microns can in fact allow them to be observed by naked eye. The second reason is to emphasize the mass of the system in determining its quantum mechanical behavior in terms of coherent matter waves.

Coherence of the matter wave, representing quantum nanomechanical motion, can be characterized by a decoherence time τ_ϕ. In simple cases, the decoherence time or the associated decoherence length may be dominated by the de Broglie wavelength, which describes the spread of a gaussian wave packet. However, a proper analysis of intrinsic decoherence mechanisms must be made for the correct estimate. The simplest approach is to follow the convention in defining decoherence of Schrodinger cat states in BEC [13, 14].

3 Potential Quantum Nanomechanical Systems

It is necessary for the physical realization of quantum nanomechanics that the appropriate conditions of quantum mechanics are satisfied. As mentioned earlier, the requirement of high resonance-mode frequency at low temperature, $\hbar\omega > k_B T$, may not be sufficient for the system to be fully quantum mechanical. The fundamental difficulty is to legislate what time scale or corresponding length scale, among those listed in Table 1, is the single characteristic length scale that determines if the macroscopic nanomechanical system is in the true quantum regime. Although there are relevant conventions in both atomic bose condensation and electronic mesoscopic physics, it is important to obtain experimental data to be able to fully identify the appropriate length and time scales. In this section, I list four different classes of experimental nanomechanical systems in which efforts are currently being made towards the observation of quantum effects.

Linear displacement and velocity of the nanomechanical systems can be detected by a number of transduction mechanisms, which allow conversion of a mechanical signal to an electrical signal. These include electrostatic detection technique in which the beam's motion is detected by measuring the change in the capacitance between an electrode on the beam and a nearby control electrode. As the distance between the two plates changes, the capacitance changes. In order to induce motion in the beam, an electric field can be applied between the two plates at or near the resonance frequency of the beam. In the optical technique, beam's displacement can be measured either directly or through an interferometric method. Because of the millikelvin temperature requirement it is difficult to employ optical techniques, as the minimum incident power from the laser will tend to increase the temperature substantially. The electrostatic technique is unsuitable because of the large parasitic capacitance between the different parts of the device and the surroundings. A variation of the electrostatic method is the coupled-SET (Single-Electron Transistor) technique in which the change in the capacitance between the two electrodes due to the motion of the beam is detected by a single-electron transistor. In this configuration, one of the electrode plates is used as a gate of the SET transistor. The change in the gate voltage is measured by detecting the change in the source-drain current of the SET. In spite of its sophistication, it is difficult with this technique to detect gigahertz-range motion in a straight beam, as the change in capacitance generated by the motion at these frequencies is very small.

3.1 Straight-Beam Oscillators

A straightforward approach to the quantum regime involves measurement of displacement or energy of a straight-beam nanomechanical structure in the gigahertz range at a temperature $k_B T < hf$. Although such submicron structures with expected gigahertz-range frequencies are now routinely fabricated in laboratories, motion at frequencies in the gigahertz range has not been detected with equal ease.

The fundamental problem in straightforward miniaturization of beam or cantilever oscillators is the increase in the stiffness constant along with the increasing frequency, which is required for getting into the quantum regime. For a straight beam in the thin-beam approximation, stiffness constant increases as $w(t/L)^3$, or $1/L^3$ if the cross-sectional dimensions w and t are kept constant. A high spring constant, typically in the range of 1000–10000 N/m, results in undetectably small displacements, typically in the range of 1–10 fm (femtometer), corresponding to a force of 1 pN. However, experimental considerations such as nonlinearity and heating require the range of force to be even smaller than that. Therefore, the problem of detecting motion in the quantum regime translates to the problem of detecting femtometer-level displacements at gigahertz frequencies, assuming that the structure cools to millikelvin-range temperature. In straight-beam oscillators, thermal phonon wavelength λ_{th} becomes orders of magnitude larger than the cross-sectional dimensions, which prevents the central part of the beam from cooling to the required millikelvin temperature.

3.2 Multi-Element Oscillators

Design of structures for the detection of quantum motion at gigahertz frequencies therefore is a two-fold problem. First, the normal-mode frequencies have to be in the gigahertz range. Second, the structure in these gigahertz modes must have a much lower spring constant "k_{eff}" to generate a detectable displacement or velocity. This cannot be achieved with simple beams as "k_{eff}" and "ω" are coupled by trivial dispersion relations. The problem is to find a structure with certain modes in which "k_{eff}" and "ω" can be decoupled. However, decoupling of "k_{eff}" and "ω" cannot be achieved in single-element structures.

One type of multi-element structure, comprising of two coupled but distinct components [1] has been experimentally studied. Small identical paddles serve as the frequency-determining elements, which generate gigahertz-range natural frequencies because of their sub-micron dimensions. The paddles are arranged in two symmetric arrays on both sides of a central beam, which acts as the displacement-determining element. Because of its multipart design, the structure displays many normal modes of vibration, including the fundamental mode and numerous complex modes. By design, there exists a class of collective modes at high frequencies, apart from all other normal modes. In the collective modes, the sub-micron paddles move in phase to induce relatively large amplitude of motion along the central beam at the same frequency.

In recent experiments, the antenna structure has been studied in detail at low temperatures by the magnetomotive technique [1, 2]. It exhibits the expected classical behavior at the low frequency modes. A class of high-frequency collective modes are observed in the range of 480 MHz—3 GHz. At temperatures corresponding to high thermal occupation number k_BT/hf, the high-frequency gigahertz modes show the expected classical behavior, equivalent to the linear Hooke's law.

In the quantum regime, $N_{th} \equiv k_BT/hf \leq 1$, the gigahertz modes show discrete transitions in contrast to the classical behavior of the same modes at higher temperatures, $N_{th} \gg 1$. The 1.5 GHz mode at 110 mK ($N_{th} \to 1$) displays discrete transitions as a function of driving force or magnetic field [1]. While the transitions do not always occur at exactly the same field values from sweep to sweep, the jump size remains unchanged, suggesting that the oscillator switches between two well-defined states. Although these reproducible discrete jumps could indicate transition to quantum behavior, it is difficult to gain more insight into the nature of the two states from the data.

A higher frequency resonance mode at 1.88 GHz was studied down to a cryostat temperature of 60 mK [15], deeper in the quantum regime, corresponding to $N_{th} \sim 0.66$. Figure 3 shows a four-state discrete velocity response in the form of a staircase as a function of continuous driving force. In frequency domain, the response displays clear gaps in the growth of the resonance peak as the magnetic

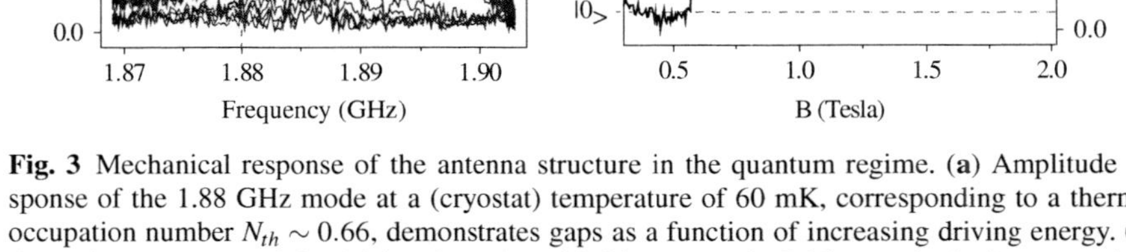

Fig. 3 Mechanical response of the antenna structure in the quantum regime. (**a**) Amplitude response of the 1.88 GHz mode at a (cryostat) temperature of 60 mK, corresponding to a thermal occupation number $N_{th} \sim 0.66$, demonstrates gaps as a function of increasing driving energy. (**b**) A continuous sweep of the driving force (provided by magnetic field) at a single frequency 1.88 GHz shows discrete jumps

field is increased in equal increments of 0.2 tesla. The lineshape remains almost Lorentzian, which excludes the possibility of a standard classical nonlinear effect.

The response is highly reproducible with a strong temperature dependence. Above the mixing-chamber temperature of 1.2 K, corresponding to $N_{th} \sim 14$, the discrete transitions abruptly disappear, and the mode response varies smoothly with applied drive, consistent with a classical behavior. In addition, low megahertz frequency modes of the same oscillator at the lowest measured refrigerator temperature of 60 millikelvin demonstrate expected classical dependence. Further investigations are currently underway to fully characterize this remarkable effect.

3.3 Tunneling Two-State Oscillator in a Double-Well Potential

A different approach to the experimental realization of quantum effects in the motion of a nanomechanical system involves quantum tunneling of the entire structure between two physically distinct states. A nanomechanical beam oscillator can be driven in the nonlinear regime where the equation of motion is given by the standard Duffing oscillator expression. The amplitude response of the oscillator changes from the standard Lorentzian form in the linear regime to an asymmetric hysteretic form in the Duffing regime where the oscillation amplitude is multi-valued. Hysteresis is demonstrated in frequency sweep, as the oscillator follows two different states depending on whether frequency is swept forward or backward through the bistable regime.

Recent experiments have demonstrated controlled switching between the two nonlinear states of the oscillator in the classical regime [8]. The addition of a slowly-varying driving force modulates the double-well potential, associated with nonlinearity, and enables the system to go over the potential barrier following the sub-threshold modulation signal. This classically coherent transition between the two states can also be enhanced by the application of white noise to the system, which results in stochastic resonance for a given range of noise power [9]. As temperature is decreased, another pathway for transition between the two states opens up as the probability for the system to directly tunnel through the barrier increases. Quantum mechanical tunneling of the macroscopic nanoscale oscillator is a fundamentally different realization of quantum nanomechanics. Towards this end, there have been a few theoretical analyses [16–18]. However, new experiments are clearly needed for a better definition of this problem.

3.4 Coupled Nano-Electro-Mechanical Systems

Currently, most of the theoretical activities have been focused on coupled-NEMS (nano-electro-mechanical systems). In this setup, motion of a nanomechanical resonator is analyzed through its coupling to a measurement appartus provided by either a single electron transistor (SET) or a cooper-pair box. Recently, two experimental groups have demonstrated that it is possible to detect mechanical motion of oscillators using the SET-detection technique [19,20]. Although the initial experiments (performed on oscillators with relatively low frequencies in the megahertz range, and hence high thermal occupation number N_{th}) have demonstrated relatively high displacement detection sensitivity, it is not clear if at higher gigahertz-range frequencies the detection sensitivity will be as high. Nevertheless, this technique, motivated essentially by qubit experiments in mesoscopic physics, offers the possibility of detecting quantum motion in straight-beam oscillators.

4 Endnote

The field of quantum nanomechanics is off to a good start. However, new experiments are needed to build a solid phenomenology since this is a new territory in respect to our intuition of large quantum systems. It is still not known if quantum mechanics in its current form remains a valid description of systems as large as the nanomechanical structures with billions of atoms [6]. There are further complications due to finite dissipation and finite decoherence. Therefore, it is important to manage our expectation of what we ought to observe in experiments. Perhaps it is prudent to be guided by phenomenology as we build up the conceptual and theoretical framework.

Experiments with fundamentally different measurements of displacement, velocity, acceleration and other mechanical properties are needed to be performed on nanomechanical systems with high frequencies at low temperatures. It is important to include multiple actuation and detection approaches to address the issues of eigen selection, quantum non-demolition and back action. Materials choice also becomes an important concern, hence repetition of the same measurements on structures of different materials can also elucidate fundamental quantum effects. Beyond the obvious materials such as silicon, silicon carbide, gallium arsenide, carbon nanotube and carbon 60, it is important to explore doped and undoped diamond [21], aluminum nitride, graphene [22] and other new materials for nanomechanics.

On the theoretical side, a bootstrap approach to developing the appropriate framework will be needed. Proper definitions of relevant length scales and time scales must be done along new experimental data. Furthermore, basic theoretical analyses must be done to understand the classical dynamics and the energy spectrum of extended mechanical objects of relevant geometry. A lot of progress is being currently made in multi-element oscillators, where exact calculations of the energy spectrum for driven and undriven periodic structures have been done. Similar calculations are needed to define and understand measurable quantum properties of realistic systems studied in experiments. Lastly, it is important to continue the debates and discussions [23, 24] which are fundamentally important to developing a new field.

Acknowledgments The experimental part of the work has been done in collaboration with numerous members of my group, including Alexei Gaidarzhy, Robert Badzey, Guiti Zolfagharkhani, Yu Chen, Matthias Imboden, Josef-Stefan Wenzler, Tyler Dunn and Diego Guerra. This work is supported by National Science Foundation (grant no. DMR-0449670).

References

1. A. Gaidarzhy, G. Zolfagharkhani, R. L. Badzey and P. Mohanty, Phys. Rev. Lett. 94, 030402 (2005).
2. A. Gaidarzhy, G. Zolfagharkhani, R. L. Badzey and P. Mohanty, Appl. Phys. Lett. 86, 254103 (2005).
3. M. Bocko and R. Onofrio, Rev. Mod. Phys. 68, 755 (1996).
4. Ulrich Weiss, Quantum Dissipative Systems (2nd Edition) (World Scientific, 1999).
5. A. J. Leggett, in *Chance and Matter* -Les Houches XLVI (eds J. Souletie et al.) (Elsevier Science Publishers, 1987).
6. A. J. Leggett, J. Phys. Cond. Matt. 14, R415 (2002).
7. R. Phillips, Crystals, Defects and Microstructures: Modeling Across Scales (Cambridge University Press, 2001).
8. R. L. Badzey, G. Zolfagharkhani, A. Gairdazhy, and P. Mohanty, Appl. Phys. Lett. 85, 3587 (2004).
9. R. L. Badzey and P. Mohanty, Nature 437, 995 (2005).
10. P. Mohanty, G. Zolfagharkhani, S. Kettemann, and P. Fulde, Phys. Rev. B 70, 195301 (2004).
11. L. D. Landau and E. M. Lifshitz, Theory of Elasticity (Pergamon, 1959).
12. S. P. Timoshenko and J. N. Goodier, Theory of Elasticity (McGraw-Hill, 1971).
13. Diego A.R. Dalvit, J. Dziarmaga, and W. H. Zurek, Phys. Rev. A 62, 13607 (2000); *ibid.* cond-mat/0001301; cond-mat/0006349.
14. Y. P. Huang and M. G. Moore, cond-mat/0508659.
15. A. Gaidarzhy et al., AIP Conference Proceedings 850, 1677 (2006).
16. S. M. Carr, W. E. Lawrence and M. N. Wybourne, Phys. Rev. B 64, 220101 (2001).
17. V. Peano and M. Thorwart, Phys. Rev. B 70, 235401 (2004).
18. P. Werner and W. Zwerger, Europhys. Lett. 65, 158 (2004).
19. R. G. Knobel and A. N. Cleland, Nature 424, 291 (2003).
20. M. D. LaHaye et al., Science 304, 74 (2004).
21. M. Imboden et al., Appl. Phys. Lett. 91, 203503 (2007).
22. J. Bunch et al., Science 315, 490 (2007).
23. K. Schwab et al., Phys. Rev. Lett. 95, 248901 (2005).
24. A. Gaidarzhy et al., 95, 248902 (2005).

Coupled-Core Fluxgate Magnetometer

Andy Kho, Visarath In, Adi Bulsara, Patrick Longhini, Antonio Palacios, Salvatore Baglio and Bruno Ando

Abstract A fluxgate magnetometer is a magnetic field sensor that is used to detect relatively low intensity magnetic fields. It belongs to a class of nonlinear systems known as the overdamped bistable system whose dynamics can be described by the generic form $\dot{x}(t) = -\nabla U(x)$. It is well known that an overdamped system will not oscillate on its own. It is usually driven with a known periodic signal with an amplitude large enough to overcome the potential barrier height. Instead of this approach, we will show how similar oscillations can be generated with a carefully selected coupling scheme, coupling parameters, and initial conditions [1, 2], thus eliminating the need for expensive driving signals. We will show that the coupled-core fluxgate magnetometer can not only detect DC magnetic fields, but it can detect and classify AC magnetic fields as well.

1 Introduction

Fluxgate magnetometers have long been used as room temperature magnetic field sensors for detecting low intensity (below .01 mT) magnetic fields. Since it is an overdamped system, the normal operation of the device is to drive it with a known periodic signal sufficient to overcome the potential barrier height and quantify the response of the signal by looking at the second harmonics of the power spectral density (PSD). This usually requires a large driving signal in the kHz frequencies and complex electronics. However, advances in time domain readout schemes [3] have afforded a less complex readout strategy as well as a drive signal with significantly less amplitude and lower frequency (a few hundred Hz). We have successfully designed, built, and tested working prototypes of this type of magnetometer using Residence Time Detection (RTD) readout.

A. Kho (✉)
SPAWAR Systems Center San Diego, 53560 Hull St, Code 71730, San Diego CA 92152
e-mail: andy.kho@navy.mil

V. In et al. (eds.), *Applications of Nonlinear Dynamics*, Understanding Complex Systems,
DOI 10.1007/978-3-540-85632-0_4,

The fluxgate magnetometer can be treated as a nonlinear dynamic system by assuming the core as approximately single-domain. In order to improve the sensitivity of the device and add new detection regimes by exploiting the dynamics of the system, we have developed a carefully selected coupling scheme [1, 2]. The topology we propose is a unidirectionally coupled elements with cyclic boundary conditions. Our analysis shows that N (taken to be odd, although the oscillatory behavior is also seen for N large and even) would, in fact, oscillate when a control parameter – in this case the coupling strength – exceeds a critical value. Note that energy conservation dictates that at least one of the elements must have an initial state different from the others, for the oscillations to occur; with identical initial conditions, the overdamped dynamics will rapidly collapse into the stable steady states (potential energy minima) of each element. By substituting the driving function with a coupling term λ, the following dynamical equation is obtained:

$$\dot{x}_i = -x_i + \tanh(c(x_i + \lambda x_{i+1} + \varepsilon)), \tag{1}$$

where $x_i(t)$ represents the normalized magnetic flux at the output of unit i, c a temperature-dependent system parameter, and ε is an externally applied DC "target" magnetic flux. The individual dynamics in (1) are derived via a mean field description of the domain dynamics in a ferromgnetic core, effectively, the core is treated as a "single-domain" entity; it is, in fact, possible to realize this, in practice, by meticulous fabrication [4–7]. It is important to stress that this behavior is quite general; it has been demonstrated in coupled overdamped Duffing elements [1, 2], and applied to the analyses of the frequency-selective properties of interacting neural networks [8].

The oscillatory behavior of the system is controlled by the coupling strength λ. When $\lambda = 0$, there are no oscillations because of the overdamped nature of the system. However, oscillatory behavior occurs at a critical coupling, λ_c; note that in our convention, $\lambda < 0$ so that oscillations occur for $|\lambda| > |\lambda_c|$. λ_c is given by the expression:

$$\lambda_c = -\varepsilon - \sqrt{(c-1)/c} + \frac{1}{c}\tanh^{-1}\sqrt{(c-1)/c}. \tag{2}$$

The individual elemenatal oscillations are separated in phase by $2\pi/N$, and have period $T_i = \frac{N\pi}{(c(c-1))^{1/4}}\left[\frac{1}{\sqrt{\lambda_c-\lambda}} + \frac{1}{\sqrt{\lambda_c-\lambda+2\varepsilon}}\right]$; these oscillations can be experimentally produced at frequencies ranging from a few Hz to high kHz. Hence, the frequency increases as λ increases (ie. gets more negative) as shown in Fig. 1. The coupling parameter λ can be used to tune the device to the desired operating frequency. Interestingly, the *summed* output oscillates at period $T_\Sigma = T_i/N$ and its amplitude (as well as that of each elemental oscillation) is *always supra*threshold i.e., the emergent oscillations correspond to switching events between the stable steady states of each core. Just like the single fluxgate magnetometer, when the coupled-core version is subjected to a DC magnetic field where $\varepsilon \neq 0$, the residence time shifts so the RTD method can still be used for magnetic field detection. Though,

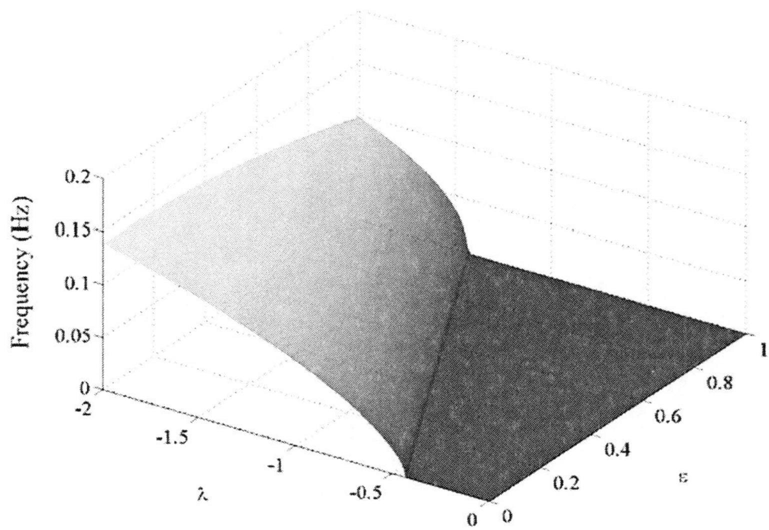

Fig. 1 Frequency scaling of a coupled-core fluxgate as a function of coupling strength λ and magnetic field strength ε

with the ability to tune the device with the coupling parameter λ, the device can be made much more sensitive by operating it near λ_c. For DC magnetic field detection, the readout electronics is exactly the same as with the single fluxgate.

Unlike the single fluxgate, however, the coupled-core fluxgate changes frequency in the presence of different magnetic field strengths, (see Fig. 1). The frequency decreases with increasing magnetic field strength according to the square root scaling equation. This adds another way to detect DC signals other than the RTD method.

Another new feature of the coupled-core system is the ability to detect and lock onto AC magnetic fields of the form $h(t) = \varepsilon sin(\omega t)$ (see Fig. 2). If the device is tuned just past λ_c so that it is oscillating *apriori*, it will oscillate out of phase at its own natural frequency. As a small AC magnetic field is introduced, the system continues to oscillate out of phase at its own natural frequency, but the frequency of the AC magnetic field is modulated on top of that. As the AC magnetic field is increased slightly, the system switches to a state where it is no longer oscillating at its own natural frequency but is instead oscillating out of phase at the AC signal divided by the number of sensor cells. Finally, as the AC magnetic field is increased even more, the system switches to oscillating in phase at the same frequency of the AC magnetic field.

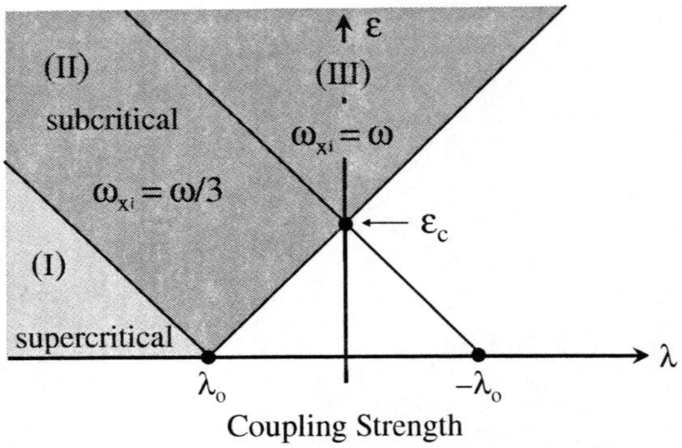

Fig. 2 Theoretical Phase Diagram: Oscillatory behavior of the coupled fluxgate model when applying a AC magnetic field $h(t)$. In the supercritical regime, the oscillations form a traveling wave pattern. In the subcritical regime, with $h(t)$ small, the system oscillates about one of the steady states ± 1, while with $h(t)$ large, the system oscillates between two steady states. In both cases the oscillation form a traveling wave and their frequency is exactly $\omega/3$. For ε greater than a critical value, all three waves are in-phase with each other and frequency synchronized with the external signal $h(t)$ in region (III)

2 Experimental Setup

The minimum number of cores that can be used for the above described coupling topology is $N = 3$ cores. For the sake of simplicity, this case was used for the experiments. The equations describing the dynamics are:

$$\begin{aligned} \dot{x}_1 &= -x_1 + \tanh(c(x_1 + \lambda x_2 + \varepsilon)) \\ \dot{x}_2 &= -x_2 + \tanh(c(x_2 + \lambda x_3 + \varepsilon)) \\ \dot{x}_3 &= -x_3 + \tanh(c(x_3 + \lambda x_1 + \varepsilon)). \end{aligned} \tag{3}$$

The parameters c and λ are assumed to be identical in simulations and as closely matched as possible in experiments.

The fluxgate sensors in the experiments are constructed in one of two ways. The PCB (printed circuit board) technology based fluxgates are constructed using a three-layer architecture (see Fig. 3(a)). The two outer layers consist of PCB with copper traces printed on them and the middle layer is the ferromagnetic material with a high permeability constant (Metglas 2714A). There are two sets of traces printed on each of the PCBs. One set is for the excitation coil, and the other set is for the sensing coil. At both ends of the traces are small holes (0.5 mm). The traces and holes are drawn in such a way that when the PCBs are aligned, they complete the coils for the sensor. The middle layer is a small sheet of the ferromagnetic material cut to the right dimensions, which can either be a single slit or racetrack configuration. The sensor is constructed by sandwiching the ferromagnetic material between the two PCB boards and aligning the holes on the two PCBs. A small jumper is passed through the holes from one board to the other and soldered in place to make

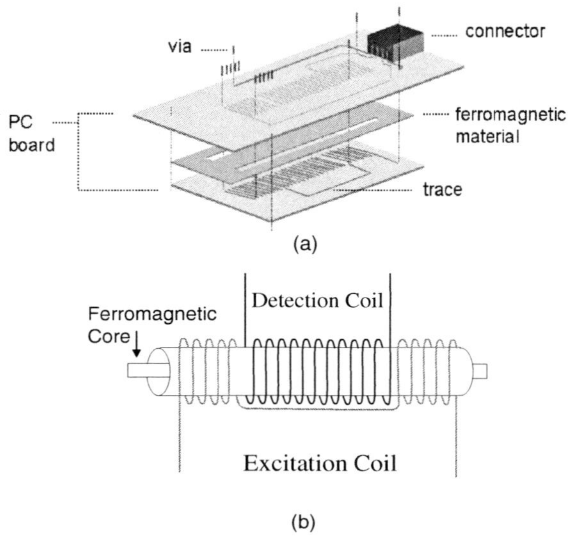

Fig. 3 The sensors for the fluxgate magnetometers are made from one of two ways. (**a**) shows a 3-layer PCB version and (**b**) shows the version with magnet wire wrapped around a small tube

the connections. It is important that the ferromagnetic material is not soldered or heated beyond the curie temperature. The two PCB boards are fused in this manner to complete the windings for the sensing and excitation coils. The interface connector for the sensor is a standard 4-pin Molex connector used for easy connection of the sensor coils to the electronics.

The tube version of the sensor is made with a core 4 mm in diameter and 10 cm in length (see Fig. 3(b)). 38 AWG magnet wire is used to wrap around the tube to make the loops for the sensor. The excitation coil is wrapped around the core first to a length of 5 cm and the detection coil is wrapped around the excitation coil to a length of about 1.5 cm. These coils are once again connected to a 4-pin Molex connector. The ferromagnetic material is slipped inside of the core and held securely in place. Using this type of core allows for more loops, thus reducing the power required to drive and read the sensor.

The electronic coupling circuitry is shown in the flowchart in (Fig. 4). The (voltage) readout of one fluxgate signal (i.e. the derivative signal of the flux detected by the sensing coil) is amplified by an instrumentation amplifier with a high gain and input impedance. Following this, the signal is integrated with a "leaky" integrator to convert the derivative signal from the sensing coil back to the "flux" variable so that the experimental system closely conforms to the model. Even though the "leaky" integrator removes a lot of the DC component from the signal, the remaining DC component can cause the operational amplifiers to saturate. A high pass filter following the "leaky" integrator prevents this from happening. The signal then passes through an amplifier which gives the ability to tune the coupling parameter λ.

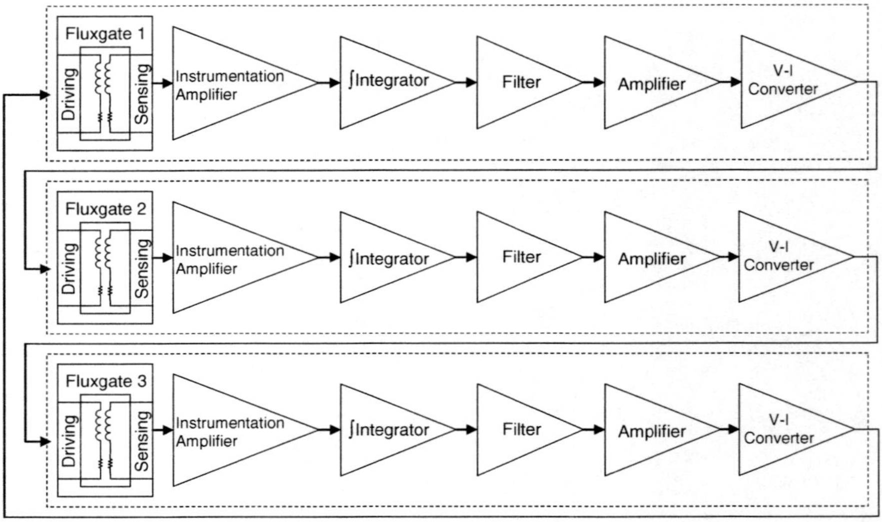

Fig. 4 Circuit flowchart of a coupled core fluxgate magnetometer

Finally, the signal passes through a voltage-to-current converter (V–I converter) to drive the primary coil of the adjacent fluxgate. The setup is the same for the remaining two cores and all values of the coupling circuit parameters are closely matched from one set to the other. The coupling circuitry is placed on a PCB to reduce noise as well as circuit parasitics. In addition, high speed and precision operational amplifiers are used to minimize the time delay in order to conform closely to the model since knowledge of state variable x_i is known instantaneously. Bifurcation analysis of the system and simulation results with delayed coupling terms shows no behavior change for delay times up to 10% of the oscillation frequency.

The readout signal is taken from the voltage output of the integrator. The voltage outputs from the three sensors are summed together with summing circuitry. Since the signals are 120° out of phase and non-sinusoidal, the resultant sum signal is 3 times the frequency of the individual signals. The signal is then passed through a Schmmitt Trigger, which outputs 0 V or +5 V corresponding to a Low and High, respectively. The output from the Schmitt Trigger is then passed to a PIC microcontroller. For every period of the sum signal the PIC microcontroller, with help from hardware counters, counts how long the Schmitt Trigger output is in the Low and High states. The Low and High states correspond to the lower (RT−) and upper (RT+) residence times, respectively. In practice, both the upper and lower residence times are accumulated for about 60 oscillation cycles. The number must be a multiple of 3 because of the inability to perfectly match all three sensor units. A mean is then computed for both, and the two means are subtracted. The difference of the two values is proportional to the external magnetic field. Likewise, the two numbers could also be added together to get the frequency which can also be used to quantify the external magnetic field strength.

3 Results and Conclusions

The oscillations from the experimental setup match very well qualitatively with the numerical simulations as shown in Fig. 5. The experimental setup cannot be exactly matched up with the simulation dynamics because the material parameter value, c, and the time constant, τ, cannot be easily extracted from the actual device. Also, in the experiments, the amplitude of the oscillations are set by amplifiers where as in the numerical simulations $\tau = 1$.

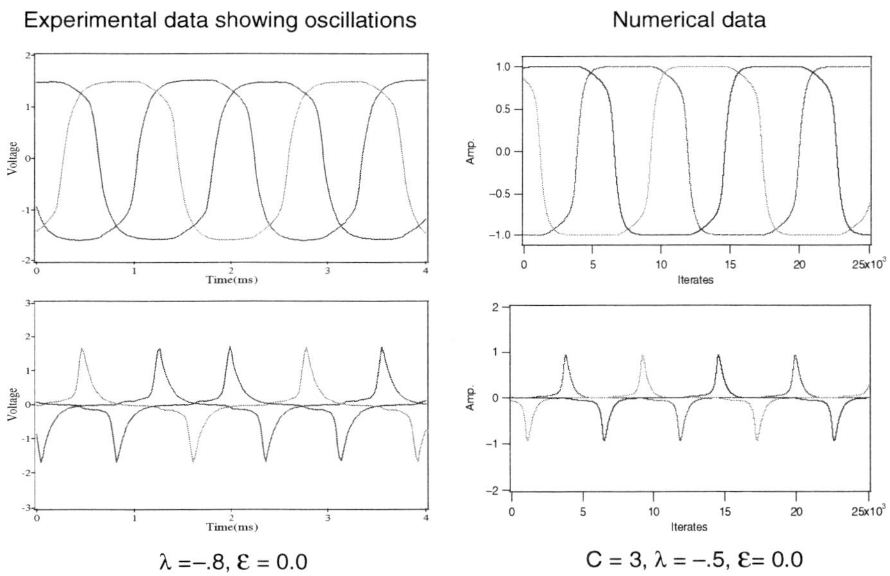

Fig. 5 Time series data of coupled-core fluxgate magnetometer. The two top plots are the flux signals and the bottom two plots are the derivative (observable)

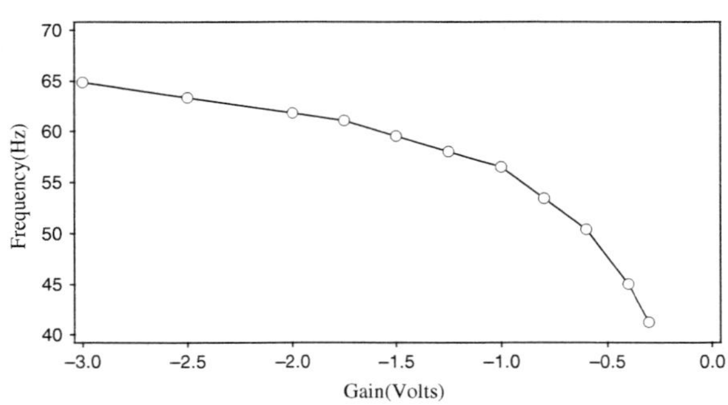

Fig. 6 Experimental results of frequency scaling appear to agree very well with theoretical results and with estimates obtained from computer simulations

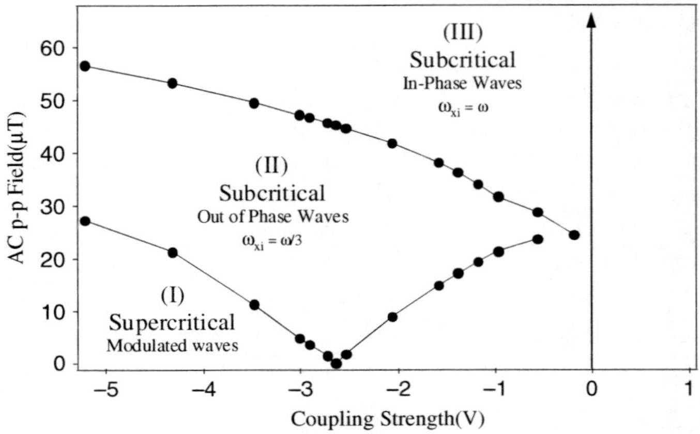

Fig. 7 Chart showing experimental data of the different oscillating regimes when the coupled-core fluxgate is subjected to an AC magnetic field

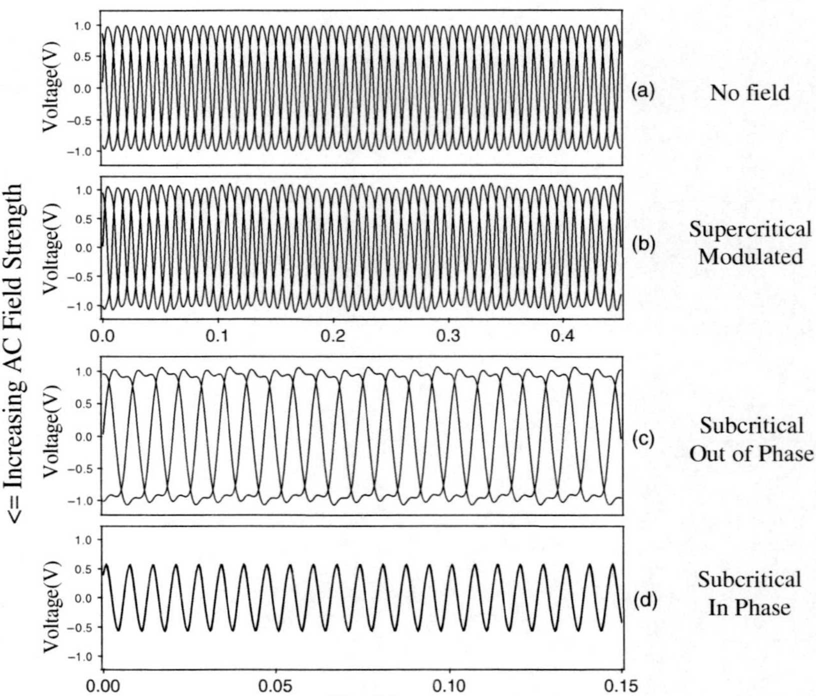

Fig. 8 Time series showing experimental data of the waveformes subjected to an AC magnetic field at an increasing magnetic field strength

Figure 6 shows that the frequency scaling from the experiment matches well with the numerical simulations. The experimental data clearly shows the characteristic square root scaling. As the coupling strength λ is increased (getting more negative), the frequency increases just as the theory predicts. The theory suggests that the frequency will continue to increase as λ is increased indefinitely, but in a real system, the operational amplifiers limit the actual practical tuning range of λ.

The response of the coupled core fluxgate magnetometer to an AC magnetic field also matches very well with theoretical prediction. All three oscillation regimes are seen in the experiment as shown in Fig. 7. The data clearly shows how the oscillations change as the AC field changes. Figure 8 shows experimental data of the time series of the different oscillation regimes. In (a), there is no AC magnetic field. The system is oscillating out of phase at its own natural frequency according to Eq. (2). As the AC field strength is increased (b), a modulation signal is introduced into the system; however, the system continues to oscillate out of phase at its own natural frequency. With an even larger AC magnetic field (c), the modulation stops and the system is no longer oscillating at its own natural frequency. Instead, it is oscillating out of phase at $\frac{1}{3}$ the frequency of the AC target signal. When the AC magnetic field is increased even more (d), the system no longer oscillates out of phase and locks onto the exact frequency of the target AC magnetic field.

Another possible way to operate the device is to set the coupling strength to the point just before where natural oscillations begin. That way the system is not oscillating *apriori*. When an AC magnetic field is detected by the sensor, it locks onto $\frac{1}{N}$ of the target signal. The coupling strength needs to be carefully chosen close to λ_c so that the maximum sensitivity is achieved.

Acknowledgments We gratefully acknowledge support from the Office of Naval Research (Code 331) and the SPAWAR internal research funding (S&T) program. A.P. was supported in part by National Science Foundation grants CMS-0625427 and CMS-0638814.

References

1. V. In, A. Bulsara, A. Palacios, P. Longhini, A. Kho, J. Neff, Phys. Rev. **E68**, 045102(R) (2003).
2. A. Bulsara, V. In, A. Kho, P. Longhini, A. Palacios, W-J. Rappel, J. Acebron, S. Baglio, B. Ando, Phys. Rev. **E70**, 036103 (2004).
3. A. Bulsara, C. Seberino, L. Gammaitoni, M. Karlsson, J. Robinson, Phys. Rev. E **67**, 016120 (2003).
4. B. Ando, S. Baglio, A. Bulsara, L. Gammaitoni; in Proceedings of the IEEE International Symposium on Circuits and Systems, Scottsdale, Arizona 2003.
5. B. Ando, S. Baglio, A. Bulsara, V. Sacco; Proceedings of the 24th IEEE Instrumentation and Measurement Conference, Como, Italy Pg. 1419 (2004).
6. B. Ando, S. Baglio, A. Bulsara, V. Sacco; IEEE Trans. Instr. Measur. **54**, 1366 (2005).
7. B. Ando, S. Baglio, A. Bulsara, V. Sacco, IEEE Sensors **5**, 895 (2005).
8. V. In, A. Palacios, P. Longhini, J. Neff, B. Meadows, Phys Rev. Lett. **91**, 244101 (2003).

Data Assimilation in the Detection of Vortices

Andrea Barreiro, Shanshan Liu, N. Sri Namachchivaya, Peter W. Sauer
and Richard B. Sowers

Abstract We develop new algorithms for target detection in multi-sensor environments. These methods are applied to study point vortex motion based on Lagrangian tracer information. First we solve analytically the nonlinear filtering problem for the special case of equal strength vortices. Recently developed methods, the particle filters that are based on importance sampling Monte Carlo simulations, are used for the detection of vortices in the the general case. Unlike the well-known extended Kalman filter, it is applicable to highly nonlinear systems with non-Gaussian uncertainties.

1 Introduction

Nonlinear filtering methods are used to dynamically integrate the computational and measurement aspects of real-time applications. The range of subjects in which nonlinearity and noise play a significant role is enormous. In this paper, we are interested in the development and application of prediction techniques to a specific *fluid mechanics* problem – vortex-driven tracer dynamics. Because of low viscosity vortices are long-lived structures. We study the conditional law of the vortices on the basis of the tracer observations.

The main focus of filtering is to combine computational models with sensor data to predict the dynamics of large-scale evolving systems. Filtering deals with recursive estimation of a signal or state of a random dynamical system from noisy measurements. When the signal and the observation model is linear and Gaussian, the filtering equation is linear as well and it is given by the well-known Kalman-Bucy filter. Otherwise the filter has a more complicated nonlinear structure. The signal that is represented by a Markov process cannot be accessed or observed directly and is to be "filtered" from the trajectory of the observation process which

A. Barreiro (✉)
Mathematics, University of Illinois, Urbana, IL 61801, e-mail: abarreir@math.uiuc.edu.

V. In et al. (eds.), *Applications of Nonlinear Dynamics*, Understanding Complex Systems,
DOI 10.1007/978-3-540-85632-0_5,

is statistically related to the signal. Suppose we make a forecast about the behavior at a future time of a complex system with some uncertainties (randomness) and there is near-continuous data available from remote sensing instrumentation networks. As new information becomes available through observations, it is natural to ask how to best incorporate this new information into the estimation and prediction.

The optimal estimate is given by the conditional expectation and can be generated by a recursive equation, called filter, driven by the observation process. Sensor data usually contain noise, and mathematical models are limited in accuracy due to model uncertainties. But, when used together, the resulting prediction of the state of large-scale dynamical systems must be superior to using either models or data alone.

2 Modeling: Vortex-Driven Tracer Dynamics

To obtain the point-vortex model, one starts with the 2-D vorticity equations. If viscous and external forces are neglected, the vorticity-transport equation corresponding to the Euler equations is [9, 12]:

$$\frac{\partial \omega}{\partial t} + (\mathbf{u}\cdot\nabla)\,\omega = (\omega\cdot\nabla)\,\mathbf{u} \quad \text{with} \quad \nabla\cdot\omega = \mathbf{0}, \tag{1}$$

where the vorticity vector $\omega \equiv \nabla \times \mathbf{u}$. These equations simplify for two-dimensional flows in (x_1, x_2) plane to

$$\frac{\partial \omega}{\partial t} + (\mathbf{u}\cdot\nabla)\,\omega = 0 \quad \text{with} \quad \nabla^2\psi = \omega, \tag{2}$$

where $\omega \equiv \omega_{x_3}$ is the vorticity component normal to the (x_1, x_2) plane and ψ is the stream function defined by

$$u_{x_1} = -\frac{\partial \psi}{\partial x_2} \qquad u_{x_2} = \frac{\partial \psi}{\partial x_1}$$

If ω consists of isolated, well-separated vortices, then a reasonable approximation is to consider the vortices as singularities or "point" vortices. In this case we express the vorticity field as

$$\omega(\mathbf{x},t) = \sum_{i=1}^{n} \Gamma_i \delta\left(\mathbf{x}_t^i - \mathbf{x}\right) \quad \text{with} \quad \mathbf{x}_0^i = \mathbf{x}^i, \tag{3}$$

where $\Gamma_i \neq 0$ is the circulation of vortex i. By inserting (3) in the Euler equations (1), and using the divergence free constraint along with Biot-Savart law, one obtains

$$\dot{\mathbf{x}}_t^i = \sum_{j,j\neq i}^{n} \frac{\Gamma_j}{2\pi} \frac{\left(\mathbf{x}_t^j - \mathbf{x}_t^i\right)^{\perp}}{|\mathbf{x}_t^j - \mathbf{x}_t^i|^2} \quad \text{with} \quad \mathbf{x}_0^i = \mathbf{x}^i. \tag{4}$$

Point vortex models that account for viscous effects also exist. Chorin [3] introduced the first random point vortex method to simulate viscous incompressible flows. Later Marchioro and Pulvirenti in [10] considered a continuous-time random vortex method with Gaussian random walks replaced by independent Brownian motions and proved a corresponding mean field type result. It was shown by Marchioro and Pulvirenti [10] and Agullo and Verga [1] that a stochastic vortex dynamics model approximates the evolution of vorticity with viscosity, in the same way in which the deterministic vortex dynamics simulates the Euler equations. Vortex dynamics with viscosity are governed by set of Langevin or stochastic differential equations:

$$\dot{\mathbf{x}}_t^i = \sum_{j,j\neq i}^{n} \frac{\Gamma_j}{2\pi} \frac{\left(\mathbf{x}_t^j - \mathbf{x}_t^i\right)^{\perp}}{|\mathbf{x}_t^j - \mathbf{x}_t^i|^2} + \sqrt{2\nu}\xi_t^i \quad \text{and} \quad \mathbf{x}_0^i = \mathbf{x}^i, \tag{5}$$

where $\xi^i(t) \equiv (\xi_1^i(t), \xi_2^i(t))$ are zero mean white noise processes and Eq. (5) show that the velocity of each vortex is the sum of two terms, namely, the fluid velocity at the vortex position and a diffusive (stochastic) perturbation proportional to the fluid viscosity.

Lagrangian meters, such as ocean drifters and floats, provide a substantial part of ocean data which are used to reconstruct mean large-scale currents, estimate the rate of relative dispersion and give insight into the formation, movement and interactions of coherent structures such as point vortices and eddies. Based on the near-continuous data available from these instrumentation networks and to lower computational costs, we would like to develop more practical techniques required to analyze and interpret the data for dispersion modeling.

Trajectories of a Lagrangian tracer contain quantitative information about the dynamics of the of the underlying flow [6]; a tracer is advected according to

$$\dot{\mathbf{y}}_t^i = \mathbf{J} \sum_{j}^{n} \frac{\Gamma_j}{2\pi} \frac{\mathbf{y}_t^i - \mathbf{x}_t^j}{|\mathbf{y}_t^i - \mathbf{x}_t^j|^2} + \sqrt{2\nu}\eta_t^i \quad \text{and} \quad \mathbf{y}_0^i = \mathbf{y}^i. \tag{6}$$

The coupling between the dynamical model of the vortices and the tracer allows us to extract maximal information about the vortices by tracking the tracer. We can also correct the model variables on the fly using data from the tracers.

The theory of nonlinear filtering forms the framework in which problems of data assimilation for the nonlinear models will be treated. This can also be achieved using the idea of reduced dimensional nonlinear filtering as explained in Park et al. [13], where dimensional reduction made it possible in principle to reduce the cost of such filtering algorithms by a considerable factor.

3 Analytical Results: Continuous Signal and Discrete Observations

The theoretical aspect of data assimilation will be accomplished by constructing nonlinear filter equations based on continuous dynamics and discrete observation. Consider, for example, a two-vortex problem. Its state variables are the positions and velocities of the vortices. In the deterministic two-vortex problem, the distance between the vortices r is a constant of motion and the pair of point vortices rotate rigidly about the center of vorticity with a constant angular frequency. We use (x_1,x_2) and (x_3,x_4) to represent the position coordinates of the first and second vortices respectively. The dynamics of the signal process $X_t \equiv \{x_t^1,x_t^2,x_t^3,x_t^4\}$ is governed by

$$dx_t^1 = -\frac{\Gamma_2}{2\pi}\frac{(x_t^2-x_t^4)}{r^2}dt+\sqrt{2\nu}dW_t^1, \quad dx_t^2 = \frac{\Gamma_1}{2\pi}\frac{(x_t^1-x_t^3)}{r^2}dt+\sqrt{2\nu}dW_t^2 \tag{7}$$

$$dx_t^3 = -\frac{\Gamma_2}{2\pi}\frac{(x_t^4-x_t^2)}{r^2}dt+\sqrt{2\nu}dW_t^3, \quad dx_t^4 = \frac{\Gamma_1}{2\pi}\frac{(x_t^3-x_t^1)}{r^2}dt+\sqrt{2\nu}dW_t^4, \tag{8}$$

where $r=\sqrt{(x^1-x^3)^2+(x^2-x^4)^2}$. For various practical reasons, the initial value of the signal $X_0=\bar{x}$ is unknown, but the distribution of the initial condition x is given by $p(x)$. Now we introduce *relative* and "center of mass" coordinates as

$$x^r = x^3-x^1, \quad x^c = \frac{\Gamma_1 x^1+\Gamma_2 x^3}{\Gamma_1+\Gamma_2} \tag{9}$$

$$y^r = x^4-x^2, \quad y^c = \frac{\Gamma_1 x^2+\Gamma_2 x^4}{\Gamma_1+\Gamma_2}. \tag{10}$$

Then by Itô lemma

$$\begin{pmatrix} dx_t^r \\ dy_t^r \\ dx_t^c \\ dy_t^c \end{pmatrix} = \begin{pmatrix} \frac{\Gamma_1+\Gamma_2}{2\pi}\frac{y^r}{r^2} \\ -\frac{\Gamma_1+\Gamma_2}{2\pi}\frac{x^r}{r^2} \\ 0 \\ 0 \end{pmatrix} dt + \sqrt{2\nu}\begin{pmatrix} -1 & 1 & 0 & 0 \\ 0 & 0 & -1 & 1 \\ \frac{\Gamma_1}{\Gamma_1+\Gamma_2} & \frac{\Gamma_2}{\Gamma_1+\Gamma_2} & 0 & 0 \\ 0 & 0 & \frac{\Gamma_1}{\Gamma_1+\Gamma_2} & \frac{\Gamma_2}{\Gamma_1+\Gamma_2} \end{pmatrix}\begin{pmatrix} dW_t^1 \\ dW_t^2 \\ dW_t^3 \\ dW_t^4 \end{pmatrix} \tag{11}$$

Defining

$$z=\{x^r,y^r,x^c,y^c\}, \quad \tau=\Gamma_1+\Gamma_2, \quad \text{and} \quad \kappa_i=\frac{\Gamma_i}{\Gamma_1+\Gamma_2}$$

the generator of the Markov process is given by

$$\begin{aligned}(\mathscr{L}f) &= \nu\left(\kappa_1^2+\kappa_2^2\right)\left(\frac{\partial^2 f}{\partial z_3^2}(z)+\frac{\partial^2 f}{\partial z_4^2}(z)\right)+2\nu\left(\kappa_2-\kappa_1\right)\left(\frac{\partial^2 f}{\partial z_1\partial z_3}(z)+\frac{\partial^2 f}{\partial z_2\partial z_4}(z)\right) \\ &+\frac{\tau}{2\pi}\frac{1}{z_1^2+z_2^2}\left(-z_2\frac{\partial f}{\partial z_1}+z_1\frac{\partial f}{\partial z_2}\right)+2\nu\left(\frac{\partial^2 f}{\partial z_1^2}(z)+\frac{\partial^2 f}{\partial z_2^2}(z)\right)\end{aligned} \tag{12}$$

for $f \in C^2(\mathbb{R}^4)$. The probability density is governed by the forward Kolmogorov equation

$$\frac{\partial P}{\partial t} = \mathscr{L}^* P \tag{13}$$

where $\mathscr{L}^*$ is the adjoint operator.

3.1 Continuous Signal: Two-Vortices with Equal Strengths

In this section, we consider a special case of the signal process. Our signal process is given by the two-vortex dynamics given in (8) with equal vortex strengths, that is, $\kappa_1 = \kappa_2 = 1/2$. Then the generator (12) of the signal process is separable into the "relative" and "center of mass" coordinates. The center of mass (x^c, y^c) and relative distance (x^r, y^r) evolve independently; if one solves the forward equation (13) with initial conditions such that $p(z, t_0) = p_c(x^c, y^c, t_0) p_r(x^r, y^r, t_0)$, then $p(z, t) = p_c(x^c, y^c, t) p_r(x^r, y^r, t)$ for all time $t > t_0$. This suggests that the evolution equation – here a PDE with four spatial dimensions – can be simplified by considering the evolution of (x^c, y^c) and (x^r, y^r) separately.

The evolution equation for the density of (x^c, y^c) is simply the heat equation. So we are left with the evolution of the density of (x^r, y^r), which in polar coordinates is given by the generator

$$(\mathscr{L}^{r,\theta} f) = \frac{\tau}{2\pi} \frac{1}{r^2} \frac{\partial f}{\partial \theta} + \nu \left(\frac{\partial^2 f}{\partial r^2} + \frac{1}{r} \frac{\partial f}{\partial r} + \frac{1}{r^2} \frac{\partial^2 f}{\partial \theta^2} \right). \tag{14}$$

We can rescale time and redefine ν so that the constant $\tau/2\pi$ is removed from the equations.

We use a Green's function approach to find a solution from an arbitrary initial condition. Let $\rho(r, \theta, t)$ be a probability distribution evolving according to this law. Then

$$\rho(r, \theta, t) = \int ds\, s \int_0^{2\pi} d\phi\, P(r, \theta, t; s, \phi) \rho(s, \phi, 0), \tag{15}$$

where $P(r, \theta, t; s, \phi)$ is the solution to

$$\frac{\partial P}{\partial t} = (\mathscr{L}^{r,\theta})^* P \tag{16}$$

with initial condition $P(r, \theta, 0; s, \phi) = \delta(r - s)\delta(\theta - \phi)$. Without loss of generality we can set $\phi = 0$.

An exact solution is available for this initial value problem, due to Agullo and Verga [1]; we state the final result here.

$$P(r, \theta, t; s, 0) = \frac{1}{4\pi \nu t} \sum_{p \in Z} e^{ip\theta} e^{-(r^2+s^2)/(4\nu t)} I_{\mu_p}\left(\frac{rs}{2\nu t}\right) \tag{17}$$

where $I_m(z)$ is the modified Bessel function of the first kind with order m and argument z and $\mu_p^2 = p^2 + ip/\nu$, and the root should be chosen so that $Re(\mu_p) \leq 0$. In terms of the variables (x^c, y^c, x^r, y^r) our probability density is

$$P(x^r, y^r, x^c, y^c, t) = p_r(x^r, y^r, t) p_c(x^c, y^c, t) \tag{18}$$

where

$$p_r(x^r, y^r, t) = \frac{1}{4\pi\nu t} \int\int d\xi d\eta \, e^{-(|\bar{x}^r|^2 + |\bar{\xi}|^2)/(4\nu t)} \times \left[\sum_{p \in Z} e^{ip\tan^{-1}(y^r/x^r) - ip\tan^{-1}(\eta/\xi)} I_{\mu_p}\left(\frac{|\bar{x}^r||\bar{\xi}|}{2\nu t}\right) \right] p_r(\xi, \eta, 0), \tag{19}$$

$$p_c(x^c, y^c, t) = \frac{1}{\pi\nu t} e^{-((x^c)^2 + (y^c)^2)/(\nu t)} \tag{20}$$

and $\bar{x}^r \equiv (x^r, y^r)$ and $\bar{\xi} \equiv (\xi, \eta)$.

We use this formula to observe the evolution of a pair of vortices, initially at a distance $r = 1$ and $\theta = 0$, see Fig. 1. We evaluate the modified Bessel functions using a freely available code developed for Coulomb functions [14], of which the Bessel functions are a subset. We show the distribution of the vortex positions at $t = 1$ and $t = 5$, see Fig. 1, while the mean value of estimated position of the vortices is shown in Fig. 2. These can be compared to the histograms presented in Fig. 3 of [1].

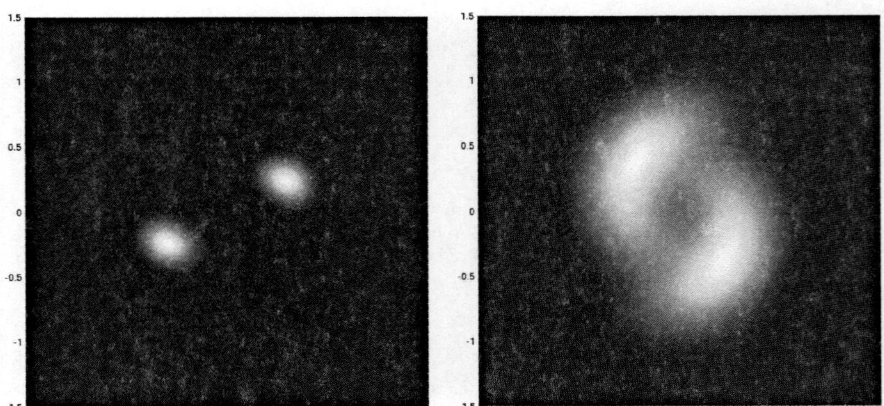

Fig. 1 The left figure shows the superimposed distributions of (x^1, x^2) and (x^3, x^4) at $t = 1$. The right figure shows $t = 5$

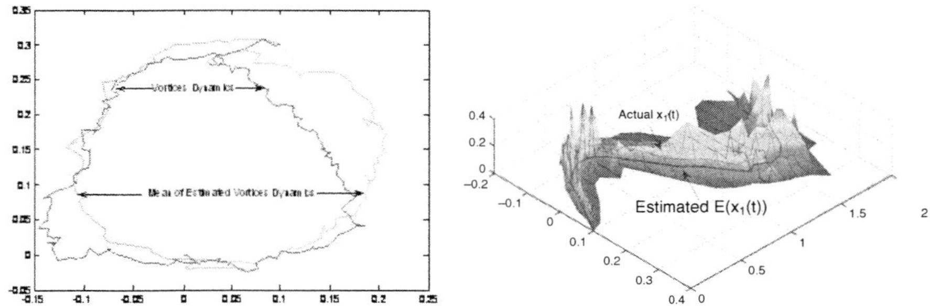

Fig. 2 The left figure shows the mean value of estimated position of the vortices by tracking the single tracer. The right figure shows the conditioned pdf of the position

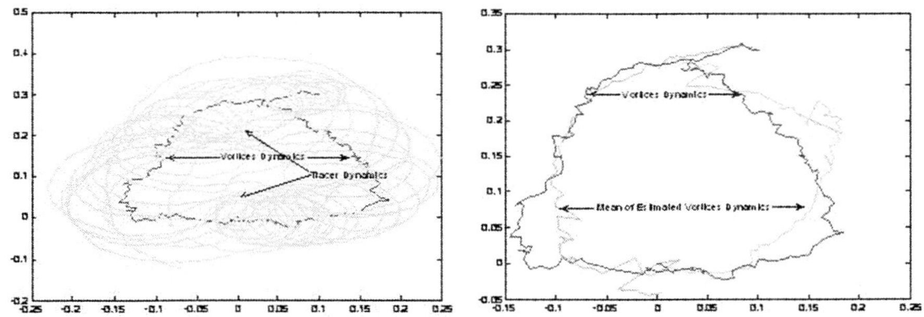

Fig. 3 The left figure shows the vortex-tracer dynamics.The right figure shows that with two or more tracers, the extraction results can be improved

3.1.1 Nonseparable Probability Densities

In general, particularly after the incorporation of an observation, the probability density will no longer be separable into radial and center coordinates as in (18). In this general case, the forward Kolmogorv equation (13) must be solved by first dividing the density into separable components with a spectral transform. We apply the cosine tranform to P, that is

$$P(x^r,y^r,x^c,y^c,0) = \sum_{k,l} \hat{P}_{k,l}(x^r,y^r,0)\cos(k\pi x^c/L)\cos(l\pi y^c/L) \tag{21}$$

if x^c,y^c are defined on $(0,L)$ (if x^c and y^c are defined on another interval, such as $(-L/2,L/2)$, then the argument of the cosine functions would be shifted appropriately).

We choose the cosine tranform as most natural to a probability distribution, where we wish to have "no-flux" boundaries (ideally, we would resolve the probability distribution on a large enough grid that the probability would be negligible near the boundaries).

For each pair k,l we compute $\hat{P}_{k,l}(x^r,y^r,t)$ to be the solution of (16), with initial condition $\hat{P}_{k,l}(x^r,y^r,0)$. This solution is given explicitly by Eq. (19).

The final answer is given by applying the diffusion operator to the (x^c, y^c) coordinates that is appropriate to the Fourier mode, and taking the inverse transform to yield

$$P(x^r, y^r, x^c, y^c, t) = \sum_{k,l} e^{-\nu(k^2+l^2)\pi^2 t/2L^2} \hat{P}_{k,l}(x^r, y^r, t) \cos(k\pi x^c/L) \cos(l\pi y^c/L). \quad (22)$$

3.2 Discrete Observations: Tracer Advection

The observations are defined by the tracers and are taken at discrete time instants t_k. The model that we will use for y is the following:

$$y_k^i = h_i(z_k, y_{k-1}) + v_k^i, \qquad z_k = z_{t_k}, \quad y_k^i = y_{t_k}^i, \qquad i = 1 \dots 2n \quad (23)$$

The sensor functions are given by first-order approximation to the tracers' equations of evolution (6); in the case of one tracer, we have

$$h_1(z_k, y_{k-1}) = y_{k-1}^1 + \Delta t \left(\frac{-(y_{k-1}^2 - (y^c - y^r/2))}{r_{1,2}^2} + \frac{-(y_{k-1}^2 - (y^c + y^r/2)}{r_{3,4}^2} \right) (24)$$

$$h_2(z_k, y_{k-1}) = y_{k-1}^2 + \Delta t \left(\frac{y_{k-1}^1 - (x^c - x^r/2)}{r_{1,2}^2} + \frac{y_{k-1}^1 - (x^c + x^r/2)}{r_{3,4}^2} \right) \quad (25)$$

where

$$r_{1,2} \equiv |(y^1, y^2) - (x^1, x^2)|^2 = |(y^1, y^2) - (x^c - x^r/2, y^c - y^r/2)|,$$
$$r_{3,4} \equiv |(y^1, y^2) - (x^3, x^4)|^2 = |(y^1, y^2) - (x^c + x^r/2, y^c + y^r/2)|$$

give the distances of the tracer from the first and second vortices respectively. $v_k = v_{t_k}$ is a $\mathbb{R}^2$-valued white Gaussian noise process independent of z_k, that is $v_k \sim N(0, R_{v_k})$. Here $z = (x^1, x^2)$, (x^3, x^4) are the signal variables, and (y^1, y^2) are the observation variables; corresponding formulas would be used for additional tracers.

Once again the observation σ-field

$$\mathscr{F}_t^y = \sigma\{y_l \sup_{0 \le t \le T} 0 \le t_l \le t\} = \{y_l \sup_{0 \le t \le T} l = 1, 2, \cdots, n; n\tau \le t\}$$

where τ is the sampling intervals. $\mathscr{F}_t^Y$ contains all the information available upto time instant t. To solve the filtering problem, for each $t \ge 0$, we would like to find the conditional **pdf** called the *posterior density*. Assume that the conditional probability distribution of the state z_t, given the observation up time t, denoted by

$$\pi_t(dz) = \mathbb{P}\left(z_t \in dz | \mathscr{F}_t^y\right)$$

has a conditional **pdf** $p(z, t | \mathscr{F}_t^y)$.

3.3 Nonlinear Filters

Hence, *between observarions*, the conditional **pdf** $p(z,t|\mathscr{F}_t^y)$ is governed by the Kolmogrov's forward equation [7, 8], that is,

$$\begin{aligned} \frac{\partial}{\partial t}p(z,t|\mathscr{F}_{t_k}^y) &= \mathscr{L}^* p(z,t|\mathscr{F}_{t_k}^y), \quad t_k < t < t_{k+1}, \\ \text{with} \quad \lim_{t \to t_k} p(z,t|\mathscr{F}_{t_k}^y) &= p(z,t_k|\mathscr{F}_{t_k}^y). \end{aligned} \tag{26}$$

where $(\mathscr{L})^*$ is the adjoint of the operator given in (12). This implies that in the discrete observation case, once we know the initial condition at $t = t_k$ given by $p(z,t_k|\mathscr{F}_{t_k}^y)$, we can compute the conditional **pdf** $p(z,t|\mathscr{F}_{t_k}^y)$ using the explicit solution (18) at any time $t > t_k$. However, at time $t = t_{k+1}$, we get more information from the observation y_{k+1}, which has to be used to *update* this conditional **pdf** at $t = t_{k+1}$.

The natural question is how to determine the initial conditional **pdf** $p(z,t_k|\mathscr{F}_{t_k}^y)$ at $t = t_k$ given in (26), knowing the previous evolution $p(z,t_k|\mathscr{F}_{t_{k-1}}^y)$ evaluated at $t = t_k$, and the new information y_k. Then, by Bayes' rule we have

$$p(z,t_k|\mathscr{F}_{t_k}^y) = \frac{p\left(y_k|\{z,t_k\},\mathscr{F}_{t_{k-1}}^y\right) p\left(z,t_k|\mathscr{F}_{t_{k-1}}^y\right)}{p\left(y_k|\mathscr{F}_{t_{k-1}}^y\right)} \tag{27}$$

The denominator in (27) is just the normalization of the numerator and can be calculated from

$$p\left(y_k|\mathscr{F}_{t_{k-1}}^y\right) = \int p\left(y_k|\{z,t_k\},\mathscr{F}_{t_{k-1}}^y\right) p\left(z,t_k|\mathscr{F}_{t_{k-1}}^Y\right) dz. \tag{28}$$

The conditional **pdf** on the right-hand side of (27)

$$p\left(z,t_k|\mathscr{F}_{t_{k-1}}^Y\right)$$

is given by (18). Since $\{v_k\}$ is a white noise and z_k is independent of v_k, the conditional **pdf**

$$p\left(y_k|\{z,t_k\},\mathscr{F}_{t_{k-1}}^y\right),$$

from the observation y_k can be simplified somewhat. The observation y_k at time k conditioned on z_k, is independent of all other measurements but y_{k-1}.

$$p\left(y_k|\{z,t_k\},\mathscr{F}_{t_{k-1}}^y\right) = p\left(y_k|\{z,t_k\},y_{k-1}\right) = p\left(y_k|z_k,y_{k-1}\right). \tag{29}$$

Once $p\left(y_k|z,y_{k-1}\right)$ is determined, we can get the desired map for the conditional density at an instant of observation.

Consider the observation Eq. (23). Since v_k is Gaussian and y_k is linear in v_k, for a given value of $z_k = z$ and $y_{k-1} = y$

$$p_y\left(y_k|z_k = z, y_{k-1} = y\right) = p_{v_k}\left(y_k - h(z,y,t_k)\right)|\frac{\partial v_k}{\partial y_k}| = p_{v_k}\left(y_k - h(z,y,t_k)\right)$$

Since $v_k \sim N(0, R_k)$ we can explicitly write

$$p\left(y_k|z,y\right) = \frac{1}{(2\pi)^{\frac{m}{2}}|R_k|^{\frac{1}{2}}}\exp\left\{-\frac{1}{2}\left(y_k - h(z,y,t_k)\right)^T R_k^{-1}\left(y_k - h(z,y,t_k)\right)\right\} \quad (30)$$

We can summarise the results by combining the Eqs. (18), (27), (29) and (30) as follows. The conditional **pdf** $p(z,t|\mathscr{F}_t^y)$ satisfies the following partial differential equation

$$\frac{\partial}{\partial t}p(z,t|\mathscr{F}_t^y) = \mathscr{L}^* p(z,t|\mathscr{F}_t^y), \quad t_k < t < t_{k+1}, \quad (31)$$

with the initial conditions at t_k given by the updating equations

$$p(z,t_k|\mathscr{F}_{t_k}^Y) = C_k\, \psi_k(z)\, p(z,t|\mathscr{F}_{t_{k-1}}^y), \quad (32)$$

where

$$\psi_k(z) = \mathbb{E}\left\{-\frac{1}{2}\left(y_k - h(z,y_{k-1},t_k)\right)^T R_k^{-1}\left(y_k - h(z,y_{k-1},t_k)\right)\right\},$$

C_k is a normalizing factor and $\mathscr{F}_{t_{k-1}}^Y$ the information on Y up to the instant in time right before t_k. The Eq. (31) has an explicit solution given by (18). The first Eq. (31) is the Kolmogorov's forward equation which is used to compute predictions between measurements, while the second Eq. (32) is used to update the information about the state via Bayes' rule.

4 Numerical Results: Particle Filters

One of the recent, more efficient and most popular classes of filtering methods is called particle methods. Importance sampling Monte Carlo offers powerful approaches to approximating Bayesian updating in sequential problems. Specific classes of such approaches are known as particle filters. The particles in these methods refer to independent samples generated with the Monte Carlo method, and they include sequential Monte Carlo, ensemble Kalman filter and interacting particle filters. The popularity of particle methods is attested by the recent surge of papers in this area. Particle algorithms are techniques for implementing a recursive Bayesian filter by MC simulations [5] (see for example, Arulampalam et al. [2]). In all particle methods, we evolve the particles between measurements by a set of random samples with associated weights and update the ensemble using Bayes' rule at the measurement time based on these samples and weights. Particle methods are very flexible and easy to implement; also they are ideally suited for a parallel computing architecture. This method has recently given rise to extensive mathematically rigorous studies, see for instance [4, 11] for the nonlinear filtering problem.

The idea is based on the Importance Sampling technique, that is, we can calculate an expected value

$$\mathbb{E}[f(z_k)] = \int f(z_k)p(z_k|y_{1:k})dz_k = \int f(z_k)\frac{p(y_{1:k}|z_k)\,p(z_k)}{p(y_{1:k})\,q(z_k|y_{1:k})}q(z_k|y_{1:k})dz_k \quad (33)$$

by using a known and simple proposal distribution $q(\cdot)$. This can be further simplified to

$$\mathbb{E}[f(z_k)] = \int f(z_k)\frac{w_k(z_k)}{p(y_{1:k})}q(z_k|y_{1:k})dz_k, \text{ where } w_k(z_k) = \frac{p(y_{1:k}|z_k)\,p(z_k)}{q(z_k|y_{1:k})} \quad (34)$$

is defined as the filtering non-normalized weight at step k. Hence,

$$\mathbb{E}[f(z_k)] = \frac{\mathbb{E}_q[w_k(z_k)\,f(z_k)]}{\mathbb{E}_q[w_k(z_k)]} = \mathbb{E}_q[\hat{w}_k(z_k)\,f(z_k)],$$

$$\text{where}\hat{w}_k(z_k) = \frac{w_k(z_k)}{\mathbb{E}_q[w_k(z_k)]} \quad (35)$$

These procedures rely on the simulation of samples or ensembles of the unknown quantities and the calculation of associated weights for the ensemble members. Hence, using Monte-Carlo sampling from the distribution $q(z_k|y_{1:k})$ we can write

$$\mathbb{E}[f(z_k)] \approx \sum_{i=1}^{N} \hat{w}_k(z_k^i)\,f(z_k^i), \quad \text{where} \quad \hat{w}_k(z_k^i) = \frac{w_k(z_k^i)}{\sum_{i=1}^{N} w_k(z_k^i)} \quad (36)$$

In addition if our proposal distribution $q(\cdot)$ satisfies the Markov property, it can be shown that $\hat{w}_k(z_k^i)$ satisfies a cursive relationship. The basic ideas of particle filters are: (1) represent the required posterior density function by a set of random samples with associated weights; (2) compute estimates based on these samples and weights.

In principle, armed with these algorithms, we should be able to handle a large class of nonlinear filtering problems. The problem of this method is that for high dimensional systems, these stochastic algorithms are usually slow and computational complexity grows too quickly with dimension. In extreme cases, after a sequence of updates the particle system can collapse to a single point or to several particles with so much internal correlation that summary statistics behave as if they are derived from a substantially smaller sample. To compensate, large numbers of particles are required in realistic problems. Hence, the method is not always implementable in real time nonlinear applications when the state space is too large.

4.1 Data Fusion

The results presented thus far are, in general, well understood in terms of single sensor filtering theory. However, when there are multiple sensors, then the problem of combining information from them arises. We consider some approaches generally proposed in the literature and discuss some criticisms associated with them.

To begin with, we assume that M sensors are available and the observations from the kth are given by the vector $y^k \in \mathbb{R}^m$ (i.e., the number of observations m is the same for all sensors). What is now required is to compute the global posterior distribution $p(z|y^1, y^2, \cdots, y^M)$, given the information contributed by each sensor. We shall assume that each sensor provides either a local posterior distribution $p(z|y^k)$, or a likelihood function $p(y^k|z)$.

Since the information is received from different sensors, the natural question to ask in tackling the problem of fusion, is how relevant and how reliable is the information from each sensor. One of the ways to address this problem is by attaching a weight to the information provided by each sensor.

On the other hand when each information source has common prior information, i.e. information obtained from the same origin, the situation is better described by the *independent likelihood pool*, which is derived as follows. According to Bayes' theorem for the global posterior, we obtain

$$p(z|y^1, y^2, \cdots y^M) = \frac{p(y^1, y^2, \cdots y^M|z)p(x)}{p(y^1, y^2, \cdots y^M)}. \tag{37}$$

For a system of tracers it is reasonable to assume that the likelihoods from each tracer $p(y^M|z), m = 1, 2, \cdots, M$ are independent since the only parameter they have in common is the state x of the vortices., that is,

$$p(y^1, y^2, \cdots y^M|z) = p(y^1|z)p(y^2|z)\cdots p(y^M|z).$$

Thus, the Independent Likelihood Pool is defined by the following equation

$$p(z|y^1, y^2, \cdots y^M) = p(z)\Pi_{m=1}^{M} p(y^M|z). \tag{38}$$

As may be seen from the above both the Independent Opinion Pool and the Independent Likelihood Pool more accurately describe the situation in multi-sensor systems where the conditional distribution of the observation can be shown to be independent. However, in most cases in sensing the Independent Likelihood Pool is the most appropriate way of combining information since the prior information tends to be from the same origin. If there are dependencies between information sources the Linear Opinion Pool should be used.

Acknowledgments The authors would like to acknowledge the support of AFOSR and the National Science Foundation. Any opinions, findings, and conclusions or recommendations expressed in this paper are those of the authors and do not necessarily reflect the views of the National Science Foundation. Part of the work was done while the authors were visiting the Mathematical Sciences Research Institute (MSRI). The authors gratefully acknowledge the hospitality of MSRI during its Spring 2007 program in Dynamical Systems. They would also like to acknowledge Professor Chris Jones of University of North Carolina, Chapel Hill for helpful discussions at the initial stage of this work during our stay at MSRI.

References

1. Agullo, O., Verga, A.: Effect of viscosity in the dynamics of two point vortices: Exact results. Physical Review E **63(5)**, 056304 (2001)
2. Arulampalam, M.S., Maskell, S., Gordon, N., Clapp, T.: A tutorial on particle filters for online nonlinear/non-Gaussian Bayesian tracking. IEEE Transactions on Signal Processing **50(2)**, 174–188 (2002)
3. Chorin, A.J.: Numerical study of slightly viscous flow. Journal of Fluid Mechanics **57(4)**, 785–796 (1973)
4. Crisan, D., Gaines, J., Lyons, T.: Convergence of a branching particle method to the solution of the Zakai equation. SIAM Journal of Applied Mathematics **58(5)**, 1568–1590 (1998)
5. Gordon, N., Maskell, S., Kirubarajan, T.: Efficient particle filters for joint tracking and classification. IProceedings of SPIE Signal and Data Processing of Small Targets **4728**, 439–449 (2002) **Journal or contributed?**
6. Ide, K., Kuznetsov, L., Jones, C.K.R.T.: Lagrangian data assimilation for point vortex systems. Journal of Turbulence **3**, 053 (2002)
7. Jazwinski, A.H.: Stochastic Processes and Filtering Theory. Academic Press, New York (1970)
8. Kallianpur, G.: Stochastic filtering theory. Applications of Mathematics, 13. Springer-Verlag, New York-Berlin (1980)
9. Lugt Hans, J.: Vortex flow in nature and technology. Wiley (1983)
10. Marchioro, C., Pulvirenti, M.: Hydrodynamics in two dimensions and vortex theory. Communications in Mathematical Physics **84**, 483–503 (1982)
11. Del Moral, P.: Nonlinear filtering using random particles. Theory of Probability and its Applications **40**, 690–701 (1976)
12. Newton, P.K.: The N-Vortex Problem: Analytical Techniques. Applied Mathematical Sciences Series, Vol. 145. Springer-Verlag (2001)
13. Park, J., Namachchivaya, N. Sri and Sowers, R.B.: A Problem in Stochastic Averaging of Nonlinear Filters, (submitted)
14. Thompson, I.J., Barnett, A.R: Coulomb and Bessel Functions of Complex Argument and Order. Journal of Computational Physics **64**, 490–509 (1986)

The Role of Receptor Occupancy Noise in Eukaryotic Chemotaxis

Wouter-Jan Rappel and Herbert Levine

Abstract Chemotacting eukaryotic cells are able to translate a gradient of occupied receptors into cell motion. For small concentrations and shallow gradients, fluctuations in the number of occupied receptors can become important. Here, we present an effective way to numerically simulate the correlations of these fluctuations. Furthermore, we apply our previously developed formalism ([10]) to a simple version of the local excitation, global inhibition model.

1 Introduction

Eukaryotic chemotaxis is involved in a variety of important biological processes. Examples of chemotaxis can be found in wound healing, cancer metastasis and embryogenesis. During chemotaxis, cells crawl in a direction that is determined by the external ligand gradient. Exactly how the cells are able to determine this direction has been the subject of intense recent research, including a number of theoretical studies [1,4–7,9,11,12] Here, we will extend our recent investigation on the effect of sensing noise on the process of directional sensing [10]. In particular, we will discuss an alternative way to numerically produce correlated noise sequences and will apply these sequences to a specific directional sensing model [3].

2 Receptor Occupancy Fluctuations

Recent experiments have demonstrated that *Dictyostelium discoideum* cells, a social amoeba, are able to chemotax in extremely shallow gradients [13]. These gradients

W.-J. Rappel (✉)
Center for Theoretical Biological Physics, University of California San Diego,
La Jolla, CA 92093-0319 USA

V. In et al. (eds.), *Applications of Nonlinear Dynamics*, Understanding Complex Systems,
DOI 10.1007/978-3-540-85632-0_6, © Springer-Verlag Berlin Heidelberg 2009

were produced in microfluidic devices, ensuring a constant linear profile. At the minimum gradient for motion ($\sim 10^{-3}$nM/μm) we can estimate difference between the number of bound receptors at the front and at the back of the cell. Assuming that the number of receptors on a cell is 70000, that the disassociation constant is $K_d = 30$ nM and that a typical cell diameter is 10 μm we find that this difference is approximately 20. The total number of bound receptors depends on the background concentration and thus on the location of the cell in the microfluidic channel but can measure as little as a few hundreds. These small numbers immediately raise the question of the importance of fluctuations in the directional sensing process.

In a recent study, we have quantified the fluctuations in the receptor occupancy using a Monte Carlo simulation package, MCell3, which explicitly tracks diffusing particle [10]. MCell3 has been highly optimized and can handle complex geometries and receptor/ligand binding reactions [14]. In our simulations, we took the cell to be spherical with radius R and covered uniformly with receptors. At the start of each simulation, we populated one half of the spherical cell with bound receptors while the receptors of the remaining half were all empty. The number of bound receptors was taken from the equilibrium value:

$$N_{bound} = \frac{N}{2}\frac{c}{c+K_d} \tag{1}$$

where N is the total number of receptors and where c is the ligand concentration.

The cell was placed in a computational box that was much larger than R and had absorbing boundaries. The simulation ended when all particles had been absorbed on the box boundary. From the number of occupied receptors at each half as a function of time we were able to calculate the auto- and crosscorrelations. To ensure a statistically significant result we ran 100 simulations. We found that, consistent with our result in Ref. [17], the autocorrelation C_{ff} and C_{bb} (the index f denotes front and b denotes back) was a decaying exponential with a time constant and amplitude given by $\tau_a = 1/k_- + N/(8\pi D_l R K_d)$ and $A_a = NcK_d/(2c+2K_d)$, respectively. Here, k_- is the off rate of the ligand (measured to be around $1s^{-1}$ [15]) and D_l is its diffusion constant. The crosscorrelation was given by $C_{fb} = \frac{A_c}{\tau_c} t e^{-t/\tau_c}$ where the amplitude and time constant were found by fits to the numerical data.

3 Application to a One-Dimensional Geometry

As in Ref. [10], we will now consider a directional sensing model in which a diffusive cytosolic inhibitor B is linearly generated by the external signal S, which is of course directly related to the number of bound receptors:

$$\begin{aligned} \frac{\partial B}{\partial t} &= D\nabla^2 B \quad \text{in the cytosol} \\ D\frac{\partial B}{\partial n} &= k_a S - k_b B \quad \text{on the membrane} \end{aligned} \tag{2}$$

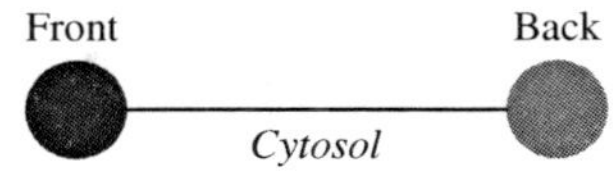

Fig. 1 Schematic representation of our one dimensional cell

The signal at the front and the back can thus be written as the sum of a deterministic and a stochastic part:

$$\begin{aligned} S_f &= S_{f,0} + \eta_f \\ S_b &= S_{b,0} + \eta_b \end{aligned} \tag{3}$$

where η_f and η_f are the receptor occupancy fluctuations described in Sect. 2. In particular, their auto- and crosscorrelations, given by $< \eta_f(t)\eta_f(t') >= C_{ff}(t-t')$, $< \eta_b(t)\eta_b(t') >= C_{bb}(t-t')$ and $< \eta_f(t)\eta_b(t') >= C_{fb}(t-t')$, were directly computed using the MCell simulations.

To facilitate analytical progress, we will consider a simplified one dimensional geometry shown in Fig. 1. This geometry contains a front and a back, both considered to be points, connected by a line of length L, representing the interior, or cytosol, of the cell. In this case, one can calculate the fluctuation spectra of the diffusive inhibitor as a function of the correlation spectra of the receptor occupancy. This is done in Fourier space where the noise is diagonal in frequency and we can write

$$\tilde{B}(x,\omega) = \tilde{b}_e(\omega)\frac{\cosh(\kappa \mathrm{x})}{\cosh(\kappa \mathrm{L}/2)} + \tilde{b}_o(\omega)\frac{\sinh(\kappa \mathrm{x})}{\sinh(\kappa \mathrm{L}/2)} \tag{4}$$

where $\kappa = \sqrt{-i\omega/D}$ and where we have introduced the tilde to denote a quantity in Fourier space. Furthermore,

$$\tilde{b}_e = \frac{k_a(\tilde{\eta}_f + \tilde{\eta}_b)}{2(k_b + D\kappa \tanh(\kappa \mathrm{L}/2))} \tag{5}$$

and

$$\tilde{b}_o = \frac{k_a(\tilde{\eta}_f - \tilde{\eta}_b)}{2(k_b + D\kappa \coth(\kappa \mathrm{L}/2))} \tag{6}$$

leading to

$$\begin{aligned} \tilde{B}_f &= \tilde{b}_e + \tilde{b}_o \\ \tilde{B}_b &= \tilde{b}_e - \tilde{b}_o \end{aligned} \tag{7}$$

We can then express the correlation spectra in B as functions of the correlation spectra in S. The final expressions are:

$$\tilde{N}_{ff}(\omega) = \frac{k_a^2}{4}\left(\tilde{C}_{ff}(\omega)\left[\frac{1}{k_b + D\kappa\tanh(\kappa L/2)} + \frac{1}{k_b + D\kappa\coth(\kappa L/2)}\right]^2 + \right.$$
$$\tilde{C}_{bb}(\omega)\left[\frac{1}{k_b + D\kappa\tanh(\kappa L/2)} - \frac{1}{k_b + D\kappa\coth(\kappa L/2)}\right]^2 +$$
$$\tilde{C}_{fb}(\omega)\left[\frac{1}{k_b + D\kappa\tanh(\kappa L/2)} + \frac{1}{k_b + D\kappa\coth(\kappa L/2)}\right]$$
$$\left.\left[\frac{1}{k_b + D\kappa\tanh(\kappa L/2)} - \frac{1}{k_b + D\kappa\coth(\kappa L/2)}\right]^* + c.c.\right)$$
$$\tilde{N}_{bb}(\omega) = \frac{k_a^2}{4}\left(\tilde{C}_{bb}(\omega)\left[\frac{1}{k_b + D\kappa\tanh(\kappa L/2)} + \frac{1}{k_b + D\kappa\coth(\kappa L/2)}\right]^2 + \right.$$
$$\tilde{C}_{ff}(\omega)\left[\frac{1}{k_b + D\kappa\tanh(\kappa L/2)} - \frac{1}{k_b + D\kappa\coth(\kappa L/2)}\right]^2 +$$
$$\tilde{C}_{fb}(\omega)\left[\frac{1}{k_b + D\kappa\tanh(\kappa L/2)} + \frac{1}{k_b + D\kappa\coth(\kappa L/2)}\right]$$
$$\left.\left[\frac{1}{k_b + D\kappa\tanh(\kappa L/2)} - \frac{1}{k_b + D\kappa\coth(\kappa L/2)}\right]^* + c.c.\right)$$
$$\tilde{N}_{fb}(\omega) = \frac{k_a^2}{4}\left(\tilde{C}_{fb}(\omega)\left[\frac{1}{k_b + D\kappa\tanh(\kappa L/2)} + \frac{1}{k_b + D\kappa\coth(\kappa L/2)}\right]^2 + \right.$$
$$\tilde{C}^*_{fb}(\omega)\left[\frac{1}{k_b + D\kappa\tanh(\kappa L/2)} - \frac{1}{k_b + D\kappa\coth(\kappa L/2)}\right]^2 +$$
$$\tilde{C}_{ff}(\omega)\left[\frac{1}{k_b + D\kappa\tanh(\kappa L/2)} + \frac{1}{k_b + D\kappa\coth(\kappa L/2)}\right]$$
$$\left[\frac{1}{k_b + D\kappa\tanh(\kappa L/2)} - \frac{1}{k_b + D\kappa\coth(\kappa L/2)}\right]^* +$$
$$\tilde{C}_{bb}(\omega)\left[\frac{1}{k_b + D\kappa\tanh(\kappa L/2)} - \frac{1}{k_b + D\kappa\coth(\kappa L/2)}\right]$$
$$\left.\left[\frac{1}{k_b + D\kappa\tanh(\kappa L/2)} + \frac{1}{k_b + D\kappa\coth(\kappa L/2)}\right]^*\right)$$

4 Numerical Generation of the Noise

One way the noise can be generated, as detailed in [10], is to use the spectrum of the noise in Fourier space:

$$\tilde{C}_{ff}(\omega) = \frac{2A_a/\tau_a}{1/\tau_a^2 + \omega^2}; \qquad \tilde{C}_{fb}(\omega) = \frac{2A_c(1/\tau_c^2 - \omega^2)/\tau_c}{(1/\tau_c^2 + \omega^2)^2} \tag{8}$$

The correlation spectrum for B is then also calculated in Fourier space, followed by an inverse Fourier transform, after which the remaining equations for the directional sensing model are calculated in real space.

Another way to generate the required noise traces is by starting with

$$\tilde{\eta}_f(\omega) = f(\omega)\tilde{\xi}_f + g_1(\omega)\tilde{\xi}_1 + g_2(\omega)\tilde{\xi}_2$$
$$\tilde{\eta}_b(\omega) = f(\omega)\tilde{\xi}_b + g_2(\omega)\tilde{\xi}_1 + g_1(\omega)\tilde{\xi}_2 \tag{9}$$

where $\tilde{\xi}_f$, $\tilde{\xi}_b$, $\tilde{\xi}_1$, $\tilde{\xi}_2$ are independent standard Brownian motions. Then, the spectra can be expressed in terms of f, g_1 and g_2:

$$\tilde{C}_{ff}(\omega) = |f|^2 + |g_1|^2 + |g_2|^2$$
$$\tilde{C}_{fb}(\omega) = g_1 g_2^* + g_1^* g_2 \tag{10}$$

Choosing $g_1(\omega)$ and $g_2(\omega)$ to be

$$g_1(\omega) = \frac{\sqrt{A_c/\tau_c}}{1/\tau_c + i\omega}$$
$$g_2(\omega) = \frac{\sqrt{A_c/\tau_c}}{1/\tau_c - i\omega} \tag{11}$$

we find that f needs to be

$$f = \frac{u}{1/\tau_a + i\omega} + \frac{v}{1/\tau_c + i\omega} \tag{12}$$

where

$$u = \frac{B - 1/\tau_a}{1/\tau_c - 1/\tau_a}\sqrt{2(A_a/\tau_a - A_c/\tau_c)}$$
$$v = \frac{B - 1/\tau_c}{1/\tau_a - 1/\tau_c}\sqrt{2(A_a/\tau_a - A_c/\tau_c)} \tag{13}$$

and

$$B = \sqrt{\frac{A_a/\tau_c - A_c/\tau_a}{A_a\tau_c - A_c\tau_a}} \tag{14}$$

Thus, in Fourier space we have

$$\begin{aligned}\tilde{\eta}_f(\omega) &= \frac{u}{1/\tau_a + i\omega}\tilde{\xi}_f + \frac{v}{1/\tau_c + i\omega}\tilde{\xi}_f + \frac{\sqrt{A_c/\tau_c}}{1/\tau_c + i\omega}\tilde{\xi}_1 + \frac{\sqrt{A_c/\tau_c}}{1/\tau_c - i\omega}\tilde{\xi}_2 \\ &= \tilde{Y}_{f1} + \tilde{Y}_{f2} + \tilde{U}_{f1} + \tilde{U}_{f2}\end{aligned}$$

$$\begin{aligned}\tilde{\eta}_b(\omega) &= \frac{u}{1/\tau_a + i\omega}\tilde{\xi}_b + \frac{v}{1/\tau_c + i\omega}\tilde{\xi}_b + \frac{\sqrt{A_c/\tau_c}}{1/\tau_c + i\omega}\tilde{\xi}_2 + \frac{\sqrt{A_c/\tau_c}}{1/\tau_c - i\omega}\tilde{\xi}_1 \\ &= \tilde{Y}_{b1} + \tilde{Y}_{b2} + \tilde{U}_{b1} + \tilde{U}_{b2}\end{aligned}$$

In time domain, the Y's and U's follow simple stochastic processes:

$$\begin{aligned} dY_{f1} &= -k_1 Y_{f1} dt + u dB_f \\ dY_{f2} &= -k_2 Y_{f2} dt + v dB_f \\ dU_{f1} &= -k_2 U_{f1} dt + \sqrt{A_c/\tau_c} dB_1 \\ dU_{f2} &= k_2 U_{f2} dt - \sqrt{A_c/\tau_c} dB_2 \end{aligned} \tag{15}$$

$$\begin{aligned} dY_{b1} &= -k_1 Y_{b1} dt + u dB_b \\ dY_{b2} &= -k_2 Y_{b2} dt + v dB_b \\ dU_{b1} &= -k_2 U_{b1} dt + \sqrt{A_c/\tau_c} dB_2 \\ dU_{b2} &= k_2 U_{b2} dt - \sqrt{A_c/\tau_c} dB_1 \end{aligned} \tag{16}$$

which are straightforward to simulate. The only complication arises from the fact that U_{f2} and U_{b2} are not stable processes. This problem is circumvented by simulating U_{f2} and U_{b2} backwards from time T to time 0, while all others processes are simulated forwards starting from time 0. To reach equilibrium, a time frame of length T_w is dropped for both the starting part and the ending part.

5 A Specific Directional Sensing Model

In our previous work, we applied the above formalism to a recently developed directional sensing model [10]. Here, we will apply our formalism to a simplified version of the local excitation, global inhibition (LEGI) model. This model was originally proposed by Parent and Devreotes [8] and examined in detail by Levchenko and Iglesias [4]. In this model, the external stimulus produces a local activator as well as a globally diffusing inhibitor, as described above. Key element in their model is perfect adaptation which ensures that a gradient can be "sensed" for arbitrary background concentration.

The version of the LEGI model we will examine couples the equations for B (see Eq. (2)) to two additional variables on the membrane [3]:

$$\begin{aligned} \frac{\partial E}{\partial t} &= k_e S - k_{-e} E \\ \frac{\partial R}{\partial t} &= k_r E - k_{-r} B R \end{aligned} \tag{17}$$

The variable R is the read-out component which is a measure of the internal asymmetry following an external gradient. A precise coupling of this variable to chemotactic motion would require extensive knowledge of the signal transduction pathways that are involved in cell motion. In the absence of this, we define a phenomenological measure for the chemotactic response by calculating the quantity

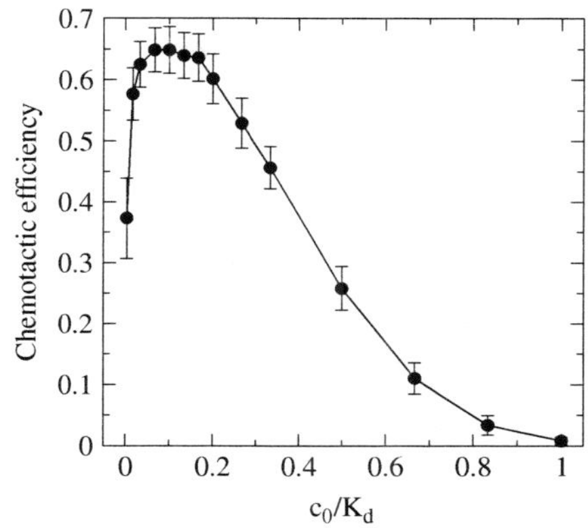

Fig. 2 The chemotactic efficiency as a function of the background concentration for a simplified LEGI model. Parameter values for the LEGI model are: $k_e = 0.1$, $k_{-e} = 0.5$, $k_a = 0.4$, $k_b = 0.4$, $k_r = 0.25$, $k_{-r} = 0.1$, $D_l = 200\,\mu m^2/s$, $R = 5\mu m$ and $D = 20\,\mu m^2/s$. Motility parameters were chosen as follows: $T_{dec} = 30$ s, $T_{int} = 10$ s and $\theta = 0.007$

$$G(t) = \frac{\int_{-\infty}^{t} e^{(t-s)/T_{int}} (R_f - R_b)\, ds}{\int_{-\infty}^{t} e^{(t-s)/T_{int}} (R_f + R_b)\, ds} \tag{18}$$

once every $t = T_{dec}$. This is translated into motion using the following rule: the cell takes positive unit step (i.e. into the direction of the gradient) if $G > \theta$, a negative unit step if $G < \theta$ and will not move otherwise. The chemotactic efficiency CE is then defined as the total distance covered by the cell divided by the number of decision intervals and can take on values between -1 and $+1$.

The CE as a function of the midpoint concentration c_0 is plotted in Fig. 2 for typical system parameters. Motivated by our results in [10], which showed that for biological realistic parameter values $A_c \ll A_a$, we have neglected the contribution of the crosscorrelations. The gradient steepness was chosen such that $c_f = 1.01c_0$ and $c_b = 0.99c_0$. The error bars in the figure correspond to the standard deviation and were obtained by running 50 simulations of 5000 s each. As can be seen from these results, the CE reaches a maximum for very small values of c_0 ($c_0 \sim 0.1\ K_d$). The CE in the balanced inactivation model [10], as well as in other recent theoretical studies [16] and experiments on neutrophils [2], displayed similar qualitative behavior. How the exact location of the maximum depends on the model parameters and phenomenological motility model is the subject of future work.

As a consequence of our findings, the minimum gradient for onset of chemotaxis should also be a function of the background concentration. This should be testable using the microfluidic set up of Ref. [13]. Instead of setting up a linear gradient that varies between $c = 0$ at one end of the microfluidic channel and a non-zero value at the other end, it should be possible to set the concentration to nonzero values at both ends of the channel. In this way, a systematic study of the interplay between background concentration and gradient steepness and the chemotactic efficiency can be performed.

6 Summary

In this paper we have detailed how one can generate fluctuations in the number of occupied receptors with the desired cross- and autocorrelation. Furthermore, we have applied our methodology described in [10] to a simplified LEGI model. We would like to point out, however, that more complicated directional sensing models in which the resulting equations are not diagonal in frequency space can still be examined using our formalism through direct numerical integration.

Acknowledgments Numerical assistance from Kai Wang and John Lewis is gratefully acknowledged.

References

1. A. Gamba, A. de Candia, S. D. Talia, A. Coniglio, F. Bussolino, and G. Serini, *Diffusion-limited phase separation in eukaryotic chemotaxis*, PNAS, 102 (2005), pp. 16927–16932.
2. P. Herzmark, K. Campbell, F. Wang, K. Wong, H. El-Samad, A. Groisman, and H. R. Bourne, *Bound attractant at the leading vs. the trailing edge determines chemotactic prowess*, Proc Natl Acad Sci USA, 104 (2007), pp. 13349–13354.
3. B. Kutscher, P. Devreotes, and P. A. Iglesias, *Local excitation, global inhibition mechanism for gradient sensing: an interactive applet*, Sci STKE, 2004 (2004).
4. A. Levchenko and P. A. Iglesias, *Models of eukaryotic gradient sensing: application to chemotaxis of amoebae and neutrophils*, Biophys J, 82 (2002), pp. 50–63.
5. H. Levine, D. A. Kessler, and W. J. Rappel, *Directional sensing in eukaryotic chemotaxis: a balanced inactivation model*, Proc Natl Acad Sci USA, 103 (2006), pp. 9761–9766.
6. M. Meier-Schellersheim, X. Xu, B. Angermann, E. J. Kunkel, T. Jin, and R. N. Germain, *Key role of local regulation in chemosensing revealed by a new molecular interaction-based modeling method*, PLoS Comput Biol, 2 (2006), p. e82.
7. A. Narang, K. K. Subramanian, and D. A. Lauffenburger, *A mathematical model for chemoattractant gradient sensing based on receptor-regulated membrane phospholipid signaling dynamics*, Ann Biomed Eng, 29 (2001), pp. 677–691.
8. C. A. Parent and P. N. Devreotes, *A cell's sense of direction*, Science, 284 (1999), pp. 765–770.
9. M. Postma and P. J. M. van Haastert, *A diffusion-translocation model for gradient sensing by chemotactic cells*, Biophys J, 81 (2001), pp. 1314–1323.
10. W.-J. Rappel and H. Levine, *Receptor noise and directional sensing in eukaryotic chemotaxis*, Submitted, (2008).
11. W. J. Rappel, P. J. Thomas, H. Levine, and W. F. Loomis, *Establishing direction during chemotaxis in eukaryotic cells*, Biophys J, 83 (2002), pp. 1361–1367.
12. R. Skupsky, W. Losert, and R. J. Nossal, *Distinguishing modes of eukaryotic gradient sensing*, Biophys J, 89 (2005), pp. 2806–2823.
13. L. Song, S. M. Nadkarni, H. U. Bödeker, C. Beta, A. Bae, C. Franck, W.-J. Rappel, W. F. Loomis, and E. Bodenschatz, *Dictyostelium discoideum chemotaxis: threshold for directed motion*, Eur J Cell Biol, 85 (2006), pp. 981–989.
14. J. R. Stiles and T. M. Bartol, *Monte Carlo methods for simulating realistic synaptic microphysiology using MCell*, in Computational Neurobiology: Realistic Modeling for Experimentalists, E. de Schutter, ed., CRC Press, Boca Raton, FL, 2001.

15. M. UEDA, Y. SAKO, T. TANAKA, P. N. DEVREOTES, AND T. YANAGIDA, *Single-molecule analysis of chemotactic singaling in Dictyostelium cells*, Science, 294 (2001), pp. 864–867.
16. M. UEDA AND T. SHIBATA, *Stochastic signal processing and transduction in chemotactic response of eukaryotic cells*, Biophys J, 93 (2007), pp. 11–20.
17. K. WANG, W. J. RAPPEL, R. KERR, AND H. LEVINE, *Quantifying noise levels of intercellular signals*, Phys Rev E, 75 (2007), p. 061905.

Applications of Forbidden Interval Theorems in Stochastic Resonance

Bart Kosko, Ian Lee, Sanya Mitaim, Ashok Patel and Mark M. Wilde

Abstract Forbidden interval theorems state whether a stochastic-resonance noise benefit occurs based on whether the average noise value falls outside or inside an interval of parameter values. Such theorems act as a type of screening device for mutual-information noise benefits in the detection of subthreshold signals. Their proof structure reduces the search for a noise benefit to the often simple task of showing that a zero limit exists. This chapter presents the basic forbidden interval theorem for threshold neurons and four applications of increasing complexity. The first application shows that small amounts of electrical noise can help a carbon nanotube detect faint electrical signals. The second application extends the basic forbidden interval theorem to quantum communication through the judicious use of squeezed light. The third application extends the theorems to noise benefits in standard models of spiking retinas. The fourth application extends the noise benefits in retinal and other neuron models to Levy noise that generalizes Brownian motion and allows for jump and impulsive noise processes.

1 Forbidden Interval Theorems for Stochastic Resonance

Stochastic resonance (SR) occurs in a nonlinear system when noise benefits the system [3, 17, 33]. The noise benefit can take the form of an increase in mutual information or a signal-to-noise ratio or correlation or a decrease in an error measure. But when will such a noise benefit occur?

Forbidden interval theorems answer that SR question for several nonlinear systems. The theorems act as a type of SR *screening device* because they can give sufficient or necessary conditions for an SR noise benefit. We here restrict noise benefits

---

B. Kosko (✉)
Department of Electrical Engineering, University of Southern California, Los Angeles, California 90089, USA, e-mail: kosko@usc.edu

V. In et al. (eds.), *Applications of Nonlinear Dynamics*, Understanding Complex Systems, DOI 10.1007/978-3-540-85632-0_7,

to improvements in the system's input-output Shannon mutual information given an increase in the variance or dispersion of the exogenous noise. But forbidden interval theorems do not show how to find such a noise benefit. Nor do they indicate the magnitude of such a noise benefit if it occurs. They merely indicate whether such a noise benefit exists in theory for a given combination of parameters.

Stochastic learning algorithms can often find the actual noise benefit that a forbidden interval theorem predicts. These adaptive algorithms can find the local noise benefits if we take enough samples from the process and if we apply a sufficiently robust SR learning algorithm of the form $\sigma_{k+1} = \sigma_k + \mu_k \phi(\frac{\partial I}{\partial \sigma})$ where I is mutual information or some other performance measure, σ is the noise intensity, and ϕ is a statistical "robustifier" such as signum or other impulse suppressor [18, 23, 24]. Both the learning algorithms and the forbidden interval theorems apply to many other system performance measures other than mutual information.

The simplest forbidden interval theorem is the strongest because it gives both necessary and sufficient conditions for an SR noise benefit while it requires only a threshold nonlinearity. It applies to a threshold neuron or other threshold system with threshold θ for subthreshold bipolar signal amplitudes A and $-A$: $-A < A < \theta$. The additive noise n has either a finite mean $E[n]$ or comparable location parameter a in the case of infinite-variance stable noise. Then the simplest forbidden interval theorem fully characterizes the presence or absence of a noise benefit because it gives both necessary and sufficient conditions for this SR effect: SR occurs if and only if $E[n] \notin (\theta - A, \theta + A)$. So a noise benefit occurs just in case the mean

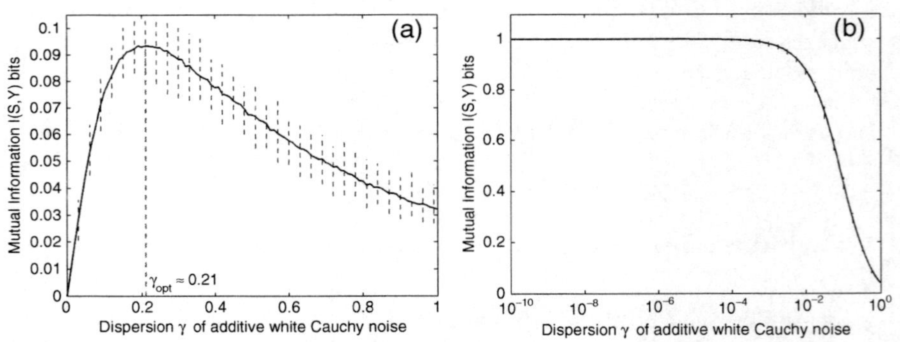

Fig. 1 Forbidden interval effects. The graphs show the smoothed input-output mutual information of a threshold system (1) as a function of the dispersion of additive white alpha-stable noise n_t with $\alpha = 1$ (infinite-variance Cauchy noise). The system has threshold $\theta = 0.5$. The bipolar input Bernoulli signal s_t has amplitude $A = 0.4$ with success probability $p_A = \frac{1}{2}$. Each trial produced 10,000 input-output samples $\{s_t, y_t\}$ that estimated the probability densities to obtain the mutual information. The vertical dashed lines show the absolute deviation between the smallest and largest outliers in each sample average of 100 outcomes. (**a**): Stochastic resonance (SR) when the Cauchy noise n_t has location $a = 0$ and thus a lies outside the forbidden interval: $a = 0 \notin (\theta - A, \theta + A) = (0.1, 0.9)$. The system has a nonzero noise optimum at $\gamma_{opt} \approx 0.21$ and thus shows the SR effect. (**b**): No SR when the Cauchy noise n_t has location $a = 0.2$ that lies in the forbidden interval: $a = 0.2 \in (\theta - A, \theta + A) = (0.1, 0.9)$. The system is optimal when $\gamma \to 0$ and thus *does not* show the SR effect: The mutual information $I(S,Y)$ is maximum because it equals the input entropy $H(S) = 1$

or location-scale noise does not fall in an interval that depends on the threshold and signal amplitudes (see Fig. 1). A reviewer of [20] referred to this interval as a "forbidden interval." We liked the colorful term and have used it ever since. The theorems below show just how complicated these interval conditions can become due to the complexity of the system's nonlinear dynamics and the nature of the noise or diffusion processes.

2 Proof Strategy: *What Goes Down Must Go Up*

This section presents sufficient and necessary forbidden interval theorems for SR in threshold systems. We use the discrete-time threshold system [4, 10, 16]

$$y_t = \operatorname{sgn}(s_t + n_t - \theta) \tag{1}$$

where $\theta > 0$ is the system's threshold, y_t is the system's output, s_t is the bipolar input Bernoulli signal with amplitude $A > 0$ and with success probability $p_A = \frac{1}{2}$, n_t is the additive white noise with probability density $p(n)$, and the signum function is

$$\operatorname{sgn}(x) = \begin{cases} 1 & \text{if } x \geq 0 \\ -1 & \text{if } x < 0 \end{cases} . \tag{2}$$

Note that the theorems below hold for any two-symbol alphabet set $\mathscr{Y}$ for the output y_t. The Shannon mutual information $I(S,Y)$ has the form [8]

$$I(S,Y) = H(Y) - H(Y|S) = \sum_{s\in\mathscr{S}} \sum_{y\in\mathscr{Y}} P_{SY}(s,y) \log \frac{P_{SY}(s,y)}{P_S(s)P_Y(y)} . \tag{3}$$

The idea behind forbidden interval theorems is that we can cast the proof in terms of showing that a limit must go to zero as the noise intensity goes to zero.

Theorem 1. *Suppose that the threshold system* (1) *has noise probability density function $p(n)$ and that the input signal is subthreshold $A < \theta$. Suppose that there is some statistical dependence between input random variable S and output random variable Y (so that $I(S,Y) > 0$). Suppose that the noise mean $E[n]$ does not lie in the signal-threshold interval $(\theta - A, \theta + A)$ if $p(n)$ has finite variance. Then the threshold system* (1) *exhibits the nonmonotone SR effect in the sense that $I(S,Y) \to 0$ as $\sigma \to 0$.*

Proof. Assume $0 < P_S(s) < 1$ to avoid triviality when $P_S(s) = 0$ or 1. We show that S and Y are asymptotically independent: $I(\sigma) \to 0$ as $\sigma \to 0$. Recall that $I(S,Y) = 0$ if and only if S and Y are statistically independent [8]. So we need to show only that $P_{SY}(s,y) = P_S(s)P_Y(y)$ or $P_{Y|S}(y|s) = P_Y(y)$ as $\sigma \to 0$ for all signal symbols $s \in \mathscr{S}$ and $y \in \mathscr{Y}$. The two-symbol alphabet set $\mathscr{S} = \{0,1\}$ gives

$$P_Y(y) = \sum_{s\in\mathscr{S}} P_{Y|S}(y|s)P_S(s) = (P_{Y|S}(y|0) - P_{Y|S}(y|1))P_S(0) + P_{Y|S}(y|1) . \tag{4}$$

So we need to show only that $P_{Y|S}(y|0) - P_{Y|S}(y|1) = 0$ as $\sigma \to 0$. This condition implies that $P_Y(y) = P_{Y|S}(y|1)$ and $P_Y(y) = P_{Y|S}(y|0)$. We assume for simplicity that the noise density $p(n)$ is integrable. The argument below still holds if $p(n)$ is discrete and if we replace integrals with appropriate sums.

Consider $y =$ "0." Then

$$P_{Y|S}(0|0) - P_{Y|S}(0|1) = \int_{-\infty}^{\theta+A} p(n)dn - \int_{-\infty}^{\theta-A} p(n)dn = \int_{\theta-A}^{\theta+A} p(n)dn \quad . \quad (5)$$

Similarly for $y =$ "1":

$$P_{Y|S}(1|0) - P_{Y|S}(1|1) = \int_{\theta+A}^{\infty} p(n)dn - \int_{\theta-A}^{\infty} p(n)dn = -\int_{\theta-A}^{\theta+A} p(n)dn \quad . \quad (6)$$

The result follows if we can show that

$$\int_{\theta-A}^{\theta+A} p(n)dn \to 0 \qquad \text{as } \sigma \to 0 \quad . \quad (7)$$

Denote the mean of the noise by $\mu = E[n]$ and the variance by $\sigma^2 = E[(n-\mu)^2]$. Then $\mu \notin (\theta - A, \theta + A)$ by hypothesis.

Now suppose that $\mu < \theta - A$. Pick $\varepsilon = \frac{1}{2}(\theta - A - \mu) > 0$. So $\theta - A - \varepsilon = \theta - A - \varepsilon + \mu - \mu = \mu + (\theta - A - \mu) - \varepsilon = \mu + 2\varepsilon - \varepsilon = \mu + \varepsilon$. Then

$$P_{Y|S}(0|0) - P_{Y|S}(0|1) = \int_{\theta-A}^{\theta+A} p(n)dn \leq \int_{\theta-A}^{\infty} p(n)dn \leq \int_{\theta-A-\varepsilon}^{\infty} p(n)dn \quad (8)$$

$$= \int_{\mu+\varepsilon}^{\infty} p(n)dn = P(n \geq \mu + \varepsilon) = P(n - \mu \geq \varepsilon) \leq P(|n-\mu| \geq \varepsilon) \quad (9)$$

$$\leq \frac{\sigma^2}{\varepsilon^2} \qquad \text{by Chebyshev's inequality} \quad (10)$$

$$\to 0 \qquad \text{as } \sigma \to 0 \quad . \quad (11)$$

A symmetric argument shows that for $\mu > \theta + A$

$$P_{Y|S}(0|0) - P_{Y|S}(0|1) \leq \frac{\sigma^2}{\varepsilon^2} \to 0 \quad \text{as } \sigma \to 0 \quad . \quad \text{QED} \quad (12)$$

The next forbidden interval shows that a structurally similar interval condition holds for *all* types of infinite-variance stable noise [19]. Stable models apply to a quantum alpha-stable noise source. Stable models apply to diverse physical phenomena that include impulsive interrupts in phone lines, underwater acoustics, low-frequency atmospheric signals, and gravitational fluctuations [25]. Symmetric alpha-stable noise [25, 30] results from an impulsive noise source and describes a family of thick-tailed bell-curve densities per (13) below. The parameter α (which differs from the coherent state α in Sect. 4) lies in $(0,2]$ and governs the thickness of the distribution's tail: $\alpha = 1$ corresponds to the thick-tailed Cauchy random variable and $\alpha = 2$ corresponds to the familiar thin-tailed Gaussian random variable. The bell curve's tail thickness increases as α decreases. The generalized central limit theorem states that all and only normalized stable random variables converge in distribution to a stable random variable [2].

Theorem 2. *Suppose $I(S,Y) > 0$ and the threshold system (1) uses alpha-stable noise with location parameter $a \notin (\theta - A, \theta + A)$. Then the system exhibits the nonmonotone SR effect if the input signal is subthreshold.*

Proof. Again the result follows if $\int_{\theta-A}^{\theta+A} p(n)dn \to 0$ as $\gamma \to 0$. The characteristic function $\varphi(\omega)$ of alpha-stable density $p(n)$ has the exponential form

$$\varphi(\omega) = \begin{cases} \exp\left\{ia\omega - \gamma|\omega|^{\alpha}\left(1+i\beta\operatorname{sign}(\omega)\tan\frac{\alpha\pi}{2}\right)\right\} & \text{for } \alpha \neq 1 \\ \exp\left\{ia\omega - \gamma|\omega|(1-i\frac{2}{\pi}\beta\ln|\omega|\operatorname{sign}(\omega))\right\} & \text{for } \alpha = 1 \end{cases}. \tag{13}$$

This reduces to a simple complex exponential in the zero-dispersion limit: $\lim_{\gamma\to 0}\varphi(\omega) = \exp\{ia\omega\}$ for *all* α's, skewness β's, and location a's. So Fourier transformation gives the corresponding density function in the limiting case ($\gamma \to 0$) as a translated delta function $\lim_{\gamma\to 0} p(n) = \delta(n-a)$. Then

$$P_{Y|S}(0|0) - P_{Y|S}(0|1) = \int_{\theta-A}^{\theta+A} p(n)dn = \int_{\theta-A}^{\theta+A} \delta(n-a)dn = 0 \tag{14}$$

because $a \notin (\theta-A, \theta+A)$. QED

Similar proofs give converses in Theorems 3 and 4 [20].

Theorem 3. *Suppose that the threshold system* (1) *has noise probability density function* $p(n)$ *and that the input signal* S *is subthreshold. Suppose that the noise mean* $E[n]$ *lies in the signal-threshold interval* $(\theta-A, \theta+A)$ *if* $p(n)$ *has finite variance. Then the threshold system* (1) *does not exhibit the nonmonotone SR effect in the sense that* $I(S,Y)$ *achieves its maximum when* $\sigma \to 0$*:* $I(S,Y) = H(Y) = H(S)$ *when* $\sigma \to 0$.

Theorem 4. *Suppose that the threshold system* (1) *has subthreshold input signal and use alpha-stable noise with location parameter* $a \in (\theta-A, \theta+A)$*. Then the threshold system* (1) *does not exhibit the nonmonotone SR effect:* $I(S,Y)$ *achieves its maximum when* $\gamma \to 0$*:* $I(S,Y) = H(Y) = H(S)$ *when* $\gamma \to 0$.

3 SR in a Carbon Nanotube Signal Detector

A carbon nanotube signal detector benefits from small amounts of added electrical noise in accord with the forbidden interval Theorems 1 and 2 above [21]. Our experiments used a carbon nanotube field-effect transistor to detect noisy subthreshold electrical signals. Two new SR hypothesis tests in [21] also confirmed the SR effect in the nanotube transistor. Three measures of detector performance showed the SR effect: Shannon's mutual information, the normalized correlation measure, and an inverted bit error rate compared the input and output discrete-time random sequences. The nanotube detector had a threshold-like input-output characteristic in its gate effect (see Fig. 2). It produced little current for subthreshold digital input voltages that fed the transistor's gate. Three types of synchronized white noise corrupted the subthreshold Bernoulli sequences that fed the detector. Gaussian, uniform, and impulsive Cauchy noise combined with the random input voltage sequences to help the detector produce random output current sequences.

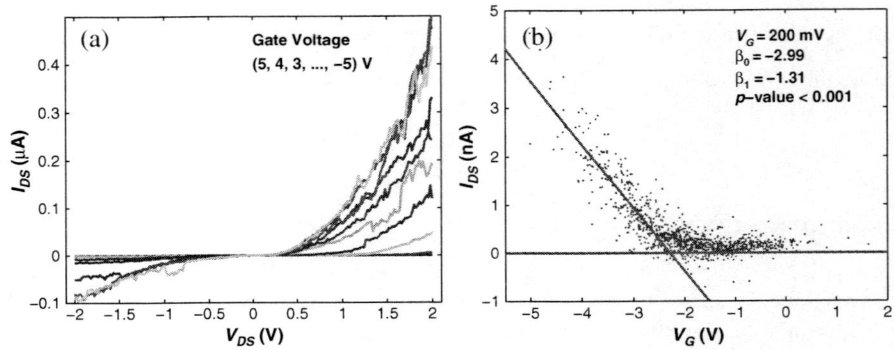

Fig. 2 Less than ideal carbon nanotube transistor gate effect. (**a**) plots the drain-source current versus the drain-source voltage for different gate voltages. (**b**) plots the experimental input-output pairs and shows the transistor's current-voltage I-V_G characteristics in response to noisy input signals. Linear regression fit the data to the transistor equation, extrapolated the nonlinearity, and estimated the threshold voltage $V_T \approx -2.3$ V

The experiments observed the SR effect by measuring how well an output sequence matched its input sequence. Shannon's mutual information used histograms to estimate the probability densities and computed the entropies. The correlation measure was a scalar inner product of the input and output sequences. The inverted bit error rate computed how often the bits matched between the input and output sequences. The observed nanotube SR effect was robust: it persisted even when infinite-variance Cauchy noise corrupted the signal stream.

Simulations and lab experiments both showed that the above forbidden interval theorems apply to threshold detectors. The simulations modeled transistors in general. They used a threshold-like ramp function that often models an ideal transistor's current-to-gate-voltage (I-V_G) characteristics: $Y = G(S - V_T)$ where Y is the output current, S is the input voltage, V_T is the threshold voltage, and G is a nonzero gain for suprathreshold inputs and zero otherwise. The negative threshold voltage V_T gives the forbidden voltage intervals of the form $(V_T + x, V_T - x)$ for threshold voltage $V_T = -2.3$ for $x = -2$, -1.8, -1.6, and -1.4 volts. So $E[n] = 0$ fell outside all intervals.

The experiments tested single-walled carbon nanotube transistors. The prototype transistors had non-ideal characteristics such as the I-V curves in Fig. 2. The laboratory data generated SR-curves Fig. 3(b), (c), (d) that qualitatively agreed with those generated in simulation Fig. 3(a).

Experiments confirmed the SR prediction: noise helped a pristine (undoped) single-walled carbon nanotube transistor [32] detect subthreshold signals. The experiments applied different Bernoulli input sequences that used different combinations of subthreshold gate voltages as their ON/OFF symbols. Synchronized Gaussian, uniform, and infinite-variance Cauchy noise added to the input sequences and helped the nanotube transistor detect the subthreshold input. The performance

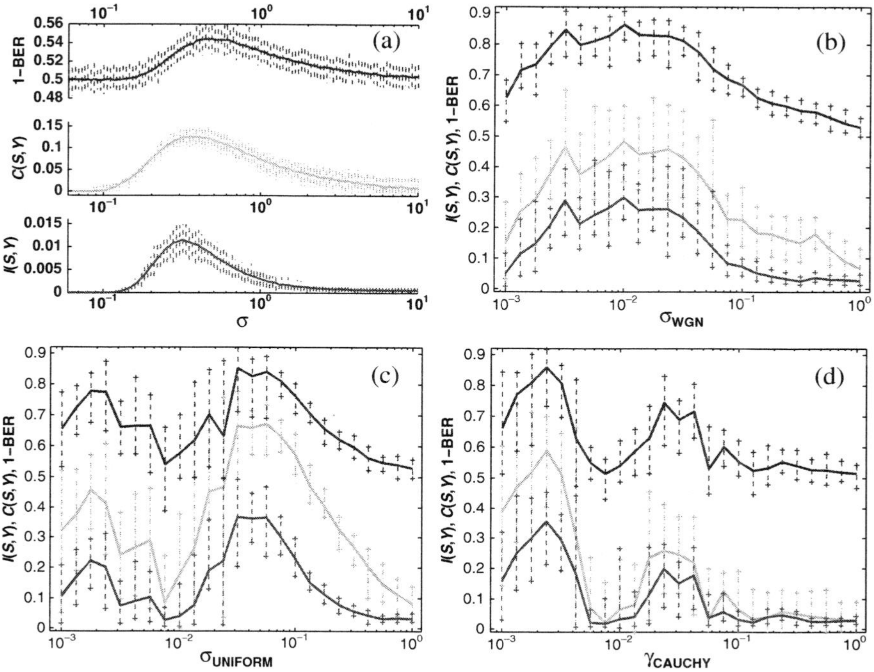

Fig. 3 Noise-enhanced detector performance. (**a**) A simulated threshold detector exhibits the signature SR modes with additive white Gaussian noise. The experiments found the SR effect for three different measures, for three different types of noise, and for multiple combinations of voltage-symbols. (**b**) shows the SR effect for Gaussian noise and binary symbols $(-1.6, -1.4)$ V, (**c**) for uniform noise and binary symbols $(-2.0, -1.8)$ V, and (**d**) for impulsive Cauchy noise and binary symbols $(-2.0, -1.8)$ V. The subthreshold symbols were more positive than the threshold voltage ($V_T = -2.3$ V) consistent with p-type semiconductors. Each pair of voltage symbols had a 0.2V separation because sensitivity analysis showed that the separation gave complete SR modes within the range of noise levels. A linear regression of the transistor's gate effect estimated the threshold voltage and aided the selection of the subthreshold ON/OFF symbols. (**b**) The SR mode of the mutual-information curve is six times the value at minimal noise. The SR mode of the correlation-measure curve is three times the value at minimal noise. The SR mode of the inverted bit-error-rate curve shows a 40% improvement over the value at minimal noise. (**d**) Cauchy noise. We passed impulsive or infinite-variance white Cauchy noise through the nanotube detector to test whether it was robust to occasional large noise spikes. We chose the highly impulsive Cauchy noise for this task. Not all Cauchy experiments produced a measurable SR effect

measures were Shannon's mutual information $I(S,Y)$ in (3), an input-output correlation measure $C(S,Y)$, and an inverted bit error rate $1-BER$.

Both the experimental and simulated nanotube detectors had correlated SR curves. The correlation coefficient R measured the strength of the correlation. The simulated SR curves in Fig. 3(d) had correlation coefficients of $R = 0.9367$ for $I(S,Y)$ and $C(S,Y)$, $R = 0.8265$ for $I(S,Y)$ and $1-BER$, and $R = 0.9541$ for $C(S,Y)$ and $1-BER$. The experimental SR curves in Fig. 3(a) had correlation coefficients of $R = 0.9830$ for $I(S,Y)$ and $C(S,Y)$, $R = 0.9774$ for $I(S,Y)$ and $1 - BER$, and $R = 0.9877$ for $C(S,Y)$ and $1 - BER$. The correlations were statistically significant for p-value < 0.001.

We observed the nanotube SR effect in Fig. 3(b) as one of four such successful combinations of input binary values with the parameter choices ON = -1.6 V and OFF = -1.4 V. Figures 3(c) and (d) each represents a selection from successful combinations. [21] contains additional details of the experimental setup and parameters. This SR effect occurred despite the nanotube instabilities that caused fluctuations in the stochastic I-V_G curve in Fig. 2(b). [21] discusses the generation of the stochastic current-voltage curve in the presence of hysteresis.

The nanotube experiments produced the SR effect for Shannon's mutual information [31], an input-output correlation measure [6, 7], and an inverted bit error rate that measured how well the output sequences matched the input Bernoulli sequences. The input signal S was a sequence of random binary voltages that produced a random output sequence in the form of a transistor current. Histograms of the sequences estimated the probability density functions that computed the entropies. The correlation measure normalized the zero-lag value ($l = 0$) of the cross-correlation sequence

$$r_{SY}(l) = \sum_{k=1}^{N} s(k)y(k-l) \tag{15}$$

of the two sequences with subtracted means. These two measures did not assume that the nanotube detector had a special structure and did not impose a threshold scheme on the experiment. But the inverted bit error rate $1 - BER$ decided whether each output y_i was a "0" or "1" by applying a threshold scheme: a Baye's discriminant function in a two-class minimum-distance classifier [9] that used complete knowledge of the input. The rate $1 - BER$ measured how often the input and output bits agreed: $1 - BER = 1 - N_{error}/N = N_{correct}/N$ where N_{error} counted the number of bits that differed between the length-N input and output sequences and $N_{correct}$ counted the number of bits that agreed.

The experimental observation of nanotube SR used the nonlinear field effect of carbon nanotubes. A semiconductor single-walled carbon nanotube (SWNT) can change its conductivity in response to an external electric field in a gate effect [32]. The SR experiments used a chemical-vapor-deposition (CVD) grown SWNT [15]. The reference contains the details of the fabrication.

The semiconductor SWNT forms a Schottky diode at the interface with metal so that a metal-nanotube-metal contact forms a field-effect transistor (FET) with an adjacent gate electrode [35]. The typical current-voltage I-V_G characteristics

$$I = \begin{cases} G(V_G - V_T) & \text{for } V_G < V_T \\ 0 & \text{else} \end{cases} \tag{16}$$

indicate that the pristine semiconductor nanotubes act as hole-doped semiconductors at room temperatures and that the nanotube devices are p-type FETs [32]. The transconductance G is negative and the gate voltage $V_G < V_T$ is suprathreshold for p-type FETs. Equation (16) modeled the ideal p-type FET in the simulated experiments.

We provide a terse summary of the laboratory setup— [21] contains the details. Each of the nanotube experiments applied 25 sampled noise levels that ranged from 0.001 to 1 standard deviation σ (dispersion γ for infinite-variance Cauchy) linearly in logarithmic scale. The noisy input S was a *synchronized* Bernoulli sequence $s_i = b_i + n_i$ of the sum of random subthreshold binary values b_i and additive white noise n_i of three types. So there was no *timing* noise in the pulse train as in the FHN neuron model [28, 29]. Synchronization allows the nanotube systems to implement a variety of algorithms from signal processing and communications.

The experiments updated the noisy input symbols s_i about once every 10 ms. A 200 mV drain-source voltage biased the nanotube at room temperature in vacuum. The experiments measured and averaged ten samples of the detector output at 100 kilosamples/s near the end of each symbol interval to estimate the output symbols y_i. This allows the transistor output to settle. A gate voltage is subthreshold if it is more positive than a p-type FET's threshold voltage and produces picoamp current in an OFF state. The experiment tested whether noise could enhance subthreshold signals to produce measurable currents.

The detector consisted of a single-walled semiconductor carbon nanotube bridging two electrodes [15]. The CVD technique combined with e-beam lithography to grow a single-walled nanotube that was 3–5 μm long and less than 2 nm in diameter between two electrodes. The gap between the electrodes was approximately 3 μm wide but the single-walled nanotube was not straight as it spanned the gap. Atomic force microscopy examined the detector and showed that the nanotube had a diameter $d < 2$ nm that was consistent with a single-walled nanotube.

A PC-based National Instruments PCI-MI0-16XE-10 data acquisition (DAQ) board converted the noise-corrupted signal S from digital to analog (DA) and converted the conditioned noisy output Y from analog to digital (AD). The AD-DA conversion has a 16-bit resolution and a 10 μs rise time. A DL 1211 current-voltage preamplifier conditioned the detector output current before data acquisition sampled it as a voltage. The amplifier converts a small current (10 nA) into a large voltage (1 V) with the 10^{-8} A/V gain setting. The analog voltage has a maximal time delay of 0.1 ms with the 0.1 ms rise-time setting.

Two new hypothesis tests verified that the SR curves were nonmonotonic and confirmed the nanotube SR effect. A chi-square test and a Kolmogorov-Smirnov test both rejected the similarity between a monotonically decreasing β-probability density function and each of the three SR curves with p-value < 0.001. The statistical tests were goodness-of-fit tests that treated the SR curves as candidate pdfs and compared them against the benchmark pdf. The parameters of the benchmark β-pdf were $\alpha = 0.5$ and $\theta = 5$ among many others that we tested [21].

4 Forbidden Interval Theorem for Quantum Communication

Quantum optics and "squeezed light" [22] extend the basic or classical forbidden interval theorems above for detecting subthreshold signals [34]. The new quantum forbidden interval theorem shows that a noise benefit occurs if and only if the position quadrature of the noise mean or location does not lie in a given interval. The result holds for all types of finite-variance noise and for all infinite-variance stable noise.

The quantum forbidden-interval theorem applies to the quantum-optical communication system of Fig. 4. This system assumes weak or subthreshold signals and further assumes that noise corrupts these signals. Alice sends a squeezed displaced vacuum as a binary signal to Bob. Then Bob decodes the binary message by position-quadrature homodyning and thresholding. Figure 5 shows the non-monotonic signature of the predicted SR noise benefit in the quantum-optical communication system.

We develop the quantum-optical model in the Heisenberg picture. Suppose Alice possesses a vacuum mode. Let $\hat{x}$ denote the position-quadrature operator of this vacuum state. This operator collapses to a zero-mean $1/2$-variance Gaussian random variable X if Alice measures her state. Suppose that Alice does not measure it. Suppose instead that she sends her mode through a squeezer. Suppose further that she can control the strength of squeezing with a squeezing parameter r. Her position-quadrature operator $\hat{x}$ evolves under the squeezer to become $\hat{x}e^{-r}$. She encodes a random message bit $S \in \{0,1\}$ by displacing her state by $\alpha \in \mathbb{C}$ if $S = 1$ or by $-\alpha$ if $S = 0$. Her operator evolves under the displacement to become $\hat{x}e^{-r} + (-1)^{S+1}\alpha_x$ where $\alpha_x = Re\{\alpha\}$. She sends her state to Bob over an additive noisy bosonic channel [12]. A noisy bosonic channel affects any annihilation operator $\hat{a}_{in}$ by $\hat{a}_{out} = \hat{a}_{in} + \nu$. Annihilation operator $\hat{a}_{out}$ represents the output mode. Complex random variable ν represents the noisy effects of the bosonic channel. Random variable ν need not be Gaussian for the SR effect to occur—it can have finite variance or possess an alpha-stable distribution [25].

Bob receives the state $\hat{x}e^{-r} + (-1)^{S+1}\alpha_x + \nu_x$ from the noisy channel where $\nu_x = Re\{\nu\}$. Bob performs a position-quadrature homodyne detection so that the state collapses to the random variable $(-1)^{S+1}\alpha_x + N$ where $N = Xe^{-r} + \nu_x$ sums both noisy random variables. Bob thresholds the result of this homodyne detection with threshold θ to retrieve a bit Y. This bit Y should be the message bit S that Alice first sent.

Random variables Xe^{-r} and ν_x are independent because random variable Xe^{-r} originates from vacuum fluctuations and because ν_x represents Bob's loss of knowledge due to the state's propagation through a noisy quantum channel. The density

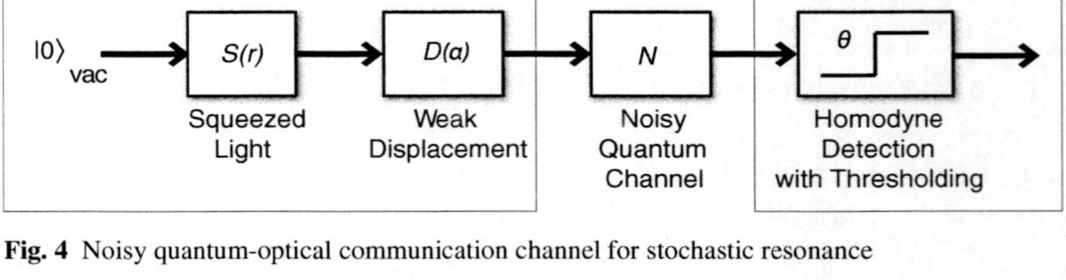

Fig. 4 Noisy quantum-optical communication channel for stochastic resonance

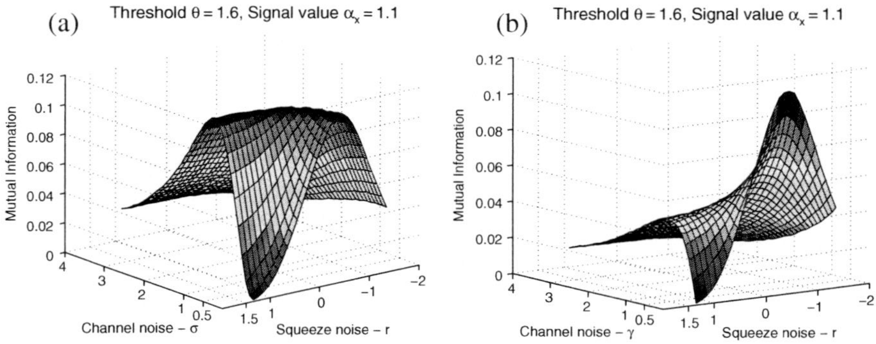

Fig. 5 Quantum SR noise benefits for (**a**) Gaussian noise and (**b**) Cauchy noise

$p_N(n)$ of random variable N is $p_N(n) = (p_{Xe^{-r}} * p_{v_x})(n)$ where $p_{Xe^{-r}}(n)$ is the density of a zero-mean $(e^{-2r}/2)$-variance Gaussian random variable, $p_{v_x}(n)$ is the density of v_x, and $*$ denotes convolution.

The quantum forbidden interval theorem below gives necessary and sufficient conditions for the nonmonotone SR effect to occur under both finite-variance and infinite-variance alpha-stable noise. We state the parameters for the finite-variance case without parentheses and the parameters for the infinite-variance case with parentheses.

Theorem 5. *Suppose the channel noise's position quadrature has finite variance σ_x^2 (dispersion γ_x) and mean μ_x (location a_x). Suppose the input signal's position quadrature α_x is subthreshold: $\alpha_x < \theta$. Suppose there is some statistical dependence between input signal S and output signal Y so that the mutual information obeys $I(S,Y) > 0$. Then the quantum communication system exhibits the nonmonotone SR effect if and only if the position quadrature of the noise mean does not lie in the forbidden interval: $\mu_x \notin (\theta - \alpha_x, \theta + \alpha_x)$ (or $a_x \notin (\theta - \alpha_x, \theta + \alpha_x)$ in the stable case). The nonmonotone SR effect is that $I(S,Y) \to 0$ as $\sigma_x \to 0$ (or $\gamma_x \to 0$) and as squeezing parameter $r \to \infty$.*

Proof. The finite-variance proof for sufficiency and necessity follows the proof method in [19] and [20] respectively if we use $p_N(n)$ as the noise density. The infinite-variance proof for sufficiency and necessity follows the stable proof method in [19] and [20] respectively if we use $p_N(n)$ as the noise density and if v_x is an alpha-stable random variable. QED

Figure 5 shows simulation instances of Theorem 5. Figure 5 displays the full "inverted-U" curve for realistic squeezing values [36]. The theorem guarantees only that the nonmonotone SR effect occurs. It does not give the optimal combination of channel noise and squeezing or guarantee a large increase in mutual information.

One criticism is that the theorem is not realistic because it requires infinite squeezing and thus requires infinite energy to produce the SR effect. But the theorem guarantees that the SR effect occurs for some finite squeezing. The simulations in

Fig. 5 display the full joint σ^2 and r nonmonotone SR signature for experimentally plausible squeezing values and for realistic channel noise levels.

5 Forbidden Interval Theorem for Spiking Retinal Neurons

A different forbidden interval theorem guarantees an SR noise benefit for spiking retinal neurons [26]. Figure 6 below shows an SR noise benefit in a spiking retinal neuron. The neuron should emit more spikes when the brightness contrast level is low rather than high. The right amount of Gaussian noise helps the neuron discriminate between two levels of brightness contrast. The retinal neuron emits too few spikes if no noise corrupts the Bernoulli sequence of contrast levels. The neuron also emits too many spikes and emits many of them at the wrong time if too much noise corrupts the sequence.

The retina model of Fig. 6 is a noisy version of a common Wiener-type cascade model [5, 14]:

$$r(t) = r_0 h\left[\int_{-\infty}^{\infty} f(z)\{S(t-z)+n_1(t)\}dz + n_2(t)\right] \tag{17}$$

where S is the input stimulus defined below, r is the instantaneous Poisson spike rate that gives the exponential interspike-interval density function as $p(t) = r(t)$ $\exp[-\int_0^t r(\tau)d\tau]$, f is a band-pass linear filter function, and h is a memoryless monotone–nondecreasing function. Here n_1 denotes the combined stimulus and photoreceptor noise and n_2 denotes the combined ion-channel noise and the synaptic noise.

The input stimulus S is Michelson's visual contrast signal: $S = (L_c - L_s)/(L_c + L_s)$. L_c is the amount of light that falls on the center of the ganglion cell's receptive field. L_s is the light that falls on its surround region. The sigmoid-shaped memoryless function h approximates the spike threshold and saturation level. We define h as a piecewise-linear approximation of a sigmoidal nonlinearity [37]:

$$h(x) = \begin{cases} \theta_2 - \theta_1 & \text{if } x > \theta_2 \\ x - \theta_1 & \text{if } \theta_1 \le x \le \theta_2 \\ 0 & \text{if } x < \theta_1 \end{cases} \tag{18}$$

and so

$$r(w(t)) = \begin{cases} r_0(\theta_2 - \theta_1) & \text{if } w(t) > \theta_2 \\ r_0(w(t) - \theta_1) & \text{if } \theta_1 \le w(t) \le \theta_2 \\ 0 & \text{if } w(t) < \theta_1 \end{cases} \quad . \tag{19}$$

The subthreshold contrast signal $S(t) \in \{A, B\}$ is a random Bernoulli sequence with $P(S(t) = A) = p$ and $P(S(t) = B) = 1 - p$. The time duration of each signal value A and B in $S(t)$ is much larger than the time constant of the linear filter $f(t)$. We define $v(t)$ as the filtered output of the contrast signal $S(t)$ without noise $n_1(t)$ and such that

$$v(t)|_{S(t)=A} = v_1 \quad \text{and} \quad v(t)|_{S(t)=B} = v_2 \tag{20}$$

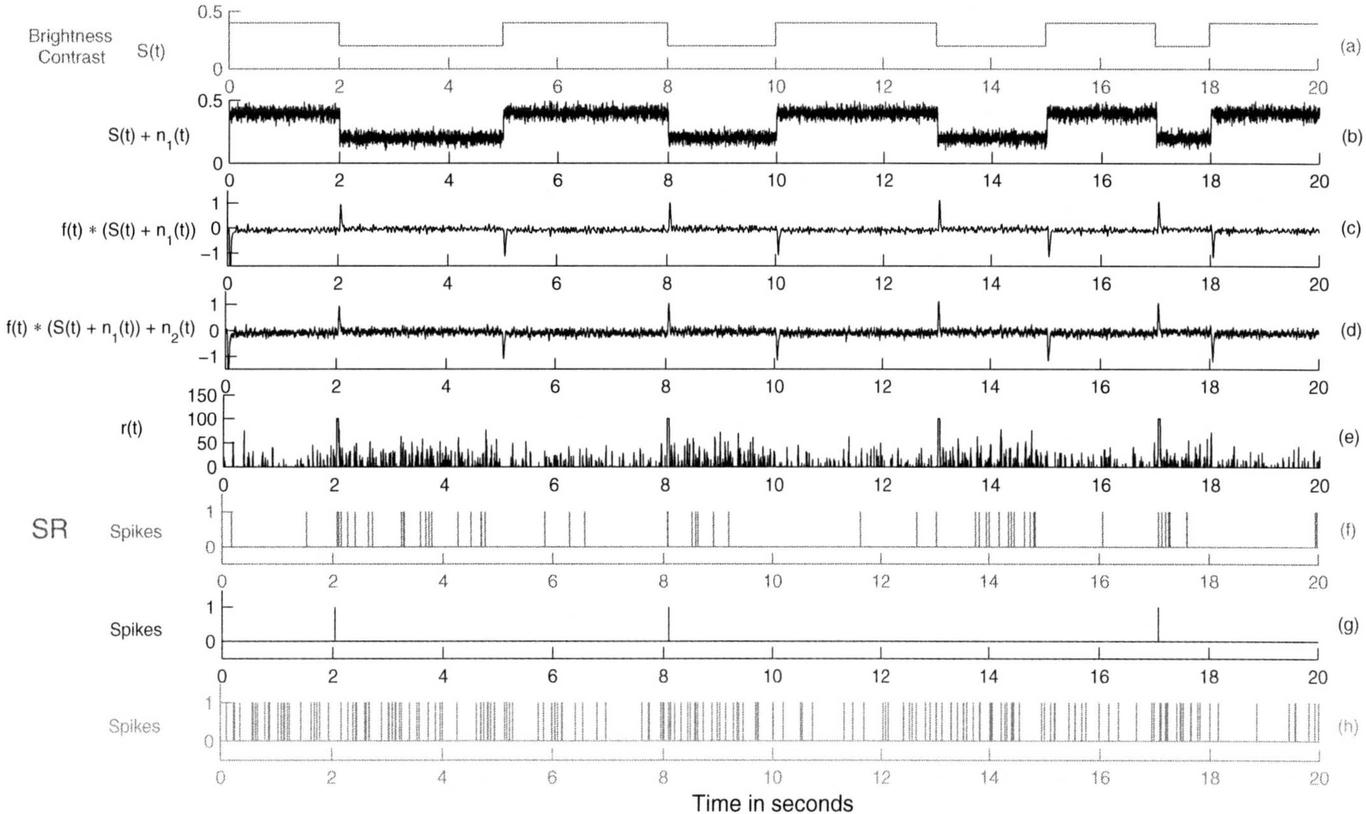

Fig. 6 Stochastic resonance in a spiking retinal neuron. The neuron should emit more spikes when the brightness contrast level is low rather than high. Noise improves the discrimination of subthreshold contrast stimuli in the retina model (17), (18), (19). (**a**) Bernoulli contrast signal S as a function of time t. (**b**) Contrast signal S plus Gaussian white noise n_1 with variance $\sigma_1 = 0.03$. (**c**) Signal in plot (**b**) filtered with f in (1). (**d**) Filtered noisy signal in (**c**) plus noise n_2 (synaptic and ion-channel noise) with variance $\sigma_2 = 0.06$. (**e**) Noisy spike rate $r(t)$. (**f**) SR effect: Output Poisson spikes that result from the noisy spike rate $r(t)$. (**g**) Output spikes in the absence of noise. (**h**) Output spikes in the presence of too much noise

in steady-state where $v_1 > v_2$ and $\max(v_1, v_2) < \theta_1 < \theta_2$. So the input signal $S(t)$ is subthreshold. We measure the average spike rate for each symbol only when the corresponding value of $v(t)$ is in steady-state. Theorem 6 below gives necessary and sufficient conditions for an SR noise effect in the retina neuron model (17), (18), (19) for either noise source n_1 or n_2. It states that standard spiking retinal models benefit from additive white noise if and only if a joint noise mean or location parameter does not fall in a forbidden interval of threshold-based values. Theorem 6 holds for all finite-variance noise and for all impulsive or infinite-variance stable noise [26].

Theorem 6. *Suppose that the noise sources n_1 and n_2 in the retina model* (17), (18), (19) *are white and have finite-variance (or finite-dispersion in the stable case) probability density functions $p_1(n)$ and $p_2(n)$ with corresponding variances (dispersions) σ_1^2 and σ_2^2 (γ_1 and γ_2). Suppose that the input signal S is subthreshold ($v_2 < v_1 < \theta_1 < \theta_2$) and that there is some statistical dependence between the input contrast random variable S and the output random variable R so that $I(S,R) > 0$. Then the retina model* (17), (18), (19) *exhibits the nonmonotone SR effect in the sense that $I(S,R) \to 0$ as σ_1^2 and σ_2^2 (or γ_1 and γ_2) decrease to zero if and only if the mean sum $E[n_1]\int f(\tau)d\tau + E[n_2]$ (or the location parameter sum in the stable case) does not lie in the interval $(\theta_1 - v_1, \theta_2 - v_2)$. The only-if part holds in the sense that the system performs better without noise than with it when the interval condition fails.*

6 Forbidden Interval Theorems for Levy Noise Diffusions

The most complex forbidden interval theorems apply to nonlinear stochastic differential equations with additive Levy diffusions. Levy processes generalize Brownian motions to allow for jumps and other impulsive behavior [1]. Levy processes include not only Brownian processes but also compound Poisson processes, infinite-variance α-stable processes, generalized inverse Gausssian processes, and generalized hyperbolic processes. A random process L_t is a Levy process if it has *independent increments* $L_t - L_s$, if it is *stationary* ($L_t - L_s$ has the same distribution as L_{t-s}), and if it is *continuous in probability* ($L_s \to L_t$ in probability if $s \to t$). A Levy process L_t has a drift component, a Brownian (Gaussian) component, and a jump component. Figure 7 shows sample paths from four different types of Levy processes.

The forbidden interval results of Theorems 7 and 8 below show that a broad class of additive white Levy noise [27] with finite second moments can benefit a wide range of noisy feedback continuous and sensory spiking neuron models of the general form

$$dX_t = b(X_{t^-})dt + c(X_{t^-})dL_t \tag{21}$$

$$Y_t = g(X_t) \quad . \tag{22}$$

Here $c(X_{t^-})$ is a bounded Levy diffusion term, dL_t is a white Levy noise with noise scale κ, and $b(X_{t^-})$ is a Lipschitz continuous drift term that has the additive net

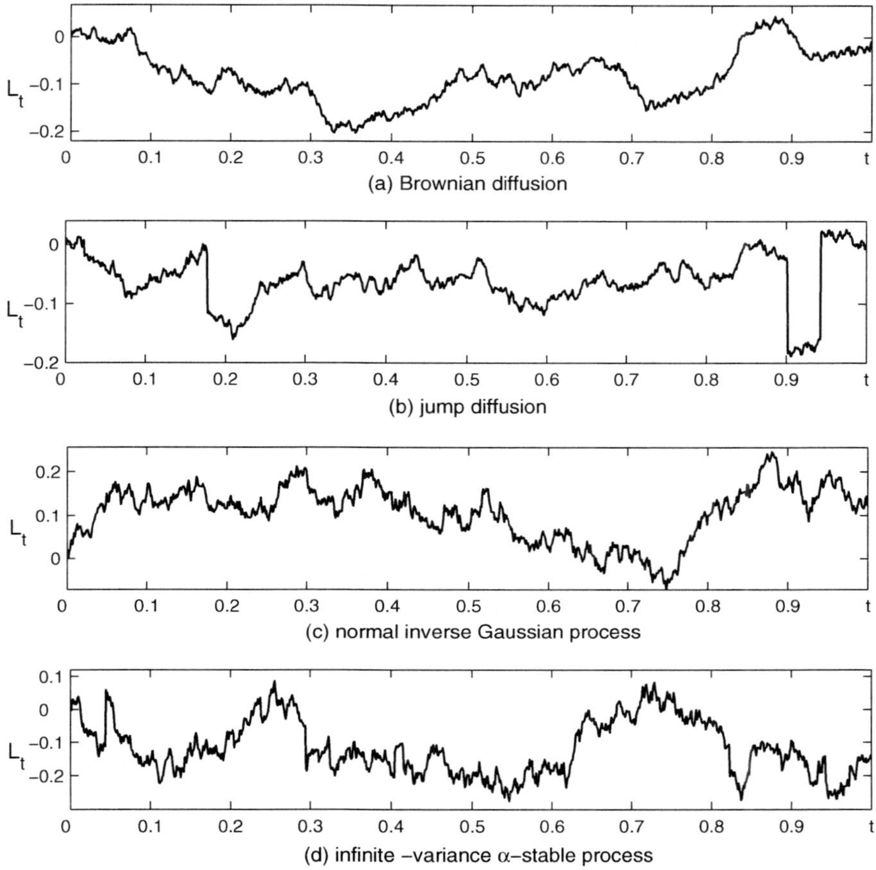

Fig. 7 Sample paths from one-dimensional Levy processes: (**a**) Brownian motion with drift $\mu = 0.1$ and variance $\sigma = 0.15$, (**b**) jump diffusion with $\mu = 0.1$, $\sigma = 0.225$, Poisson jump rate $\lambda = 3$, and uniformly distributed jump magnitudes in the interval $[-0.2, 0.2]$ (and so with Levy measure $\nu(dy) = (3/0.4)dy$ for $y \in [-0.2, 0.2]$ and zero else in [27]]), (**c**) normal inverse Gaussian (NIG) process with parameters $\alpha = 20$, $\beta = 0$, $\delta = 0.1$, and $\mu = 0$, (**d**) infinite-variance α-stable process with $\alpha = 1.9$ and dispersion $\kappa = 0.0272$ ($\mu = 0$, $\sigma = 0$, and $\nu(dy)$ is of the form $\frac{k}{|y|^{1+\alpha}}dy$ [27])

excitatory or inhibitory input forcing signal S_t—either s_1 or s_2. The neuron feeds its activation or membrane potential signal X_t back to itself through the drift term $b(X_{t^-})$ and emits the (observable) thresholded or spike signal Y_t as output. Here g is a static transformation. We use the threshold $g(X_t) = 1$ if $X_t > 0$ and zero else for continuous neuron models. We use a related threshold g in spiking neuron models where g determines the spike occurrence. The drift term $b(X_{t^-})$ is $-X_{t^-} + f(X_{t^-}) + S_t$ for continuous neuron models where the neuronal signal function $f(x)$ includes common signal functions such as the logistic, bistable, linear-threshold, and Gaussian or "radial basis" [16]. The drift term $b(X_{t^-})$ of sensory spiking neurons includes popular spiking neuron models such as the FitzHugh-Nagumo (FHN) model, the leaky integrate-and-fire model, and the reduced Type I neuron model. There exist θ_1 and θ_2 for continuous neuron models such that the input S_t is subthreshold when $\theta_1 \leq s_1 < s_2 \leq \theta_2$ while there exists B for spiking neuron models such that S_t is subthreshold when $S_t < B$. The values of θ_1, θ_2, and B depend on the model parameters.

Equation (21) is shorthand for the system of stochastic differential equations

$$dX_t^i = b^i(X_{t^-})dt + \sum_{j=1}^{m} c_j^i(X_{t^-})dL_t^j \qquad \text{for} \quad i = 1,..,d \tag{23}$$

with initial condition X_0^i. Here $X_t = (X_t^1,...,X_t^d)'$, $b(X_t) = (b^1(X_t),...,b^d(X_t))'$, and c is a $d \times m$ matrix with rows $c^i(X_t) = (c_1^i(X_t),...,c_m^i(X_t))$. The functions b^i: $\mathbb{R}^d \to \mathbb{R}$ are locally or globally Lipschitz measurable functions. The functions c_j^i: $\mathbb{R}^d \to \mathbb{R}$ are bounded globally Lipschitz measurable functions such that $|c_j^i|^2 \leq H_j^i \in \mathbb{R}^+$. The L_t^j terms are independent Levy processes for $j = 1, ...,m$.

Levy noise has advantages over standard Gaussian noise in neuron models despite its increased mathematical complexity. A Levy noise model more accurately describes how the neuron's membrane potential evolves than does a simpler diffusion model because the more general Levy model includes not only pure-diffusion and pure-jump models but jump-diffusion models as well. Neuron models with additive Gaussian noise are pure-diffusion models. These neuron models rely on the classical central limit theorem for their Gaussian structure and thus they rely on special limiting-case assumptions of incoming Poisson spikes from other neurons. These assumptions require at least that the number of impinging synapses is large and that the synapses have small membrane effects due to the small coupling coefficient or the synaptic weights [11]. The Gaussian noise assumption may be more appropriate for signal inputs from dendritic trees because of the sheer number of dendrites. But often fewer inputs come from synapses near the post-synaptic neuron's trigger zone and these inputs produce impulses in noise amplitudes because of the higher concentration of voltage-sensitive sodium channels in the trigger zone [13]. Engineering applications also favor the more general Levy model because physical devices may be limited in their number of model-neuron connections and because real signals and noise can often be impulsive. Adding Levy noise to enhance faint signals could apply to a variety of signal and image processing problems that include low-light imaging in satellites and other sensor devices, night vision, artificial vision and olfaction, neural prosthetics, infrared imaging, impulsive signal detection, and some types of pattern recognition.

Theorems 7 and 8 below generalize the forbidden interval Theorem 6 for continuous and sensory spiking neuron models [26] to a broad class of Levy noise that may depend on the neuron's membrane potential. The next two forbidden interval theorems require the following Ito-theoretic lemma and corresponding limitation that the Levy process have finite second moments [27].

Lemma. *Let* $b^i : \mathbb{R}^d \to \mathbb{R}$ *and* $c_j^i : \mathbb{R}^d \to \mathbb{R}$ *in* (23) *be measurable functions that satisfy the respective local and global Lipschitz conditions*

$$\|b^i(z) - b^i(y)\| \leq C_n\|z-y\| \quad \textit{when } \|z\| \leq n \textit{ and } \|y\| \leq n \tag{24}$$
$$\|c^i_j(z) - c^i_j(y)\| \leq K_1\|z-y\| \textit{ for all } z \textit{ and } y \in \mathbb{R}^d \tag{25}$$
$$\textit{and} \qquad |c^i_j|^2 \leq H^i_j \qquad \textit{for} \quad i = 1,...,d \textit{ and } j = 1,...,m. \tag{26}$$

Suppose $dX_t = b(X_t)dt + c(X_t-)dL_t$ *and* $d\hat{X}_t = b(\hat{X}_t)dt$ *where* dL_t *is a Levy noise with* $\mu = \mathbf{0}$ *and finite second moments. Then for every* $T \in \mathbb{R}^+$ *and for every* $\varepsilon > 0$*:*

$$E[\sup_{0\leq t\leq T} \|X_t - \hat{X}_t\|^2 > \varepsilon] \to 0 \quad \textit{as} \quad \sigma^j \to 0 \textit{ and } \nu^j \to 0 \textit{ for all} \quad j = 1,...,m, \tag{27}$$

and hence

$$P(\sup_{0\leq t\leq T} \|X_t - \hat{X}_t\|^2 > \varepsilon) \to 0 \quad \textit{as} \quad \sigma^j \to 0 \textit{ and } \nu^j \to 0 \textit{ for all } j = 1,...,m \tag{28}$$

since mean-square convergence implies convergence in probability.

Theorem 7. *Suppose that the continuous neuron models of the form* (21) *and* (22) *have a bounded globally Lipschitz Levy diffusion term* $c(X_{t^-}) \leq H$ *and that the additive Levy noise has drift velocity* μ. *Suppose also that the input signal* $S(t) \in \{s_1, s_2\}$ *is subthreshold:* $\theta_1 \leq s_1 < s_2 \leq \theta_2$ *and that there is some statistical dependence between the input random variable S and the output spike-rate random variable R so that* $I(S,R) > 0$. *Then such continuous neuron models exhibit the nonmonotone SR effect in the sense that* $I(S,R) \to 0$ *as the Levy noise parameters* $\sigma \to 0$ *and* $\nu \to 0$ *if* $\theta_1 - s_1 \leq H\mu \leq \theta_2 - s_2$.

Theorem 8. *Suppose that the spiking neuron models of the form* (21) *and* (22) *have a locally Lipschitz drift term* $b(X_{t^-})$ *and a bounded globally Lipschitz Levy diffusion term* $c(X_{t^-}) \leq H$. *Suppose also that the additive Levy noise has drift velocity* μ *and that the input signal* $S(t) \in \{s_1, s_2\}$ *is subthreshold:* $S(t) < B$. *Suppose there is some statistical dependence between the input random variable S and the output spike-rate random variable R so that* $I(S,R) > 0$. *Then such spiking neuron models exhibit the SR effect in the sense that* $I(S,R) \to 0$ *as the Levy noise parameters* $\sigma \to 0$ *and* $\nu \to 0$ *if* $H\mu < B - s_2$.

These forbidden interval theorems still require that the Levy process have a finite second moment. Simulations show that often we can drop this condition in the case of infinite-variance stable processes and still produce an SR noise benefit [27]]. But it is an open question whether some form of forbidden interval theorem holds for more general Levy diffusions.

References

1. D. Applebaum, "Levy Processes–From Probability to Finance and Quantum Groups," *Notices of American Mathematical Society*, vol. 51, no. 11, pp. 1336–1347, December 2004.
2. L. Breiman, *Probability*, Addison-Wesley, 1968.
3. A. R. Bulsara and L. Gammaitoni, "Tuning in to Noise," *Physics Today*, vol. 49, pp. 39–45, March 1996.

4. A. R. Bulsara and A. Zador, "Threshold Detection of Wideband Signals: A Noise-Induced Maximum in the Mutual Information," *Physical Review E*, vol. 54, no. 3, pp. R2185–R2188, September 1996.
5. D. Chander and E. J. Chichilnisky, "Adaptation to Temporal Contrast in Primate and Salamander Retina," *Journal of Neuroscience*, vol. 21, no. 24, pp. 9904–9916, December 2001.
6. J. J. Collins, C. C. Chow, A. C. Capela, and T. T. Imhoff, "Aperiodic Stochastic Resonance," *Physical Review E*, vol. 54, no. 5, pp. 5575–5584, November 1996.
7. J. J. Collins, C. C. Chow, and T. T. Imhoff, "Aperiodic Stochastic Resonance in Excitable Systems," *Physical Review E*, vol. 52, no. 4, pp. R3321–R3324, October 1995.
8. T. M. Cover and J. A. Thomas, *Elements of Information Theory*, John Wiley & Sons, 1991.
9. R. O. Duda and P. E. Hart, *Pattern Classification and Scene Analysis*, Wiley, New York, 1973.
10. L. Gammaitoni, "Stochastic Resonance and the Dithering Effect in Threshold Physical Systems," *Physical Review E*, vol. 52, no. 5, pp. 4691–4698, November 1995.
11. W. Gerstner and W. M. Kistler, *Spiking Neuron Models: Single Neurons, Populations, Plasticity*, Cambridge University Press, 2002.
12. A. S. Holevo and R. F. Werner, "Evaluating Capacities of Bosonic Gaussian Channels," *Physical Review A*, vol. 63, pp. 032312, 2001.
13. E. R. Kandel, J. H. Schwartz, and Thomas M. Jessell, *Principles of Neuroscience*, McGraw-Hill, 4th revised edition, 2000.
14. K. J. Kim and F. Rieke, "Temporal Contrast Adaptation in the Input and Output Signals Signals of Salamander Retinal Ganglion Cell," *Journal of Neuroscience*, vol. 21, no. 23, pp. 287–299, December 2001.
15. J. Kong, H. Soh, A. Cassell, C. F. Quate, and H. Dai, "Synthesis of Individual Single-Walled Carbon Nanotubes on Patterned Silicon Wafers," *Nature*, vol. 395, pp. 878–881, 1998.
16. B. Kosko, *Neural Networks and Fuzzy Systems: A Dynamical Systems Approach to Machine Intelligence*, Prentice Hall, 1991.
17. B. Kosko, *Noise*, Viking/Penguin, 2006.
18. B. Kosko and S. Mitaim, "Robust Stochastic Resonance: Signal Detection and Adaptation in Impulsive Noise," *Physical Review E*, vol. 64, no. 051110, pp. 1–11, October 2001.
19. B. Kosko and S. Mitaim, "Stochastic Resoannce in Noisy Threshold Neurons," *Neural Networks*, vol. 16, no. 5–6, pp. 755–761, June-July 2003.
20. B. Kosko and S. Mitaim, "Robust Stochastic Resonance for Simple Threshold Neurons," *Physical Review E*, vol. 70, no. 031991, pp. 1–10, September 2004.
21. I. Lee, X. Liu, C. Zhou, and B. Kosko, "Noise-Enhanced Detection of Subthreshold Signals with Carbon Nanotubes," *IEEE Transactions on Nanotechnology*, vol. 5, no. 6, pp. 613–627, November 2006.
22. R. Loudon and P. L. Knight, "Squeezed Light," *Modern Optics*, vol. 34, pp. 709–759, 1987.
23. S. Mitaim and B. Kosko, "Adaptive Stochastic Resonance," *Proceedings of the IEEE: Special Issue on Intelligent Signal Processing*, vol. 86, no. 11, pp. 2152–2183, November 1998.
24. S. Mitaim and B. Kosko, "Adaptive Stochastic Resonance in Noisy Neurons Based on Mutual Information," *IEEE Transactions on Neural Networks*, vol. 15, no. 6, pp. 1562–1540, November 2004.
25. C. L. Nikias and M. Shao, *Signal Processing with Alpha-Stable Distributions and Applications*, John Wiley and Sons, 1995.
26. A. Patel and B. Kosko, "Stochastic Resonance in Noisy Spiking Retinal and Sensory Neuron Models," *Neural Networks*, vol. 18, pp. 467–478, August 2005.
27. A. Patel and B. Kosko, "Stochastic Resonance in Continuous and Spiking Neuron Models with Levy Noise," *IEEE Transactions on Neural Networks*, 2008, to appear.
28. X. Pei, K. Bachmann, and F. Moss, "The Detection Threshold, Noise and Stochastic Resonance in the Fitzhugh-Nagumo Neuron Model," *Physics Letters A*, vol. 206, pp. 61–65, October 1995.
29. X. Pei, L. Wilkens, and F. Moss, "Noise-Mediated Spike Timing Precision from Aperiodic Stimuli in an Array of Hodgkin-Huxley-Type Neurons," *Physical Review Letters*, vol. 77, no. 22, pp. 4679–4682, November 1996.

30. S. I. Resnick, *Heavy-Tail Phenomena: Probabilistic and Statistical Modeling*, Springer, 2007.
31. N. G. Stocks, "Information Transmission in Parallel Threshold Arrays," *Physical Review E*, vol. 63, no. 041114, 2001.
32. S. J. Tans, R. M. Verschueren, and C. Dekker, "Room Temperature Transistor Based on a Single Carbon Nanotube," *Nature*, vol. 393, pp. 49–52, 1998.
33. K. Wiesenfeld and F. Moss, "Stochastic Resonance and the Benefits of Noise: From Ice Ages to Crayfish and SQUIDs," *Nature*, vol. 373, pp. 33–36, January 1995.
34. M. Wilde and B. Kosko, "Quantum Forbidden-Interval Theorems for Stochastic Resonance with Squeezed Light," in *Proceedings of the 8th International Conference on Quantum Communication, Measurement, and Computing*, November 2006, pp. 553–556.
35. J. W. G. Wildoer, L. C. Venema, A. G. Rinzler, R. E. Smalley, and C. Dekker, "Electronic Structure of Atomically Resolved Carbon Nanotubes," *Nature*, vol. 391, pp. 59–62, 1998.
36. L.-A. Wu, H. J. Kimble, J. L. Hall, and H. Wu, "Generation of Squeezed States by Parametric Down Conversion," *Physical Review Letters*, vol. 57, pp. 2520, 1986.
37. Y. Yu, B. Potetz, and T. S. Lee, "The Role of Spiking Nonlinearity in Contrast Gain Control and Information Transmission," *Vision Research*, vol. 45, no. 5, pp. 583–592, March 2005.

Smart Materials and Nonlinear Dynamics for Innovative Transducers

B. Andò, A. Ascia, S. Baglio, N. Pitrone, N. Savalli, C. Trigona, A.R. Bulsara and V. In

Abstract The exploitation of smart material properties and nonlinear dynamics behavior is the focus of this work, with emphasis on the realization of innovative transducers. Both ferromagnetic and ferroelectric properties are taken into account to develop sensors and actuators for various application fields. In particular, ferromagnetic materials have been used to realize sensors adopted in several innovative measurement solutions such as volcanic ash detection or magnetic bead-based estimation; they have also yielded innovative sensing strategies to estimate fluid viscosity with a very low invasive approach. The use of *smart* magnetic fluids has been considered, for their shock resistance and their intrinsic feature to be shapeless, for the development of suitable sensors and actuators. Ferrofluidic pumps, gyroscopes and inclinometers have been developed and characterized. Finally, a nonlinear dynamical system based on ferroelectric capacitors coupled into a unidirectional ring circuit is considered with particular interest for developing novel electric field sensors. The focused approach is based on the exploitation of circuits made up by the ring connection of an odd number of elements containing a ferroelectric capacitor, which under particular conditions exhibits an oscillating regime of behavior. For such a device a weak, external, target electric field interacts with the system thus inducing perturbation of the polarization of the ferroelectric material; this, the target signal can be indirectly detected and quantified via its effect on the system response.

1 Introduction

The present work summarizes a series of research achievements on the exploitation of both smart material characteristics and nonlinear dynamics for the development of innovative transducers. The paper is organized to deal with two classes of materials having magnetic and ferroelectric properties. In particular, in the case of

B. Andò (✉)
Facoltà di Ingegneria, Univ. degli Studi di Catania, DIEES, V. A. Doria 6, 95125 Catania, Italy

V. In et al. (eds.), *Applications of Nonlinear Dynamics*, Understanding Complex Systems,
DOI 10.1007/978-3-540-85632-0_8,

magnetic materials, nonlinear ferromagnetic properties play a key role in the development of magnetometer; further, the magnetic properties of ferrofluids have been exploited to realize different transducers, such as a static accelerometer (inclinometer), a vibrating gyroscope, and a mechanical actuator (micro-pump). In addition, ferroelectric materials properties have been exploited here into a novel nonlinear oscillator to realize innovative electric field sensors based on the dynamics of the chain of nonlinear coupled circuits.

The magnetic transducers discussed in the following sections include two different types of Residence Times Difference (RTD) based fluxgate magnetometers [1–3] exploiting ferromagnetic materials such as Metglas® and amorphous microwire . They have common driving setups and readout strategies, but are different in terms of geometry and, consequently, static and dynamic performances. These devices have been to realize a measurement system to monitor volcanic ash fall out following eruptions of Mt. Etna.

The research activity on magnetic materials properties includes ferrofluidic materials and has lead us to develop different transducers. The ferrofluid property to align to the magnetic field lines, or to concentrate where the field intensity is higher, has been exploited to realize proof-masses for inertial sensors, or suitable occlusions for mechanical actuators (valves or pistons). Therefore, in both cases, the opportunity of dynamically change the proof mass properties in inertial sensors and, therefore, their operating conditions (operating range and sensitivity), together with the possibility of recovering from system faults are among the distinguishing features of this approach [4–6].

Ferromagnetic materials have also been exploiteded for the development of unidirectionally coupled, overdamped nonlinear elements, to realize a novel class of electric field sensors. This approach has, recently, proven to be successful in the development of a coupled core fluxgate magnetometers, wherein the active elements are unidirectionally coupled ferromagnetic cores. At the same way, the development of a coupled core ferroelectric, static E-field meter, has confirmed that these systems are underpinned by emergent oscillatory behaviors, stemming from unidirectional coupling, when a control parameter crosses a threshold value. The emergent oscillations occur even in the absence of an applied "target" signal, and their frequency depends on the parameters of each individual element (i.e. the potential energy function when isolated), as well as the coupling strength. Then, the onset of a quasi-static target signal makes the system asymmetric and leads to changes in the frequency of oscillation, as well as other dynamic characteristics [7] including the "residence times" in the stable stationary states of the active non linear core. These changes are used to quantify the symmetry-breaking signal, i.e. the amplitude of the target electric field.

2 Magnetic Materials

In this section we discuss the use of suitable materials with interesting magnetic properties. We will deal with soft ferromagnetic materials having a very sharp hysteresis loop and large saturation field. These materials come in the form of ribbons and wires, hence the device to be discussed will exploit both the magnetic features and the mechanical properties. Thin ribbon layers and tiny microwires allow one to fabricate small devices with noticeable performances that can also be arranged in suitable geometries to accomplish specific sensing goals. Moreover, their small dimensions affords sensors that can be profitably adopted when invasivity is an issue. An example of the latter is the ferromagnetic viscosimeter.

Magnetic fluids have also been considered, since they adapt and change their physical properties. Ferrofluids, in aqueous solution, have been adopted to realize sensors, as inclinometer or vibrating gyroscopes, as well as actuators, and pumps. In both cases, the significant advantage of having mechanical transducers without any moving mechanical parts must be highlighted; the smart arrangement of the systems plays the key role. Permanent magnets provide, with the actual prototypes, the energy required by the system both in the case of sensors (e.g. allowing one to impose the driving force for the vibrating gyroscope and the static force for the inclinometer), and actuators (allowing one to impose the forces needed to realize valves or pistons). Moreover, since the dynamic performances of the devices depend on externally imposed currents, they can be tuned depending on applications or operating conditions. An inductive-based readout strategy is considered for the developed sensors.

2.1 RTD Fluxgate Magnetometer Based on Innovative Materials

Residence Times Difference (RTD) Fluxgate magnetometers represent an innovative solution to sense dc magnetic field or field perturbations. These magnetometers have always been of interest to the technical and scientific communities to sense weak magnetic field with a resolution of few microtesla at room temperature. These non linear devices find applicability in fields such as space, biomedics, vehicle navigation, security, military sensing, geomagnetic field measurement, and assorted proximity sensors applications.

Recently, RTD fluxgates based on an innovative core material have been proposed as competitive devices to the traditional second harmonic architectures [1–3]. Low cost, small dimensions, high sensitivity, low noise floor, low power consumption, and an intrinsic digital form of the output signal are the main advantages given by the innovative readout strategy carried out in the time domain.

Typically the RTD fluxgate is based on a two coil architecture (excitation and detection coils) shown in Fig. 1a. The coils are wound around a suitable ferromagnetic core showing a sharp hysteretic input-output characteristic which allows one to infer that switching between the two stable states of the magnetization occurs almost instantaneously (the formal assumption is to neglect the device dynamics, since the device time-constant is, typically, very small) when the applied magnetic field exceeds the coercive field level H_c. A periodic driving current, I_e, is forced in the excitation coil and generates a periodic magnetic field, H_e parallel to the main

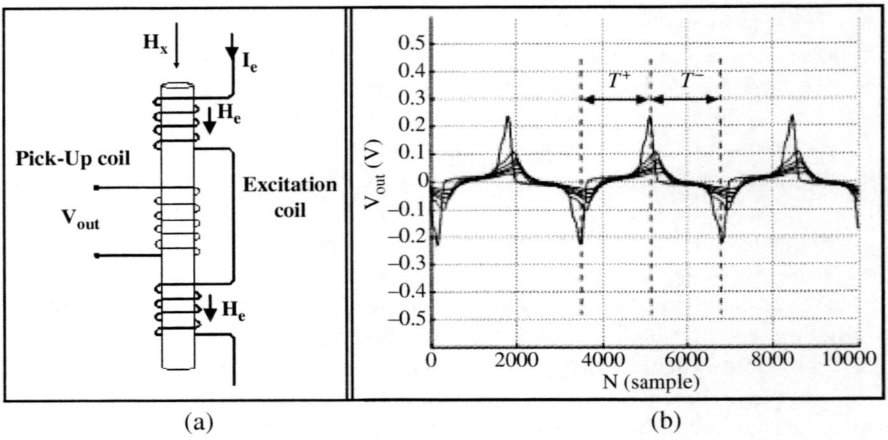

Fig. 1 (**a**) RTD fluxgate structure; (**b**) Output waveform

axis of the core. A target dc field H_x is applied in the same direction of H_e; the secondary coil is used as pick-up (detection) coil such that the output voltage V_{out} is proportional to the first derivative of the magnetization and contains information on the target magnetic field (Fig. 1b).

The phenomenological model for the dynamical response of a hysteretic ferromagnetic core, can be described via a bistable potential energy function $U(x)$ [7].

The difference between the two residence times is directly correlated to the target field H_x. In the next section two different structures of an RTD fluxgate magnetometer are presented. The first technological approach is based on the use of a metalized FR4 layer 1.6 mm thick, the second system is based on the use of 100 μm ferromagnetic amorphous wires (FeSiB).

2.1.1 Ferromagnetic Foil: FR4-Fluxgate Magnetometer

The FR4-Fluxgate magnetometer developed in PCB technology is based on a Metglas® 2714 As cast (cobalt based) foil core [8] with a non-crystalline structure and DC permeability of ~80.000. The technological approach is based on the use of a metalized FR4 layer 1.6 mm thick, a three layer architecture with a ferromagnetic middle layer is adopted, and a patterned metglas foil is embedded between two FR4 PCB layer [9, 10]. A FR4 fluxgate prototype is shown in Fig. 2. A complete characterization of such a prototype was performed in [11], with a sensitivity of 0.7 μs/nT and resolution of 1 nT @ 80 Hz and 20 mA being the main device characteristics.

The FR4 fluxgate prototype can be used for detection of volcanic particles that have an intrinsic ferromagnetic nature. Typically, volcanos release carbon dioxide, sulphur dioxide, various other gas molecules, and small jagged rocks, minerals, and other particles having diameter ranging from under 0.001 mm to a few millimeters.

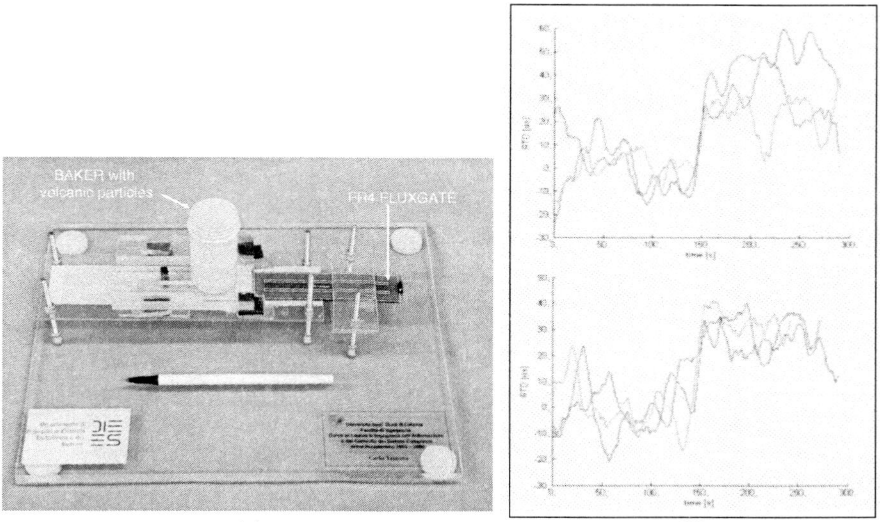

(a) (b)

Fig. 2 (**a**) FR4-Fluxgate prototype; (**b**) Evolution (moving average) of RTD for ∼7500 and ∼5000 particles respectively, for three similar family samples

Many measurement systems for lava flow monitoring have been developed, based on analysis of the magnetic properties of the lava composition [12, 13]. Hence, the intrinsic magnetic properties of the erupted magma can be detected with the magnetometer. Figure 4a shows the experimental setup for the detection of volcanic particles (from the 2001 eruption of Mt. Etna) adopting the FR4 fluxgate, with a sinusoidal bias current @ 80 Hz. A plastic beaker is used to hold the volcanic particles, which have an average diameter of 0.5 mm, and a permanent magnet is used to polarize the ferromagnetic particles.

The analysis has been conducted adopting different quantities of volcanic particles. A LabView® real time acquisition system linked with the fluxgate conditioning circuit is adopted to evaluate the RTD stemming from the magnetic particles. About 5000 particles is the minimum number of volcanic particles detected. Figure 2b shows the Residence Times Difference with and without the presence of magma, for different numbers of magnetic particles.

This experiment demonstrates the possibility of using the FR4 fluxgate magnetometers to estimate the concentration of magnetic particles in a solution. Moreover, the same magnetometer prototype can be used to analyze the intrinsic magnetic properties of erupted magma to estimate the concentration of volcanic magnetic particles with average diameters of 0.5 mm.

2.1.2 Amorphous FeSiB: Microwire Fluxgate Magnetometer

Microwire fluxgate magnetometers are based on 100 μm FeSiB amorphous ferromagnetic material. The microwire ferromagnetic cores are produced by the rapid cooling of alloys: 80% Fe, Ni or Co, and 20% P, Si, Al, C, B. In particular, FeSiB microwires are obtained using the in-water quenching technique with a typical diameter range of 80–160 μm and cylindrical structure. Typically, the solidification process induces two magnetic domain regions: (1) an inner core, easy axis parallel to the wire axis; (2) an outer shell with radial easy axes [14]. The internal stress accompanying the solidification process can be reduced through an annealing process at 350°–400°.

Our highly sensitive magnetometer exploits the properties of such a magnetoelastic material core to detect and measure the presence of a few magnetic particles, for biomedical and security applications. Such a fluxgate is based on the two-coil structure (excitation coil and detection coil) wound around the ferromagnetic core having a hysteretic input-output characteristic. The RTD has been adopted as the readout strategy. Such a magnetometer shows interesting physical characteristics in terms of magnetic performance and flexibility with high spatial resolution. The simplified process description can be summarized as:

- 100 μm wire-coils (excitation and detection) diameter are wound around a plastic structure (∼1.2 mm).
- A cylindrical glass-structure (∼1 mm) is used to contain the 100 μm FeSiBf fluxgate core.
- The magnetic core is centered with respect to the cylindrical plastic structure.

Our research activity on this class of transducers has started with confirming and characterizing their non-linear behavior, supported by models developed for our earlier fluxgate magnetometers, through the characterization phase. Figure 3 shows the wirecore fluxgate. The primary coil has 1 layer, ∼900 windings, the internal coil radius is ∼100 μm, while the coil resistance is 30 Ω; the secondary coil has four layers, each layer has ∼900 windings, the internal coil radius is ∼100 μm, and the coil resistance is 30 Ω.

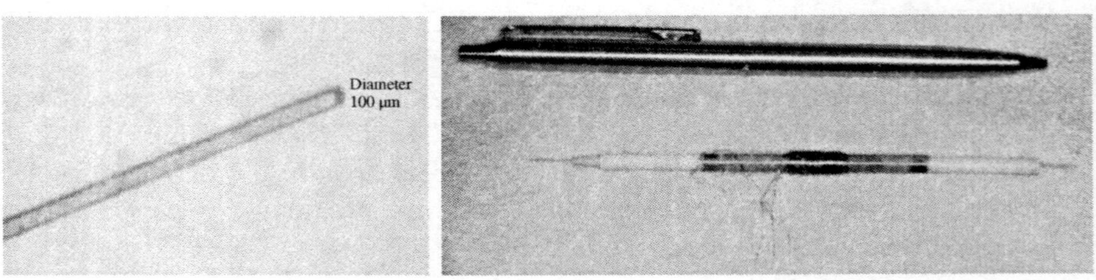

(a) (b)

Fig. 3 (**a**) Picture of the 100 μm FeSiB amorphous ferromagnetic microwire; (**b**) Wirecore fluxgate magnetometer

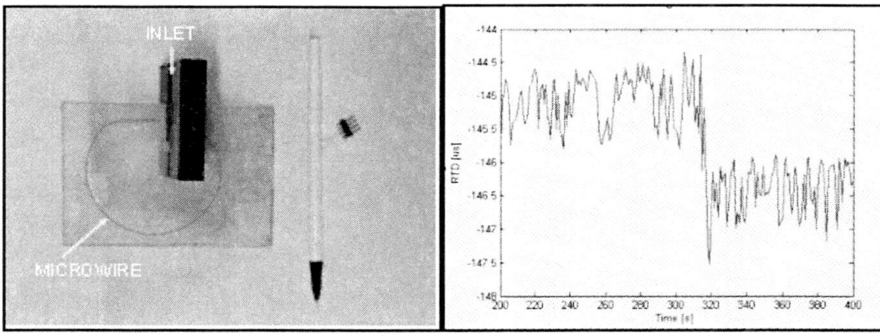

(a) (b)

Fig. 4 (**a**) Experimental setup for detection of volcano particles; (**b**) RTD evolution for Microwire-Fluxgate

In Fig. 4 we show an alternative microwire-Fluxgate magnetometer architecture specific to the detection of the presence of very low concentrations of volcaic particles. The device consists of 1000+1000 coils (primary and secondary) that is wound around a ferromagnetic microwire core.

The analysis has been conducted adopting a real time acquisition system to evaluate the magnetic contributions of the volcanic particles (about 1 mg with different particle diameters).

The primary coil has 1 layer, ~1000 windings, the coil length is 0.06 m, the internal radius is ~100 μm, while the coil resistance is 6.5 Ω; the secondary coil has 1 layer, ~1000 windings, the coil length is 0.06 m, the internal radius is ~100 μm, and the coil resistance is 6.5 Ω.

2.1.3 Amorphous FeSiB Microwire Viscosimeter

The amorphous magnetic micro-wires also present special features that can be exploited for technological applications such as active vibration measurement, micropositioning systems, microactuators, and hysteretic transformers. The main properties that characterize such devices are hysteresis, anisotropy, magnetostriction, material composition.

The FeSiB microwires are used as viscosity sensors. When an actuated magnetic core is immersed in a liquid, the liquid motion changes as a function of viscosity. The magnetic device consists of a glass pipe wherein a microwire (4 cm long) is contained in a liquid e.g. air, water, alcohol. The sensing system is composed of a primary excitation coil and two detection coils, wound around the glass pipe. A Linear Variable Displacement Transducers (LVDT) readout strategy is adopted. Figure 5 shows the schematic system and the technological characteristics of the magnetic system. The experimental set-up consists of a magnetic viscosimeter with a sinusoidal input @ 6 kHz, and a 3 Hz excitation frequency.

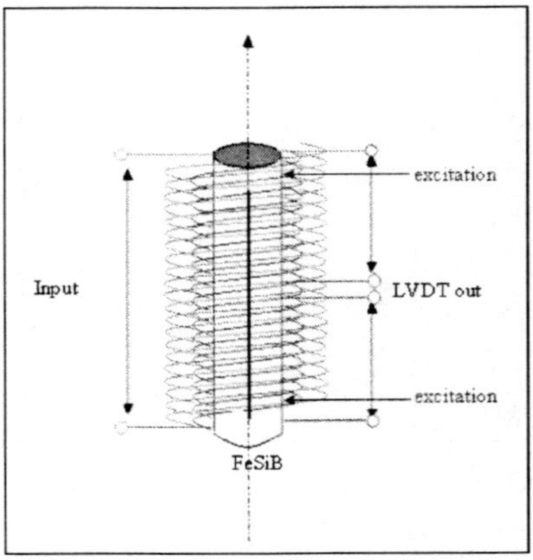

Primary coil	Characteristics	
	Value	*units*
Layer number	~12	-
Coils number in one layer	~150	-
Coil length	0.030	m
Coil internal radius	~0.0099	m
Resistance	65	Ω
Secondary coil	Characteristics	
	Value	*units*
Layer number	~9	-
Coils number in one layer	~75	-
Coil length	0.015	m
Coil internal radius	~0.0081	m
Resistance	20	Ω

Fig. 5 Schematic representation of the experimental set-up for the "microwire viscosimeter"

The higher frequency signal is used to drive the differential transformer, while the one at lower value is used to control the movements of the microwire. The experiment has been conducted by observing, for a fixed driving signal at 3 Hz, the LVDT output that is proportional to the wire displacement for different liquids. Figure 6 shows the viscosity and output voltages for different reference fluids.

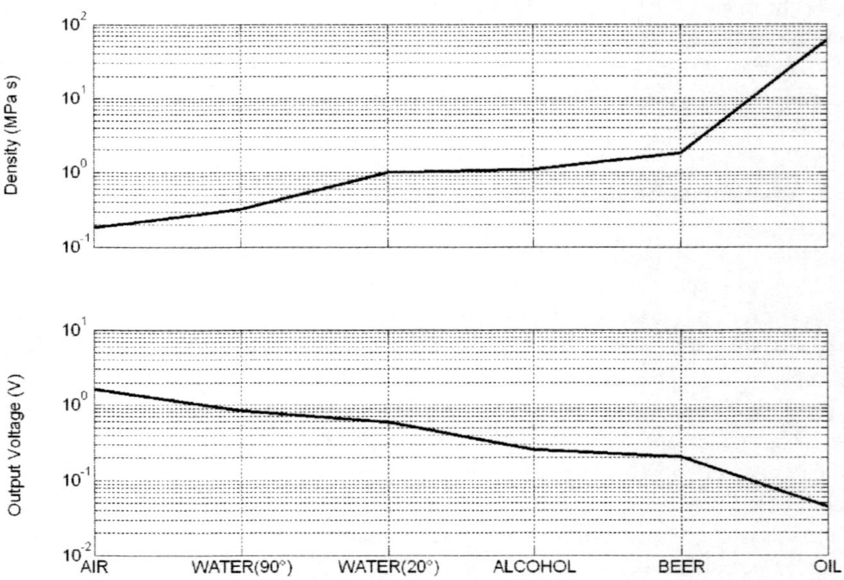

Fig. 6 Viscosity and LVDT output voltage for different "reference" fluids

2.1.4 Ferrofluidic Transducers

The use of magnetic fluids has been considered, because of their shock resistance and their intrinsic feature to be shapeless, for the development of a class of sensors and actuators. Magnetic fluids show, in fact, interesting patterns coming from hydrodynamic instabilities, such as lines, labyrinths and various other structures which can be exploited to produce actuation. Interest in these materials derives from the possibility to implement valuable and efficient conversion of elastic energy into mechanical energy.

Moreover their use for biomedical applications is, today, assuming a fundamental role in the development of efficient solutions for diagnosis, therapy, and analysis. Specifically, ferrofluids that are colloidal suspensions of magnetic particles show a diameter ranging in the order of nanometers and in contrast with magneto-rheological fluids, they maintain their liquid status in the presence of a strong magnetic field. In addition, a relatively small magnetic field is required to bring the material magnetization to the saturation state. The higher the magnetization strength, the greater is the magnetic pressure exerted by the fluid. The possibility of controlling the material characteristics (such as viscosity) by an external magnetic field has also afforded the development of smart devices with tunable operating ranges for an extended set of applications. Ferrofluidic pumps, gyroscopes and inclinometers have been developed and characterized, see Figs. 7, 8 and 9. Although the nonlinear properties of this material have not been exploited during the realization of transducer prototypes for the pump, gyroscope, and accelerometer, the research activity on this field is actually focusing on the possibilities of exploiting the nonlinearities in various ways.

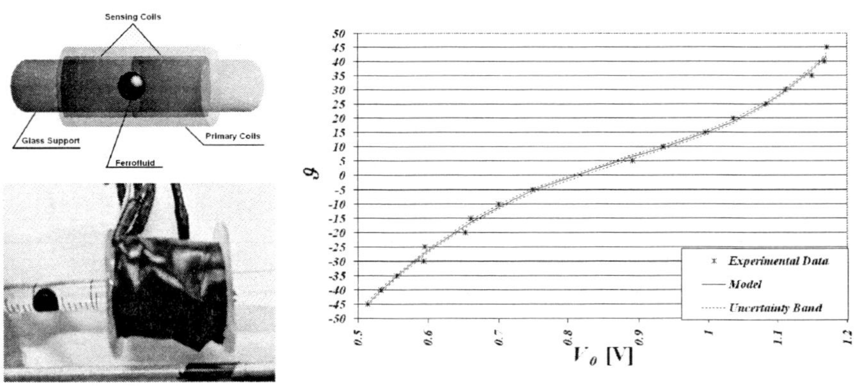

(a) (b)

Fig. 7 (**a**) 3D scheme and photograph of the ferrofluidic inclinometer; (**b**) Experimental and theoretical transduction function

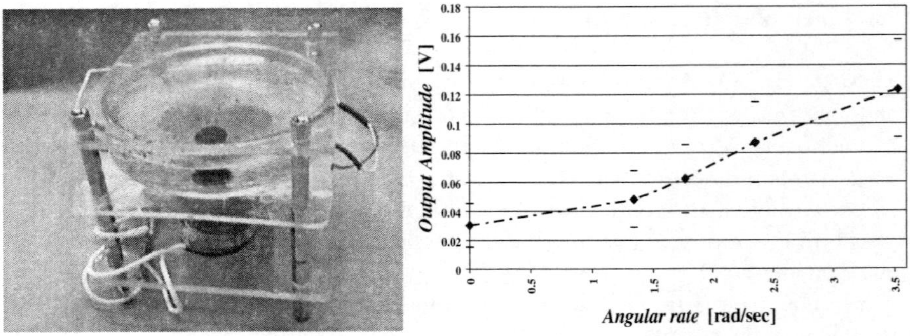

(a) (b)

Fig. 8 (**a**) Photograph of the ferrofluidic gyroscope prototype; (**b**) Experimental calibration curve

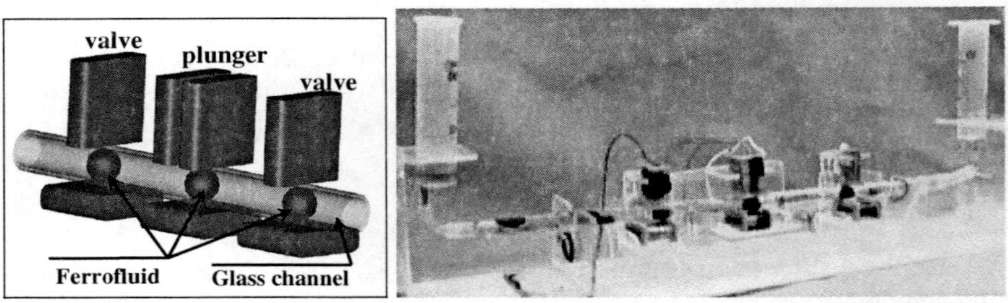

(a) (b)

Fig. 9 (**a**) 3D model of a ferrofluidic pump; (**b**) Image of the experimental prototype

2.1.5 Device Prototypes

Although, in the past, ferrofluids have been mainly conceived as damper fluids with controlled viscosity [15], they offer the opportunity to develop novel transducer architectures such as inertial sensors whose specifics are controllable by manipulating the ferrofluidic core properties (viscosity, volume, etc.). The possibility of actively controlling such properties is valuable in the development of smart transducers, with reliability and flexibility representing other important features. The main advantages are the tunable operating range and sensitivity, absence of mechanics, robustness against shocks, biocompatibility.

Briefly, the device consists of a glass pipe containing water and a ferrofluidic mass subjected to a magnetic field which is generated through an external coil wound around the pipe. As an inclinometer, the device exploits the displacement of the ferrofluidic mass as a measure of the device tilt [16]. The mass position, in the steady state regime, is predicated by the equilibrium between the magnetic force and the gravitational force, and is monitored by a differential transformer also wound around the glass pipe. The main advantages of the proposed solution derive from the adoption of such a ferrofluidic aggregate as the seismic mass, resulting in high robustness against shock with respect to devices containing a lumped magnetic material within a coil. Actually, if a tilt higher than the maximum allowed inclination is applied to the system, the ferrofluidic mass escapes from the primary coil, overcoming the allowable displacement range and dropping to the container. This is a non-destructive event and when the applied tilt returns in the specified range the ferrofluidic mass retrieves an operative position inside the driving coil and the device becomes, again, operative.

In the case of ferrofluidic gyroscopes, a different setup has been considered to realize the operating conditions of a vibrating-type gyroscope. Two electromagnets allow us to excite the ferrofludic mass along the driving axis, whereas two sensing coils (planar coils), on the bottom of the baker containing the ferrofluid, allow us to detect the position of the ferrofluidic mass during its oscillations. When the external target (angular velocity) acts on the system, inducing rotation about the z axis, part of the energy is transferred to the sense axis (y axis), due to the Coriolis force.

The resultant trajectory describes a path (resulting from the composition of the two actions along the driving and the sensing axes) which unbalances the output voltage of the differential transformer stage (made up by the two coils and a suitable conditioning circuit). In particular, since the mass describes a trajectory different from the imposed one, it will semi-periodically (with a period proportional to that of the drive signal) be closer to one of the sensing coil, as shown in Fig. 8.

Both these transducers have been experimentally realized and characterized. Moreover, in the case of a ferrofluidic inclinometer, the behavior of the system has been experimentally validated, with particular reference to its main characteristic and to the magnetic force imposed along the working axis. Then, a linear and tunable (equivalent) elastic constant of such an inertial system, can be obtained.

Finally, in the case of the ferrofluidic pump the possibility of conveying controlled amounts of liquid is of great interest for many applications ranging from lab-on-chip devices to *in vivo* implants. The ability to handle small volumes of fluids is important to decrease the consumption of expensive reagents, and valuable drug delivery systems can be developed if suitable pumps are available [17]. The proposed transducer implements a volumetric pump without moving parts. The pump is realized through the use of three ferrofluidic masses (one plunger and two valves) and a set of electromagnets for the realization of the active components of the device. The main idea is to move a small amount of liquid in a glass channel by imposing a suitable actuation sequence on the plunger and the valves.

The advantages of the proposed strategy, as compared with conventional architectures, are summarized as follows. The use of ferrofluidic masses to implement the active components of the pump and the absence of mechanical moving parts together minimize mechanical stress, thus increasing the life-time of the device, and making the pump robust against external shocks.

The device specifics (e.g. the valve pressure drop) can be tuned on the basis of the application, by exploiting the electromagnetically tunable viscosity of the ferrofluid samples adopted to implement the valves and the plunger. The pump flow rate is easily controllable by tuning the frequency of the pumping sequence which is not constrained by mechanical actions (sliding parts) and can easily be implemented by controlling the magnetization of the electromagnets.

3 Ferroelectric Materials and E-Field Sensors

A nonlinear dynamical system based on ferroelectric capacitors coupled into a unidirectional ring circuit has been considered for developing novel electric field sensors. The conceived devices exploit the synergetic use of bistable ferroelectric materials, micromachining technologies that allow us to address charge density amplification, and implement novel sensing strategies based on coupling non-linear elemental cells [18]. The weak external (dc) target electric field interacts with the system's elementary cells inducing a perturbation of the polarization state in the ferroelectric material of each cell; the perturbation can be indirectly detected and quantified via its effect on the oscillation frequency (of the polarization parameter), and on the asymmetry of the coupled system output signals; this is analogous to the coupled core magnetometer strategy, described earlier.

Simulation results have shown that for a coupling factor (related to the external field strength), between the ring cells greater than the critical one, a change in the harmonic content of the permanent oscillations generated in the coupled system occurs. Advanced simulation tools have been used for modeling a system including electronic components and non linear elements as the conceived micro-capacitors. Moreover, Finite Element Analysis (FEM) has allowed us to steer the capacitor electrodes design toward optimal geometries and to improve the knowledge of effects of the external target E-Field on the electric potential acting on the ferroelectric material.

An experimental characterization of the whole circuit, including three cells coupled in a ring configuration has also been carried out in this case. The results confirm the increasing of the circuit oscillation frequency as a function of the coupling factor, as expected from the mathematical and numerical models. A theoretical development of the underlying dynamics has been given in [21].

3.1 System Description

Each active element is realized by a micromachined capacitor whose core is a ferroelectric material that can be polarized through an imposed bias field. The polarization state is altered when the target electric field is applied, resulting in a distortion of the polarization vs. electric field hysteresis loop. In addition, a suitable external receptor allow for amplification of the target field to the sensing element. Each nonlinear ferroelectric capacitor can be modeled by the following differential equation:

$$\tau \dot{P} = aP - bP^3 + cE \tag{1}$$

The over dot denotes the time derivative, P represents the material polarization, a, b and τ denote material dependent system parameters governing its bistable behavior, finally c is a coefficient that relates the action of the external electric field E applied to the dielectric sample. Thus, the potential energy function for the given material is expressed as:

$$U(P,t) = -\frac{a}{2}P^2 + \frac{b}{4}P^4 - c\,\varepsilon P \tag{2}$$

The presence of a target signal results in the asymmetry of $U(P,t)$, and detection techniques are aimed at quantifying this asymmetry. For a single capacitor with a very high coercive field, one must apply an auxiliary electric field of amplitude strong enough to overcome the coercive field and set up periodic (clocked by the frequency of the bias field) switching events between the stable attractors of the potential. The electric field E is, then, the result of the contributions coming from an auxiliary electric field (the bias field) and from an external electric field (the target field); it is this target field that modifies the oscillation characteristic of the network in the case of the coupled system.

3.2 Elementary Cell – Circuital Realization

The ferroelectric capacitors investigated in this work have been preliminarily realized at the Penn State University [19]. Starting from a silicon substrate a common silver electrode has been evaporated, the ferroelectric material has been deposited over the bottom electrode, and finally several top electrodes have been spotted over the top surface of the ferroelectric in order to realize both the capacitors, and the external connections. A microscope picture of the ferroelectric sample is shown in Fig. 10a.

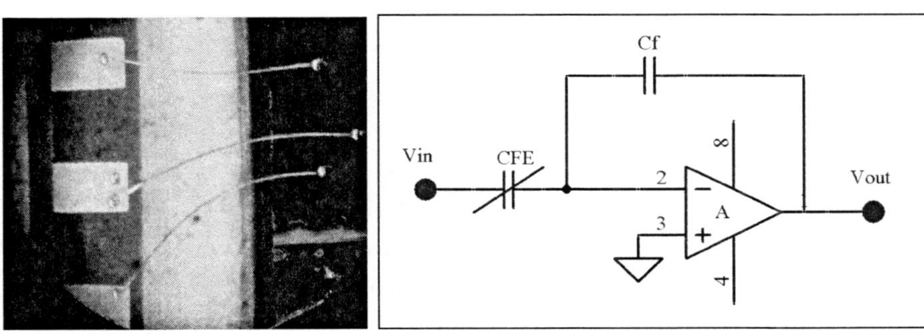

(a) (b)

Fig. 10 (**a**) Picture of the prototype ferroelectric capacitor. The top electrodes are visible; (**b**) Sawyer Tower circuit, V_{out} is proportional to the polarization in the C_{FE} capacitor

In order to evaluate the two quantities characterizing the ferroelectric device a suitable circuit has been considered (shown in Fig. 10b); it is based on a charge amplifier to realize the topology known as "Sawyer-Tower circuit" [18]. In this circuit, the polarizationP_{FE}, of the capacitor dielectric is proportional to the circuit output voltage. The definitions of the polarization as a function of the electric field and of the electric flux as a function of the capacitance charge yield the following equation:

$$V_{out} = -\frac{A}{C_f} P_{FE} \tag{3}$$

which expresses the proportionality between the circuit output and the polarization of the ferroelectric capacitor, where A_{FE} is the armature area.

The hysteretic behavior of the sample material has been confirmed experimentally [18]. These experimental observations were used to identify the parameters of the analytical model reported in Eq. (1) of the system.

3.3 PSpice Model

In order to perform numerical simulations of the system, a circuital dynamic model has been developed for the ferroelectric capacitors. Excellent agreement has been achieved between our measurements and the corresponding circuit simulations. In particular, the circuital modeling of a ferroelectric capacitor through PSPICE has been developed using a behavioral representation. The method also optimizes the process of establishing the parameter values germane to a particular hysteresis loop.

Consider the model that schematizes the capacitor as a *"displacement current"* generator, shown in Fig. 11a , driven by a voltage difference that can be derived from

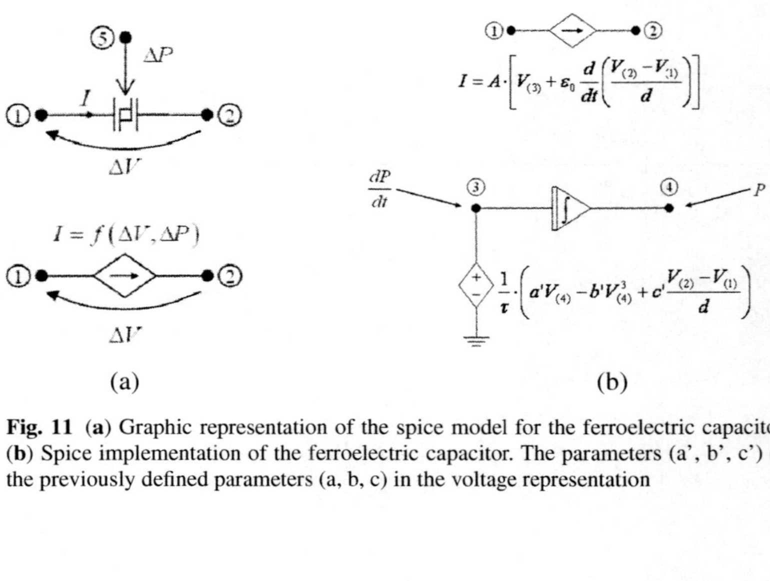

Fig. 11 (**a**) Graphic representation of the spice model for the ferroelectric capacitor *FCAP_Pin*; (**b**) Spice implementation of the ferroelectric capacitor. The parameters (a', b', c') correspond to the previously defined parameters (a, b, c) in the voltage representation

two fundamental equations. Models underpinning the device behavior can then be realized as follows:

$$D = P + \varepsilon_0 E = \frac{Q}{A} \tag{4}$$

where D, P and E are, respectively, the normal components to the capacitor electrodes of the vectors for displacement, electric polarization, and electric field respectively. Rearranging Eq. (4), it is possible to derive:

$$Q = A\,(P + \varepsilon_0 E) \tag{5}$$

Next, using Eq. (5), an expression for the current is given by:

$$I = \frac{dQ}{dt} = A\left(\frac{dP}{dt} + \varepsilon_0 \frac{dE}{dt}\right) \tag{6}$$

which is our first "constitutive" equation. In the second equation we have to take into account the dynamic behavior of the ferroelectric material:

$$\tau \cdot \frac{dP}{dt} = aP - bP^3 + cE \tag{7}$$

where a, b, c and τ are the model parameters and E is electric field amplitude. A parallel plate capacitor, with a plate separation d, has been considered to evaluate the electric field amplitude E:

$$E = \frac{V_{(2)} - V_{(1)}}{d} \tag{8}$$

The previous model may also take into account an external perturbation, to the polarization of the ferroelectric capacitor, induced by the target field through the sensing electrode. An auxiliary input allows for introducing such a perturbation, where the voltage at this node (ΔP) is expressed in units (C/m^2) and summed to the actual value of P. Thus, we can write Eq. (7) as:

$$\frac{dP}{dt} = \frac{1}{\tau} \cdot \left[a\,(P + \Delta P) - b\,(P + \Delta P)^3 + cE\right] \tag{9}$$

Hence, Eq. (9) models the representation of the PSPICE circuit displayed in Fig. 11b.

3.3.1 Dynamic Cooperative Behavior in the Coupled Capacitor System

As pointed out earlier, operating the E-field sensors as a single device, via a reference applied signal to induce switching between the stable polarization states is problematic due to the high coercive fields that are typical in ferroelectric materials. In a recent paper [20], it has been demonstrated, however, that coupling an odd number of overdamped bistable elements in a ring, with unidirectional coupling and cyclic boundary conditions, can lead to oscillatory behavior when the coupling strength exceeds a critical value.

Typically, this behavior is dictated by symmetry conditions and is generated by Hopf bifurcations; it appears to occur in any coupled system of overdamped bistable elements, none of which would oscillate when isolated and undriven, subject to the appropriate choice of parameters and operating conditions (albeit through different bifurcations mechanisms). The practical importance of this effect lies in the potential sensitivity enhancement when the system is "tuned" very close to the oscillations threshold [20].

The circuit implementation of this system comprises three elementary cells (one of which is shown in Fig. 10a), each with a ferroelectric capacitor active element. The complete coupled circuit device is shown in Fig. 12a, with the jth element modeled by:

$$\tau \frac{dP_j}{dt} = aP_j - bP_j^3 + cE_j \tag{10}$$

where j=1...N. From Eq. (3) the relationship between the output voltage and the polarization of the jth element is given by:

$$P_j = -V_{out}\frac{C_f}{A} \tag{11}$$

with

$$E_j = \frac{V_{out\,j-1} - V_{out\,j}}{d}\left(1 + \frac{R_{G1}}{R_{G2}}\right) \tag{12}$$

Using relation (11), Eq, (12) becomes:

$$E_j = \frac{A}{C_f d}\left(1 + \frac{R_{G1}}{R_{G2}}\right)(P_j - P_{j-1}) \tag{13}$$

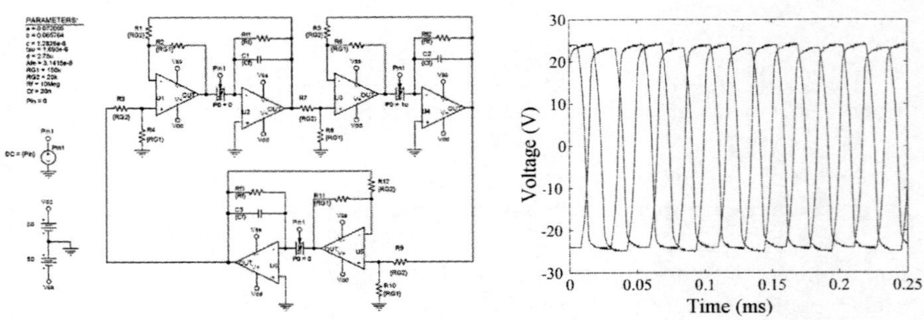

(a) (b)

Fig. 12 (**a**) Coupled system implemented with analog electronics; (**b**) Time series obtained in the coupled capacitor system

Hence, the coupled dymamics in Fig. 12b has the following form (for N=3):

$$\begin{aligned}\tau \dot{P}_1 &= aP_1 - bP_1^3 + \lambda (P_1 - P_3)\\ \tau \dot{P}_2 &= aP_2 - bP_2^3 + \lambda (P_2 - P_1)\\ \tau \dot{P}_3 &= aP_3 - bP_3^3 + \lambda (P_3 - P_2)\end{aligned} \tag{14}$$

Where,

$$\lambda = c\left(1 + \frac{R_{G1}}{R_{G2}}\right)\frac{A}{C_f d} \tag{15}$$

is the coupling coefficient.

The experimental results follow the theoretically predicted out of phase behavior as shown in Fig. 12b, with calculations yielding the critical coupling (at the onset of oscillations) [7, 18, 19, 21]

$$\lambda_c = \frac{a}{2} + \frac{3}{2}\frac{cE_{ext}}{1+\sqrt{2a/b}} \tag{16}$$

where E_{ext} is the external target dc electric field. The system oscillates for values $\lambda > \lambda$ c. The frequency dependence of the oscillations as a function of the coupling gain can be theoretically computed, and is found to be in very good agreement with experiments and simulations (Fig. 13).

The circuit has been validated, by using an experimental calibration chamber, with respect to the dependence on the coupling gain and on the external field. Moreover, preliminary experimental results, reported in Fig. 13b, show the variation (reduction) of the output signal frequency as a function of the external electric field.

The measurement setup included two 3 m by 3 m facing electrodes, used to impose the target field to be measured, and the apparatus made up by the microcapacitors and the charge collector plate placed in between the two electrodes.

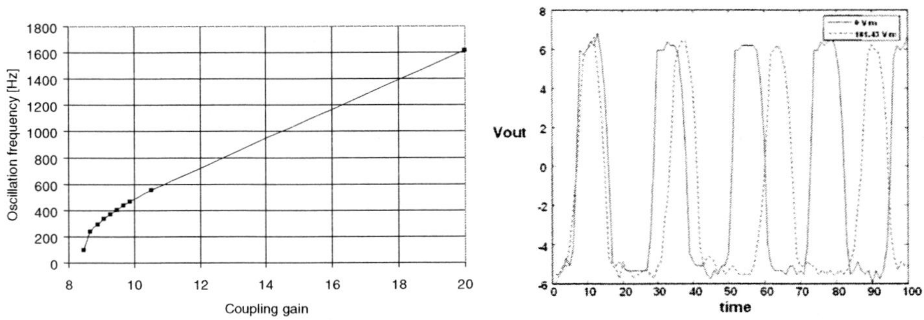

(a) (b)

Fig. 13 (**a**) Coupled system oscillation frequency vs. the coupling gain; (**b**) Experimental output signal as a function of the external electric field in the test chamber: a reduction in the output signal frequency is observed as a response to an increase in the target signal

4 Concluding Remarks

Several different applications have been presented in this paper, ranging from magnetic field sensors applied to the monitoring of volcanic particles to ferrofluidic pumps and to ferroelectric E-field sensors. Despite the ample fan of areas addressed this work has a common denominator that relies on the exploitation of material properties for applications to smart transducers. Both magnetic and ferroelectric materials have been taken into account here. In particular the nonlinear dynamics that characterize hysteretic behaviors have been extensively considered together with some magnetorheologic properties shown by suitable suspensions of paramagnetic particles in oil.

References

1. Andò, S. Baglio, A. Bulsara, V. Sacco (2005). Theoretical and experimental investigations on residence times difference fluxgate magnetometers, Measurements (Elsevier) 38 (2). p. 89-112.
2. B. Andò, S. Baglio, A. Bulsara, V. Sacco (2005). "Residence Times Difference" Fluxgate Magnetometers, *Sensors Journal, IEEE* 5 (5). pp. 895-904.
3. B. Andò, S. Baglio, A.R. Bulsara, V. In, V. Sacco, "PCB Fluxgate Magnetometers with a Residence Times Difference (RTD) Readout Strategy: The Effects of Noise", in press (Dec. 2007) IEEE transaction Instrumentation and Measurements.
4. B. Andò, A. Ascia, S. Baglio, N. Pitrone, Development of Novel Ferrofluidic Pumps, EMBS 2006, pp.2828–2831, 2006.
5. B. Andò, A. Ascia, S. Baglio, G. Franco, N. Savalli, A Novel Ferrofluidic Gyroscope, Eurosensors 2006.
6. S. Baglio, P. Barrera, N. Savalli, Novel Ferrofluidic Inertial Sensors, Proceeding of Instrumentation and Measurement Technology Conference, 2006. IMTC 2006. April 2006 Page(s):2368–2372.
7. In V. Bulsara, A. Palacios, A,. Longhini, P. Kho, A. Neff J (2003). Phys. Rev. E68:045102(R).
8. www.metglas.com.
9. B. Andò, S. Baglio, V. Caruso, V. Sacco, A. Bulsara (2006). Multilayer based technology to build RTD fluxgate magnetometer, IFSA, Sensors & Transducers Magazine, Vol.65, issue3, pp. 509-514.
10. B. Andò, S. Baglio, V. Caruso, V. Sacco (2006). Investigate the optimal geometry to minimize the demagnetizing effect in RTD-Fluxgate, IEEE IMTC 2006, proceedings of the IEEE-Instrumentation and Measurement Technology Conference, pp. 2175–2178.
11. B. Andò, A. Ascia, S. Baglio, A.R. Bulsara, V. In, Towards the Optimal Reading of RTD Fluxgate, Elsevier Sensors and Actuators, in press.
12. D. Renk, H. de Wall, U. Martin, K. Nemeth, Volcanic Influence of titanomagnetite composition on the magnetic anisotropy in a dyke-sill complex in Hungary, Massey University, Visual Geosciences, January 2006.
13. Flow direction and magnetic mineralogy of lava flows from the central parts of the Peninsula of Baja California, Mexico, Bulletin of Volcanology, Springer Berlin / Heidelberg, Volume 66, Number 5/July, 2004, Thursday, February 19, 2004.
14. H. Chiriac, J. Yamasaki, T.A. Ovari, M. Takajo (1999). Magnetic domain structure in amorphous glass-covered wires with positive magnetostriction, Magnetics, IEEE Transactions on, Vol. 3, issue 5, pp. 3901–3903.

15. J. Popplewell, R.E. Rosenweig, "Magnetorheological fuid composites", J. Phys. D, Appl. Phys., vol. 29, pp. 2297–2303, Jan. 1996.
16. B. Andò, A. Ascia, S. Baglio, N. Savalli, A novel Ferrofluidic inclinometer, IEEE Transaction On Instrumentation and Measurements, Volume 56, Issue 4, August 2007, Page(s): 1114–1123.
17. Q.A. Pankhurst, J. Connolly, S.K. Jones and J. Dobson, Applications of magnetic nanoparticles in biomedicine, J. Phys. D: Appl. Phys. 36 (2003) R167–R181, IOP Journ. Phys. D: Appl. Physics.
18. S. Baglio, B. Andò, A.R. Bulsara, V. In, N. Savalli, *Electric field detectors in a coupled ring configuration: preliminary results*, SPIE 2006 – Smart Materials, Nano & Micro-Smart Systems, 10–13 December 2006, Adelaide, Australia.
19. Visarath In, Private communication.
20. Bulsara A, In V, Kho A, Longhini, P, Palacios A, Rappel W-J, Acebron J, Baglio S, Ando B, Phys. Rev. E70:036103 (2004).
21. Bulsara A. et al., Phys. Rev. E73, 066121 (2006).

Dynamics in Non-Uniform Coupled SQUIDs

Patrick Longhini, Anna Leese de Escobar, Fernando Escobar, Visarath In, Adi Bulsara and Joseph Neff

Abstract Recently there is growing interest in Superconducting Quantum Interference Filters (SQIFs) in the science community. Much in the way that antenna array technology has moved from uniform arrays to non-uniform, sparsely populated arrays, SQIF devices open the trade space from traditional uniform SQUID arrays to unconventional SQIF structures making these mesoscopic quantum devices (for RF reception) a feasible concept. Computational modeling of SQIFs exhibits a non-periodic voltage response vs. external flux where an anti-peak is present only around the zero applied magnetic field. Improved dynamics range is investigated in this work by considering inductive coupling, number of loops in an array, and unconventional grating structures. We show that changes in these parameters can lead to a significant performance enhancement.

1 Introduction

For identically coupled SQUIDs the averaged voltage response $\langle V \rangle$ versus the external magnetic flux is Φ_0-periodic, where $\Phi_0 = \frac{h}{2e}$ is the elementary flux quantum. Interesting, when loop sizes are chosen to be different we can find a voltage response such that it becomes a unique function around the zero magnetic field [1–4]. This research explorers variations in system parameters including: inductive coupling (i.e. chain, uni- and bi-directional), unconventional grating and number of element in superconducting SQUID arrays (i.e. SQIFs) for achieving a new generation of low-noise amplifiers (LNA), plus application in other devices (Sect. 5). The focus of this research covers numerical modeling of the coupled SQUID arrays. There are theoretical indications that variations in coupling could provide very high dynamic

P. Longhini (✉)
Space and Naval Warfare Systems Center, 53560 Hull Street, San Diego, CA 92152-5001, USA
e-mail: longhini@spawar.navy.mil

V. In et al. (eds.), *Applications of Nonlinear Dynamics*, Understanding Complex Systems,
DOI 10.1007/978-3-540-85632-0_9,

range, but as of yet a numerical model of this doesn't exist such that large number arrays and fabrication parameters can be explored in the modeling.

2 Background

The SQUID magnetometer is a popular and extremely useful device which uses the interaction between magnetic flux and the Josephson Junction. Magnetic flux modulates the current passing through the junction. This modulation is detected and amplified by some feedback electronics. There are two main types of SQUIDs, named according to the applied bias, i.e. AC (or RF) and DC SQUIDs. DC SQUIDs are easier to analyze and understand, but until the past decade, AC SQUIDs were more popular as they were easier to construct and use. There are many variants of each type, many differing only in the type of shunting, e.g. capacitive and inductive.

Many nonlinear dynamical systems are highly sensitive to small external perturbations, especially when they are near the onset of a bifurcation. Moreover, previous work on DC SQUID arrays coupled in a uni-directional or bi-directional ring configurations were shown to be beneficial to detecting external signals by analyzing their Signal to Noise Ratios [5]. To better understand the dynamics of the non-uniform SQUID array our interest lies in the following attributes. (1) Flux coupling between neighbor and nearest neighbor SQUID loops exploring all possibilities of coupling (i.e. uni- and bi-direction, both chain and ring configuration), (2) SQUID loop parameters (loop size, orientation, and number of elements), (3) Voltage response vs.

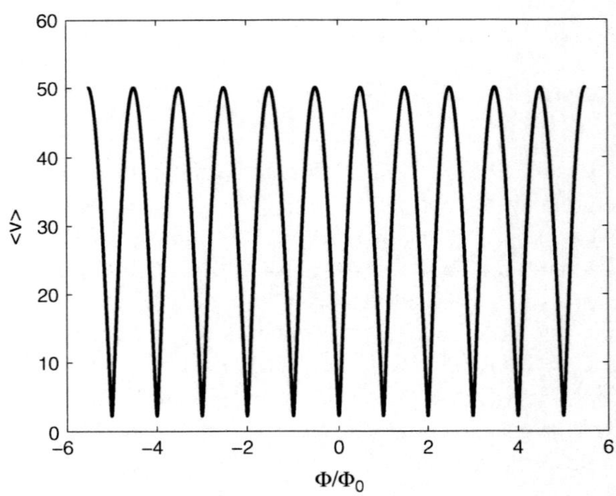

Fig. 1 Displays the average voltage response $\langle V \rangle$ vs. the external magnetic field for a uniform SQUID array, where $N = 50$, $\beta_{k_max} = 1.0$, and $I_b = 1.001 I_c$. One period of voltage variations corresponds to an increase of one flux quantum

External flux, (4) Bias currents, inductive couplings, vs. voltage response, and (5) Periodicity in the voltage response in Gaussian SQUID arrays.

By devising a theoretical model for DC SQUID loops (which vary in size, inductive coupling, and many other parameter as described above) we study the time averaged voltage response function $\langle V \rangle$ for both networks of parallel and series SQUID arrays. However, in this paper we focus only on the series arrays. Thus, in bi-directional coupling the flux in the nth SQUID loop is dependent on the flux in the $(n+1)$th and also in the $(n-1)$th. Exploration of this parameter (inductive coupling) as well as others parameters are the key to discovering new dynamical phenomena for improvement in the desired device.

Figure 1 displays the conventional uniform SQUID arrays with a voltage response that is Φ_0-periodic function. Compared to the non-uniform SQUID array such that the loop sizes are chosen to be different. Hence, the voltage response is unique around the zero magnetic field (Fig. 2). We would like to be able to design the shape of the voltage response function, optimizing dynamic range and power tolerance as well as sensitivity, ideally in an open loop configuration without the use of phase locked loops or additional amplifiers. This device can be used in an RF system as an LNA providing gain to the signal off the antenna, allowing a much smaller antenna to be used, or we anticipate eventually, as the antenna itself while still providing gain.

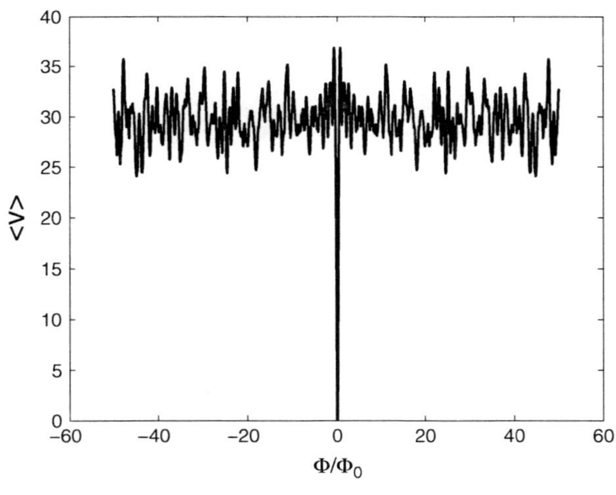

Fig. 2 Numerical simulations of the voltage response vs. external magnetic flux for an unconventional grating SQUID array. $N = 50$, $\beta_{k_{max}} = 1.0$, and $I_b = 0.98 I_c$

3 Simulation Details

Modeling single DC SQUID using the Josephson relations linking the voltage and the quantum phase difference across the ith junction, we arrive at following equation,

$$\tau\dot{\varphi}_i = J + (-1)^i/\beta(\varphi_1 - \varphi_2 - 2\pi x_e) - sin(\varphi_i) \tag{1}$$

for $i = 1, 2$ and where β is the nonlinearity parameter which has dependents on loop size, τ is a rescaling of the time constant, $J = I_b/I_c$ is the normalized biased flux and $x_e = \Phi_e/\Phi_0$ is the external flux normalized to the flux quantum.

The dynamics of the series SQUID array as shown in Fig. 3 is described by the following differential equations

$$\begin{aligned} \dot{\varphi_{1j}} &= J_1 + (-1)^j/\beta_1(\varphi_{11} - \varphi_{12} - 2\pi x_e \lambda I_2) - sin(\varphi_{1j}) \\ \dot{\varphi_{kj}} &= J_k + (-1)^j/\beta_k(\varphi_{k1} - \varphi_{k2} \cdots \\ &\qquad -2\pi x_e \lambda (I_{k-1} + I_{k+1})) - sin(\varphi_{kj}) \\ \dot{\varphi_{Nj}} &= J_N + (-1)^j/\beta_N(\varphi_{N1} - \varphi_{N2} \cdots \\ &\qquad -2\pi x_e \lambda I_{N-1}) - sin(\varphi_{Nj}) \end{aligned} \tag{2}$$

where

$$I_{k+1} = \frac{I_{c,k+1}}{\beta_{k+1}}[\varphi_{k+1,1} - \varphi_{k+1,2} - 2\pi x_e] \tag{3}$$

and $j = 1, 2$, N is the number of SQUID loops, λ is the normalized inductive coupling coefficient, and β_k is the nonlinear parameter related to the size of each loop.

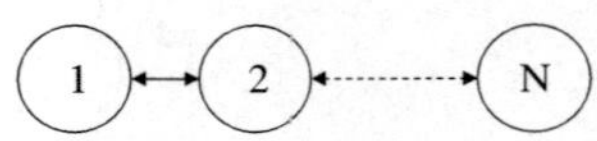

Fig. 3 Network of series SQUID array

4 Results

In a serial array of unconventional grating SQUID (SQIF) when accounting for the changes in parameters such as the flux coupling between neighboring SQUID loops the voltage to flux response changes with variation of coupling parameter λ. We define dynamic range as the ratio of the largest to smallest signal that can be measured at one time (usually measured in the Decibel scale). For example, measure a voltage of 10 volts accurately, and at the same time measure a voltage of 0.00001 volts, then the dynamic range is computed as 20 log(10/0.00001) = 120 dB. Hence, 120 dB means that the analyzer can accurately measure signals that are different in scale by a factor of as much as one million.

The dynamic range of a SQIF is related to ΔV_{max} which is the maximum voltage swing as show in Fig. 4. The dynamic range of a typical SQIF with 50 loops is ≈ 7 times (or 17 dB) larger than for a conventional single loop SQUID.

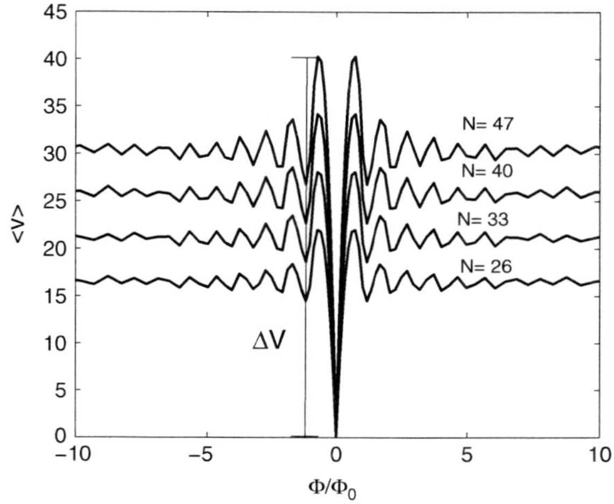

Fig. 4 Demonstrates the benefit of increasing the number of elements, by increasing the voltage response. Varying number of SQUID loops N in the series arrangement. The maximum voltage swing ΔV_{max} increases as the number of loops increase

Reference [1–3] Hence, increasing the number of element in an the array of SQUIDs increasing the dynamic range. Consider I_b near critical current I_c, as λ is varied the voltage response remains the same around the anti-peak. However, for values $\phi \neq 0$, $\bar{V}$ increases as the coupling (λ) increases, where $\bar{V}$ is the mean value of the voltage response excluding the anti-peak. Figure 5 displays that the maximum voltage

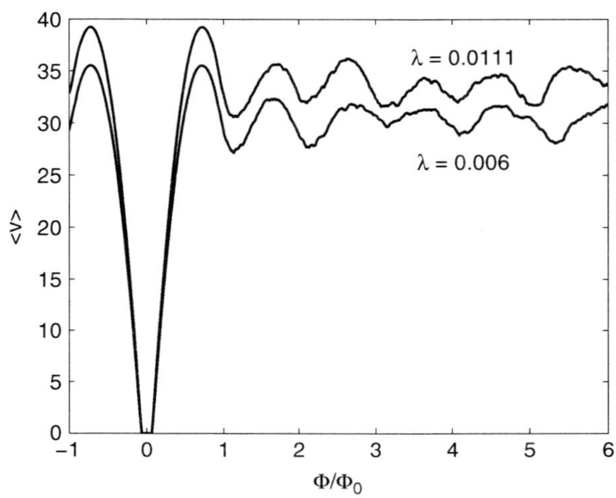

Fig. 5 Varying coupling strength for $I_b = 0.98I_c$. As the coupling strength increases the maximum voltage response increases. Number of loops: N = 100

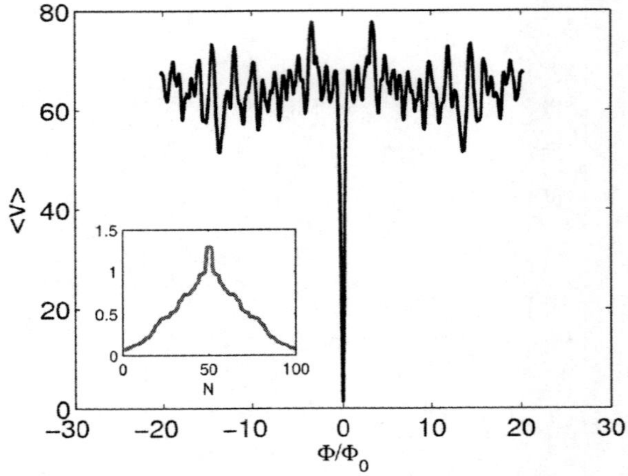

Fig. 6 Displays voltage response arrangement of triangle grating structures. Parameters used N = 100

response can also increase due to the coupling parameter λ. Therefore, we have shown that the dynamic range can increase due to the number of loops in a series and also by the increasing of inductive coupling between the neighboring loops. Suppose we are looking at how the arrangement of the loops play a role in the voltage response. In Fig. 6 we arrange the loop sizes from small to large and then back to small. The size of the loop is shown in the inserted plot on Fig. 6. Different arrangement can change the structure of the voltage response as shown in Fig. 7. In the latter case the gap between the anti-peak increases with this arrangement. For comparison the total area of the loops are the same in both Figs. 6 and 7.

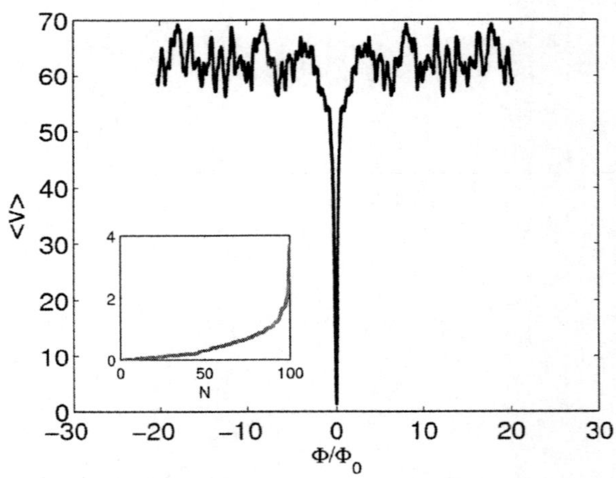

Fig. 7 Varying SQUID loops size from small to large loop sizes

5 Application

By modeling the dynamics of a SQIF, we demonstrated that their exists a non-periodic voltage response vs. external flux where an anti-peak is present only around the zero applied magnetic field. From Sect. 4 the dynamics range improves by changes in the following parameters: inductive coupling, number of loops coupled in series, and unconventional grating structures, i.e. vary each loop size in the array. Also, these devices have a large intrinsic bandwidth, can be operated in an unshielded environment, and can be used as robust, high sensitivity, low noise amplifiers (LNA) for rf-signals.

The use of SQUIDs, the most sensitive magnetic sensing device known, allows for the high sensitivity. The novel antenna is composed of an electrically small antenna that drives current into a current sensing SQUID amplifier. This configuration can be packaged in a unit much smaller than present antennas and can be used in place of two or three antenna systems that would normally be required to cover this frequency band. A wide range of magnetic field detection applications that require high-levels of sensitivity benefit from the SQUID array devices, such as: biomedical tracking of magnetic particles (e.g., MRI machines commonly used for diagnosing multiple sclerosis, brain tumors, and spinal infections) [6], geological equipment (e.g., NASA explorers, and remote sensing equipment for oil and mineral prospecting), homeland defense (e.g. detection of explosives, and building security applications) [7].

Applied as extremely high dynamic range, very low noise LNA, these devices will increase the link margin for system designers. The increased margin can be applied as reduction in antenna size, increased range or decreased transmit power as the system designer decides. While the requisite first step is to understand and apply these devices as LNA, beyond application as LNA, these devices offer a frequency insensitive, wide band, magnetic detector. Application of these devices ability to detect and amplify will result in a new generation of electrically small antennas.

6 Conclusion

Improved dynamic range by increasing the coupling parameter. Other contribution factors for increasing dynamic range include bias current, loop size distribution and number of elements coupled. Application in replacing current antennas with smaller single SQIF device.

The rich behavior provided by the nonlinear dynamics of coupled devices, numerically display improve sensitive. Motivation of this work stems from previous results with coupled e-field and fluxgate sensors [8, 9] where both have shown interesting dynamic behavior and dramatic improvement when compared to their un-coupled counterparts. Thus, by demonstrate a more general SQUID array, where the inductive coupling, size and orientation between loop area (a_n, for $n = 1, \dots, N$) vary, a next generation highly sensitive coupled SQUID array device can be realized. Extending work to 2-D arrays (large number of loops). Develop the theoretical model, to include coupling, fabrication parameters

Acknowledgments We acknowledge support from the Office of Naval Research (Code 30) and SPAWAR internal (S&T) program.

References

1. Ch. Häussler, J. Oppenländer, and N. Schopohl; J. Appl. Phys., Vol. 89, Issue 3, pp. 1875–1879 (2001).
2. Ch. Haussler, T. Trauble, J. Oppenlander, and N. Schopohl; IEEE Trans. Appl. Superconductivity, Vol. 11 Issue. I, (2001).
3. J. Oppenlnder, Ch. Hussler, and N. Schopohl; Phys. Rev. B **63**(2) 024511 (2000).
4. J. Oppenlnder, Ch. Hussler, and N. Schopohl; J. Appl. Phys., Vol. 86, Issue 10, pp. 5775–5779 (1999).
5. A. Palacios, J. Aven, P. Longhini, V. In, and A. Bulsara; Phys. Rev. E 74, 021122 (2006).
6. J.M. Talmadge, J. Gao, and J.G. Eden; Bio. Imaging: Macro to Nano, 2004. IEEE Int. Sym.; Vol. 2 p 1514–1515, April (2004).
7. Hideo Itozaki; Supercond. Sci. Technol. 16 pg. 13401343, (2003).
8. V. In, A. Bulsara, A. Palacios, P. Longhini, A. Kho, and J. Neff, Phys. Rev. E 68, 045102(R) (2003).
9. A. Bulsara, Phys. Rev. E 70, 036103 (2004).

Applications of Nonlinear and Reconfigurable Electronic Circuits

Joseph Neff, Visarath In, Christopher Obra and Antonio Palacios

1 Introduction

In the search for small, fast and low power electronic circuits the microelectronic engineering community is increasingly faced with understanding the physical and non-ideal properties of the circuit at the transistor level. This is especially true with regard to the use of modern fabrication processes where transistor gate length is measured in tens of nanometers and physical properties such as leakage current and non-ideal sub-threshold behavior become more prevalent. The purpose of this work is to present the construction of some simple electronic circuits that possess nonlinear and dynamical properties and to illustrate their utility in a few applications. Specifically, the construction of oscillating, pattern forming and mixed signal reconfigurable circuits is discussed. In these examples, the nonlinear and dynamical properties of the circuits play a critical role in their operation. Experimental results from fabricated circuits are presented. Useful analytical tools are also discussed in the context of constructing and understanding the operation of the circuits. Similarities in the construction of several circuits lead to the development of a generic analog and nonlinear electronic fabric, similar to field-programmable gate arrays. Experimental results demonstrating configurability are shown. The array is constructed using non-volatile floating-gate based transistors. This type of proposed circuit construction and operation takes advantage of the nonlinear properties of the circuit and illustrates how many undesirable properties can actually provide utility.

J. Neff (✉)
SPAWAR Systems Center-San Diego, Code 71730, San Diego, CA, USA
e-mail: jdneff@spawar.navy.mil

V. In et al. (eds.), *Applications of Nonlinear Dynamics*, Understanding Complex Systems,
DOI 10.1007/978-3-540-85632-0_10,

2 Nonlinear and Dynamical Circuits

In this section we introduce a few basic CMOS circuits that can be used as building blocks for creating higher order dynamical systems. The section starts with the introduction of a nonlinear transconductor (a device that ideally converts a voltage level to a current level) and then introduces its use in a bistable sub-circuit. The bistable sub-circuit is used as a building block for an example oscillating system. The oscillator uses a ring topology, similar to traditional ring oscillator, but with additional tunable features.

2.1 A Nonlinear Transconductor

In analog circuit design, a transconductor is used to refer to a device that converts a voltage level to a current level. In this case we use a traditional operational-transconductance-amplifier, or OTA, as one of the key building blocks for our dynamical circuits. An ideal OTA converts linearly an input voltage to an output current. The ideal OTA has an infinitely high input impedance, meaning that it draws no current at the input, and an infinitely low output impedance, meaning that it can output any current level, regardless of the resistive load. Along with accepting the fact that these ideal qualities are impossible to obtain, we also accept that the transconductance itself is characteristically nonlinear. In fact, the nonlinear qualities play in important role in the proper function of the systems described in this paper. A common theme in this paper is that, in many, cases the inherent properties of the devices used to construct a system, be they linear or nonlinear, can be used to create an efficient design.

Figure 1 illustrates a simple OTA design using five transistors (in practice a nine transistor design is used that has better output characteristics). The example OTA is constructed of two components that are fundamental to analog circuit design: the differential pair and the current mirror. The differential pair consists of the three bottom transistors. We use (1) to empirically describe the currents I_+ and I_- in the two branches of the differential circuit. The currents are functions of the two voltage inputs V_+ and V_-. In (1) and (2) I_b is set by the bias voltage V_b and $c = constant$ for $V_b < V_{thresh}$ or $c = \sqrt{\frac{\beta}{I_b}}$ for $V_b > V_{thresh}$. V_{thresh} is known as the *threshold voltage* and is a parameter that depends on the manufacturing process. With the addition of the current mirror and the use of Kirchoff's current law, we obtain (2) to describe the OTA. A more detailed description of modeling an OTA operating in subthreshold can be found in [1] A symbol typically used to represent an OTA in circuits is shown in the inset.

$$I_{\pm} = \frac{I_b}{2}\{1 \pm tanh(c(V_+ - V_-))\} \tag{1}$$

$$I_{out} = I_b tanh(c(V_+ - V_-)) \tag{2}$$

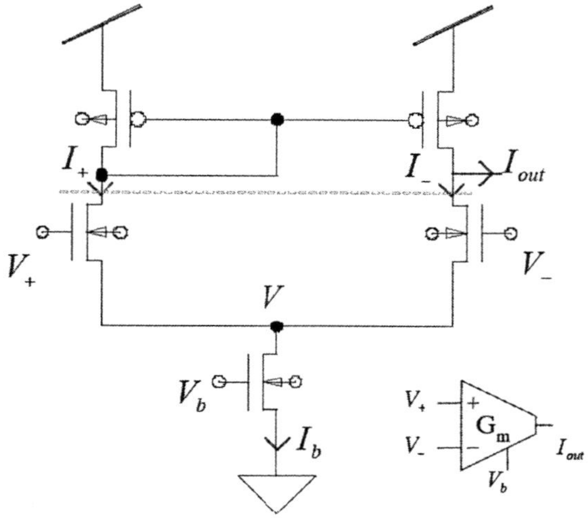

Fig. 1 A simple ordinary-transconductor-amplifier, or OTA. The OTA converts a differential input voltage to a single-ended output current level. V_b. is a bias voltage that sets the over-all output current level. The transconductance of the OTA is nonlinear function well described by the empirical Eq. (1). The OTA is well suited for constructing higher order dynamical systems due to its high input impedance and low output impedance

2.2 A Bistable Circuit

The OTA presented inf Fig. 1 can be used, along with two additional transistors (or resistors) and a capacitor, to create a simple bistable circuit (Fig. 2). The bistable circuit is similar to the over-damped Duffing oscillator adopted in [2] In general terms, the bistable circuit consists of a linear function, which is provided by resistors or short channel transistors, and a nonlinear function, which is provided by the OTA. A symbol used to represent the circuit is shown in the inset. Using a transistor as a resistive device is not unusual to circuit design. This circuit demonstrates simplicity by using a few components and by taking advantage of the available properties of the short channel transistor as well as the fundamental nonlinear transconductance provided by the OTA.

This circuit takes advantage of the short channel effect where the depletion of charge carriers in the conductive channel of the transistor increases approximately linearly with an increase in drain-source voltage. Equation (3) is typically used to represent the source-drain current of the short channel transistor in above-threshold operation. In (3), V_e is known as the Early voltage and I_0 is determined by the gate-source voltage, the channel width/length, and the conductive properties of the channel. The effect is more pronounced as the minimum channel length gets smaller with modern processes.

$$I_{ds} = I_0 \left(1 + \frac{V_{ds}}{V_e} \right) \tag{3}$$

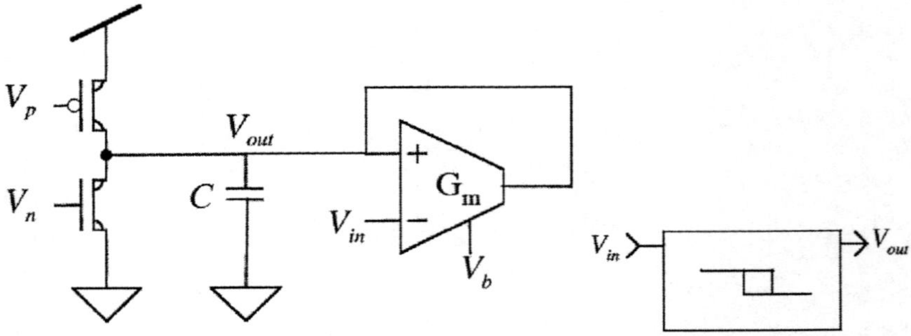

Fig. 2 A simple bistable circuit constructed of short channel transistors, a capacitor, and an OTA. The short channel transistors are used as linear devices and are adjustable by changing their gate voltages V_{bp} and V_{bn}

Using (2) and (3), a simple ordinary differential equation can be obtained to describe the bistable circuit. For the short-channel transistors, we use $I_p = I_{0p}(1+\frac{Vdd-V}{V_{ep}})$ and $I_n = I_{0n}(1+\frac{V-V_{ss}}{V_{en}})$, where $I_{0p} = I_{0p}(V_p), I_{0n} = I_{0n}(V_n)$. Using Kirchoff's law and the equation for a charging capacitor we obtain (4).

$$C\dot{V} = I_0 - gV + I_b tanh(c(V - V_{in})) \tag{4}$$

In (4), $I_0 = I_{0p} - I_{0n} + \frac{V_{dd}}{V_{ep}}$, $g = \frac{1}{V_{ep}} + \frac{1}{V_{en}}$ and $I_b = I_b(V_b)$, and $V_{ss} = 0$. To simplify the problem, we set $I_{0p} = I_{0n}$ via the bias voltages V_p and V_n. We also assume $V_{ep} = V_{en} = V_e$, which can be approximately achieved by careful design of the actual circuit. Linear stability analysis reveals (ignoring the nonlinear term) a stable fixed point at $V_0 = \frac{V_{dd}}{2}$. In this case it may be useful to introduce a change of variables $V' = V - \frac{V_{dd}}{2}$ so that the fixed-point and related bifurcations are centered at $V' = 0$. For $V_{ss} = -V_{dd}$ the fixed point solution is $V_0 = 0$. These considerations are important in analog circuit design where sub-circuits and systems can operate at different voltage supply levels and input and coupling signals need to be placed optimally.

To obtain bistability, the condition $g < I_s c$ has to be met. With that condition met, the circuit will demonstrate hysteresis in it's output V_{out} with regard to the input voltage V_{in} and so it can be said that the circuit possesses the quality of memory. In that sense, the cell can be compared functionality to a Static Random Access Memory cell (SRAM). The circuit compares in complexity in size and number of transistors as well, with an SRAM cell using about six transistors and the bistable cell in Fig. 2 using seven to eleven transistors. For the circuit in Fig. 2, a single control voltage V_{in} is used to switch between the two states, whereas a typical SRAM circuit uses a complimentary input to set the value along with an assert line to load the value. Other than these observations, no effort is made to compare the benefits of one memory circuit design over another. The purpose here is to illustrate how this circuit, which is nonlinear and dynamical in nature (so is an SRAM circuit), can be used to construct higher order systems. It is worthwhile to note here that a fully differential design can be created that can be similarly described using (4).

2.3 An Oscillator with a Ring Topology

A simple oscillator can be constructed by coupling an odd number of the bistable unit cells together in a ring topology Fig. 3. Similar to a ring oscillator, which is typically constructed of more simplified inverting circuits, this oscillator demonstrates periodic oscillation in each of the voltages V_i. For identical parameters, each voltage oscillates with the same frequency ω (period T). The voltages oscillate in an out-of-phase pattern so that each adjacent oscillator in the ring is leads or lags is neighbor by a phase delay of T/N, where N is the number of oscillators in the array. The chief difference between this ring oscillator and a traditional ring oscillator is the ability to "tune" the nature of the circuit, such as the oscillating frequency, using the control voltages V_p, V_n and V_s. Using (4) to describe the unit cell, a set of coupled first order equations can be obtained. In (5), N is an odd number. A detailed analysis of the bifurcation and dynamical properties of this system is not given here, but analysis of a similarly described system can be found in [3–5] This structure forms an integral part of a pattern forming system presented in Sect. 3.2.

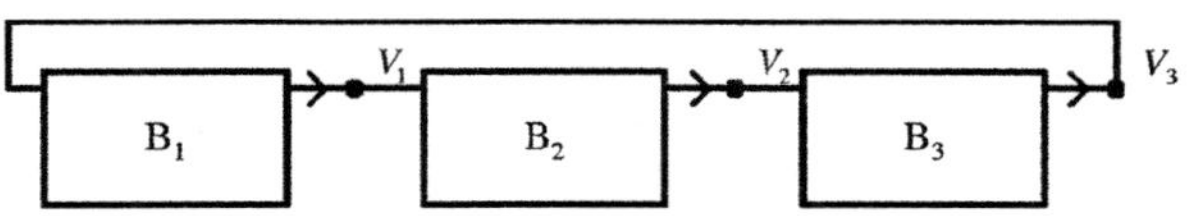

Fig. 3 A simple bistable circuit constructed of short channel transistors, a capacitor, and an OTA

$$
\begin{aligned}
C\dot{V}_1 &= I_0 - gV_1 + I_b tanh\left(c(V_1 - V_N)\right) \\
C\dot{V}_i &= I_0 - gV_i + I_b tanh\left(c(V_i - V_{i-1})\right) \; i = \{2, ..., N\}
\end{aligned} \tag{5}
$$

3 Nonlinear System

Here we use the circuit building blocks introduced in Sect. 5.2.1 and 5.2.2 to construct a higer-order system. Six bistable sub-circuits are used to construct a system that demonstrates multiple simultaneous frequency patterns that are synchronized to each other. Symmetry arguments are used to predict the behavior of the circuit and the predicted patterns are observed in an experimental system. Even though significant mismatch is present in the physical system (so exact symmetry cannot be achieved), the behaviors predicted by the ideal symmetrical system are still observed and appear to be robust. The implication is that the use of symmetry in the topological construction of coupled dynamical systems can be used to mitigate uncertainty associated with their actual construction. Furthermore, topological considerations my lead to simplified designs by relaxing constraints on the particular transfer functions (since the emphasis is no longer on implementing an ideal function such as a linear OTA). As a result, for example, coupling terms no longer need to be idealized functions, and therefor their implementations may require fewer transistors. An example of where such a design approach can be used is in the design of clock distribution networks, where timing signals across relatively large-scale circuits needs to be carefully coordinated.

3.1 Topological Symmetry

In this section we introduce the concept of symmetry in the construction of an idealized system and how symmetry arguments can be used to construct pattern forming arrays. For example, the oscillator in Sect. 2.1 possesses both the cyclic group symmetry, $\mathbf{Z}_N$, as well as temporal group $\mathbf{S}^1$ symmetry. For a more complete discussion on the role of symmetry and the group theoretic approach in understanding pattern formation in coupled systems see [6–8] Detailed discussions on applying the group theoretic approach to physical systems with practical applications can be found in [9–11]

For the sake of discussion, consider two coupled arrays of equal size and construction (in Sect. 3.3 we give an example of an actual system.) We use $X(t) = \{X_1(t),\ldots,X_N(t)\}$ to represent the state of the first array and $Y(t) = \{Y_1(t),\ldots,Y_N(t)\}$ to denote the state of the second array. At any given time t, the spatio-temporal pattern generated by the two coupled arrays is described by $P(t) = \{X(t),Y(t)\}$. Let us assume that this pattern is a periodic solution of period T with the following characteristics. On one side of the network, for instance, the X-array, the oscillators form a *traveling wave* (TW), i.e., same wave form X_0 shifted (delayed) by a constant time lag $\phi = T/N$: $X_k(t) = X_0(t+(k-1)\phi)$, $k = 1,\ldots,N$. On the opposite side, the oscillators are assumed to be *in-phase* (IP) with identical wave form Y_0, i.e., a synchronous state: $Y_k(t) = Y_0(t)$, $k = 1,\ldots,N$. Given these assumptions we can show that such a system remains unchanged under spacial $\mathbf{Z}_N$ and temporal $\mathbf{S}^1$ transformation and so possesses $\mathbf{Z}_N \times \mathbf{S}^1$ symmetry.

Assume that $P(t)$ has spatio-temporal symmetry described by the cyclic group $\mathbf{Z}_N$, i.e, the group of cyclic permutations of N objects generated by $(1,2,\ldots,N) \mapsto (N,1,\ldots,N-1)$, and by the group $\mathbf{S}^1$ of temporal shifts. Together, $\mathbf{Z}_N \times \mathbf{S}^2$ acts on $P(t)$ as follows. First, $\mathbf{Z}_N$ cyclically permutes the oscillators of both arrays:

$$\begin{aligned}
\mathbf{Z}_N \cdot X_{TW}(t) &= \{X_N(t+(N-1)\phi), X_1(t),\ldots,\\
&\qquad X_{N-1}(t+(N-2)\phi)\},\\
\mathbf{Z}_N \cdot Y_{IP}(t) &= \{Y_N(t), Y_1(t),\ldots,Y_{N-1}(t)\}.
\end{aligned}$$

Then $\mathbf{S}^1$ shifts time by ϕ so that

$$\begin{aligned}\mathbf{Z}_N \times \mathbf{S}^1 \cdot X_{TW}(t) &= \{X_N(t), X_1(t+\phi), \ldots,\\ &\qquad X_{N-1}(t+(N-1)\phi)\},\\ \mathbf{Z}_N \times \mathbf{S}^1 \cdot Y_{IP}(t) &= \{Y_N(t+\phi), Y_1(t+\phi), \ldots,\\ &\qquad Y_{N-1}(t+\phi)\}.\end{aligned}$$

Since the oscillators are identical, we get

$$\begin{aligned}\mathbf{Z}_N \times \mathbf{S}^1 \cdot X_{TW}(t) &= X_{TW}(t),\\ \mathbf{Z}_N \times \mathbf{S}^1 \cdot Y_{IP}(t) &= Y_{IP}(t+\phi).\end{aligned}$$

It follows that in order for $Y_{IP}(t)$ to have $\mathbf{Z}_N \times \mathbf{S}^1$ symmetry the in-phase oscillators must oscillate at N times the frequency of the oscillations of the traveling wave. The same conclusion is reached if the roles of the X and Y arrays are interchanged.

3.2 An Example Pattern Forming System

An experimental circuit was constructed to demonstrate multifrequency pattern formation using the ring oscillator as a starting point. The circuit possesses the $\mathbf{Z}_N \times \mathbf{S}^1$ symmetry described in the Sect. 3.1. The circuit is constructed of two ring oscillators (Sect. 2.3) coupled together in such a way that each sub-circuit in each ring is coupled to all the sub-circuits in the opposite ring. The system is illustrated in Fig. 4. In the figure the inset illustrates how the bistable circuit is modified to include the coupling term. In this case we rely on the symmetry of the system, which is described in Sect. 3.1, to dictate the topology of the coupling and find that only one additional transistor is needed for each coupling term. This is significant since in circuit design the fewer the transistors you use the better you can predict the behavior of the physical system (primarily due to mismatch and inaccuracies in the transistor model).

A set of coupled first-order differential equations can be obtained for the system presented in Fig. 4 using (1) and (5). In (6) the two arrays are identified as X and Y. The equations generalize the system of six coupled ordinary differential equations that describe the multifrequency system.

$$\begin{aligned}C\dot{V}_{X,Y_1} &= I_0 - gV_{X,Y_1} + I_b tanh\left(c(V_{X,Y_1} - V_{X,Y_N})\right) + \frac{I_b}{2} tanh\left(c(V_{Y,X_1} - V_{Y,X_N})\right)\\ &\quad + \frac{I_b}{2} tanh\left(c(V_{Y,X_2} - V_{Y,X_1})\right) + \frac{I_b}{2} tanh\left(c(V_{Y,X_3} - V_{Y,X_2})\right)\\ C\dot{V}_{X,Y_i} &= I_0 - gV_{X,Y_i} + I_b tanh\left(c(V_{X,Y_i} - V_{X,Y_{i-1}})\right) + \frac{I_b}{2} tanh\left(c(V_{Y,X_1} - V_{Y,X_N})\right)\\ &\quad + \frac{I_b}{2} tanh\left(c(V_{Y,X_2} - V_{Y,X_1})\right) + \frac{I_b}{2} tanh\left(c(V_{Y,X_3} - V_{Y,X_2})\right)\, i = \{2, \ldots, N\}\end{aligned} \tag{6}$$

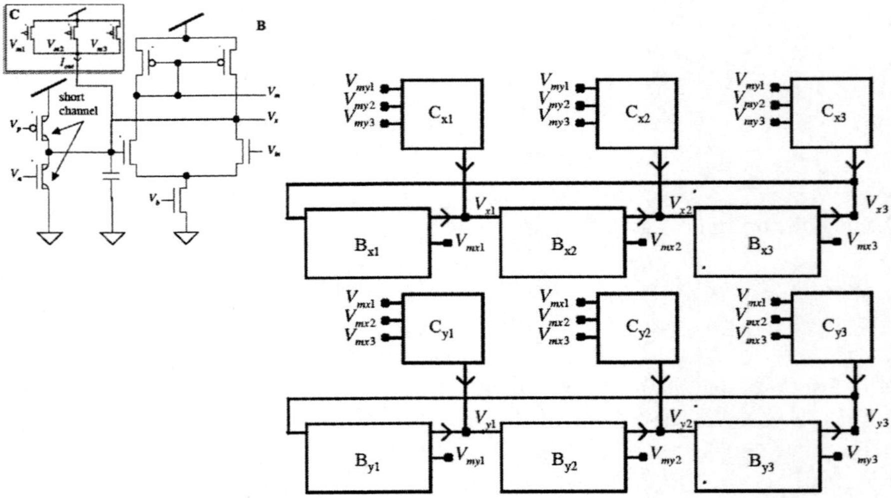

Fig. 4 A six-cell mltifrequency generating network. Each bistable circuit is labeled **B** with the associated coupling circuit labeled **C**. The system consists of two arrays, each with three uni-directionally coupled cells arranged in a ring. This coupling topology favors the traveling wave patterns described in Sect. 3.1. In addition to the unidirectional coupling, each cell in each array receives a coupling input from all the cells in the opposite array. The symmetries within this coupling topology result in the in-phase and traveling wave multifrequency solutions

3.3 Experimental Results

A circuit was constructed using the TSMC 0.35 μ CMOS process though the MOSIS foundry service [12] The mask layout for the design is shown in Fig. 5. Very little effort was made to optimize the design for a particular performance. As a result, we observed a significant amount of mismatch between the performance of each sub-circuit in the system. However, the operation of the circuit is apparently robust

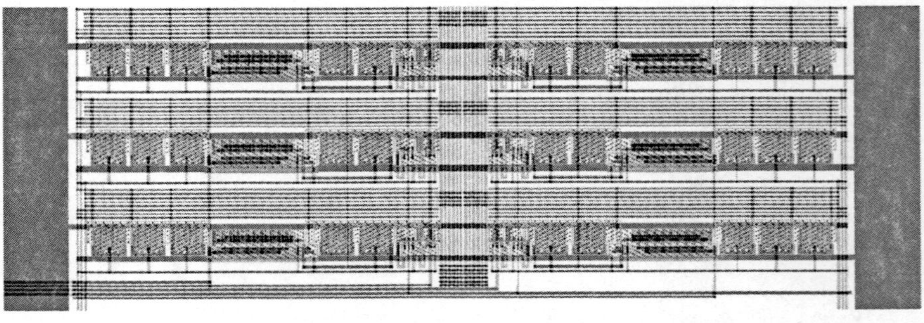

Fig. 5 Mask layout for the multifreqency circuit. The circuit was constructed using the TSMC 0.35 μ CMOS process. The *X* array is on the left-hand side of the layout and the *Y* array is on the right-hand-side

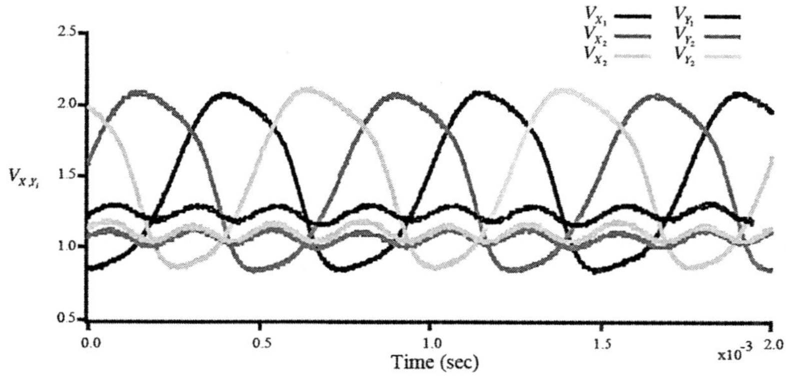

Fig. 6 Experimental results of a six-cell mltifrequency generating network. The figure shows the time-series results for the V_X and V_Y arrays. The figure shows the X array operating with a traveling wave solution, with each variable oscillating at a frequency ω, while the Y array operates with an in-phase solution oscillating at a frequency of ωN. Differences in the solutions are due to mismatch between the physical devices

to mismatch. Figure 6 shows experimental results. The top and bottom plots are the time-series results for the $V_{x1,i}$ and $V_{x2,i}$ variables respectively. The figure shows the bottom array operating with a traveling wave solution, with each variable oscillating at a frequency ω, while the top array operates with an in-phase solution oscillating at a frequency of $N\omega$. Depending on the initial conditions of the system (start-up condition) another solution can be observed with the $V_{x1,i}$ and $V_{x2,i}$ swapping roles. These solutions are predicted by the group theoretic approach given in Sect. 3.1. An additional pattern is possible where both arrays oscillate with an out of phase solution, as shown in Fig. 6. In nonlinear systems the coexistence of multiple solutions is not uncommon, and the group theoretic approach is a convenient method for predicting what patterns are possible with a networked system that has symmetry. It is important to consider symmetry does not imply stability in any of the set of possible solutions. In these experiments, the two patterns are stable. Details on the construction and testing the circuit are saved for a future discussion.

4 Generalized Reconfigurable Array

In this section we attempt to generalize the pattern forming system presented in Sect. 3.2 by creating a globally connected reconfigurable circuit that takes advantage of some of the properties of the floating-gate transistor. As an intermediate example, we present a circuit that possesses properties similar to the bistable circuit presented in Sect. 2.2. The circuit is generalized to a globally coupled reconfigurable system. A brief discussion on programming the system is given and some preliminary results demonstrating programmable stability are presented. Full demonstration of a reconfigurable pattern forming system is reserved for future work.

4.1 Floating-Gate Basics

A floating-gate transistor is a transistor that uses a true floating-gate and capacitive coupling for it's input. For older CMOS processes that possess two poly-silicon layers it is simple to create a floating-gate transistor. This is done by using the poly-1 material that forms the gate to also form one side of a poly-1/poly-2 capacitor. The input (metal-1) is then connected to the poly-2 layer to create a capacitively coupled input. Similar structures can be created in modern processes (which don't typically have two polly layers) by using metal layers or other structures. One of the benefits of this approach is that, by constructing multiple capacitors that share the same floating node, a multiple-input transistor can be created. For a multiple-input transistor, the effective coupling of an input is the ratio of the input capacitance to the total capacitance of the floating node.

As an example, we modify the sub-threshold equation for a transistor by making V_g a capacitively coupled input. We use (7) to model a multiple input floating-gate pMOS transistor. In (7) I_{ds} is the drain-source current of the transistor, I_Q is set by the relationship $I_Q = I_0 e^{\frac{Q}{C_T}}$, where Q is the floating charge on the gate, and V_g is a capacitively coupled input given by $V_g = V_{tun}\frac{C_{tun}}{C_T} + \sum_{i=1}^{N} V_i \frac{C_i}{C_T}$.

Given that the input to the floating-gate transistor can be understood, we are left with having to deal with, or otherwise program, the charge on the floating node. This is a significant topic in itself, so only a brief discussion will be given. In short, the charge on a floating gate is changed by two methods often referred to as electron tunneling and hot-electron injection. Electron tunneling is achieved via a relatively high voltage that is capacitively coupled to the gate. When the tunneling input-to-gate voltage is high enough, electrons tunnel from the gate to the input, lowering the voltage of the gate. Since the tunneling voltage is much higher than the standard supply level, a global programming scheme is typically used, allowing for less complicated circuitry.

For hot-electron injection, electrons with the proper energy within the conduction band of the channel are able to tunnel from the conduction layer onto the floating-gate. This is usually achieved by operating the transistor at drain-source voltages that are slightly higher than the typical operating voltages of a specific process, and by biasing the transistor gate voltage near the threshold voltage.

Because hot-electron injection depends on both the drain-source voltage as well as the gate-source voltage, a floating-gate transistor can be conveniently programmed using a row-collumn selection scheme. For example, consider an array of floating-gate transistors with one input each (and one tunneling junction each) arranged in a two-dimesntional array, so that the transistor drains are connected together in rows and the transistor inputs are connected together in columns. This type

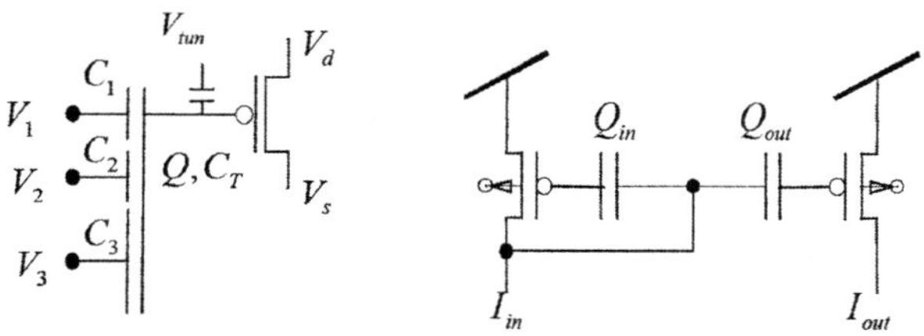

Fig. 7 A symbol for a multiple input floating-gate transistor (*left*) and a floating-gate based current mirror (*right*)

of structure allows a single transistor in the array to be programmed selectively by setting the drain (row) and gate (column) inputs appropriately [13] A symbol for a floating-gate pFET is shown in Fig. 7 (left).

$$I_{ds} = I_Q e^{\frac{\kappa V_g}{U_t}} \tag{7}$$

The ideal current mirror can be modified to provide a programmable gain between the input and output branches. The programmable current mirror consists of two single input floating-gate transistors as illustrated in Fig. 7 (right). We use (8) to describe the programmable current mirror. In (8) $W = \frac{I_{Q_{out}}}{I_{Q_{in}}}$.

$$I_{out} = W I_{in} \tag{8}$$

4.2 Generalized Array

Using the floating-gate transistor we can develop a programable system reproduces the behavior of the bistable circuit presented in Sect. 2.2. The programmable bistable circuit is shown in Fig. 8.

We use (9) to describe the programable bistable circuit, with $I_0 = I_b(W_+ + W_-) - I_l$, $W = W_+ - W_-$, $W_\pm = \frac{I_\pm}{I_0}$ and $\xi_i(t)$ is a generic noise term. The effect of noise is not considered in this work. However, because the circuit can be programed to operate in sub-threshold operation, it may prove useful in demonstrating the effects of noise in nonlinear circuits used for computation [14], in particular with regard to ultra-low-power electronics, where the effects of channel noise and other sources of noise become significant.

$$C\dot{V} = I_0 - gV + I_b W tanh\left(c(V - e(t))\right) - \xi(t) \tag{9}$$

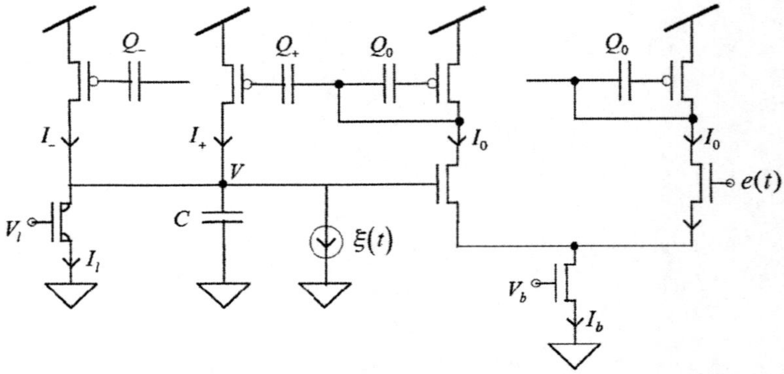

Fig. 8 A programmable bistable circuit using floating-gate transistors. The floating-gate transistors on the right-hand side of the figure are used to normalize the operation of the array by setting them equal. The transistor on the right-hand side are used to program the globally coupled array

Due to the convenient nature of the physical layout of the system, it is easy to generalize the design to arbitrarily large arrays. An arbitrarily large $N \times N$ system is given by (10).

$$C\dot{V}_i = I_{0_i} - gV_i + I_b \sum_{j=1}^{N} W_{ij} tanh(c(V_j - e(t))) - \xi_i(t) \tag{10}$$

An experimental system was constructed that implements the programable bistable system as a globally coupled 4×4 array. The test chip was manufactured using the TSMC 0.35 μ CMOS process. In the physical layout the floating-gate transistors are arranged in a rectangular array, which allows for efficient use of space and supports the row-collumn programming structure as described above. Support circuitry along the periphery of the array is used to switch between the programming and operational modes. There are established methods for programming floating-gate transistors and is a significant topic on it's own [15–17] We use a current-clamp technique to program each floating-gate in the rectangular array.

As an example of programmability, each of the four elements in the array are selectively programmed to demonstrate bistable or mono-stable behavior. Cross coupling terms (off-diagonl) are set to zero (Fig. 9) We use LabView along with National Instruments data acquisition cards and custom interface boards in our experimental setup. Our interface allows us to switch between operating and programming modes by controlling the power supply voltages. Figure 9 shows preliminary results from the experimental 4×4 array. The diagonal transistors in the array are programmed to create alternately bistable and mono-stable sub-circuits. Specifically, bistable behavior is determine by the condition $g < W_{ii} I_b c$. The response of the four sub-circuits are shown as functions of a common input signal, $e(t)$, which is swept

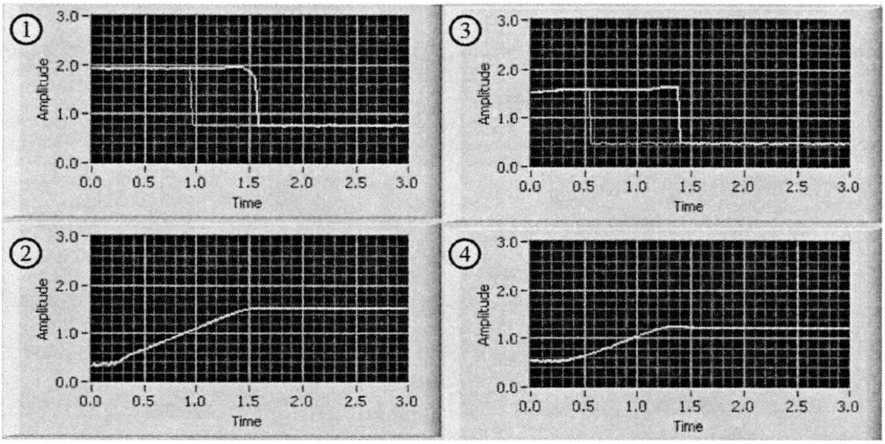

Fig. 9 Experimental results of a programmed 4×4 array. Each of the four elements in the array were programmed to demonstrate bistable and monostable behavior alternatly (all cross-coupling terms were set to zero)

from zero to V_{dd} and back again. Cells (1) and (3) show hysteresis, while cells (2) and (4) do not.

5 Conclusions

In this paper we have illustrated the use of a few simple circuits and devices, namely the differential pair, the current mirror, the short-channel transistor and some floating-gate based circuits, to create circuits and systems that demonstrate memory, pattern formation and reconfigurability. A common theme in the discussion is that simple circuits with utility can be created using a minimalist approach by taking advantage of the inherent properties of the devices, such as the linear behavior of the short-channel transistor, and the nonlinear behavior of the differential pair. In addition, by desing sysmetry into the circuit, which can be guided by a group-theoretic approach, higher-order pattern forming systems can be created that appear to be rhobust to fabrication errors (mismatch). The discussion ends with a brief introduction to programmable floating-gate transistors and how they can be used to create a reconfigurable system that offers programmable bistability and has the potential to generalize the pattern forming system to one that is arbitrarily large and that can take advantage of programmable all-to-all coupling. Future tests will involve larger arrays and will focus on demonstrating reconfigurable multi-frequency pattern-forming systems, programmable logic functions, and low-power computation.

Acknowledgments This work was funded in part by the In-House Laboratory Independent Research Program (ILIR) at SPAWAR Systems Center-San Diego and the Office of Naval Research. I would like to than Adi. Bulsara for his generous and continuous support.

References

1. C. Mead, N. *Analog VLSI and Neural Systems* Cambridge Nonlinear Sciences Series **12**, Addison-Wesley, Reading, MA, 1989.
2. A. Bulsara, V. In, A. Kho, P. Longhini, A. Palacios, W. Rappel, J.Acebron, S. Baglio and B. Ando. *Phys. Rev. E.* **70** (2004).
3. V. In, A. Bulsara, A. Palacios, P. Longhini, A. Kho and J. Neff. *Phys. Rev. E.* **68** (2003).
4. V. In, A. Bulsara, A. Palacios, P. Longhini and A. Kho. *Phys. Rev. E.* **72** (2005).
5. A. Bulsara, J. Lindner, V. In, A. Kho, S. Baglio, V. Sacco, B. Ando, P. Longhini, A. Palacios and W. Rappel. *Phys. Lett. A.* **353** (2006).
6. M. Golubitsky, I.N. Stewart, and D.G. Schaeffer. *Singularities and Groups in Bifurcation Theory: Vol. II.* Appl. Math. Sci. **69**, Springer-Verlag, New York, 1988.
7. M. Golubitsky, I. Stewart. In: *Pattern Formation in Continuous and Coupled Systems*, (M. Golubitsky, D. Luss and S.H. Strogatz, eds.) IMA Volumes in Mathematics and its Applications **115**, Springer, New York, (1999) 65–82.
8. M. Golubitsky and I. Stewart. In: *Geometry, Mechanics, and Dynamics*, (P. Newton, P. Holmes, and A. Weinstein, eds.) Springer, New York, (2002) 243.
9. V. In, A. Kho, J. Neff, A. Palacios, P. Longhini and B. Meadows. *Phys. Rev. Lett.* **91** (2003).
10. A. Palacios, R. Gonzalez, P. Longhini, N. Renz, V. In, A. Kho, J. Neff, B. Meadows and A. Bulsara. *Phys. Rev. E.* **72** (2005).
11. P. Longhini, A. Palacios, V. In, J. Neff, A. Kho and A. Bulsara *Phys. Rev. E.* **76** (2007).
12. *www.mosis.org*
13. M. Kucic, A. Low, P. Hasler and J. Neff. *IEEE Trans. Circuits Sys* **48** (2001).
14. N. Stocks. *Phys. Rev. Let.* **84** (2000).
15. A. Bandyopadhyay, G. Serrano and P. Hasler. *IEEE Journal of Solid-State Circuits* **41** (2006).
16. D. Graham, E. Farquhar, B. Degnan, C. Gordon and P. Hasler. *Proce. IEEE Int Symposium Circuits Sys* **54** (2007).
17. D. Graham and P. Hasler. *Proce. of the IEEE International Symposium on Circuits and Systems* (2007).

Multi-Phase Synchronization and Parallel Power Converters

Toshimichi Saito, Yuki Ishikawa and Yasuhide Ishige

Abstract This review paper studies nonlinear dynamics of parallel dc–dc converters and its practical applications. First, we introduce a single dc–dc converter which is a building block of the parallel system. Using a simple model of piecewise constant vector field, we show that the converters exhibit rich bifurcation phenomena including chaos and complicated superstable periodic orbits. We then introduce paralleled converters using a switching rule based on the winner-take-all principle. The switching can realize multi-phase synchronization that is crucially important to realize current sharing with smaller ripple for the low-voltage high-current capabilities. Presenting simple test circuit, typical phenomena are confirmed experimentally.

1 Introduction

The power electronics is a key technology with three main aims:

(1) conversion of electrical energy from one form to another by simple control;
(2) achievement of high conversion efficiency;
(3) minimization of the mass of power converters.

The power converters can be classified into four basic types: dc–dc, dc–ac, ac–dc and ac–ac power converters. The power converters are usually controlled by some nonlinear switching and the nonlinear dynamics can be used effectively to realize efficient power conversion [1, 2].

In a variety of systems, we have investigated paralleled systems of dc–dc converters in which multi-phase synchronous phenomena play crucial role [3–9]. Figure 1 illustrates an example of the parallel converters: N pieces of buck converters are

T. Saito (✉)
HOSEI University, Tokyo, Japan, e-mail: tsaito at k. hosei.ac.jp

V. In et al. (eds.), *Applications of Nonlinear Dynamics*, Understanding Complex Systems, 133
DOI 10.1007/978-3-540-85632-0_11,

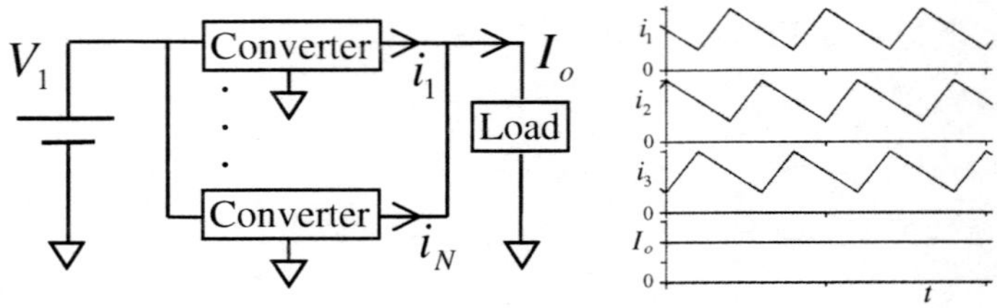

Fig. 1 Concept of paralleled buck converters and current share

interleaved between a dc input V_1 and a load. Motivations of study of such systems include the following. First, the converters share the output current [5]. The sharing is suitable for lower voltages with higher current capabilities in the next generation of micro-processors. The sharing is also effective to improve reliability and fault tolerance. Second, ripple reduction of the output current is possible [3, 4]. It is convenient to reduce size and losses of the filtering stages; and also can decrease switching and conduction losses and EMI levels. In order to realize ripple reduction with current sharing, multi-phase synchronization of inductor currents are important and some effective phase control techniques have been studied: sliding mode control [4], digital logic control [3], wireless PWM control [7] and so on. Third, the paralleled converters are nonlinear switched dynamical systems having rich periodic/chaotic phenomena and related bifurcation phenomena [9–13]. However, precise/theoretical analysis of stability and bifurcation phenomena is not easy because of complex nonlinearity.

This review paper studies nonlinear dynamics of a class of parallel buck converters. First, we introduce a single buck converter which is a building blocks of the parallel system. Using a return map we can clarify that the converters exhibit rich bifurcation phenomena including chaos and complicated superstable periodic orbits (SSPOs). We then introduce parallel converters using a switching rule based on the winner-take-all (WTA) principle [9, 10]. The WTA is a nonlinear function of multi variables that is applied effectively to various systems including self-organizing neural networks and multiplex communications [14]. In our paralleled converters the WTA switching can cause multi-phase synchronization automatically. The synchronous state is important to realize current sharing with smaller ripple for low-voltage high-current capabilities. In order to analyze nonlinear dynamics we use very simple models having piecewise constant (PWC) vector field [15–17]. The PWC models have piecewise linear (PWL) trajectories and can be justified in the case where the switching clock is much faster than time constant of the load. The PWC models are well suited for precise analysis and we can give parameter condition for existence and stability of the synchronous phenomena. Typical nonlinear phenomena are confirmed using a simple test circuit where the WTA rule is implemented by digital circuits.

2 Buck Converter

Figure 2 shows a basic buck converter with current mode control (CMC) that is a building block of the parallel system. Under the CMC, when the switch is on, the inductor current i rises, and when it reaches a reference value I_{ref}, the switch S is turned off. When S is off D turns on, and i decays. It is turned on by the arrival of the next rising edge of a free running clock signal of period T. While the switch is off, if the inductor current decays to zero, the system enters the third state in which both S and D do not conduct. Thus there can be three possible states:

State 1: S conducting, D blocking and $0 < i < I_{\mathrm{ref}}$
State 2: S blocking, D conducting and $0 < i < I_{\mathrm{ref}}$
State 3: S and D both blocking and $i = 0$

If the operation of the converter includes State 3, it is said to be operating in DCM, otherwise it is said to be operating in CCM. For simplicity, we assume that the voltage regulation is achieved in high frequency modulation ($T << RC$). This assumption enables us to replace the RC with the constant voltage source V_2 ($< V_1$). The circuit dynamics is described by Eq. (1).

$$L\frac{di}{dt} = \begin{cases} V_1 - V_2 & \text{for State 1} \\ -V_2 & \text{for State 2} \\ 0 & \text{for State 3} \end{cases} \qquad \begin{array}{l} \text{Switching Rules:} \\ \text{State 1} \rightarrow \text{State 2: when } i = i_{\mathrm{ref}} \\ \text{State 2} \rightarrow \text{State 3: when } i = 0 \\ \text{State 2 or 3} \rightarrow \text{State 1: when } t = nT. \end{array} \tag{1}$$

where all the circuit elements are assumed to be ideal. Using the following dimensionless variables and parameters:

$$\tau = \frac{t}{T}, \; x = \frac{i}{I_{\mathrm{ref}}}, \; a = \frac{T(V_1 - V_2)}{LI_{\mathrm{ref}}}, \; b = \frac{TV_2}{LI_{\mathrm{ref}}}. \tag{2}$$

Eq. (1) is transformed into Eq. (3)

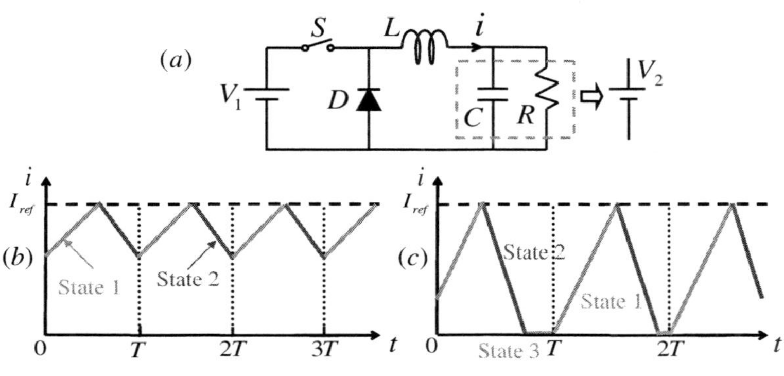

Fig. 2 PWC modeling of the buck converter and switching rule

$$\frac{dx}{d\tau} = \begin{cases} a & \text{for State 1} \\ -b & \text{for State 2} \\ 0 & \text{for State 3} \end{cases} \qquad \begin{array}{l} \text{Switching Rules:} \\ \text{State 1} \rightarrow \text{State 2: when } x = 1 \\ \text{State 2} \rightarrow \text{State 3: when } x = 0 \\ \text{State 2 or 3} \rightarrow \text{State 1: when } \tau = n. \end{array} \tag{3}$$

Note that the original five parameters (T, L, I_{ref}, V_1, V_2) of the PWC model are integrated into the dimensionless two parameters $a > 0$ and $b > 0$. Figure 3 shows typical waveforms: as b increases, the periodic orbits in CCM is changed into chaotic orbits and then to various super-stable periodic orbits in DCM (Fig. 4).

In order to analyze the dynamics we derive the 1-D return map which is a key tool to understand complex nonlinear dynamics [18–20]. Let τ_n denote n-th switching moment at which x reaches the threshold 1 and State 1 is changed into State 2. Since τ_{n+1} is determined by τ_n, we can define 1-D map $\tau_{n+1} = f(\tau_n)$. Since the system is period 1, we introduce phase variable $\theta_n = \tau_n$ mod 1 and the map can be

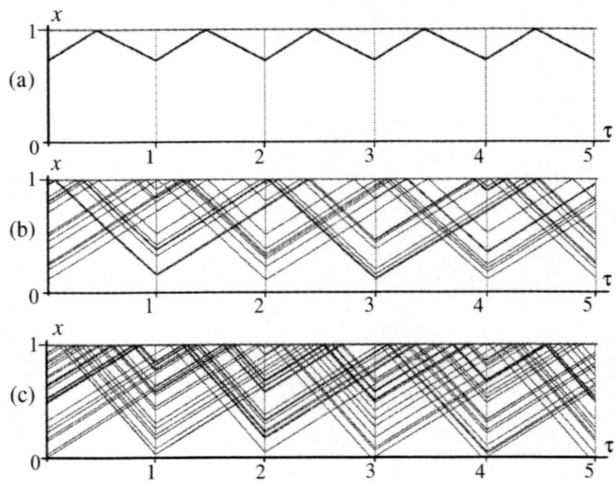

Fig. 3 Typical waveforms for $a = 0.6$. (**a**) periodic operation in CCM for $b = 0.5$, (**b**) chaos for $b = 0.9$ (superposed), (**c**) complicated periodic operation in DCM for $b = 1.1$ (superposed)

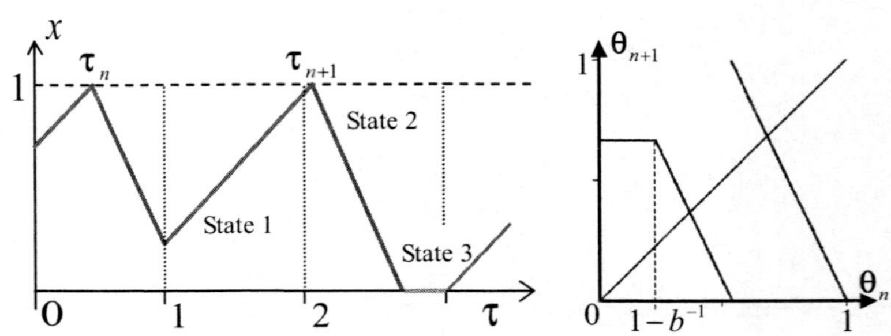

Fig. 4 Definition of the return map

reduced into the return map from $I \equiv [0,1)$ to itself: $\theta_{n+1} = F(\theta_n) = f(\theta_n) \bmod 1$. When a trajectory starts from $x = 1$ at time τ_n and returns $x = 1$ at time τ_{n+1}, there are two possibilities; Type 1: x does not reach $x = 0$ and Type 2: x reaches $x = 0$. If $b > 1$ these two types exist and the map is described by

$$\theta_{n+1} = f(\theta_n) = \begin{cases} -p(\theta_n - 1) + 1 & \text{for } 1 - b^{-1} < \theta_n \leq 1 \\ a^{-1} + 1 & \text{for } 0 < \theta_n \leq 1 - b^{-1} \end{cases} \tag{4}$$

where $p \equiv b/a$. If $0 < b < 1$, Type 2 does not exist hence the zero-slope branch can not exist. It should be noted that the return map of phase θ is simpler than the return map of sampled inductor current in [17]. In order to grasp the nonlinear dynamics we give some definitions.

Definition 1. For the map F, a point x_f is said to be a fixed point (or 1-periodic point) if $x_f = F(x_f)$. A point x_p is said to be a periodic point with period k if $x_p = F^k(x_p)$ and $x_p \neq F^l(x_p)$ for $1 \leq l < k$ where $F^l(x_p) = F(F^{l-1}(x_p))$ and $F^0(x) \equiv x$. A sequence of periodic points, $\{F(x_p), \cdots F^k(x_p)\}$, is said to be a periodic orbit. The periodic orbit is said to be unstable, stable and superstable for initial state if $|DF^k(x_p)| > 1$, $|Df^k(x_p)| < 1$ and $Df^k(x_p) = 0$, respectively, where $DF^k(x_p)$ is the slope of F^k at x_p.

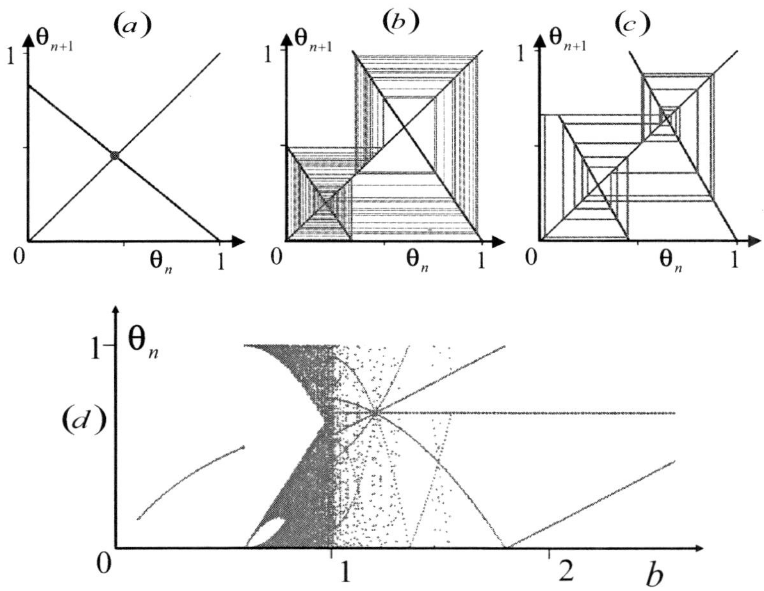

Fig. 5 Return maps for $a = 0.6$ corresponding to Fig. 3. (**a**) periodic orbit for $b = 0.5$, (**b**) chaos for $b = 0.9$, (**c**) SSPO for $b = 1.1$, (**d**) Bifurcation diagram

Definition 2. If there exists some positive integer l and $|DF^l(x_0)| > 1$ is satisfied for almost all $x_0 \in I$ then F is expanding and can not have stable periodic orbit. In this case F has positive Lyapunov exponent independently of initial values $x_0 \in I$ and the map exhibits chaotic orbit for almost all initial value in I. Such a map is said to be chaotic.

Figure 5 shows typical examples of the return map and bifurcation diagram for b. If $a > b$ the return map has contracting slope and has a stable fixed point as shown in Fig. 5 (a). If $a < b < 1$ the map has expanding slope $|p| > 1$ and exhibits chaos as shown in Fig. 5 (b). In terms of the original system parameters, it says that the buck converter loses stability when $V_2/(V_1 - V_2) = 1$. Such chaotic behavior may be suppressed, however, there exist works of EMC improvement using chaos [21]. If $b > 1$ the map has zero-slope and exhibits SSPOs as shown in Fig. 5 (c). They correspond to very robust periodic waveform in DCM. Although these SSPOs are superstable for initial state, they can be vary sensitive for parameters. It should be noted that the SSPOs becomes hard to observe as their period increases in practical systems. Because an orbit is expanding during long time interval until it reaches the "zero-slope branch" and the zero-slope can not be realized in practical systems.

3 Paralleled Buck Converters

Figure 6 shows the paralleled N buck converters. Each converter has a switch S_j and a diode D_j that can take either State 1, State 2 or State 3 defined in Sect. 2. Except for the switching to State 1, the switching rule of S_j and D_j are the same as that of the single converter defined in Eq. (1). Then the j-th converter is changed into State 1 at $t = nT$ and is connected to the input if the current i_j of the j-th converter in State 2 or State 3 is the minimum at $t = nT$ among all the currents. We regard the converter with the minimum current as a winner at time nT and refer to this switching rule as to be WTA-based. Note that co-winners can exist only if they are in State 3 at time nT. N-converters are coupled through this WTA-based switching.

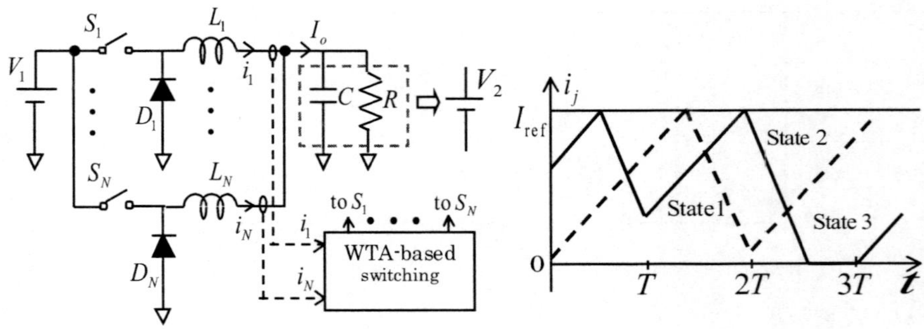

Fig. 6 Paralleled buck converters with WTA-based switching

As is in the single converter, we apply the assumption $T << RC$ and replace the load with a constant voltage source V_2. We also assume $L_1 = \cdots = L_N \equiv L$ for simplicity. The resulting PWC model is described by Eq. (5).

$$\frac{dx_j}{d\tau} = \begin{cases} a & \text{for State 1} \\ -b & \text{for State 2} \\ 0 & \text{for State 3} \end{cases} \quad \begin{array}{l} \text{Switching Rules:} \\ \text{State 1} \to \text{State 2: when } x_j = 1 \\ \text{State 2} \to \text{State 3: when } x_j = 0 \\ \text{State 2 or State 3} \to \text{State 1: when } x_j\text{=min at } \tau = n. \end{array} \tag{5}$$

where the following dimensionless parameters and variables are used.

$$\tau = \frac{t}{T},\ x_j = \frac{i_j}{I_{\text{ref}}},\ a = \frac{T}{LI_{\text{ref}}}(V_1 - V_2),\ b = \frac{T}{LI_{\text{ref}}}V_2. \tag{6}$$

where $(j = 1 \sim N$. The output current $i_o \equiv \sum_{j=1}^{N} i_j$ is normalized as

$$X(\tau) \equiv \sum_{j=1}^{N} x_j(\tau) = \sum_{j=1}^{N} \frac{1}{I_{\text{ref}}} i_j(\tau). \tag{7}$$

If X is period N, $X(\tau + N) = X(\tau)$, we measure the ripple by the ripple factor:

$$R_q = \max_{\tau} X(\tau) - \min_{\tau} X(\tau),\ \text{for } 0 \le \tau < N. \tag{8}$$

Figure 7 shows typical waveforms of the paralleled converters: the WTA-switching can distribute phases automatically and can realize multi-phase synchronization. In DCM the synchronization can be achieved if the orbit falls into State 3 once: transient to the synchronization in DCM is much faster than that of CCM.

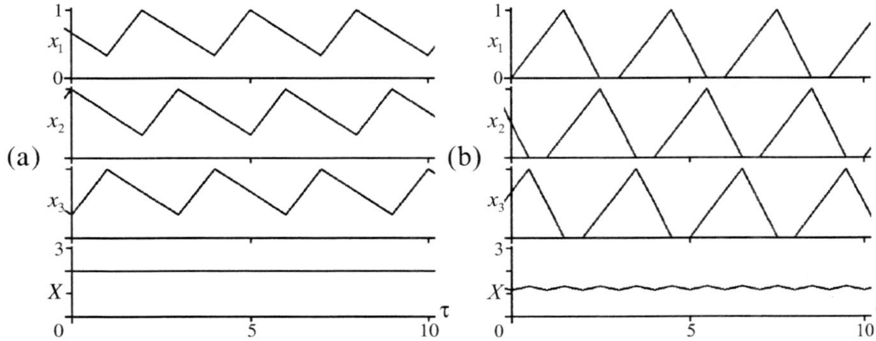

Fig. 7 Synchronous phenomena for $a = 0.67$. (**a**) 3-phase synchronization in CCM for $b = 0.33$. $R_q \doteq 0$. (**b**) 3-phase synchronization in DCM for $b = 1$. $R_q = 0.17$

However, ripple in CCM is usually much smaller than that in DCM as suggested in Fig. 7 (a). Here we give some definitions.

Definition 3. A solution $(x_{p1}, \cdots, x_{pN})$ is said to be an N-phase synchronization (N-SYN) if Eq. (9) is satisfied.

$$x_{pj}(\tau) = x_{pj}(\tau + N), \text{ for } j \in \{1, \cdots, N\}, \ x_{pj}(\tau) = \text{ winner at } \tau = p(j) \tag{9}$$

where $p(j) \in \{1, \cdots, N\}$ is a switching instant at which x_{pj} is the unique winner and $p(j) \neq p(k)$ for $j \neq k$. In the j-th converter, (S_j, D_j) is changed from State 2 (or State 3) to State 1 at time $t = p(j)T$ and repeats it with period NT. For example, 3-SYN in Fig. 7 (a) is characterized by $(p(1), p(2), p(3))$=$(1,2,3)$.

Definition 4. An N-SYN is said to be stable if a solution converges to the N-SYN for a small initial perturbation $\varepsilon_j(0)$:

$$x_j(\tau) \to x_{pj}(\tau) \ \ (\tau \to \infty) \ \ \text{for} \ \ x_j(0) = x_{pj}(0) + \varepsilon_j(0). \tag{10}$$

An N-SYN is said to be superstable if the N-SYN is achieved by some finite time:

$$x_j(\tau) = x_{pj}(\tau) \text{ for } \tau > \tau_f, \ x_j(0) = x_{pj}(0) + \varepsilon_j(0). \tag{11}$$

Performing simple geometrical calculation of the PWL solution, we obtain parameters condition for existence and stability of N-SYN [9]. First, an N-SYN exists in the CCM if

$$N < a^{-1} + b^{-1} \tag{12}$$

The N-SYN is stable for $b < a$ and is unstable if $a < b$. The condition for the stability is the same as that for a fixed point of the phase map of the single converter. When N-SYN in CCM is unstable, the system has a variety of synchronous/asynchronous phenomena. Figure 8 illustrate parameters conditions for

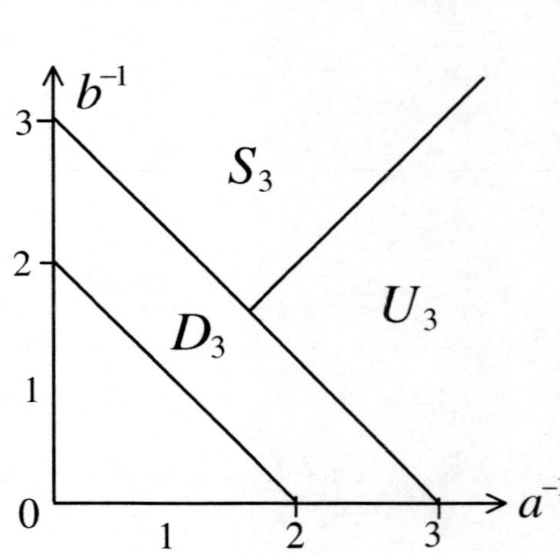

Fig. 8 Parameter condition for $N = 3$. S_3: Stable 3-SYN in CCM, U_3: Unstable 3-SYN in CCM, and D_3: Superstable 3-SYN in DCM

$N = 3$. Although the N-SYN is unstable for $a < b$, the N-SYN can be stabilized by an complementary switching rule in [10]. Second, an N-SYN exists in the DCM if

$$N - 1 < a^{-1} + b^{-1} < N \tag{13}$$

In the DCM, the N-SYN is always superstable because all the orbits traces the same route just after they fall into State 3. The superstable orbits corresponds to the zero-slope of the phase map of single converters.

4 Experiments

From viewpoints of circuit engineering, hardware experiments are important even if some theoretical and/or numerical results have been obtained. We enumerate major significance of the hardware experiments.

1. The theoretical stability must guarantee observability in a practical hardware thus the phenomena should be confirmed experimentally.
2. Although the PWC models are idealized systems, the hardware experiment bridge between ideal model and practical system design.
3. The PWC models are simple switched dynamical systems having rich nonlinear phenomena. Analysis of them may be a trigger to discover novel phenomena and/or to develop novel bifurcation theory. Historically, we know several facts

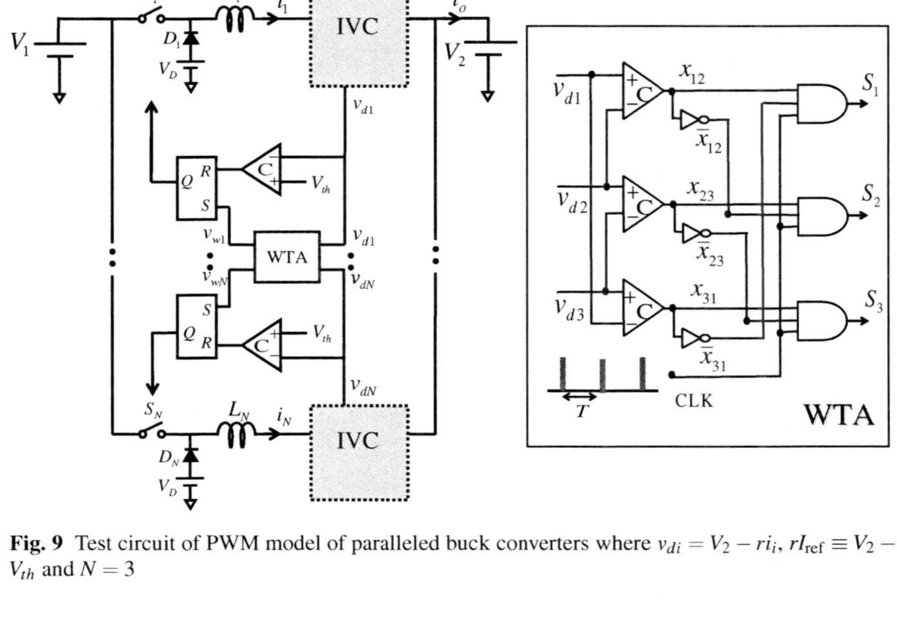

Fig. 9 Test circuit of PWM model of paralleled buck converters where $v_{di} = V_2 - ri_i$, $rI_{\text{ref}} \equiv V_2 - V_{th}$ and $N = 3$

that important phenomena are observed in hardware experiments first and then have developed into some interesting contents in nonlinear dynamical systems theory [1, 2].

Figure 9 shows the PWC model of WTA-based paralleled buck converters. Each inductor current is transformed into a voltage by the current-voltage converter IVC. The voltage is applied to the WTA circuits and a comparator. The WTA function is realized through comparators and logical gates. The outputs of the WTA and comparators are applied to set and reset terminals of each flip-flop, respectively. The outputs of the flip-flops control switches S_1 to S_N. Using this circuit we have confirmed various synchronous/asynchronous phenomena. Figure 10 (a) shows 3-SYN in CCM where almost complete ripple reduction is achieved. In the CCM the circuit can exhibit chaotic behavior for $a > b$ (Fig. 10 (b)). In the DCM the N-SYN is superstable and the transient is very fast, however, the ripple reduction is hard.

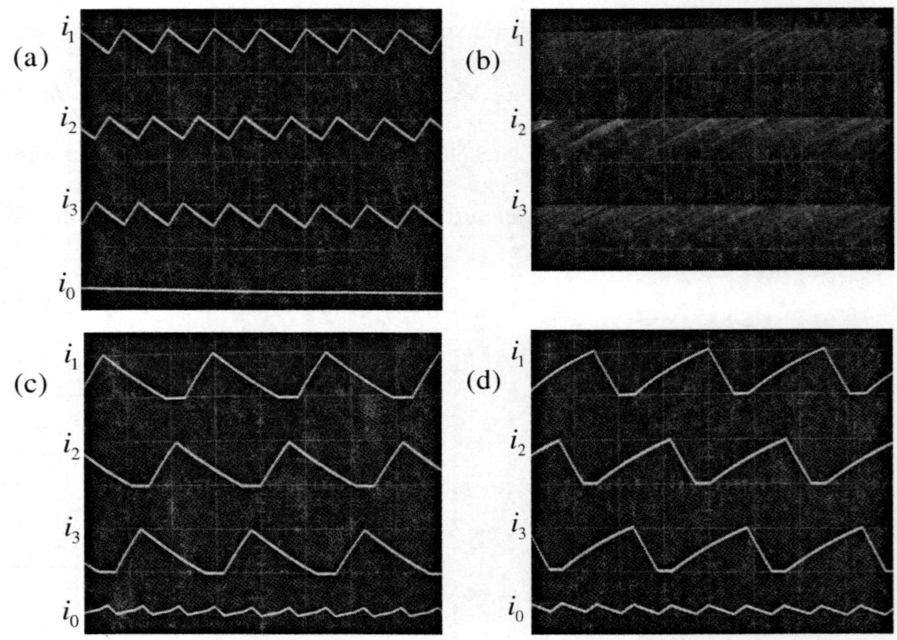

Fig. 10 Observed waveforms of PWC model of paralleled buck converters for $N = 3$, $L_j \doteq L \doteq 100$[mH] and $I_{ref} = 5$[mA]. (**a**), (**c**) and (**d**) correspond to Fig. 7 (**a**), (**b**) and (**c**), respectively. Horizontal: 0.2 ms/div, Vertical: 5 mA/div. (**a**) 3-SYN in CCM for $(a^{-1}, b^{-1}) = (1.6, 4.9) \in S_3$. $V_1 \doteq 6.3$[V], $V_2 \doteq 1.5$[V], $T \doteq 0.066$[ms]. (**b**) Chaotic orbit for $(a^{-1}, b^{-1}) = (3.5, 1.7) \in U_3$. $V_1 \doteq 6.3$[V], $V_2 \doteq 1.5$[V], $T \doteq 0.066$[ms]. (**c**) 3-SYN in DCM for $(a^{-1}, b^{-1}) = (1.4, 0.7) \in D_3$. V_1 and V_2 are as in (**a**). $T \doteq 0.17$[ms]. (**d**) 3-SYN in DCM for $(a^{-1}, b^{-1}) = (1.4, 0.7) \in D_3$. V_1 and V_2 are as in (**b**). $T \doteq 0.17$[ms]. S_3, U_3 and D_3 are shown in Fig. 3

5 Conclusions

Nonlinear dynamics of parallel dc–dc converters have been discussed. In the single converter, the dynamics can be integrated into a PWL 1D return map. In the CCM, the system exhibits periodic and chaotic behavior. In the DCM, the map has zero-slope and has complicated SSPOs. In the paralleled converters, the WTA-switching can realize N-SYN automatically. The PWC model enables us to give parameter conditions for existence and stability of the N-SYN. Based on theoretical and experimental results, we can say that current sharing with small ripple is possible for stable N-SYN in CCM, however, the N-SYN becomes unstable and may be chaotic in the CCM. The circuit can exhibit superstable N-SYN in DCM. Although the transient to N-SYN is very fast in the DCM, the ripple reduction property is less good then that in CCM. These results provide useful information for both practical and fundamental studies.

The future problems are many and include the following.

1. Detailed analysis of bifurcation phenomena. We should approach to understand an underlying scenario. The parallel system is very rich in bifurcation. In the CCM, bifurcation between N-SYN and hyperchaotic behavior is interesting.
2. Efficient hardware implementation. It is important to compare analog switching schemes with digital ones. The WTA-based switching can be realized by both analog and digital circuits. Also, there exist a variety of analog- and digital-based switching circuits for current sharing. These circuits are compared to each other by evaluating important factors: transients on the phase-shift of the modules, the sensitivity of the system to parameter variations, , conversion efficiency, and so on.
3. There exist a variety of paralleled dc–dc converters: current/voltage mode control, buck/boost type and single/muitiple input converters. The resuts of this paper should be developed into wider class of systems. s. Especially paralleled converters with multiple inputs [22, 23] are important in clean energy technology. For example, energy input from a solar panel requires adaptive conversion from time-variant multi inputs to a load.

References

1. Banerjee, S. and Verghese, G. C., eds. (2001). Nonlinear Phenomena in Power Electronics: Attractors, Bifurcations, Chaos, and Nonlinear Control. *IEEE Press*.
2. Tse, C. K. and di Bernardo, M. (2002). Complex behavior in switching power converters. *Proc. IEEE*, **90**, pp. 768–781.
3. Giral, R., Martinez-Salamero, L. and Singer, S. (1999). Interleaved converters operation based on CMC. *IEEE Trans. Power Electron.*, **14**, 4, pp.643–652.
4. Giral, R., Martinez-Salamero, L., Leyva, R. and Maixe, J. (2000). Sliding-mode control of interleaved boost converters, *IEEE Trans. Circuits Systs. I*, **47**, pp. 1330–1339.

5. Zhou, X., Xu, P. and Lee, F. C. (2000). A novel current-sharing control technique for low-voltage high-current voltage regulator module applications. *IEEE Trans. Power Electron.*, **15**, pp.1153–1162.
6. Mazumder, S. K., Nayfeh, A. H. and Borojevic, D. (2002). Robust control of parallel DC-DC buck converters by combining integral-variable-structure and multiple-sliding-surface control schemes. *IEEE Trans. Power Electron.*, **17**, pp. 428–437.
7. Mazumder, S. K., Tahir, M and Kamisetty, S. L. (2005). Wireless PWM control of a parallel DC/DC buck converter, *IEEE Trans. Power Electron.*, **20**, pp. 1280–1286.
8. Abu-Qahouq, J., Mao, H. and I. Batarseh, I. (2004). Multiphase voltage-mode hysteretic controlled dc-dc converter with novel current sharing *IEEE Trans. Power Electron.*, **19**, 1, pp. 1397–1407.
9. Saito, T., Tasaki, S. and Torikai, H. (2005). Interleaved buck converters based on winner-take-all switching. *IEEE Trans. Circuits Syst. I*, **52**, pp. 1666–1672.
10. Ishikawa, Y. and Saito, T. (2007). WTA-based Switching Strategy for Paralleled DC–DC Converters. *Proc. NOLTA*.
11. Branicky, M. S. (1998). Multiple Lyapunov functions and other analysis tools for switched and hybrid systems, *IEEE Trans. Automat. Contr.*, **43**, pp. 475–482.
12. Saito, T., Torikai, H. and Schwarz, W. (2000). Switched dynamical systems with double periodic inputs: an analysis tool and its application to the buck-boost converter. *IEEE Trans. Circuits Syst. I*, **47**, pp.1038–1046.
13. Iu, H. H. C. and Tse, C. K. (2001). Bifurcation behavior in parallel-connected buck converters, *IEEE Trans. Circuits Systs. I*, **48**, pp. 233–240.
14. Torikai, H., Saito, T. and Schwarz, W. (1999). Synchronization via multiplex pulse-train. *IEEE Trans. Circuits Syst. I*, **46**, pp.1072–1085.
15. Hamill, D. C. and Jeffries, D. J. (1988). Subharmonics and chaos in a controlled switched-mode power converter, *IEEE Trans. Circuits Systs. I*, **35**, pp. 1059–1061.
16. Deane, J. H. B., Ashwin, P., Hamill, D. C. and Jeffries, D. J. (1999). Calculation of the periodic spectral components in a chaotic DC–DC converter. *IEEE Trans. Circuits Systs. I*, **46**, pp.1313–1319.
17. Kabe, T., Parui, S., Torikai, H., Banerjee, S. and Saito, T. Analysis of current mode controlled DC-DC converters through piecewise linear models. *IEICE Trans. Fundamentals*, 2007.
18. Hamill, D. C., Deane, J. H. B. and Jeffries, D. J. (1992). Modeling of chaotic dc-dc converters by iterated nonlinear mappings. *IEEE Trans. Power Electron.*, **7**, pp. 25–36
19. Ott, E. (1993). Chaos in dynamical systems. *Cambridge*.
20. Sharkovsky, A. N. and Chua, L. O. (1993). Chaos in some 1-D discontinuous maps that appear in the analysis of electrical circuits, *IEEE Trans. Circuits Systs.* , **40**, pp. 722–731.
21. Deane, J. H. B. and Hamill, D. C. (1996). Improvement of power supply EMC by chaos. *Electron. Lett.* **32**, 12, p. 1045.
22. Matsuo, H., Lin, W., Kurokawa, F., Shigemizu, T. and Watanabe N. (2004). Characteristics of the Multiple-Input DC-DC onverter. *IEEE Trans. Ind. Electron.*, **51**, pp. 625–631.
23. Solero, L., Lidozzi, A. and Pomilio, J. A. (2005). Design of multiple-input power converter for hybrid vehicles. *IEEE Trans. Power Electron.*, **20**, pp. 1007–1016, 2005.

Coupled Nonlinear Oscillator Array (CNOA) Technology – Theory and Design

Ted Heath, Robert R. Kerr and Glenn D. Hopkins

Abstract This presentation[1] reviews the Georgia Tech Research Institute's (GTRI) contributions to the field of nonlinear oscillator antennas. Over the last seven years, GTRI, in collaboration with SPAWAR Systems Center in San Diego, has advanced the state of the art in the analysis, capability and design of coupled oscillator arrays. A theoretical framework describing the dynamics of coupled oscillators and its relevance to beam steering, beam shaping, monopulse and null steering is presented. Descriptions of 1- and 2-dimensional array designs are provided, including critical design issues encountered during their development. Performance assessments via range chamber testing and on-board diagnostic measurements are included. This paper concludes with a brief discussion of outstanding problems, directions of further study and possible insertion points for this technology.

1 Introduction

Survival requires is the ability to react to opportunities or threats present in one's surroundings; this necessitates some means by which one can make (and process) observations of the environment. Depending on the species, animals achieve this through sight, sound, smell or a combination thereof. Humans cannot exempt themselves from this fundamental need. Whether for protection from invasion, avoidance of hazardous weather, or to enable safe travel, knowledge of one's current environment remains imperative.

T. Heath (✉)
Sensors and Electromagnetic Applications Laboratory, Georgia Tech Research Institute, 7220 Richardson Road, SE, Smyrna, GA 30080

[1] This review is a modified version of that presented at the 2006 IEEE Aerospace Conference, "Nonlinear oscillator array antenna development at GTRI," Big Sky, MT, March 4–11, 2006.

V. In et al. (eds.), *Applications of Nonlinear Dynamics*, Understanding Complex Systems, 145
DOI 10.1007/978-3-540-85632-0_12,

Interrogation of one's environment can either be passive or active (or both). Sight is a common example of a passive means of observing the environment; the eye collects light scattered from objects. Bats, by contrast, utilize an active method: sonar. The advantage sonar provides for bats is the ability to hunt in absolute darkness via a source under their complete control. Animals relying on sight require a light source typically beyond their control (i.e., daylight, moonlight).

RADAR (RAdio Detection and Ranging) is a technology by which radio waves are transmitted and, by monitoring the returns scattered off of the surroundings, the existence (detection) and location (ranging) of targets of interest are established. Much like bat's sonar, RADAR's advantage is the ability to perform where optical techniques cannot by providing its own signal source (although there are passive radars) at frequencies that can "see" through certain obstructions (e.g., clouds, fog).

The transmission and reception of electromagnetic signals implies the need to control the directional sensitivity of an aperture. Electronic beam steering (a.k.a. inertia-less beam steering) alters the direction of maximum sensitivity by introducing and controlling a specified phase distribution across the aperture. This technique can improve scan speed by several orders of magnitude over that provided by mechanical means, while also allowing for rapid alteration of the sensitivity as a function of angle of arrival.

The directional sensitivity of a 1-d array of discrete antenna elements (known as the array factor), is given by:

$$E(\theta,t) = e^{-i\omega t}\sum_{n=0}^{N-1} A_n e^{-i\left(\frac{2\pi nd}{\lambda}\sin\theta - \phi_n\right)} \tag{1}$$

where (A_n, ϕ_n) are the amplitude and phase applied to the nth array element. To achieve complete constructive interference in a particular direction, Ψ_s, the required element-level phases must be

$$\Phi_n = \frac{2\pi nd}{\lambda}\sin\Psi_s \tag{2}$$

where $n = 0,1,2,\ \ldots,\ N-1$. This is a ***linear phase gradient*** as a function of array element number. This linear phase gradient effectively "shifts" the antenna pattern by compensating for geometry-induced time delays.

Typically, such a phase distribution is introduced and controlled via devices known as phase shifters. However, a phase shifter's size, weight, losses and cost can, for some applications, prohibit their use (especially at millimeter wave frequencies). As an alternative, several phase-shifterless techniques exist, most notably frequency scanning and multiple beam former networks (e.g. Rotman lenses, Butler and Blass matrices). More recently, the viability of exploiting the synchronization properties of coupled, nonlinear oscillator arrays for phase-shifterless beam steering, beam forming and signal processing has been explored.

2 Coupled Oscillator Array for Beam Steering

In 1986, Stephan and Morgan found that injecting phase shifted signals at the ends of a 1-d coupled oscillator array resulted in a linear phase gradient across the array [1]. However, the total amount of beam steering obtainable by this approach was inversely proportional to the array size. In 1993, Robert York's group at UCSB devised an alternative technique which avoided the limitations of Stephan's approach [2]. York's method did not involve external signal injection at the ends of the array, but rather small changes in bias voltage to the periphery elements. Over the last 14 years, several independent groups have verified and expanded upon York's work [3–7].

One of many significant contributions to this field by York's group was the derivation of amplitude and phase dynamical equations describing a class of coupled, nonlinear oscillators with a Van der Pol nonlinearity [8]. For a 1-dimensional array with nearest-neighbor coupling, the amplitude and phase dynamics of such an array are given by:

$$
\begin{aligned}
\frac{dA_j}{dt} &= \left(p_j - A_j^2\right)A_j + \kappa\left[A_{j+1}\cos\left(\phi_{j+1}-\phi_j+\Phi\right) + A_{j-1}\cos\left(\phi_{j-1}-\phi_j+\Phi\right)\right] \\
\frac{d\phi_j}{dt} &= \omega_j + \kappa\left[\tfrac{A_{j+1}}{A_j}\sin\left(\phi_{j+1}-\phi_j+\Phi\right) + \tfrac{A_{j-1}}{A_j}\sin\left(\phi_{j-1}-\phi_j+\Phi\right)\right]
\end{aligned}
\tag{3}
$$

where $j = 1,\ldots,N$, ω_j and κ denote the natural frequency and coupling strength, respectively. The coupling phase, Φ, represents the time delay and phase contribution of the transmission line coupling between the elements. The amplitude parameters, p_j, describe the ratio of energy gain (from negative resistance) to energy loss (from resistance) and, implicitly, the steady-state amplitude of the oscillators.

By assuming that the oscillator amplitudes rapidly (relative to the time scale of the phase dynamics) evolve to a common, steady-state value, $A_j \to A$, the amplitude and phase dynamics de-couple, leading to the Phase Model:

$$
\frac{d\phi_j}{dt} = \omega_j + \kappa\left[\sin\left(\phi_{j+1}-\phi_j+\Phi\right) + \sin\left(\phi_{j-1}-\phi_j+\Phi\right)\right] \tag{4}
$$

A majority of the work in the field of CNOA technology is based on this simplified model.

For beam steering, we seek conditions on the natural frequencies that ensure the existence of linear phase gradient solutions, $\Phi_{j+1} - \Phi_j = \theta$. Substitution of this desired solution into (4) yields

$$
\omega_j = \begin{cases} \omega - \kappa\sin(\theta+\Phi) & j=1 \\ \omega & j=2,\ldots,N-1 \\ \omega + \kappa\sin(\theta-\Phi) & j=N \end{cases} \tag{5}
$$

where ω is a common natural frequency for the interior elements. This is the key result of York's technique: to beam steer a coupled nonlinear oscillator array, ***only the two periphery elements need to be detuned***. In other words, beam steering can be accomplished via two controls rather than requiring a phase shifter at each element. This reduction in the number of controls can become significant, especially for large, 2-dimensional arrays. This result does, in fact, generalize to the 2-dimensional array: only the periphery elements need to be detuned relative to the interior elements, leading to a "row-column" type of control.

Proving the existence of the desired solutions is only half the battle; for a practical means of beam steering, such desired solutions must be stable. A linear stability analysis applied to the Phase Model reveals that perturbations about the linear phase gradient solution evolve according to:

$$\frac{d\eta_j}{dt} = a\eta_{j+1} + b\eta_j + c\eta_{j-1} \tag{6}$$

where

$$\begin{aligned} a &\equiv \kappa\cos(\theta+\Phi) \\ c &\equiv \kappa\cos(\theta-\Phi) \\ b &\equiv -(a+c) \end{aligned} \tag{7}$$

$\eta_0 = \eta_1$ and $\eta_{N+1} = \eta_N$ are the appropriate boundary conditions.

Forgoing the details in the interest of space, it can be shown that the Lyapunov exponents are

$$\lambda_n = -2\kappa\cos\Phi\cos\theta_* \left(1 - \left[\cos\frac{\pi n}{N}\right]\sqrt{1-\tan^2\Phi\tan^2\theta_*}\right) \tag{8}$$

where $n = 0, \ldots, N-1$ and θ_* denotes the desired linear phase gradient value. Most importantly, the sign of the eigenvalues are determined uniquely by $\cos\Phi\cos\theta_*$. Presuming that the coupling phase is such that $\cos\Phi > 0$, then the range of stable phase gradient values is $\theta_* \in \left(-\pi/2, \pi/2\right)$. This means that a coupled, nonlinear oscillator array has a limited scan range. In particular, for half-wavelength spacing between elements, the maximum achievable scan range is $\pm 30°$ off broadside. However, by introducing frequency doublers (or triplers) after the CNOA, a greatly enhanced scan range can be obtained. It should be noted that, due to the linearity of the stability analysis, the 2-dimensional, rectangular array with nearest-neighbor coupling case separates into two, independent 1-dimensional stability analysis problems.

3 Monopulse, Nulling and Beam Shaping

In monopulse applications such as target tracking, the generation and scanning of difference patterns (Fig. 1) is imperative for use in controlling an error-correcting closed loop servo system. Difference patterns arise from the following phase distribution:

$$\phi_j = \phi_1 + (j-1)\,\theta + h_j \tag{9}$$

where

$$h_j \equiv \begin{cases} \pi & j > N/2 \\ 0 & j \le N/2 \end{cases} \tag{10}$$

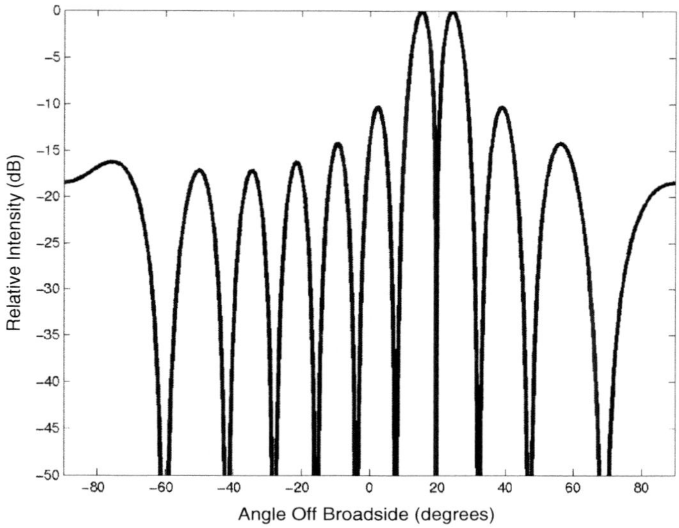

Fig. 1 A difference pattern. The central null (surrounded by the two maximal lobes) is used to maintain track on a specified target

Apart from a discontinuous jump by π between the two halves of the array, (10) is simply a linear phase gradient.

Similar to the above exercise with sum-pattern beam steering, having defined the desired phase distribution, what remains is a determination of what conditions on the natural frequencies are necessary to ensure existence, followed by a linear stability analysis. Interestingly enough, it turns out that the detunings required for difference-pattern beam steering are the same as those required for sum-pattern steering. Moreover, the subsequent linear stability analysis dynamical equations can be shown to be identical to that for sum-pattern steering. Therefore, in solving the sum pattern steering problem, we have implicitly solved the difference pattern steering problem as well.

However, there is a crucial, additional requirement for difference pattern steering: the center coupling link must be negative (e.g., a π change introduced into the coupling phase of the center coupling link). This negative (or "repulsive") coupling results in the two halves of the array becoming π out of phase relative to one another. Recently, Yan and Chu provided experimental verification of such difference pattern steering.

Often, there is a need to mitigate a source of interference by placing a null of the array factor in that direction. Several techniques (exact and approximate) exist that define the necessary array element amplitudes and phases to generate the desired pattern. A technique by Steyskal uses perturbations to the element phases to create approximate (yet, in principal, very deep) nulls. Researchers at GTRI have developed a slightly modified version of Steyskal's technique for simultaneous beam steering and null generation in a 1-dimensional array of coupled, nonlinear oscillators [9]. Once again, the same approach used for sum and difference pattern beam steering is utilized here: the required phase distribution is defined (in this case, via a modified Steyskal nulling technique), the necessary natural frequencies are determined and a linear stability analysis performed.

Unfortunately, the phase distributions required for nulling are not as straightforward as those leading to simple beam steering. In general, a steered and nulled pattern will require element-level variations about a linear phase gradient (at least for the modified Steyskal technique). Mathematically, such phase distributions can be written as:

$$\phi_j = \phi_j^{(s)} + \varphi_j \tag{11}$$

where $\phi_j^{(s)}$ is the linear phase gradient required for steering and φ_j is the phase required for nulling. Steyskal's technique provides an algorithm for computing the φ_j necessary to place an approximate null in the desired direction. It should be noted that this particular nulling technique has certain limitations, most notably (1) it is only approximate (however, considering the presence of phase errors in any realizable system, this limitation is not necessarily onerous), (2) this technique cannot produce nulls in the mainbeam, (3) this phase-only technique cannot generate two nulls placed symmetrically about the main beam and (4) the number of specified nulls must be less than the number of elements in the phased array. Despite these constraints, Steyskal's technique is attractive in that it minimizes the phase variations required for nulling. It is conceivable that large phase variations could result in a loss of stability.

As an example of this technique, Fig. 2 demonstrates the ability to steer a seven-element coupled oscillator arra -20° off broadside while simultaneously placing a null at $+22^\circ$ degrees off broadside. The pattern was generated by integrating the Phase Model dynamical equations (with the natural frequencies suitably adjusted), starting from random initial conditions. That the pattern generated by the array dynamics agrees with the desired pattern shows that this phase distribution was stable.

Owing to the increased complexity of the phase distributions required for simultaneous nulling and steering, the linear stability analysis becomes very difficult to complete in closed form. Whereas linear phase gradients produced a set of coupled perturbation dynamical equations with constant coefficients independent of element number (note that the a,b,c in the linear stability analysis of the simple sum-pattern beam steering do not depend on j), such is not the case here. At present, the best one can do is to numerically compute the eigenvalues for a specified phase distribution.

Moreover, in contrast to simple beam steering, the generation of more complex phase distributions typically requires detuning of every element in the array. Even so, this work proves that these nonlinear arrays are capable of producing more than just linear phase gradients. The principal motivation for using Steyskal's phase perturbation technique was to maintain a phase distribution close to the stable solutions required for simple sum pattern beam steering; it was reasoned that, if the phase perturbations were not large, it seemed likely that the risk of loss of stability would

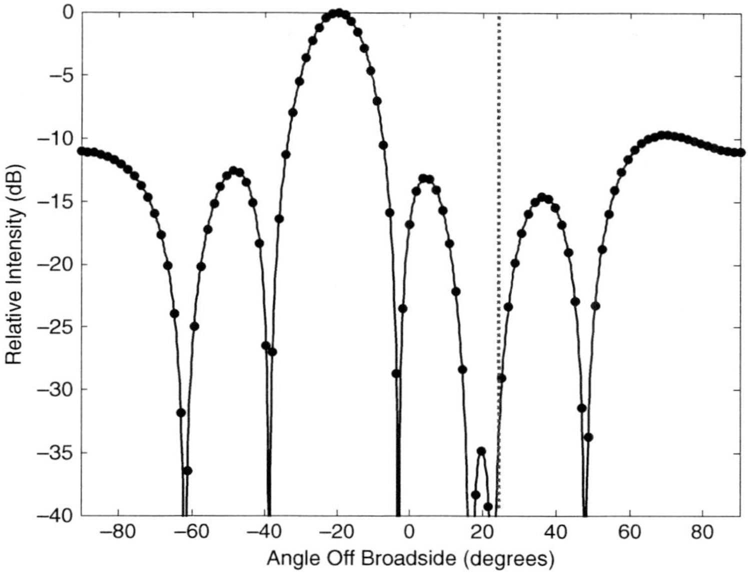

Fig. 2 Simultaneous beam steering and null generation. The red vertical line denotes the desired null direction. The dots are the array factor resulting from the CNOA dynamics; this demonstrates the capability of the CNOA to achieve the desired antenna pattern

be minimized. In the last couple of years, work by Suarez and Georgiadis have extended this nulling capability to include amplitude variations [10].

Having exploited the phase dynamics of coupled, nonlinear oscillators for beam steering and nulling, one might wonder if the amplitude dynamics could be exploited to provide beam shaping (sidelobe reduction). Work at GTRI has shown that it is indeed possible. Recall the key assumption that led to the Phase Model, namely that the oscillator amplitudes all rapidly (relative to the time scale of the phase dynamics) evolve to a common, steady-state value, $A_j \rightarrow A$. This assumption implies a uniform array which results in – 13 dB sidelobes. For many applications, such sidelobes are unacceptably high. By utilizing a non-uniform weighting across the aperture, the sidelobes can be reduced significantly, at the cost of broadening the mainbeam.

GTRI researchers proposed exploiting the amplitude dynamics of coupled, nonlinear oscillator arrays to produce the desired amplitude distribution $(A_j \rightarrow a_j)$ while exploiting the phase dynamics to achieve the desired phase distribution across the array. The amplitude parameters, p_j, allow for control of the oscillator amplitudes similar to how the natural frequencies are used to manipulate the oscillator phases. For convenience, (3) is re-written in an equivalent form using the complex quantity $z_j = A_j e^{i\phi_j}$:

$$\frac{dz_j}{dt} = \left(p_j + i\omega_j - |z_j|^2\right) z_j + \kappa e^{i\Phi}\left(z_{j+1} + z_{j-1}\right) \tag{12}$$

The approach used remains unchanged: (1) desired solutions are defined, (2) conditions for the existence of the desired solutions are determined (in this case, how the amplitude parameters and natural frequencies must be adjusted), and (3) the stability of the desired solutions is assessed.

The desired beam-steered and tapered solutions have the form: $z_j = a_j e^{i(\omega t + [j-1]\theta_*)}$ where the a_j are defined by the particular weighting scheme chosen. Substituting this desired solution into (12) leads to a set of N complex algebraic equations, the real and imaginary parts of which define how the p_j and ω_j should be adjusted, respectively. Note that the amplitude and phase dynamical equations are no longer de-coupled; consequently, simultaneous beam steering and shaping require adjustment to each element, not simply the periphery.

While the determination of the requisite p_j and ω_j is most readily obtained via (12), the linear stability analysis is easiest through (3). In this case, both the amplitudes and phases are perturbed about the desired solution. As was true for the nulling stability analysis, the resulting perturbation evolution equation coefficients are element-dependant and, therefore, a closed-form expression for the eigenvalues has yet to be found. However, computation of the eigenvalues is straightforward. Figure 3 shows the array factor generated from a 9-element array where –40 dB sidelobes were obtained via Villeneuve weighting and the mainbeam is steered off broadside. The comparison between the desired and observed pattern indicates that the nonlinear array evolved towards the desired amplitude and phase distributions.

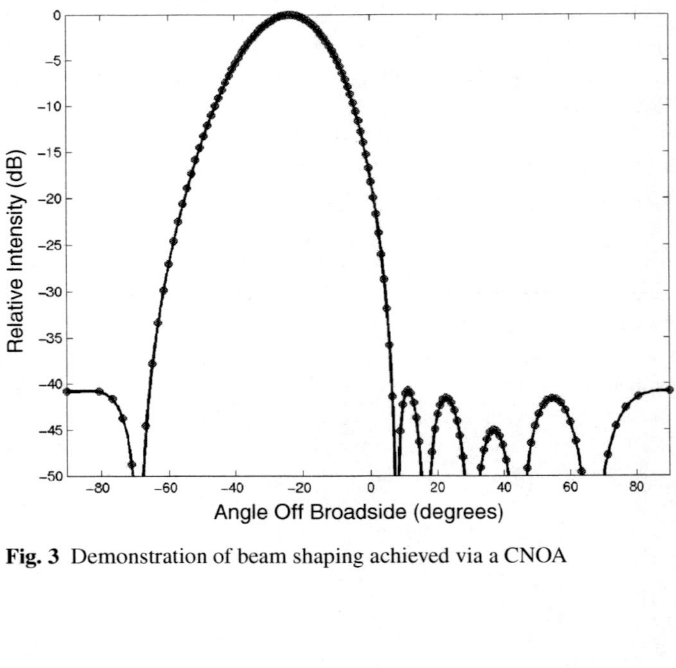

Fig. 3 Demonstration of beam shaping achieved via a CNOA

4 1-D and 2-D Array Designs

Both 1- and 2-D coupled oscillator arrays have been built at GTRI in order to provide test-beds for proof-of-principal demonstration of beam steering and pattern control. To date, all prototype hardware has been designed for testing at the GPS L1 frequency of 1.57542 GHz. In order to minimize cost, commercially available off-the-shelf (COTS), ceramic-loaded, microstrip patch radiators were procured. As a trade-off between performance and size, the Spectrum Controls part #PA18-1580-010SA radiator was chosen to populate the 2-D aperture. The element array radiators, spaced at one half of a wavelength, were individually fed with SMA coaxial connectors. Aluminum ground planes (7-gauge, measuring 36" × 18" × 0.188" for the 1-D array and 30" × 30" × 0.188" for the 2-D array) were used.

Two different unit cell oscillator designs have been developed using the Agilent Advanced Design System software (ADS). An industry-standard software package, ADS was used for schematic entry, circuit harmonic balance analysis, circuit parameter optimization and microstrip circuit layout. Owing to the fidelity of this modeling software, the simulation results were expected to faithfully predict actual circuit performance. Subsequent measurement of unit cell oscillator performance validated this expectation.

Through a method of multiple scales analysis of a Van der Pol oscillator, GTRI was able to derive a relationship between the oscillator spectral components and the Van der Pol parameters (ω, p, μ). In particular, it was found that

$$\begin{aligned} p &= \frac{1}{4} \times 10^{H_1/10} \\ \mu &= \frac{4}{p^{3/2}} \times 10^{H_3/20} \end{aligned} \tag{13}$$

where H_1 and H_3 is the power (expressed in dB) contained in the first and third harmonics, respectively. Obviously, ω is specified by the first harmonic. Ideally, a weakly nonlinear Van der Pol oscillator spectrum does not contain any even harmonics. However, any practical instantiation will possess some non-zero even harmonics. It is assumed that a particular oscillator design will be suitably "Van der Pol"-like provided this erroneous spectral content is "sufficiently" negligible.

In order to minimize the even harmonic spectral content, a differential-pair oscillator circuit topology was adopted for the 1-D array unit cell (Fig. 4). The oscillator output signal is the voltage difference between the collectors of two transistors. The two semi-circular strips act as inductors; their lower intersection is capacitively grounded, while a voltage-controlled variable capacitor connected across the gap between the upper ends of the semi-circle effectively forms two identical parallel resonant circuits. The collectors of the differential transistors are also connected to the two upper ends of the microstrip inductors at the high impedance points of these resonant circuits. The natural frequency of the oscillator is set by the control voltage on the capacitor connecting the two microstrip resonators. Two microstrip transmission lines are connected to the semi-circular inductors near the capacitive ground

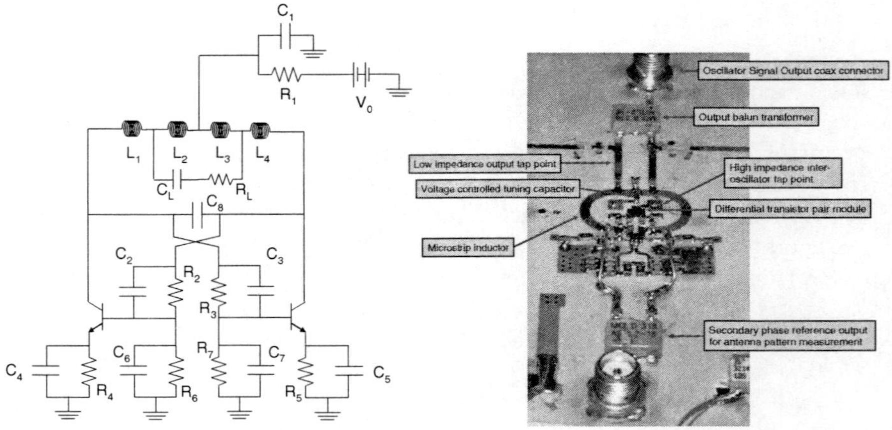

Fig. 4 Circuit schematic and instantiation of a differential-pair, cross-coupled oscillator

point (low impedance points). These two lines carry the differential oscillator output signal to an RF balun transformer, whose output is connected to a single coaxial cable. The inter-element coupling lines are provided by coaxial cables connected near the high impedance point of the microstrip inductors. In order to facilitate range pattern measurements, an external phase/frequency source for injection-locking the center oscillator of the array was included (Fig. 5).

For well-matched transistors, this configuration suppresses the even-order harmonics of the fundamental frequency. The relative amplitude of the odd order harmonics can be controlled by manipulating the DC bias of the transistors (changing the resistor values) or by adjusting the Q of the parallel L-C resonator. Through (13), this allows for specification of a desired set of Van der Pol parameters.

The unit cell design of the 2-D coupled oscillator array was based on a COTS voltage controlled oscillator (VCO), the Micronetics MW500-1109 [11]. Electronically tunable over a range of 950 to 2250 MHz and controllable via a single DC tuning voltage, this VCO possesses a 0.5" × 0.5" footprint and requires a 12VDC

Fig. 5 One-dimensional CNOA with differential-pair, cross-coupled VCOs as the unit cells

supply voltage. Other attractive features motivating its choice included (1) only a DC blocking capacitor separated the output port and the oscillator circuitry, (2) there are only three connections: one for a tuning voltage, one for a power supply and one for RF output and (3) the chip is relatively inexpensive.

Besides the VCO, the 2-D array unit cell design required additional components. The DC supply voltage was delivered at 15V to power the 12VDC voltage regulators included at each VCO. An on/off switch was added to provide the ability to calibrate each oscillator frequency individually. In order to minimize any possible undesired RF coupling between oscillators through the supply or tuning paths, RF filtering was added to both the supply and tuning voltage inputs to the VCO. A resistive divider network was added at the junction with the four coupling lines to provide the ability to reduce and, if necessary, modify the amplitudes of the couplings between oscillators. An RF attenuator and buffer amplifier was included in series with the VCO output to insure that any antenna impedance change would not load the output of the oscillators. The RF output exited the board at an SMA coaxial connector, and each of the 25 outputs was fed through phase-matched coaxial cables to the SMA inputs of the 25 COTS radiators.

A principal design goal for the 2-D array was to incorporate the capability for real-time measurement of the relative amplitude and phase between each oscillator. This was accomplished through the use of an integrated amplitude/phase detector, the Analog Devices 8302 chip. This part measures 5 mm x 6 mm and requires a 5VDC supply voltage. The AD8302 has two RF inputs, and provides two DC voltage outputs that are proportional to the relative amplitudes and relative phases of the two RF inputs.

For the 2-D array layout, there was no convenient way to package all of the circuits on a single side of the oscillator circuit board. Therefore the VCO and coupling lines were integrated on the "VCO" or front side of the board, and the AD8302 chips and most of the control and data output traces were integrated on the "Sampling" or back side of the board. As the AD8302 chips required input from two neighboring oscillator outputs, traces were coupled off the SMA RF output connectors. Resistive divider networks were utilized to provide control of the level of coupling from the RF outputs. As the AD8302 measured the relative differences between two oscillators, they were used in a serpentine arrangement such that only 24 chips were required to provide the diagnostic data for the entire array.

The prototype oscillator printed circuit measured 16" $\times$ 18" $\times$ 0.10". It was fabricated with a 0.031 inch thick fiberglass core with 0.020 inch thick Rogers 4003 substrates on top and bottom. Including the thicknesses of the six metalized layers and the two 0.0075 inch fiberglass bonding layers, the total circuit board thickness was approximately 0.092 inches. The interface to the tuning voltage control input and diagnostics output were made through a 68-pin and a 100-pin connector, respectively. Figure 6 is a photograph of the completed prototype array.

To provide a user interface to the control and diagnostic outputs, custom software was developed for use with the 2-D array. The software was LABVIEW based and incorporated a Graphical User Interface (GUI) for the array control and graphical displays of both input and output data. The GUI provided a means to control

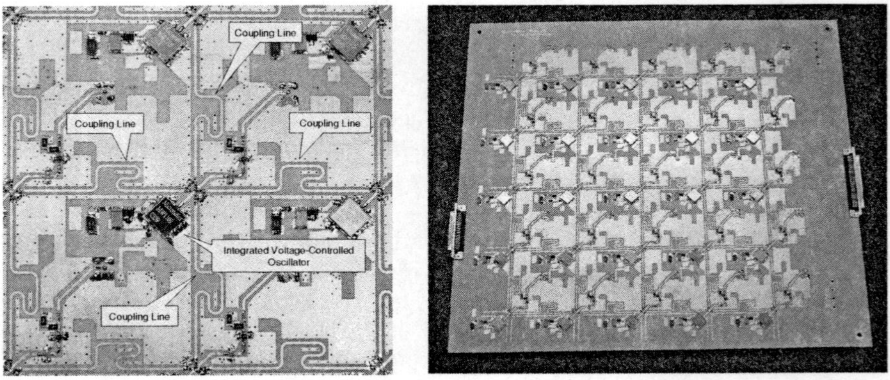

Fig. 6 Two-dimensional CNOA

the tuning voltages supplied to each of the 25 oscillators for individual calibration. Additional controls provided tuning voltages to the periphery elements to affect row/column steering of the main beam. The computer I/O was provided via two National Instruments computer cards: a 32-channel controllable DC supply, and a 64-channel Digital Acquisition (DAQ) card.

5 Experimental Results

The completed oscillator coupled array was installed in GTRI's Cobb County Planar Near-Field Range to measure its radiation pattern performance (Fig. 7). For both 1- and 2-D array, antenna pattern cuts demonstrating phase shifter-less beam steering at L1 (1575.42 MHz) were obtained.

Although the 1-D array was designed to provide full field-of-view scanning through the use of frequency-doubling, only a much more restricted scan range was observed. It was determined that unexpected coupling line behavior was the source of this discrepancy. The deviation was due in part to a combination of non-optimal tap points and line length errors. An attempt was made to model the coupling line behavior with the ADS software; unfortunately, whereas ADS was capable of accurately predicting unit cell performance, it was unable to converge to a solution for the coupled array. Thus, GTRI was unable to fully optimize the design prior to fabrication. Consequently, coupling line length modifications were accomplished through hard-wiring changes to the actual circuit. Improvements to the steady-state phase distribution were observed (Fig. 7).

For the 2-D array, the initial results of near-field range chamber testing were mixed; exceptional broadside patterns with deep nulls were observed, while steered patterns were unachievable. It was determined that the reason for the difficulty in scanning was due to the coupling of the external locking signal (required for near-field measurements) to all of the oscillators in the array. Nominally, the external

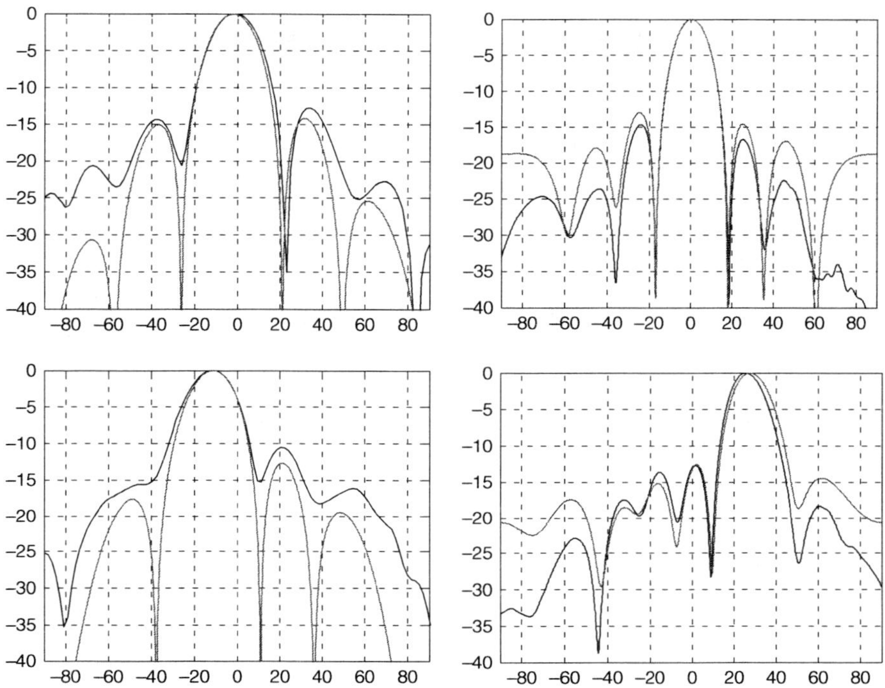

Fig. 7 One-dimensional CNOA antenna patterns measured in a GTRI anechoic chamber. The blue lines denote the ideal pattern while the black line represents the measured patterns

reference signal was to be injected into and lock the center element of the array. For the broadside pattern, the array coherence was not disrupted since all the coupling lines were one-wavelength long leading to no conflict between the oscillator interactions and the external signal (both in phase with one another). Scanned modes, however, were different in that the external signal was not in phase with the oscillator interactions, leading to an overall instability (manifested as a frequency drift) in the scanned phase distribution. The coupling lines of the array were modified through increased resistance, which appeared to mitigate the scanned pattern stability problem to some degree. It is believed that in order to completely resolve this issue, a phase-locked loop around the center array element is required; this would lock the center oscillator to the injected range reference signal without affecting the driving point impedance of the center oscillator and without introducing the external signal to the coupling lines.

Apart from the range chamber results, the on-board diagnostic circuitry incorporated into the 2-D array provided measurements of the instantaneous relative amplitudes and phases of the oscillators; this information was used to compute "pseudo-plots" of the far-field intensity patterns (Fig. 8). These results were obtained directly from bench-top (rather than range chamber) testing of the oscillator board; therefore, the stability issues arising from the injected external signal were not present.

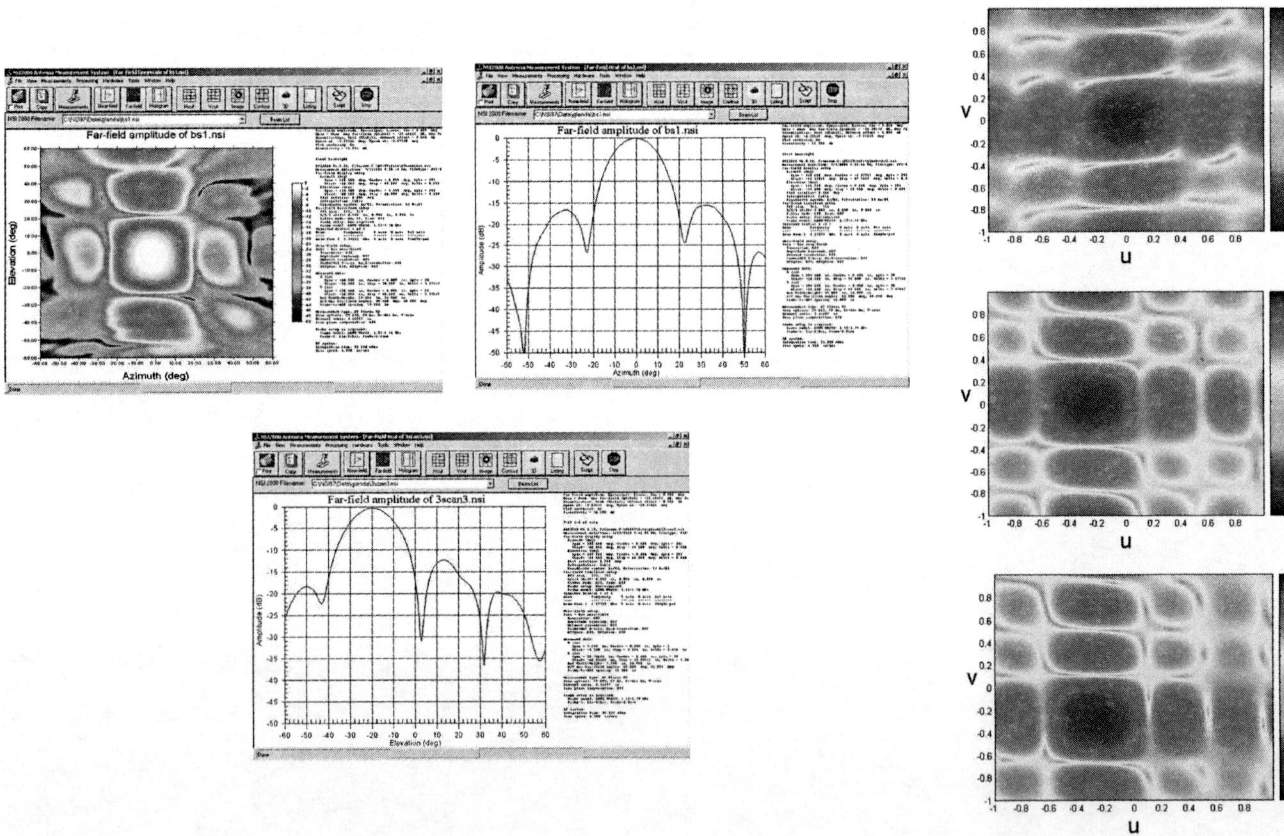

Fig. 8 Measured antenna patterns from the two-dimensional CNOA. The plots on the left were obtained via the GTRI near-field test range; the right plots were derived from the relative amplitude and phase measurements obtained from the on-board diagnostic circuitry

Consequently, the array exhibited much greater stability, enabling significant scanning in both azimuth and elevation. Based on a set of time-averaged relative amplitude and phase measurements, the oscillator array pseudo-plots incorporate the amplitude and phase errors present in the oscillator array.

As intimated in the preceding paragraphs, certain issues and limitations arise with the use of COTS oscillators: (1) the observed frequency drift was determined to be related to the large operating bandwidth of the COTS VCO; these devices were most likely designed to be used in phase-locked loops where frequency stability is ensured. To mitigate this drift, a phase-locked loop was constructed to assist in maintaining lock between the central oscillator and the reference frequency; (2) the inter-oscillator coupling was applied to the output of the COTS VCO, rather than directly at its L/C tank circuit. This led to undesirable uncertainty in understanding precisely the electrical path length from the coupling line to the L/C tank. We believe this to be a leading contributor to the limited range phase gradient and hence range of beam scan; (3) the lack of a priori knowledge of the driving point impedance of the propriety VCO design can significantly impede proper functionality. In developing the 2-D array, it became obvious that this property was crucial to achieving synchronization with a controllable phase distribution. This issue is especially relevant if a substitution of VCO type is required. Possessing different driving point impedance, use of this replacement part can lead to difficulties in stability and desired performance.

6 Outstanding Problems

The preceding sections detailed theoretical and experimental work accomplished over the last five years at GTRI. While not exhaustive, it succinctly reviews most of the most notable contributions to the technology. Efforts to improve the coupled oscillator array design and extend its capabilities are ongoing.

Much work (theory and experiment) remains to be done. At present, theory appears to have out-paced experiment. Of the capabilities described, only beam steering has been validated in the laboratory. Although validated by numerical simulations, experimental verification of beam shaping and nulling has yet to be accomplished with coupled, nonlinear oscillator arrays. Consequently, such efforts are a high priority for GTRI.

Most importantly, an answer must be found to the most pressing question any nascent technology encounters: what is its "killer app"? In other words, what does this technology provide that no other technology can? The initial development of this technology (in 1993) was as a power-, cost- and space-effective solution to collision avoidance radar. At the millimeter-wave frequencies used by such applications, phase shifter size, weight and power distribution losses can be prohibitive. However, since that time, development of low-loss, cheap MEMs phase shifters has progressed to the point where the relative advantages of nonlinear oscillator array technology is not as compelling, at least in terms of simple beam steering. Certainly, the promise of steering through a reduced number of controls (i.e., row-column steering) provided by nonlinear arrays is of interest to the radar community; however, in practice, every element must be tweaked (calibrated) to some extent.

In addition to finding its niche, a possible weakness of this technology – graceful degradation in performance due to element failure – must be assessed. By its very nature, the nonlinear array technology is holistic, crucially dependent on the interactions between elements to maintain phase coherence. How robust the coherence is when some of the array elements fail remains an open question. It seems possible that, for a 1-D array, failure of any element could be catastrophic. This should be contrasted with the conventional phase shifter approach where the presumed element-level independence ensures graceful pattern degradation. If it is shown that nonlinear arrays are not robust with respect to element failures, then, at the very least, including redundancy in a fielded design will be imperative; certainly, the technology's practical usefulness will be more difficult to justify.

These concerns should not be interpreted as sounding the death knell of nonlinear array technology. Any cutting-edge technology will have its critics and skeptics (including MEMs) and rightly so; as a result, the community must be aware of and have an answer to such questions. Every technology has its shortcomings; the relative benefits, costs and potential of nonlinear arrays have yet to be fully explored. As evidenced by GTRI's contributions, this technology possesses capabilities far greater than anticipated at its inception in 1993; what this technology is ultimately capable of remains to be seen.

References

1. K. D. Stephan and W. A. Morgan, "Analysis of Inter-injection Locked Oscillators for Integrated Phased Arrays," IEEE Trans. Antennas Propag., vol. AP-35, no. 7, pp. 771–781, July 1987.
2. P. Liao and R. A. York, "A New Phase-Shifterless Beam-Scanning Technique Using Arrays of Coupled Oscillators," IEEE Trans. Microwave Theory Tech., vol. 41, no. 10, 1810–1815, October 1993.
3. Jae-Ho Hwang and Noh-Hoon Myung, "A New Beam-Scanning Technique by Controlling the Coupling Angle in a Coupled Oscillator Array," IEEE Microwave Guided Wave Lett., vol. 8, no. 5, 191–193, May 1998.
4. C. Kykkotis, P. S. Hall and H. Ghafouri-Shiraz, "Performance of Active Antenna Oscillator Arrays Under Modulation for Communication Systems," IEE Proc. Microwave Antennas Propag., vol. 145, no. 4, 313–320, August 1998.
5. J. Shen and L. W. Pearson, "Oscillator Reproducibility Consideration in Coupled Oscillator Phase-Steering Arrays," Microwave Symp. Digest, 2000 IEEE MTT-S Int., vol. 2, 827–830.
6. Ronald J. Pogorzelski, "On the Dynamics of Two-Dimensional Array Beam Scanning via Perimeter Detuning of Coupled Oscillator Arrays," IEEE Trans. Antennas Propagat., vol. 49, no. 2, 234–242, February 2001.
7. Brian K. Meadows, et al., "Nonlinear Antenna Technology," Proc. IEEE, vol. 90, no. 5, 882–897, May 2002.
8. R. A. York and Z. B. Popovic, Eds., *Active and Quasi-Optical Arrays for Solid State Power Combining*, Series in Microwave and Optical Engineering, New York: Wiley, 1997.

9. T. Heath, "Simultaneous Beam Steering and Null Formation with Coupled, Nonlinear Oscillator Arrays," IEEE Trans. Ant. Propagat., vol. 53, no. 6, 2031–2035, June 2005.
10. A. Georgiadis and A. Suarez, "Pattern Nulling in Coupled Oscillator Antenna Arrays," IEEE Trans. Ant. Propagat., vol. 55, no. 5, pp. 1267–1274, May 2007.
11. T. Heath, R. Kerr and G. Hopkins, "Two-Dimensional, Nonlinear Oscillator Array Antenna," 2005 IEEE Aerospace Conference, March 5–12, 2005, Big Sky, MT.

Nonlinear Dynamic Effects of Adaptive Filters in Narrowband Interference-Dominated Environments

A.A. (Louis) Beex and Takeshi Ikuma

Abstract When applying certain adaptive algorithms, such as the popular LMS and NLMS algorithms, in adaptive noise canceling or adaptive equalization scenarios where a strong narrowband interference component is present, these algorithms exhibit nonlinear dynamic behaviors. The latter is expressed in filter weights – that are generally expected to converge to constants – exhibiting a dynamic component. Furthermore, the performance of these adaptive filters with dynamic weight behavior can exceed the performance of any filter of the same structure in which those weights are fixed. In adaptive noise canceling applications, the dynamic component of the weights can be unmistakably large. In adaptive equalization scenarios, however, the dynamic aspect of the weight behavior can easily be mistaken for low level noise. These various findings will be illustrated.

1 Introduction

Adaptive filters achieving better mean-square error (MSE) performance than the fixed Wiener filter of corresponding structure has been observed in adaptive noise canceling, adaptive equalization, and adaptive prediction [1–6]. In all these cases there is a narrowband signal component present and – characteristically – the improved performance is associated with the use of large step-sizes in adaptation. The latter is suggestive of dynamic weight behavior as it is coupled to tracking.

In the adaptive noise canceling application, with a different center frequency of the narrowband component in the primary channel than the center frequency of the narrowband component in the reference channel, a pronounced dynamic behavior in the weights of the adaptive filter may be observed. A hypothesized model for

A.A. (Louis) Beex (✉)
Department of Electrical & Computer Engineering – Wireless@VT – DSPRL, Virginia Tech, Blacksburg, VA 24061-0111, USA, e-mail: beex@vt.edu

V. In et al. (eds.), *Applications of Nonlinear Dynamics*, Understanding Complex Systems, DOI 10.1007/978-3-540-85632-0_13,

the origin of this behavior is based on the adaptive filter algorithm attempting to instantaneously track the underlying manifold of time-varying equivalents of a two-channel Wiener filter [6]. For an algorithm that entails more time-averaging the performance improvement disappears [7].

In the adaptive equalization application, where the equalizer is used to mitigate narrowband interference, the MSE performance can again be better than that of the Wiener filter of corresponding structure. However, in this application the dynamic weight behavior is of a much more subtle character, to the point that it can easily be thought to be a small random variation. It was recently shown that one aspect of the dynamic weight behavior is a pronounced change in the mean of the weight vector, away from the fixed Wiener weight vector [8, 9]. Again this effect happens in the presence of narrowband interference and when large step-sizes are used in adaptation.

In this paper we examine the dynamic behavior of the weights in the adaptive equalization case in more detail.

2 The LMS Algorithm

The LMS algorithm takes two input signals, a reference vector input $\mathbf{u}_n \in \mathbb{C}^M$ and a scalar primary input $d_n \in \mathbb{C}$ as shown in Fig. 1. The adapted weight vector $\mathbf{w}_n \in \mathbb{C}^M$ is updated by

$$\mathbf{w}_{n+1} = \mathbf{w}_n + \mu \mathbf{u}_n e_n^* \tag{1}$$

where μ is the step-size parameter, $*$ denotes the complex conjugate operation, and the a priori error signal is computed from

$$e_n = d_n - \mathbf{w}_n^H \mathbf{u}_n \tag{2}$$

with ^{H} denoting the Hermitian (conjugate) transpose operation.

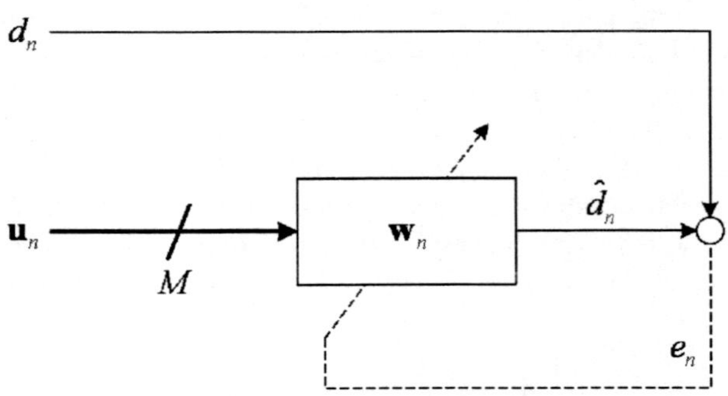

Fig. 1 General adaptive filter formulation

While the LMS algorithm is popular, due to its simplicity of operation and numerical robustness, it is fundamentally a nonlinear algorithm as its coefficients depend on input values in a quadratic fashion. As readily seen from (1), when the step-size is large there is an instantaneous change in the weights that depends on the current input vector only; for smaller and smaller step-sizes the nonlinear aspects of the algorithm vanish, for they arise from the weight vector update term.

2.1 Traditional Statistical LMS Theory

The LMS algorithm is usually analyzed according to the following decomposition of its weights:

$$\mathbf{w}_n = \mathbf{w}_o + \mathbf{v}_n \tag{3}$$

where the corresponding Wiener solution is given by

$$\mathbf{w}_o \triangleq E\{\mathbf{u}_n\mathbf{u}_n^H\}^{-1}E\{\mathbf{u}_n d_n^*\} \tag{4}$$

and $\mathbf{v}_n$ is the deviation or misadjustment caused by the adaptation. The underlying assumption here is that the adaptation tries to identify $\mathbf{w}_o$(assuming the input signals are wide-sense stationary) but the adapting weights $\mathbf{w}_n$ are subject to an unwanted deviation term $\mathbf{v}_n$ due to the adaptation taking place on the basis of data instead of on the basis of statistical averages.

One of the most popular measures, and one we utilize here, to assess the performance of the adaptive filter is the mean square error (MSE), $J \triangleq E\{|e_n|^2\}$. This performance metric is often viewed according to the weight decomposition in (3). The performance of the adaptive filter is generally believed to be bounded below by the mean square error (MSE) of the corresponding Wiener filter:

$$\begin{aligned} J_W &\triangleq E\{|d_n - \mathbf{w}_o^H\mathbf{u}_n|^2\} \\ &= \sigma_d^2 + \mathbf{w}_o^H E\{\mathbf{u}_n d_n^*\} \end{aligned} \tag{5}$$

Because it is the best possible mean square error attainable by a fixed filter, $J_W$in (5) is often referred to as the minimum mean square error. The error due to the deviations in the adapting weights is referred to as the excess mean square error.

The excess mean square error, both to characterize the transient and the steady-state behaviors of the LMS filter, has been studied usually for small step-sizes, where the dynamic behavior, if any, is largely suppressed. One such estimate for the MSE of the LMS algorithm is given by

$$J_{LMS,\infty} \approx J_W + \frac{\mu J_W}{2}\sum_{k=1}^{M}\lambda_k \quad \text{for } \mu \text{ small} \tag{6}$$

where $\{\lambda_k, k = 1,2,...,M\}$ are the eigenvalues of the input correlation matrix $E\{\mathbf{u}_n\mathbf{u}_n^H\}$ [10].

2.2 Dynamic Weight Behavior: Departure from Statistical Theory

The above classical theory largely depends on Gaussian stochastic processes at the inputs. The behavior of the LMS algorithm can be substantially different when one or both of the input signals contain or consist of narrowband spectral components.

When adaptive noise canceling is used with sinusoidal interference the dynamic behavior of the adapting weights can be strongly periodic [11]. When the interference consists of narrowband autoregressive processes the adaptive filter weights typically exhibit semi-periodic dynamic behavior [6].

In the adaptive prediction and adaptive equalization applications, when narrowband components are present, performance can again be better than that of the Wiener filter with the same structure as the adaptive filter but with its coefficients fixed based on statistical quantities. This behavior is prominent at step-sizes that are large to very large compared to what is commonly used. While the non-Wiener or nonlinear performance improvement is observed in these latter two applications, the behavior of the adaptive filter weights is not prominently dynamic (as in the adaptive noise canceling case). An indication of beneficial dynamic weight behavior results from freezing the adaptive filter weights at a locally obtained time-average value, in which case any mean-square error performance improvement is turned into a mean-square error performance loss; if one wants fixed weights the Wiener weights produce the best performance in mean-square error sense.

We next analyze the subtle dynamic weight behavior in the adaptive equalizer application.

3 Adaptive Equalizer with Narrowband Interference

An application in which the dynamic behavior of the LMS algorithm becomes beneficial is when transversal adaptive equalizers are operating in an environment with strong additive narrowband interference. In digital communication, adaptive equalization is often utilized to mitigate the effect of a time-varying or unknown channel that causes intersymbol interference. Equalizers can also be used to alleviate other forms of interference such as additive narrowband interference. The latter function of an adaptive equalizer is our focus. When an adaptive equalizer is subject to strong additive narrowband interference, the LMS adaptation can provide much better performance than the optimal time-invariant Wiener equalizer both in mean-square-error (MSE) sense and in bit-error-rate sense [3,4,12].

The simplified adaptive equalizer problem with narrowband interference is shown in Fig. 2; the system is stripped down to the minimal set of components that cause the non-Wiener effect. First, the input u_n to the equalizer is merely the sum of the communication signal $x_n$and the interference i_n. In other words, the equalizer is purely responsible to mitigate the narrowband interference and not called upon to correct channel imperfections. Also, the receiver noise is assumed small enough to be negligible. Secondly, we assume that the entire transmitted symbol sequence is

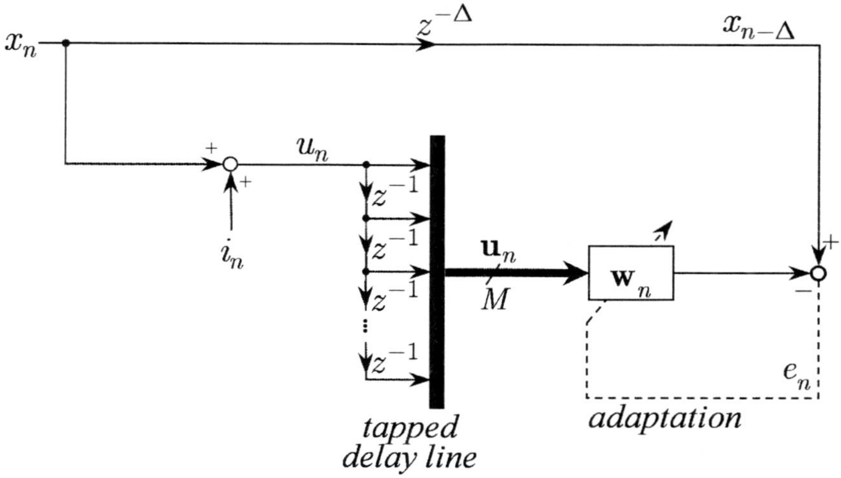

Fig. 2 Adaptive equalizer problem with narrowband interference

known a priori so that e_n is computed with the delayed version of the transmitted signal, $x_{n-\triangle}$. Typically, the adaptation error c_n is computed with the decision made by the receiver (decision-directed mode) or with the known training sequence (training mode). Our assumption is essentially that the equalizer is perpetually operating in either the training mode or in an error-free low noise environment (where the correct decision is made all the time). Lastly, we restrict ourselves here to considering complex sinusoidal interference as the narrowband interference, i.e., $i_n = \sigma_i \exp(j\omega_i n)$ with power σ_i^2 and frequency ω_i rad. While the dynamic weight behavior can be observed under stochastic narrowband interference, the current analysis does not directly extend to such stochastic interference.

Figure 3 illustrates the MSE performance of the LMS adaptive equalizer as a function of the LMS step-size based on numerical simulation[1]. The LMS MSE, which is shown relative to the Wiener equalizer MSE, depends on only two parameters: the number of filter taps M and the interference-to-signal (ISR) ratio σ_i^2/σ_x^2 with the communication signal power σ_x^2. The step-size is normalized with respect to the number of equalizer taps M and the received signal power $\sigma_u^2 = \sigma_x^2 + \sigma_i^2$. The illustrated MSE behavior of the adaptive equalizer is completely unexpected from the traditional analysis result in (6). The adaptive filter is clearly outperforming the corresponding fixed Wiener equalizer as indicated by the negative relative MSE.

In addition to the MSE-sense performance improvement, the dynamic weight behavior of the LMS equalizer interacting with narrowband interference causes an interesting side effect: a prominent shift in the time-averaged weights. This behavior is different from that in the adaptive noise cancellation application, where the

[1] Three system parameters are fixed in all the simulations: $\triangle = 0$, $\sigma_x^2 = 1$, and $\omega_i = 0.2_\pi$ rad/sample. Also, the simulations use a random sequence of 8-QAM symbols as the communication signal in all the simulations.

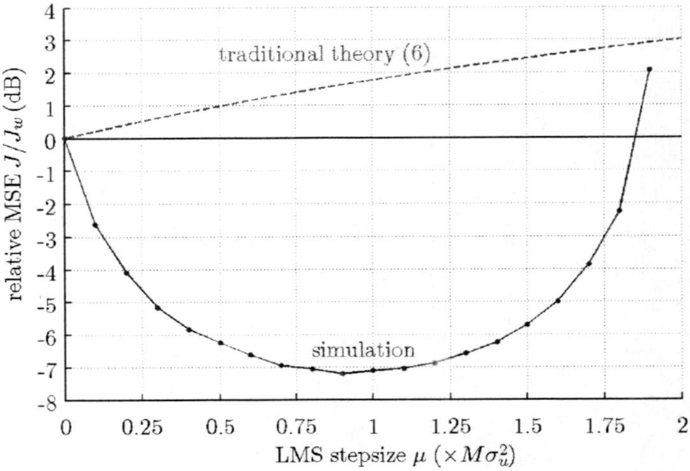

Fig. 3 MSE performance (Monte Carlo) as a function of step-size: $M = 20$, ISR = 20 dB

prominent feature is dynamic pseudo-periodic weight behavior close to the Wiener solution (which was all zeros in the purely sinusoidal case). The steady-state behavior of the LMS equalizer weights is illustrated in Fig. 4. The deviation of the LMS weights from the Wiener solution is prominently visible. The variation of the weights about their mean value is indicated by the little "clouds" of 10,000 consecutive weight values, i.e. the changes in the weights are small relative to their mean value, which deviates substantially from the Wiener weight values.

The spiral-shaped mean of the LMS equalizer weights has been analyzed for the strong sinusoidal interference case and is found to be [8]

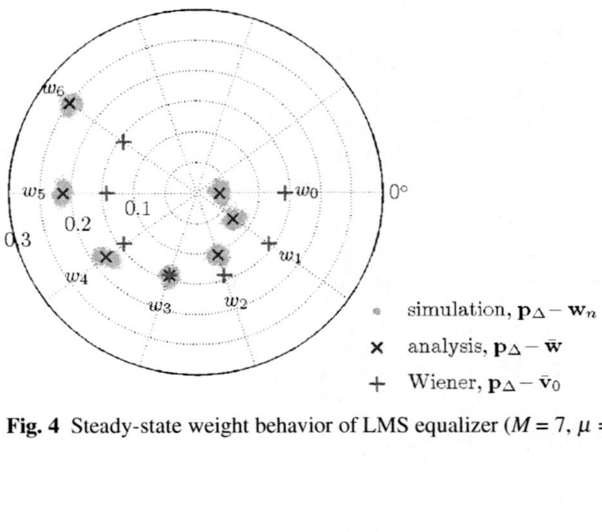

Fig. 4 Steady-state weight behavior of LMS equalizer ($M = 7$, $\mu = (M\sigma_u{}^2)^{-1}$, ISR = 20 dB

$$\overline{\mathbf{w}} = \mathbf{p}_{\triangle} - \frac{\sigma_i^2}{\lambda_{max}} (\mathbf{I} - \mu \Xi \Theta)^{-1} \mathbf{e} \tag{7}$$

where

$$\mathbf{p}_{\triangle} \triangleq [\underbrace{0 \quad \ldots \quad 0}_{\Delta} \quad 1 \quad 0 \quad \ldots \quad 0]^T, \tag{8}$$

$$\lambda_{\max} = \sigma_x^2 + M\sigma_i^2, \tag{9}$$

$$\mathbf{e} = e^{j\omega i \Delta}[1e^{-}j\omega i \ldots e^{-}j\omega i(M-1)]^T, \tag{10}$$

$$\Xi \triangleq \frac{1}{\sigma_x^2} \left[\mathbf{I} - \frac{\sigma_i^2}{\lambda_{\max}} 11^H \right], \tag{11}$$

and

$$\Theta \triangleq \frac{\sigma_i^2 M}{\sigma_x^2} \sum_{P=1}^{M-1} \mathbf{Z}^P (1 - \mu \lambda_{\max})^{p-1}, \tag{12}$$

In the above, $\mathbf{I}$ is the identity matrix, $\mathbf{1}$ is the vector of all ones, and $\mathbf{Z}$ is the lower triangular shift matrix (ones on the subdiagonal and zeros elsewhere).

The mean offset itself does not explain the performance improvement of the adaptive equalizer over the Wiener equalizer. If anything, it suggests that the LMS equalizer should underperform because it is a suboptimal solution. Instead, the time-varying portion of the weights, which is smaller than the magnitude of the mean portions, must be responsible for the MSE performance. Inspired by the time-varying subspace analysis for the adaptive noise canceller [13, 14], we propose the following dynamic weight model for LMS used in our problem:

$$\mathbf{w}_n = \overline{\mathbf{w}} + \mathbf{e}\alpha_n + \beta_n \tag{13}$$

The LMS weights consist of the fixed portion $\overline{\mathbf{w}}$ and the time-varying portion $\mathbf{e}\alpha_n + \beta_n$. Of the two time-varying components, $\mathbf{e}\alpha_n$ represents the dynamic behavior while the vector process βn is the weight misadjustment. The dynamic components of the weights are locked to each other by $\mathbf{e}$, and their collective motion α_n is a lowpass stochastic process. The misadjustment weights β_n are assumed to be zero-mean.

The model in (13) is verified in this paper by estimating α_n with

$$\hat{\alpha}_n = \frac{1}{M} \sum_{m=0}^{M-1} (w_{m,n} - \overline{w}_m) e^{j\omega_i (m-\Delta)} \tag{14}$$

where $w_{m,n}$ and $\overline{w}_m$ indicate the m-th element of the weight vectors $\mathbf{w}_n$ and $\overline{\mathbf{w}}$, respectively. Subsequently, the weights are estimated by

$$\hat{\mathbf{w}}_n = \overline{\mathbf{w}} + \mathbf{e}\hat{\alpha}_n \tag{15}$$

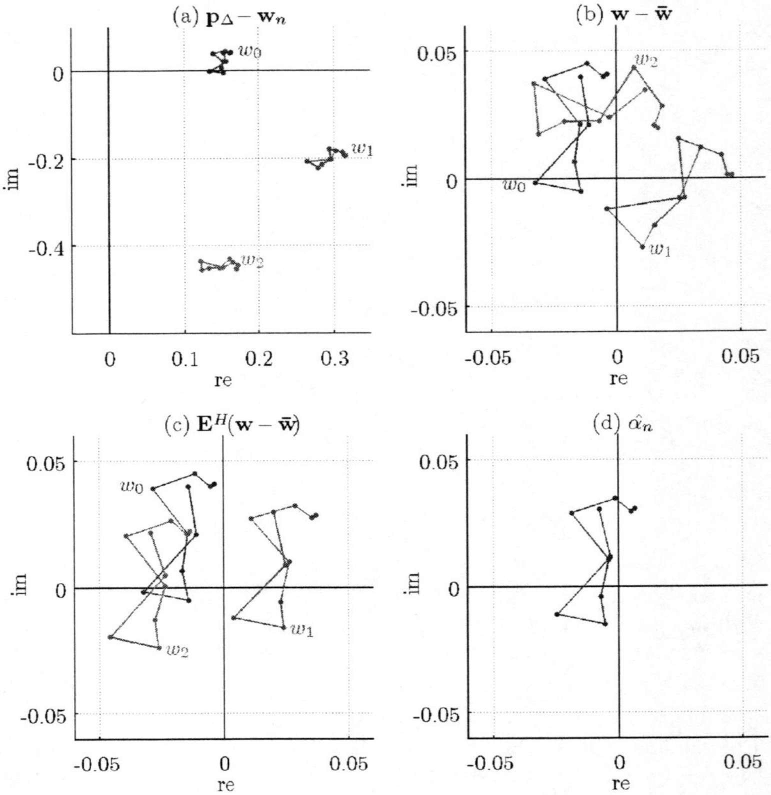

Fig. 5 Dynamic weight behavior deconstruction for the LMS equalizer ($M = 3$, $\mu = (M\sigma_u^2)^{-1}$, ISR 20 dB): (**a**) 10 samples of the LMS weight vector components, (**b**) after removal of the mean, (**c**) after undoing the rotation, and (**d**) the estimated α_n

The MSE can be computed for the estimated weights in (15) with the same input and desired signals as M_n. Figure 5 illustrates how the LMS weights relate to α_n. Figure 5(b) and 5(c) clearly illustrate the rotated nature of the time varying weights and their similarities in the signals. The observed offsets among the rotated time-varying weights in Fig. 5(c) relate to the β_n process and are not constant over time; instead, the offsets represent low-pass stochastic processes and thus change slowly over time.

Furthermore, as functions of ISR, Fig. 6 compares the experimentally evaluated MSE (relative with respect to the Wiener MSE) of the LMS error e_n with the MSE estimated using the mean weights only, and with the MSE estimated using the time-varying weight model in [15]. The latter error signals are computed as follows.

$$\bar{e}_n = d_n - \bar{\mathbf{w}}^H \mathbf{u}_n \tag{16}$$

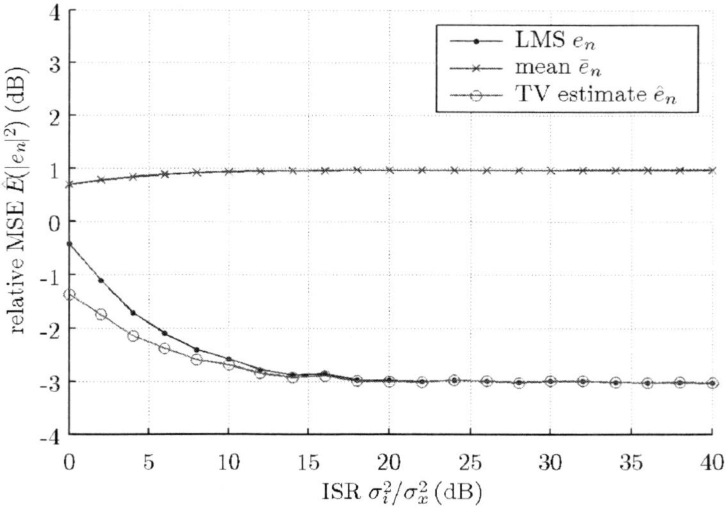

Fig. 6 Estimated MSEs of the LMS equalizer – relative to Wiener MSE – using the proposed model ($M = 7$, $\mu = (M\sigma_u{}^2)^{-1}$)

and

$$\bar{e}_n = d_n - \hat{\mathbf{w}}_n^H \mathbf{u}_n. \tag{17}$$

A 7-tap LMS equalizer with step-size $\mu = \sigma_u^{-2}$ is used in these Monte Carlo simulations. There are three notables in this result in support of the validity of our model. First, the proposed $\mathbf{e}\alpha_n$ dynamic model indeed characterizes the non-Wiener performance gain. In addition, the MSE of the estimated model is approaching the simulation from below, supporting the proposed role of β_n as the misadjustment component that causes excess error. Lastly, the difference between $e_n$and $\hat{e}_n$asymptotically disappears as the interference becomes stronger, implying that $\beta_n$vanishes as the interference increasingly dominates.

Next, the statistical properties of the $\hat{\alpha}_n$ process are evaluated. This simulation is conducted in a 20-dB ISR environment and for step-size $\mu = (M\sigma_u^2)^{-1})$. The autocorrelation functions and power spectra of $\hat{\alpha}_n$ are examined for different equalizer tap-lengths: M = 3, 7, and 20. The autocorrelation functions are shown in Fig. 7, and the power spectrum estimates are given in Fig. 8. The autocorrelation functions clearly depend on M: $\hat{\alpha}_n$ looks to be predominantly an MA(M) (M-th order moving average) process. We observe these processes to be low-pass in nature, with a bandwidth relating to the M parameter.

The last illustration in Fig. 9 shows estimates for the variance of $\hat{\alpha}_n$ in (14) and for $\hat{\beta}_{0,n}$, the 0-element of $\hat{\beta}_n \triangleq \mathbf{w}_n - \hat{\mathbf{w}}_n$, for different tap lengths and ISRs. The other elements of $\hat{\beta}_n$ behave similarly to $\hat{\beta}_{0,n}$. We observe that both time-varying components have similar variance for ISR below 30 dB; for higher ISR the variance of $\hat{\beta}_{0,n}$ starts to drop off. In Fig. 6 we saw that beyond 15 dB ISR there was no

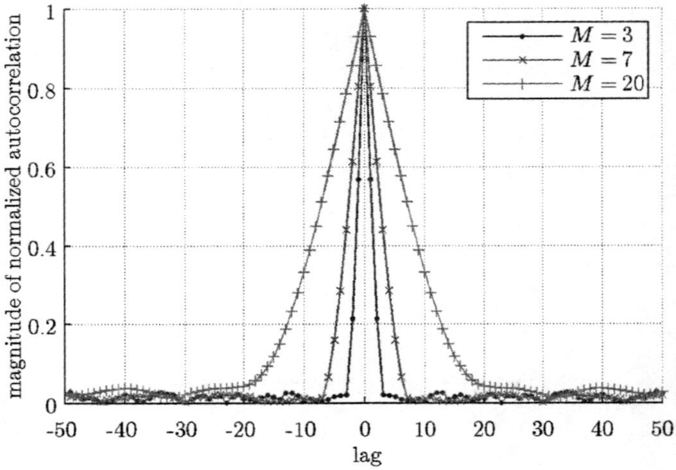

Fig. 7 Normalized autocorrelation functions of the estimated $\hat{\alpha}_n$ (ISR = 20 dB, $\mu = (M\sigma_u{}^2)^{-1}$)

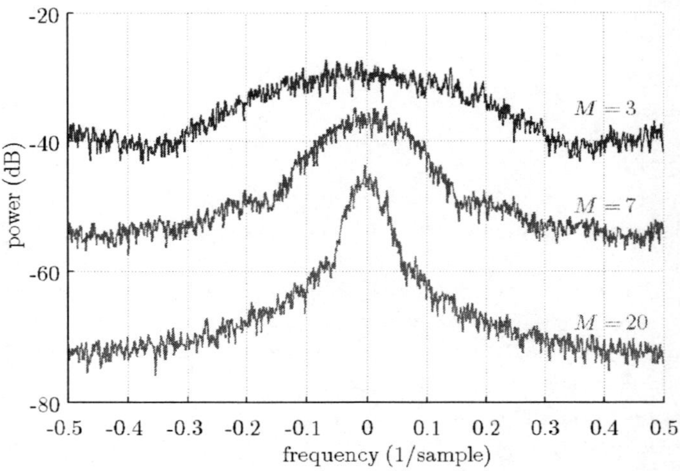

Fig. 8 Power spectrum estimates of the estimated $\hat{\alpha}_n$ (ISR 20 dB, $\mu = (M\sigma_u{}^2)^{-1}$)

visible effect $\hat{\beta}_n$ on the LMS estimation error. We verified that this is a result of $\hat{\beta}_n^H \mathbf{u}_n$ vanishing, i.e. $\hat{\beta}_n$ becomes orthogonal to $\mathbf{u}_n$ while $\mathbf{e}\hat{\alpha}_n$ does not.

The above results point to the future work refinement of the present model, in which the $\alpha_n$process is modeled explicitly as MA(M). The latter will allow us to add $\beta_n$explicitly to the model as the misadjustment process so that it can be estimated simultaneously with the MA(M) parameters. In addition, we plan to connect the time-varying weight behavior directly to the input signals.

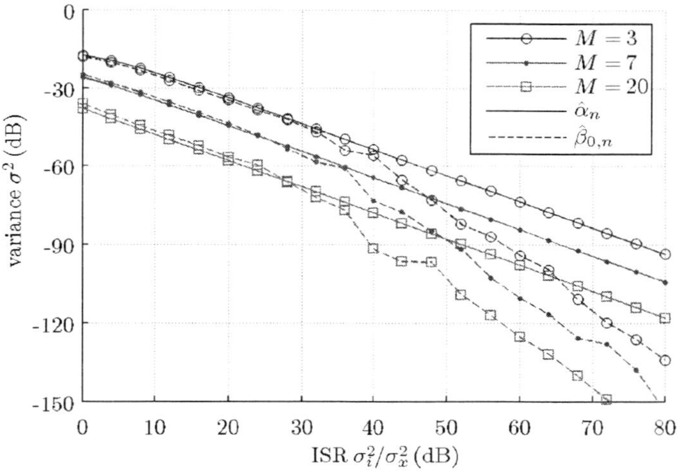

Fig. 9 Variance estimates of the two time-varying components of the proposed LMS weight model ($\mu = (M\sigma_u{}^2)^{-1}$)

4 Conclusions

We have presented the beneficial time-varying weight behavior of the popular LMS adaptive algorithm. The LMS algorithm, when applied in the presence of narrowband signals, such as the narrowband interference discussed in this paper, can outperform the Wiener mean-square error statistical framework that the algorithm is derived from. We have proposed a new model for the dynamic behavior of the LMS weights of the adaptive equalizer mitigating strong narrowband interference. The model has been shown experimentally to converge asymptotically towards the actual LMS behavior as the input becomes more interference dominated.

References

1. B. Widrow, J. R. Glover, Jr., J. M. McCool, J. Kaunitz, C. S. Williams, R. H. Hearn, J. R. Zeidler, E. Dong, Jr., and R. C. Goodlin, "Adaptive noise cancelling: Principles and applications," *Proc. IEEE,* vol. 63, pp. 1692–1716, Dec. 1975.
2. B. Widrow, J. M. McCool, M. G. Larimore, and C. R. Johnson, Jr., "Stationary and nonstationary learning characteristics of the LMS adaptive filter," *Proc. IEEE,* vol. 64, pp. 1151–1162, 1976.
3. R. C. North, R. A. Axford, and J. R. Zeidler, "The performance of adaptive equalization for digital communication systems corrupted by interference," in *27th Asilomar Conf.*, Pacific Grove, CA, 1993, pp. 1548–1553.
4. M. Reuter and J. R. Zeidler, "Nonlinear effects in LMS adaptive equalizers," *IEEE Trans. Signal Process.,* vol. 47, pp. 1570–1579, June 1999.
5. J. Han, J. R. Zeidler, and W. H. Ku, "Nonlinear effects of the LMS adaptive predictor for chirped input signals," *EURASIP J. Appl. Signal Process.,* vol. 2002, pp. 21–29, 2002.

6. A. A. Beex and J. R. Zeidler, "Steady-state dynamic weight behavior in (N)LMS adaptive filters," in *Least-Mean-Square Adaptive Filters*, S. Haykin and B. Widrow, Eds. Hoboken, NJ: John Wiley & Sons, 2003, pp. 335–443.
7. A. A. Beex and J. R. Zeidler, "Non-Wiener effects in recursive least squares adaptation," in *ISSPA 2003*, Paris, France, 2003, pp. 595–598.
8. T. Ikuma, "Non-Wiener effects in narrowband interference mitigation using adaptive transversal equalizers," Ph. D. Dissertation, Dept. Elect. Comput. Eng., Virginia Tech, Blacksburg, 2007. [Online]. Available: http://scholar.lib.vt.edu/theses/available/etd-04112007-104316/
9. T. Ikuma, A. A. Beex, and J. R. Zeidler, "Non-Wiener weight behavior of LMS transversal equalizers," in *ICASSP 2007*, Honolulu, Hawaii, 2007, pp. 1297–1300.
10. S. Haykin, *Adaptive Filter Theory*, 4th ed. Upper Saddle River, NJ: Prentice Hall, 2002.
11. J. Glover, Jr., "Adaptive noise canceling applied to sinusoidal interferences," *IEEE Trans. Acoust., Speech, Signal Process.,* vol. 25, pp. 484–491, Dec. 1977.
12. A. Batra, T. Ikuma, J. R. Zeidler, A. A. Beex, and J. G. Proakis, "Mitigation of unknown narrowband interference using instantaneous error updates," in *37th Asilomar Conf.*, Pacific Grove, CA, 2004, pp. 115–119.
13. N. J. Bershad and J. C. M. Bermudez, "Sinusoidal interference rejection analysis of an LMS adaptive feedforward controller with a noisy periodic reference," *IEEE Trans. Signal Process.,* vol. 46, pp. 1298–1313, May 1998.
14. N. J. Bershad and P. L. Feintuch, "Non-Wiener solutions for the LMS algorithm-a time domain approach," *IEEE Trans. Signal Process.,* vol. 43, pp. 1273–1275, May 1995.

Design-Oriented Bifurcation Analysis of Power Electronics Systems

Chi K. Tse

Abstract Bifurcations and chaos have been studied in many power electronics circuits and systems. Numerous results have been reported regarding the nonlinear behavior of such circuits and systems under variation of some selected parameters, e.g., period-doubling bifurcations, Hopf bifurcations, coexisting attractors, border collisions, etc. The current status of research in the identification of bifurcation behavior in power electronics can be considered mature in the sense that the salient types of bifurcation behavior, their underlying causes and the theoretical parameters affecting them have been well documented. Recently, research in this field has begun to seek possible applications that are of direct relevance to industrial power electronics. One direction is to apply some of the available research results in bifurcation behavior to the design of practical power electronics systems.The main barrier is that the abstract mathematical presentations of the available results are not directly applicable to practical design problems. Our work in this research area has been directed to bridge that gap.

1 Introduction

Much of the work in the study of nonlinear phenomena of power electronics circuits and systems has been focused on basic research into the bifurcation and chaotic behavior of power converters under variation of some selected parameters. Although the research results generated from these studies have greatly improved the understanding of the nonlinear behavior of power electronics, their use in solving practical design problems is very limited and relatively unexplored, presumably due to the fundamental difference between the approaches taken for pursuing basic research and those required for facilitating practical design. Power electronics engi-

C.K. Tse (✉)
Department of Electronic and Information Engineering. The Hong Kong Polytechnic University, Hong Kong, e-mail: encktse@polyu.edu.hk

V. In et al. (eds.), *Applications of Nonlinear Dynamics*, Understanding Complex Systems, DOI 10.1007/978-3-540-85632-0_14,

neers frequently ask: *what possible applications can the study of bifurcations and chaos offer to the practical design of industrial power electronics systems?* Much of our research efforts has been (will be) devoted to showing the practical relevance of the study of bifurcations and chaos in power electronics. Specifically, we have attempted to pursue the following tasks:

1. Re-organization of the available research results in terms of practically relevant conditions (task 1); and
2. Derivation of a design-oriented approach to analyse the bifurcation behavior of power electronics circuits and systems (task 2).

   The first task mentioned above is non-trivial and requires a complete reformulation of the problem. Specifically, currently available results are often given in terms of rather unrealistic circuit conditions and are based on idealised models which omit practical details. To make the results available for practical use, we need to include detailed circuit components and to consider practical operating scenarios. In most cases, the basic phenomena will remain the same, but the results become more significant if they are related to practical circuits (those actually in use) and are given in terms of realistic operating conditions. The second task is in effect a consolidation of existing analysis of bifurcations into a set of design-oriented procedures that can be directly applied to practical design problems.

   To gain further insights into the nonlinear behavior of power electronics circuits and systems, efforts should be spent to uncover previously unknown phenomena and their underlying causes, in very much the same way as what basic researchers did in the past, but mainly for practical power electronics circuits. Therefore, the following task is complementary to the above two tasks.
3. Pursuit of basic research into the identification of bifurcation behavior of commonly used practical power electronics systems (task 3).

2 Review of Current Research Status

Power electronics as a discipline is driven by practical needs of industry. The continual quest for better products with ever increasing standards of reliability and performance makes it necessary to consider systematic approaches for performance optimization. Practical bifurcation analysis will have a long-term impact on the design of power electronics systems because the performance of a particular power electronics system is related to the choice of design parameters which are often bifurcation-sensitive. In reality, bifurcation is to be avoided, but it is also known that designing a system too remote from bifurcation boundaries may degrade performance characteristics such as transient speed. Therefore, design-oriented bifurcation analysis should have a significant impact on the practical methodologies taken to make design trade-offs and performance optimizations.

Power electronics is a discipline spawned by real-life applications in industrial, commercial, residential and aerospace environments. Much of its development evolves around some immediate needs for solving specific power conversion problems. In the past three decades, this branch of electrical and electronic engineering has undergone an intense development in many areas of technology [1], including power devices, control methods, circuit design, computer-aided analysis, passive components, packaging techniques, and so on. The principal focus in power electronics, as reflected from topics discussed in some key conferences [2, 3], is to fulfill the functional requirements of the application for which the circuits are used. Like many areas of engineering, power electronics is mainly motivated by practical applications, and it often turns out that a particular circuit topology or system implementation has found widespread applications long before it has been thoroughly analysed. For instance, the widespread application of a simple switching converter may date back to more than three decades ago. However, good analytical models allowing better understanding and systematic circuit design were only developed in the late 1970's [4], and in-depth analytical characterization and modelling is still being actively pursued today. Despite their common occurrence in power electronics circuits, nonlinear phenomena have only recently received appropriate formal treatments. Serious basic research in this area started mainly in the late 1980's [5–7], and much of the reported work has dealt with direct application of bifurcation theory to some appropriate nonlinear models of simple converter circuits [8]. Because simplified models and idealized operating conditions are considered, these basic works uncover the cause of many so-called "strange" nonlinear phenomena in power electronics, but fall short of any design-oriented analytical result that can directly benefit practicing engineers. In brief, the basic research of nonlinear phenomena in power electronics has enjoyed good academic recognition in the past, but is incompatible with the actual practice of power electronics.

As mentioned above, serious work in this area began in the late 1980's and much of the work reported in the past has focused on fundamental study of nonlinear dynamical and chaotic behavior of power converters. The following is a summary of work conducted by research groups other than ours at Hong Kong Polytechnic University.

The occurrence of bifurcations and chaos in power electronics was first reported in the literature in the late 1980's by Hamill et al. [5]. Experimental observations regarding boundedness, chattering and chaos were also made by Krein and Bass [9] back in 1990. Although these early reports did not contain any rigorous analysis, they provided solid evidence of the importance of studying the complex behavior of power electronics and its possible benefits for practical design. Since then, much interest has been directed towards pursuing formal studies of the complex phenomena commonly observed in power electronics. In 1990, Hamill et al. [5] reported a first attempt to study chaos in a simple buck converter at the IEEE Power Electronics Specialists Conference. Using an implicit iterative map, the occurrence of period-doublings, subharmonics and chaos in a simple buck converter was demonstrated by numerical analysis, PSPICE simulation and laboratory measurements. The derivation of a closed-form iterative map for the boost converter under a current-mode control scheme was presented later by the same group of researchers [10]. This closed-form iterative map allowed the analysis and classification of bifurcations and structural instabilities of this simple converter. Since then, a number of authors have contributed to the identification of bifurcation patterns and strange attractors in a wide class of circuits and devices of relevance to power electronics. Recently, an edited book which is devoted entirely to the subject of nonlinear phenomena in power electronics has also been published [11].

The bifurcation behavior of the buck converter was studied by Chakrabarty et al. [12] who examined the bifurcation behavior under variation of a range of circuit parameters including storage inductance, load resistance, output capacitance, etc. In 1996, Fossas and Olivar [13] presented a detailed analytical description of the buck converter dynamics, identifying the topology of its chaotic attractor and studying the regions associated with different system evolutions. Various possible types of operation of a simple voltage-feedback pulse-width-modulated buck converter were also investigated through the so-called stroboscopic map obtained by periodically sampling the system states. The bifurcation behavior of dc/dc converters under current-mode control has been studied by a number of authors, e.g., Deane [14]. Typical bifurcation behavior in power electronics contains transitions whereby a "sudden jump from periodic solutions to chaos" is observed. These transitions cannot be explained in terms of standard bifurcations such as "period-doubling" and "saddle-node". In fact, as proposed by Banerjee et al. [15, 16] and Di Bernardo [17], these transitions are due to a class of bifurcations known as "border collisions", which is unique to switched dynamical systems [18].

Power electronics circuits other than dc/dc converters have also been examined in recent years. Dobson et al. [19] reported "switching time bifurcation" of diode and thyristor circuits. Such a bifurcation manifests as jumps in the switching times. Bifurcation phenomena from induction motor drives were reported separately by Kuroe [20] and Nagy et al. [21].

The occurrence of period-doubling cascades for a simple dc/dc converter operating in discontinuous mode was first reported in 1994 by the author [22, 23]. By modelling the dc/dc converter as a first-order iterative map, the onset of period-doubling bifurcations can be located analytically. The idea is based on evaluating the Jacobian of the iterative map about the fixed point corresponding to the solution undergoing the period-doubling, and determining the condition for which a period-doubling bifurcation occurs. Simulations and laboratory measurements have confirmed the findings. Formal theoretical studies of conditions for the occurrence of period-doubling cascades in a discontinuous-mode dc/dc converter were reported subsequently in [24]. For current-mode controlled converters, the author studied various types of routes to chaos and their dependence upon the choice of bifurcation parameters [25]. In 1995, the study of bifurcation phenomena was extended to a fourth-order Cuk dc/dc converter under a current-mode control scheme [26]. Moreover, when external clocks are absent and the system is "free-running" (for example, under a hysteretic control scheme), the system is autonomous and does not have a fixed switching period. A representative example is the free-running Cuk converter which has been shown [27] to exhibit Hopf bifurcation and chaos. Also, some attempts have been made to study higher order parallel-connected systems of converters which are becoming popular design choice for high-current applications [28, 29].

Very recently, we began to look into applications of bifurcation analysis. Among the various specific applications, our work has been particularly fruitful in the study of a practical power-factor-correction converter and the application of bifurcation analysis to address a difficult fast-scale instability problem. The results have been given in terms of realistic circuit parameters and conditions, clearly defining the different stability regions of the parameter space [30, 31].

3 Design-Oriented Methodologies

In the previous section, we have identified the key problem in the development of applications of nonlinear dynamics and chaos theory to practical power electronics system design. This problem boils down to the general incompatibility between theoretical results in nonlinear analysis and the practical viewpoints of power electronics practitioners. To combat this problem, our basic methodology is to re-formulate the problems in bifurcations and chaos in terms of concepts that are already familiar to the engineer.

To illustrate the idea, we take the classic current-mode controlled dc/dc converter as an example. The nonlinear analysis of this converter has uncovered interesting bifurcation patterns [14, 25], but the results have not been useful to engineers. The difficulty in using the results or recognizing the usefulness of these results can be attributed to the way in which the analysis was performed. Traditionally, nonlinear analysis involves finding nonlinear models, dealing with differential equations (or iterative maps), computing Jacobians, and analysing stability of orbits [25]. The results are often stated in terms of abstract parameters, e.g., theoretical feedback gains and reference current levels. Although period-doubling bifurcations are well explained by such analysis, the power electronics community examined the same problem many years ago using a different viewpoint, examining mainly the gradient (slope) of the inductor current [32, 33]. Effective methods for preventing period-doubling bifurcations known as "slope compensation" have already been used in practice for a long time. It transpires that if the analysis of this problem is re-formulated in terms of the familiar concept of slope compensation, results become more relevant to the design of these kinds of converters. Along this line, we have recently derived a practically relevant analysis for current-mode controlled converters [34]. This work has been further applied to power-factor-correction converters to explain a practical problem related to fast-scale instability [30, 31].

The ease of analysis and the usefulness of the results depend very much on the choice of modelling methods. Thus, the choice of methodology for model building is crucial to this project. Our basic methodology is to employ a pre-analysis simulation and/or measurement. The purpose is to get an initial idea of the possible behavior of the system under study. This is an important step because if the type of behavior is approximately known, appropriate modelling methods can be chosen to fit the likely behavior [8]. For example, it was once (incorrectly) believed that averaged modelling was useless for the study of chaos and bifurcations, and most studies were based on the more complicated discrete-time iterative mapping method. However, for systems that bifurcate from a regular high-frequency orbit to a long-period limit cycle (known as Hopf bifurcation), the expected phenomenon is essentially a low-frequency one and therefore can be captured by averaged models. If such a behavior is known from a pre-analysis simulation or experiment, then averaged models would be the best choice for analysing the system as they are simple and adequate for capturing the salient low-frequency bifurcation behavior [27].

Another important aspect of our future work that has a direct impact on the research methodology is the choice of the system for investigation. In the past, many basic research works are targeted at overly ideal systems for the sake of convenience or simplicity. The research methodology could then follow some previously developed theoretical procedures, and aim to achieve some closed-form analytical solution. For instance, in the study of border collisions, much has been done for the simple first-order case and rather detailed analytical results concerning the possible types of transitions have been reported [35, 36]. In this project, we attempt to include as many practical components as possible in order to produce results that are directly relevant to practical design. In other words, we will refrain from the commonly used methodology of studying over simplified "ideal" cases. It is expected that significant new findings can be uncovered when practical (more complex) systems are considered. For example, we have recently identified intermittent chaos in power supplies that are constructed in practical "noisy" environments [37, 38]. (Technically, the power supply under study takes into account unintended coupling of signals through the ground plane of the printed circuit board.)

Finally, as mentioned before, we should target practical systems for investigation. It is therefore unlikely that the systems under study will lend themselves to simple low-order analysis. As a consequence, we will have to resort to numerical procedures and generality may be lost since particular parameter values must be assigned to perform numerical studies. Moreover, in order for the results to remain useful for the general case, they must be presented in terms of normalized parameters [39]. Recent work that has targeted practical systems includes the study of bifurcation in power factor correction converter systems [40–43], bifurcation in SPICE simulations [44], interacting bifurcations in current-mode converters [45], and refinement of stability information in two-stage converter systems [46].

4 Example

The current-mode controlled boost converter is shown in Fig. 1(a). The system has an outer voltage loop and an inner current loop. The voltage loop, which consists

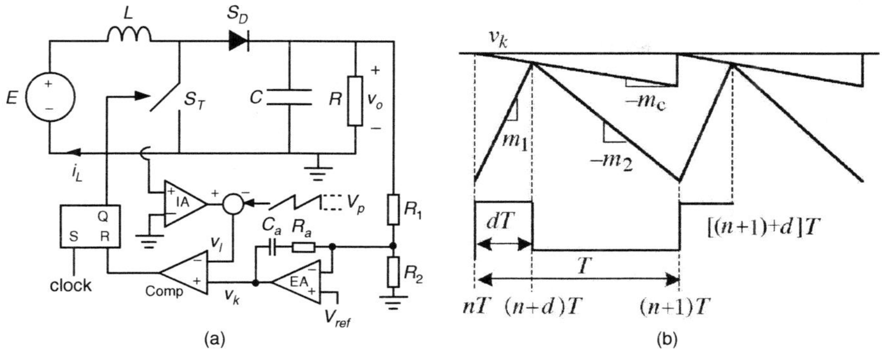

Fig. 1 (**a**) Boost converter under current-mode control; (**b**) typical waveforms

of an error amplifier (EA), a compensation network, provides the reference for the inner current loop. The inner current-loop consists of a current transformer, a compensation ramp signal and a current sense amplifier. The compensationl ramp is added to stabilize the converter if a wide range of output voltage is required [47]. The output of the two loops is then connected to the inputs of the comparator whose output is used to reset a flip-flop latch to give a pulse-width modulated waveform to control switch S_T. The operation can be briefly described as follows. The flip-flop latch is set periodically by the clock signal, turning on the switch S_T. Then, the inductor current goes up linearly, and is compared with a reference level, which is equal to the output of the error amplifier of the voltage loop minus the compensation ramp signal. When the peak inductor current reaches the reference level, the output of the comparator resets the flip-flop, thereby turning off the switch. When the switch is off, the inductor current falls almost linearly if the output capacitor is sufficiently large. The cycle repeats when the flip-flop is set again by the clock. Typical waveforms of are shown in Fig. 1(b), where m_c is the slope of the compensation ramp signal, and m_1 and m_2 are the rising and falling slopes of the inductor current with the switch S_T on and off respectively.

4.1 A Glimpse at Bifurcation Behavior

We begin with a series of typical waveforms from exact cycle-by-cycle computer simulations to show some possible bifurcation behaviors in this system.

The main parameters affecting fast-scale bifurcations are L/E and m_c, whereas those affecting slow-scale bifurcations are the voltage feedback gain and time constant, as studied previously [39]. To facilitate simulation study, we choose parameter values as listed in Table 1.

Table 1 Circuit parameters for simulation study

Component/parameter	Value
Input voltage E	3–25 V
Inductance L	120–195 μH
Capacitance C	2000 μF
Load resistance R	3–20 Ω
Switching frequency f_s	25 kHz
Reference output voltage V_{ref}	1.8 V
Voltage divider R_1, R_2	47.5 kΩ, 2.5 kΩ
Compensation network R_a, C_a	72.3 kΩ, 0.23 μF
Compensation ramp V_p	0.25 V
Inductance current sampling gain M	0.082 V/A

For a certain range of parameters, normal periodic operation, slow-scale bifurcation, fast-scale bifurcation, coexisting fast and slow-scale bifurcation, and "saturated" operation (border collision) can be observed. Typical waveforms are shown in Fig. 2. Consistent with usual understanding, we can see from these waveforms that slow-scale bifurcations can be eliminated by increasing the feedback gain or time constant, and that fast-scale bifurcations can be eliminated by increasing m_1 or m_c.

4.2 Detailed Examination: Boundaries of Operations

In this section, we take a detailed look into the qualitative behaviors of the system, and present the boundaries of stable region, slow-scale bifurcation region, fast-scale bifurcation region, and coexisting (interacting) fast and slow-scale bifurcation region in terms of selected circuit parameters.

The behavior of the system varies with E and R. Two perspectives of operating boundaries are shown in Fig. 3, where the boundaries divide regions of stable operation and "saturated" region. The transition from the stable region to the other is a Hopf type bifurcation. Parameters except R and E are kept as listed in Table 1.

The feedback gain and time constant have obvious influence on the stability, especially for the slow-scale bifurcation because of the bandwidth limitation imposed by the time constant. We keep E and R constant, and vary L/E. We can then change g and τ_f by varying R_a and C_a. Figure 4 shows the stability boundaries for different values of L/E, corresponding to the slow-scale bifurcation boundary and fast-scale bifurcation region of Fig. 3. In Fig. 4, the regions under the boundary curves corresponds to the "saturated" operation, and the operations above the boundaries depend on the values of L/E and g.

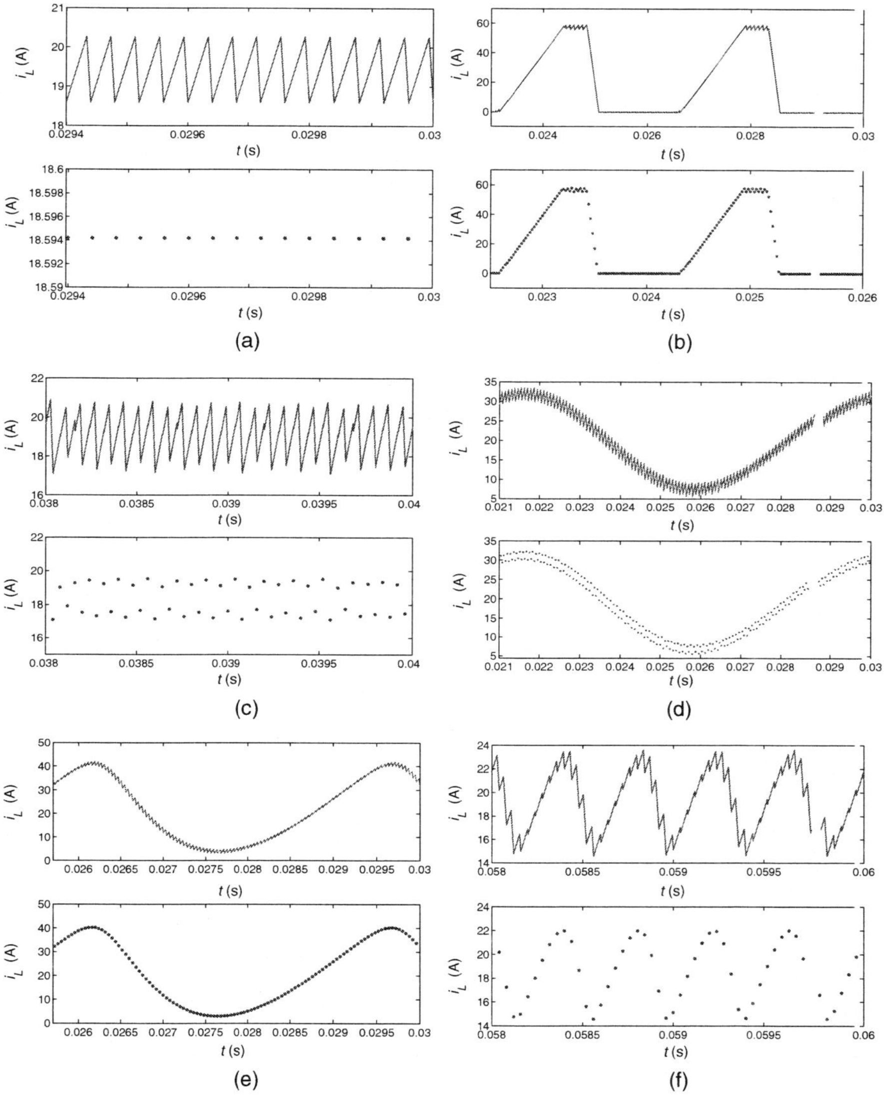

Fig. 2 Simulated behaviors for different feedback gain $g = R_a/R_1$ and time constant $\tau_f = R_1C_a$, with $1/m_1 = 19.835 \times 10^{-6}$ s/A, $m_c = 6.25 \times 10^{-3}$ V/s. Upper trace: actual waveforms, lower trace: sampled-data waveforms. (**a**) Stable periodic operation with $g = 1.522$ and $\tau_f = 0.5225$ ms, (**b**) saturated operation with $g = 1.522$ and $\tau_f = 0.4775$ ms, (**c**) fast-scale bifurcation with $g = 0.1$ and $\tau_f = 10.925$ ms, (**d**) coexisting (interacting) fast and slow-scale bifurcation with $g = 0.1$ and $\tau_f = 0.209$ ms, (**e**) slow-scale bifurcation $g = 0.8$ and $\tau_f = 0.492$ ms, (**f**) slow-scale bifurcation with $g = 2.4$ and $\tau_f = 7.125$ ms

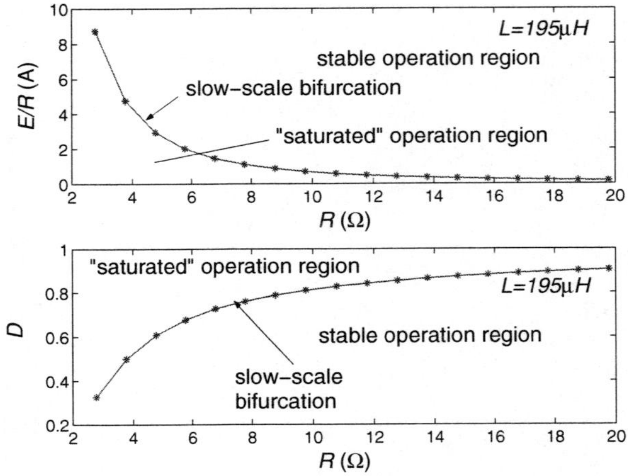

Fig. 3 Operating boundaries with varying E/R and D

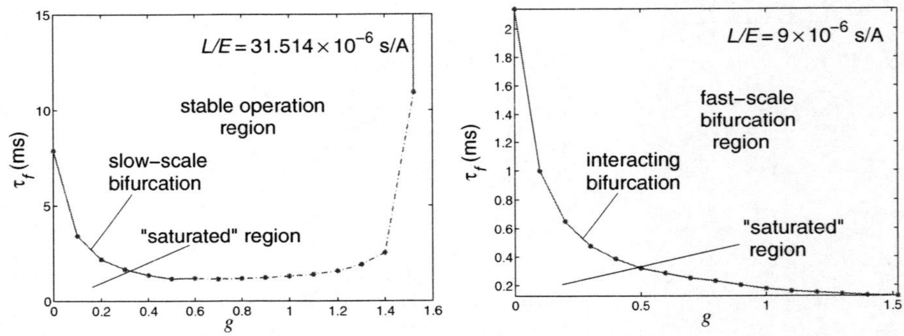

Fig. 4 Operating boundaries with varying feedback gain and time constant. $E = 6.1877$ V, $R = 10.87\Omega$

5 Conclusion

The current status of research in the identification of bifurcation behavior in power electronics can be considered mature in the sense that the salient types of bifurcation behavior, their underlying causes and the theoretical parameters affecting them have been well documented. We believe that the usefulness of research results for practical design applications rely on the way researchers formulate their problems. Study should be performed in termed of practically meaningful parameters and circuit configurations. Uncovering "new" phenomena in hypothetical power electronics systems with artificial parameter ranges will be less important, while design-oriented nonlinear analysis on practical power electronics systems will become a convincing approach for identifying the various operating regimes and the conditions under which certain regimes are possible. An obvious application is in the provision of accurate stability information for practical design of power electronics systems.

Acknowledgments This work is Supported by Hond Kong Polytechnic University Research Grant G-U284.

References

1. B.K. Bose (Ed.), Modern Power Electronics: Evolution, Technology, and Applications, New York: IEEE Press, 1992.
2. Proceedings of Annual IEEE Applied Power Electronics Conference and Exposition (APEC), New York: IEEE, since 1986.
3. Records of Annual IEEE Power Electronics Specialists Conference (PESC), New York: IEEE, since 1970.
4. R.D. Middlebrook and S. Cuk, "A general unified approach to modelling switching-converter power stages," IEEE Power Electron. Spec. Conf. Rec., pp. 18–34, 1976.
5. D.C. Hamill and D.J. Jefferies, "Subharmonics and chaos in a controlled switched-mode power converter," IEEE Trans. Circ. Syst. I, vol. 35, pp. 1059–1061, 1988.
6. J.H.B. Deane and D.C. Hamill, "Instability, subharmonics and chaos in power electronics systems," IEEE Power Electron. Spec. Conf. Rec., pp. 34–42, 1989.
7. J.H.B. Deane and D.C. Hamill, "Instability, subharmonics and chaos in power electronics systems," IEEE Trans. Power Electron., vol. 5, pp. 260–268, 1990.
8. C.K. Tse and M. di Bernardo, "Complex behavior of switching power converters," Proceedings of IEEE, vol. 90, pp. 768–781, 2002.
9. P.T. Krein and R.M. Bass, "Types of instabilities encountered in simple power electronics circuits: unboundedness, chattering and chaos," IEEE Applied Power Electron. Conf. Exp., pp. 191–194, 1990.
10. J.H.B. Deane and D.C. Hamill, "Chaotic behavior in a current-mode controlled dc/dc converter," Electron. Lett., vol. 27, pp. 1172–1173, 1991.
11. S. Banerjee and G.C. Verghese (Eds.), Nonlinear Phenomena in Power Electronics, New York: IEEE Press, 2001.
12. K. Chakrabarty, G. Podder and S. Banerjee, "Bifurcation behavior of buck converter," IEEE Trans. Power Electron., vol. 11, pp. 439–447, 1995.
13. E. Fossas and G. Olivar, "Study of chaos in the buck converter," IEEE Trans. Circ. Syst. I, vol. 43, pp. 13–25, 1996.
14. J.H.B. Deane, "Chaos in a current-mode controlled dc-dc converter," IEEE Trans. Circ. Syst. I, vol. 39, pp. 680–683, 1992.
15. S. Banerjee, E. Ott, J.A. Yorke and G.H. Yuan, "Anomalous bifurcation in dc/dc converters: borderline collisions in piecewise smooth maps," IEEE Power Electron. Spec. Conf. Rec., pp. 1337–1344, 1997.
16. G.H. Yuan, S. Banerjee, E. Ott and J.A. Yorke, "Border collision bifurcation in the buck converter," IEEE Trans. Circ. Syst. I, vol. 45, pp. 707–716, 1998.
17. M. di Bernardo, F. Garofalo, L. Glielmo and F. Vasca, "Switchings, bifurcations and chaos in dc/dc converters," IEEE Trans. Circ. Syst. I, vol. 45, pp. 133–141, 1998.
18. H.E. Nusse and J.A. Yorke, "Border-collision bifurcations for piecewise-smooth one-dimensional maps," Int. J. Bifur. Chaos, vol. 5, pp. 189–207, 1995.
19. S. Jalali, I. Dobson, R.H. Lasseter and G. Venkataramanan, "Switching time bifurcation in a thyristor controlled reactor," IEEE Trans. Circ. Syst. I, vol. 43, pp. 209–218, 1996.
20. Y. Kuroe and S. Hayashi, "Analysis of bifurcation in power electronic induction motor drive system," IEEE Power Electron. Spec. Conf. Rec., pp. 923–930, 1989.

21. Z. Suto, I. Nagy and E. Masada, "Avoiding chaotic processes in current control of ac drive," IEEE Power Electron. Spec. Conf. Rec., pp. 255–261, 1998.
22. C.K. Tse, "Flip bifurcation and chaos in a three-state boost switching regulator," IEEE Trans. Circ. Syst. I, vol. 42, pp. 16–23, 1994.
23. C.K. Tse, "Chaos from a buck switching regulator operating in discontinous mode," Int. J. Circ. Theory Appl., vol. 22, pp. 263–278, 1994.
24. W.C.Y. Chan and C.K. Tse, "On the form of control function that can lead to chaos in discontinuous-mode dc/dc converters," IEEE Power Electron. Spec. Conf. Rec., pp. 1317–1322, 1997.
25. W.C.Y. Chan and C.K. Tse, "Study of bifurcation in current-programmed boost dc/dc converters: from quasi-periodicity to period-doubling," IEEE Trans. Circ. Syst. I, vol. 44, pp. 1129–1142, 1997.
26. C.K. Tse and W.C.Y. Chan, "Chaos from a Cuk current-programmed Cuk converter," Int. J. Circ. Theory Appl., vol. 23, pp. 217–225, 1995.
27. C.K. Tse, Y.M. Lai and H.H.C. Iu, "Hopf bifurcation and chaos in a free-running current-controlled Cuk switching regulator," IEEE Trans. Circ. Syst. I, vol. 47, pp. 448–457, 2000.
28. H.H.C. Iu and C.K. Tse, "Instability and bifurcation in parallel-connected buck converters under a master-slave current-sharing scheme," IEEE Power Electron. Spec. Conf. Rec., pp. 708–713, 2000.
29. H.H.C Iu, C.K. Tse, V. Pjevalica and Y.M. Lai, "Bifurcation behavior in parallel-connected boost converters," Int. J. Circ. Theory Appl., vol. 29, pp. 281–298, 2001.
30. O. Dranga, C.K. Tse, H.H.C. Iu and I. Nagy, "Bifurcation behavior of power-factor-correction boost converters," Int. J. Bifur. Chaos, vol. 13, no. 10, pp. 3107–3114, October 2003.
31. C.K. Tse and O. Dranga, "Bifurcation analysis with application to power electronics," in Bifurcation Control: Theory and Applications, Edited by G. Chen, D.J. Hill and X. Yu, New York: Springer-Verlag, 2003.
32. B. Holland, "Modelling, analysis and compensation of the current-mode converter," Proc. Powercon 11, pp. I-2–1-I–2–6, 1984.
33. R. Redl and N.O. Sokal, "Current-mode control, five different types, used with the three basic classes of power converters," IEEE Power Electron. Spec. Conf. Rec., pp. 771–775, 1985.
34. C.K. Tse and Y.M. Lai, "Controlling bifurcation in power electronics: a conventional practice re-visited," (Invited Paper), Latin American Applied Research, vol. 31, pp. 177-184, 2001.
35. S. Banerjee, M.S. Karthik, G.H. Yuan and J.A. Yorke, "Bifurcation in one-dimensional piecewise smooth maps: theory and applications in switching circuits," IEEE Trans. Circ. Syst. I, vol. 47, 2000.
36. S. Banerjee, P. Ranjan and C. Grebogi, "Bifurcations in two-dimensional piecewise smooth maps: theory and applications in switching circuits," IEEE Trans. Circ. Syst. I, vol. 47, pp. 633–643, 2000.
37. C.K. Tse, Y. Zhou, F.C.M. Lau and S.S. Qiu, "Intermittent chaos in switching power supplies due to unintended coupling of spurious signals," IEEE Power Electron. Spec. Conf. Rec., pp. 642–647, June 2003.
38. Y. Zhou, H.H.C. Iu, C.K. Tse and J-N. Chen, "Complex intermittency in switching converters," Int. J. Bifur. Chaos, to appear.
39. C.K. Tse, Complex Behavior of Switching Power Converters, Boca Raton: CRC Press, 2003.
40. O. Dranga, C.K. Tse, H.H.C. Iu and I. Nagy, "Bifurcation Behavior of a Power-Factor-Correction Boost Converter, Int. J. Bifur. Chaos, vol. 13, no. 10, pp. 3107–3114, October 2003.
41. X. Wu, C.K. Tse, O. Dranga and J. Lu, "Fast-Scale Instability of Single-Stage Power-Factor-Correction Power Supplies," IEEE Trans. Circ. Syst. I, vol. 53, no. 1, pp. 204–213, January 2006.
42. X. Wu, C.K. Tse, S.C. Wong and J. Lu, "Fast-scale bifurcation in single-stage PFC power supplies operating with DCM boost stage and CCM forward stage," Int. J. Circ. Theory Appl., vol. 34, no. 3, pp. 341–355, May 2006.
43. D. Dai, S. Li, X. Ma and C.K. Tse, "Slow-Scale Instability of Single-Stage Power-Factor-Correction Power Supplies," IEEE Trans. Circ. Syst. I, to appear.

44. X. Wu, S.C. Wong, C.K. Tse and J. Lu, "Bifurcation Behavior of SPICE Simulations of Switching Converters: A Systematic Analysis of Erroneous Results," IEEE Trans. Power Electron., Accepted.
45. Y. Chen, C.K. Tse, S.C. Wong and S-S. Qiu, "Interaction of fast-scale and slow-scale bifurcations in current-mode controlled dc/dc converters," Int. J. Bifur. Chaos, to appear.
46. G. Chu, C.K. Tse and S.C. Wong, "A Model for Stability Study of PFC Power Supplies," IEEE Power Electron. Spec. Conf. Rec., Florida, USA, pp. 1298–1303, June 2007.
47. S.C. Wong, C.K. Tse and K.C. Tam, "Intermittent chaotic operation in switching power converters, Int. J. Bifur. Chaos, vol.14, no. 8, pp. 1971–2978, August 2004.

Collective Phenomena in Complex Social Networks

Federico Vazquez, Juan Carlos González-Avella, Víctor M. Eguíluz
and Maxi San Miguel

Abstract The problem of social consensus is approached from the perspective of nonlinear dynamics of interacting agents in a complex network. Some basic concepts, such as dynamical metastability, are discussed in the framework of the prototype voter model. In the context of Axelrod's model for the dissemination of culture we describe a co-evolutionary dynamics formulation with recent results on group formation and nonequilibrium network fragmentation and recombination transitions.

1 Introduction

Nonlinear dynamics includes a set of concepts, tools and generic phenomena that are being successfully applied to understand complex behavior of natural systems and man made devices. From this understanding new designs for improved technology are constantly being proposed. The study of social dynamics from this perspective of nonlinear phenomena is a relatively new scientific challenge attracting an increasing number of researchers coming from different disciplines. But also here, the understanding of basic mechanisms operating in the nonlinear social interactions should pave the way to innovative proposals to handle conflicts arising in the man made system which is a society.

Collective phenomena emerge from the nonlinear interactions of individual agents. The existence of thresholds or transitions between two collective states reflect an underlying nonlinear dynamics. A general class of social collective phenomena fall within what is termed consensus problems: The question is to establish when the dynamics of a set of interacting agents that can choose among several options (political vote, opinion, language, cultural features, etc.) leads to a consensus

F. Vazquez (✉)
IFISC, Instituto de Física Interdisicplinar y Sistemas Complejos (CSIC-UIB), E-07122 Palma de Mallorca, Spain, e-mail: federico@ifisc.uib.es

V. In et al. (eds.), *Applications of Nonlinear Dynamics*, Understanding Complex Systems, 189
DOI 10.1007/978-3-540-85632-0_15,

in one of these options, or when a state with several coexisting social options prevails. The former is an ordered state, while the latter is called a polarized state. An important issue is to identify nonlinear mechanisms producing a polarized state in spite of general convergent dynamics. The interaction among agents has two different ingredients. The first is the mechanism by which agents interact: the interaction rule. It can be an imitation mechanism, a social pressure mechanism (following a majority of neighbors), following a leader, etc. The second ingredient is the social network, that is, who interacts with whom. The process of formation of complex social networks is itself a nonlinear process. It turns out that the fact that a consensus is reached or not depends on nontrivial ways on both ingredients.

In this paper we address the consensus problem from the point of view of nonlinear interactions in complex networks. We first consider the prototype voter model [1, 2] of imitation dynamics for agents with two possible options. We describe a phenomena of dynamic metastability that occurs in complex networks of high dimensionality. There are long lived polarized states in which spatial domains of the two options coexist. Secondly, we consider a model proposed by Axelrod [3] for the dissemination of culture. This model can be considered as a multivariable elaboration of the voter model dynamics. This model, in a fixed network of interactions, exhibits a nonequilibrium transition between the ordered and polarized states for a critical value of the parameter measuring initial cultural diversity. We analyze here how this phenomena is modified when there is coupled dynamics (co-evolution) of the state of the agents and the links defining the social network. This results in a complex dynamics of group formation with nonequilibrium network fragmentation and recombination transitions.

2 Voter Model

The voter model is a stochastic model originated in the mathematical literature [1,2] recently considered in a variety of situations [4–8]. It is defined by a set of agents with two possible opinions or spins $s_i = \pm 1$ located at the nodes of a network. The elementary dynamical step in a *node-update* dynamics consists in randomly choosing one node and assigning to it the opinion, or spin value, of one of its neighbors, also chosen at random. A neighbor is defined as an agent connected to the one being considered by a link of the network. This interaction rule implements a simple mechanism of social imitation.

The voter model dynamics has two absorbing states, corresponding to situations in which all the spins have converged to the $s_i = 1$ or to the $s_i = -1$ consensus states. The ordering dynamics is stochastic and driven by interfacial noise. A standard order parameter to describe the ordering process [4, 5] is the average of the interface density ρ, defined as the density of links connecting nodes with different spin value. In a disordered configuration with randomly distributed spins $\rho \simeq 1/2$, while for a completely ordered system $\rho = 0$. In regular lattices of dimensionality $d \leq 2$ the system orders. This means that, in an infinite large system, there is a coarsening process with unbounded growth of spatial domains of one of the two opinions. The asymptotic regime of approach to the ordered state is characterized in $d = 1$ by a power law $\langle \rho \rangle \sim t^{-\frac{1}{2}}$, while for the critical dimension $d = 2$ a logarithmic decay is found $\langle \rho \rangle \sim (\ln t)^{-1}$ [4,9]. Here the average $\langle \cdot \rangle$ is an ensemble average.

Intuition might indicate that consensus should be reached more efficiently modifying a regular network by introducing random links that connect far away nodes of the network (that is, creating a small-world network [10]). It turns out that this intuition is wrong. For networks with random connections (and also for regular lattices with $d > 2$), the voter dynamics does not order the system in the thermodynamic limit of large systems [5,6,8,11]. Starting from a random initial condition and after an initial transient, the system falls in a partially ordered dynamical metastable state. After the initial transient ρ fluctuates randomly around an average plateau value ξ which gives the characteristic linear dimension of an ordered domain of one of the two options. In a finite system the metastable state has a finite lifetime: a finite size fluctuation takes the system from the metastable state to one of the two ordered absorbing states. In this process the fluctuation orders the system and ρ changes from its metastable plateau value to $\rho = 0$ (see Fig. 1). Note that these dynamical metastable states are different from frozen states that occur even in finite systems in other dynamical models.

Such metastable states can be characterized within a mean field approximation focusing on the link dynamics. One finds that for a random network and in the limit of infinite system size

$$\xi = \frac{\langle k \rangle - 2}{2(\langle k \rangle - 1)}, \tag{1}$$

where $\langle k \rangle$ is the average degree of the network, that is the average number of neighbors of a node. This result establishes a one parameter theory in which the degree of order reached by the system is determined only by $\langle k \rangle$ in any random

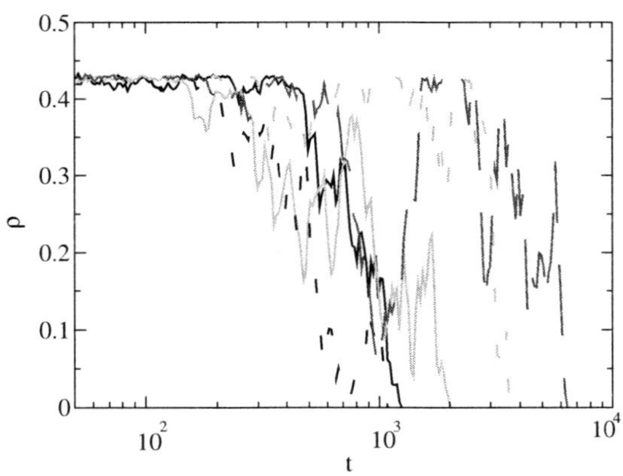

Fig. 1 Interface density evolution for several realizations of voter model dynamics in a random network

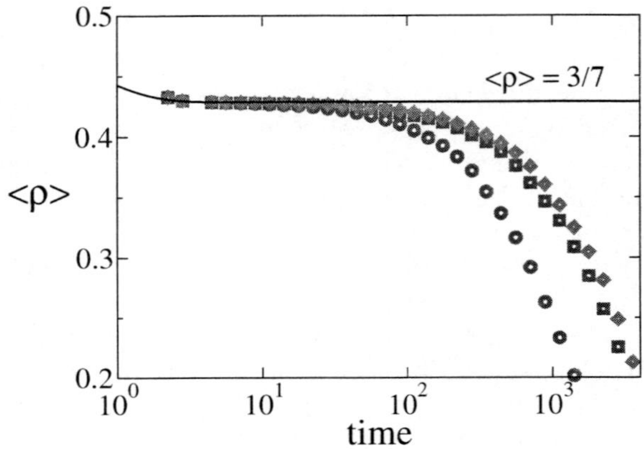

Fig. 2 Time evolution of the mean interface density for three random networks: Barabasi-Albert (*circles*), Exponential (*squares*) and Erdős-Rényi (*diamonds*). The system size is $N = 10^4$ and the average connectivity is $\langle k \rangle = 8$. The continuous line gives the mean-field result. Numerical simulations depart from this result due to finite size effects

network. Figure 2 checks this result against numerical simulations for different random networks. The exponential random network is constructed as in the Barabasi network [12] but with random instead of preferential attachment of new nodes, therefore it has an exponential degree distribution. In the Erdős-Rényi network every pair of nodes is connected by a link with probability p, giving a Poisson degree distribution with average connectivity $\langle k \rangle = Np$.

3 Axelrod Model

One of the crucial and yet unanswered questions in the social sciences is: if people become more similar in their beliefs, attitudes and behavior as they interact, why doesn't this interaction lead to homogeneity? [3].

To answer this question, social scientists have been proposing various mechanisms. A few years ago Robert Axelrod introduced a model [3] that explores the competition between consensus and polarization, in an attempt to describe how the culture disseminates within a society. He proposed a mechanism in which people become more alike when they interact (social influence) as, for instance in the voter model [1,2], but now with the possibility that the tendency to converge stops before it reaches completion. It also incorporates the idea that partners with more similarities interact more frequently than those with less similarities (homophily).

The "culture" of an individual is defined as a set of F different attributes, each of which can assume one of q possible traits, and they describe the individual's *features*. For instance, for $F = 3$ features and $q = 4$ traits per feature, we could have:

1. Language: English, Spanish, French, Hebrew;
2. Religion: Protestant, Catholic, Muslim, Jewish;
3. Food preference: pasta, meat, vegetables, tofu.

The basic mechanism is that neighbors who share at least one feature are more likely to interact, after which they agree on another feature previously not shared, thus becoming even more similar; while neighbors with no features in common do not interact.

This model has attracted the attention of many statistical physicists due to its rich dynamics and non-trivial behavior. It belongs to the class of stochastic models that, started from a given initial state, may or may not reach the consensus state. This gives rise to its most intriguing feature: an order-disorder non-equilibrium phase transition as we shall explain.

3.1 Model Definition and Results on a Square Lattice

The mathematical model consists of a population of individuals located at the nodes of a 2D square lattice that can interact only with its first four neighbors. The culture of an individual i is represented by an F-component array C_i^f $(f = 1,2\ldots,F)$ where each component can take one of q different values ($C_i^f = 0,1\ldots,q-1$). There are, therefore, q^F different possible states or type of individuals.

At each time step one individual i is chosen at random, and then it randomly picks one of its neighbors (j). Then one feature f is chosen randomly: if $C_i^f \neq C_j^f$ nothing happens but if $C_i^f = C_j^f$ another feature f' among the non-shared ones is chosen randomly and individual i adopts individual's j feature ($C_i^{f'} = C_j^{f'}$). This last step is equivalent to making i and j interact with a probability equal to the fraction of features that they share m/F, where $m = \sum_{f=1}^{F} \delta_{C_i^f, C_j^f}$ is their *overlap* and δ is the Kronecker delta function.

The model posses two competing features that lead to a very rich dynamical behavior. On the one hand, the interactions between nearest neighbors lead to more similarity producing the formation of same-culture domains. On the other hand, the incompatibility constrained between totally different neighbors prevents global consensus.

For a finite population, the system eventually freezes in a configuration that depends on q. For small q, the system coarsens as the neighbors tend to share the same culture and it freezes in a state where the average size of the largest monocultural region $\langle S_{max}\rangle$ is similar to the system size N (monocultural phase) (see Fig. 3). As q is increased beyond a critical value q_c, $\langle S_{max}\rangle/N$ suddenly drops to zero meaning that in the final state all the regions have a size much smaller than N (multicultural phase).

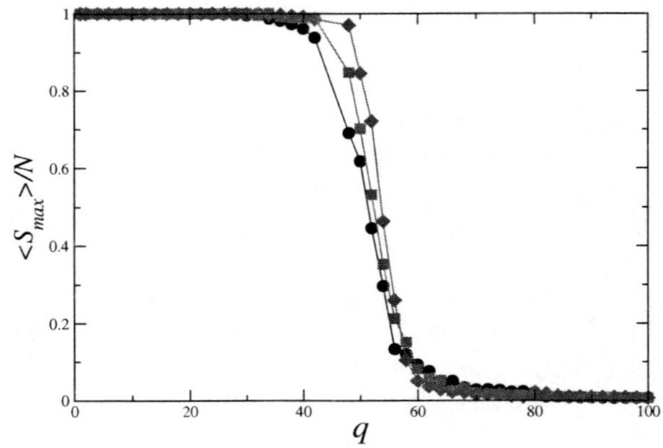

Fig. 3 Average size of the largest cultural domain $\langle S_{max} \rangle$ vs q for $F = 10$ features and system sizes $N = 900$ (*circles*), $N = 1600$ (*squares*) and $N = 2500$ (*diamonds*). The transition at the critical point $q_c \simeq 55$ becomes sharper as the system size increases

3.2 The Model in a Co-Evolving Network

In Sect. 3.1 we presented the original version of the model, where individuals are assumed to interact always with the same neighbors. However, real-life experience shows that people have a tendency to select their interacting partners according to their evolving social preferences. In this section we present a version of the model in which the interacting neighborhood of individuals changes with time. Individuals drop their ties with incompatible partners and form new ties with other like-minded individuals.

Individuals are located at the nodes of a degree-regular graph where every node has exactly $\langle k \rangle$ neighbors, and they initially take values for the features at random from 0 to $q-1$. In a Monte Carlo step, an individual i and one of its neighbors j are randomly chosen. If their overlap m is larger than zero they interact with probability m/F (as in the original model) and, in case of an interaction, the overlap is increased by one. However, if the overlap is zero, then i removes its link to j and attaches it to a randomly chosen node l with the condition that l was not already a neighbor of i.

Due to the random rewiring dynamics the original network quickly evolves into a random graph with a Poisson degree distribution with average connectivity $\langle k \rangle$. This distribution is conserved until the system freezes (see Fig. 4).

Even though the node degree distribution seems to be independent on q, the structure of the network in the final frozen state dramatically depends on q. During the evolution nodes are grouped in different communities with individuals sharing some features. Thus the network gets disconnected in different network components of like-minded individuals. Inside a component there are also groups of nodes with the same state, that we call domains.

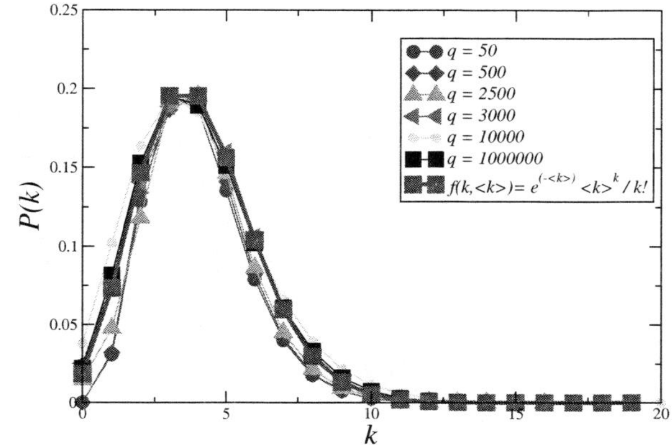

Fig. 4 Node degree distribution $P(k)$ of the co-evolving network in the final frozen state for a system with $F = 10$ features, population size $N = 10^4$ and various values of q. The system starts from random network with average degree $\langle k \rangle = 4$. $P(k)$ is very similar to a Poissonian (sketched in empty squares for comparison) for all values of q

In order to characterize the final structure of the network we calculated the average size of the largest component S as a function of q for simulations on a population of $N = 2500$ individuals and $F = 3$ features (Fig. 5).

We observe that there are two transitions in the absorbing state: a fragmentation transition from phase I to phase II and a recombination transition from phase II to phase III. In phases I and II, the system is frozen and composed by a set of disconnected components whose size depends on the degree of initial diversity q. For values of q smaller than a critical value $q_c \simeq 85$ (phase I), S is similar to the system

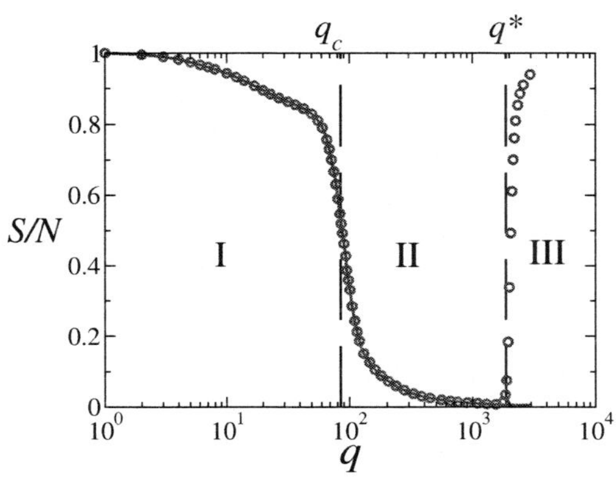

Fig. 5 Average relative size of the largest network component (*circles*) and largest domain $\langle S_{max} \rangle$ (*solid line*) in the stationary configuration vs q, for $F = 3$ features, $N = 2500$, averaged over 400 realizations. The vertical lines at $q_c = 85$ and $q^* = 1875$ indicate the transition points between the different phases

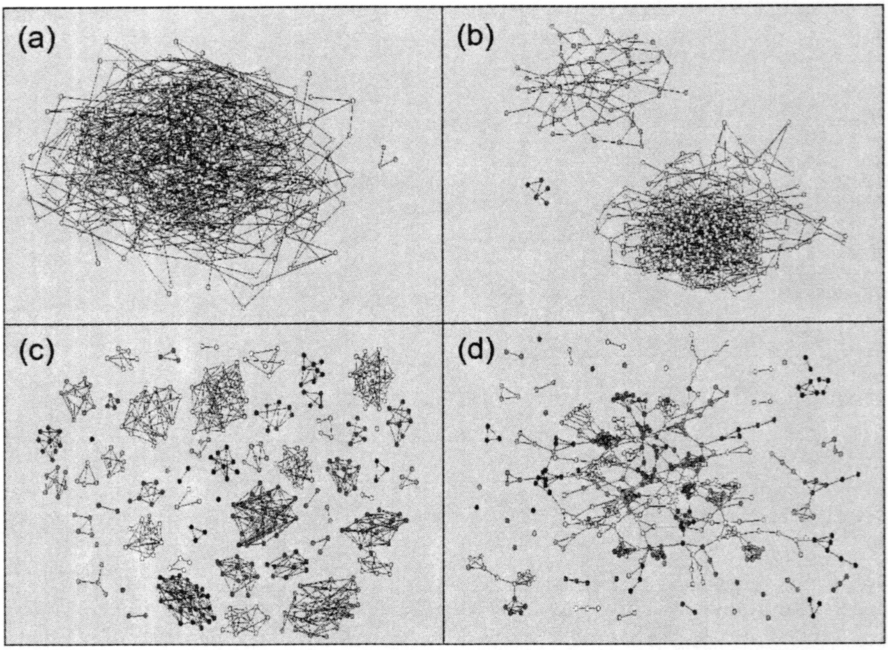

Fig. 6 Network structure in the final frozen configuration in phase *I*: $q = 3$ (**a**) $q = 20$ (**b**) and in phase *II*: $q = 100$ (**c**) for $N = 400$. (**d**) Snapshot of the network in the stationary active configuration (phase *III*) for $q = 500$

size N (Fig. 6-a and b), while for q above q_c (phase *II*), S is much smaller than N (Fig. 6-c). In analogy to the characterization of non-equilibrium phase transitions, we define q_c as the the value of q where the fluctuations in S reach a maximum value. The critical value $q_c \simeq 85$ in this co-evolving network is larger that the corresponding value $q_c \simeq 55$ obtained in a fixed 2D square lattice (Fig. 3). When q is above a transition point q^* (phase *III*), the system reaches a dynamically active configuration with many small domains. These domains have zero overlap between them and they are interconnected by links making up a large network component compared to the system size, as we observe in Fig. 6-d. Above the value $q^* \simeq \frac{NF}{\langle k \rangle}$ the initial number of pairs of nodes with at least one feature in common is smaller than the number of links in the system.

The nature of both transitions can be understood by studying the dynamical approach of the system to the final state in the three phases (see [13] for details).

3.2.1 Changing the Rewiring Rate

In the previous section we assumed that when a pair of nodes with zero overlap is chosen, the link between them is always rewired. We now consider the case in

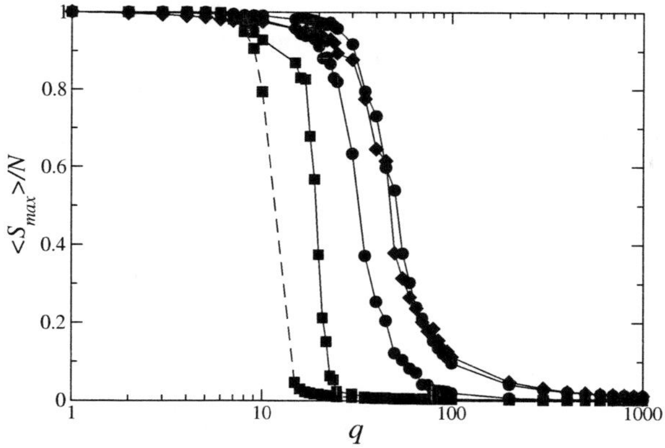

Fig. 7 Relative size of the largest domain vs q for values of the rewiring probability $p = 0, 10^{-6}, 5 \times 10^{-5}, 0.1$ and 1 from left to right, $N = 1026$ and $F = 3$ features. S was measured for every value of p at the same observation time $\tau = 10^8$

which the rewiring happens with probability p. Varying p is equivalent to change the relative time scales at which the copy and the rewiring dynamics occur. In the limit of p going to zero we expect the system to behave as in the original Axelrod model, where the network is fixed. On the other limit, when p is one we recover the co-evolving model studied before. Thus, we should see that the transition point q_c shifts to higher values of q as p is increased from zero. Figure 7 shows that the critical point for values of p above 0.1 is very close to the critical value $p_c \simeq 85$ for the $p = 1$ case, when S is measured at a fixed time $\tau = 10^8$. To investigate this

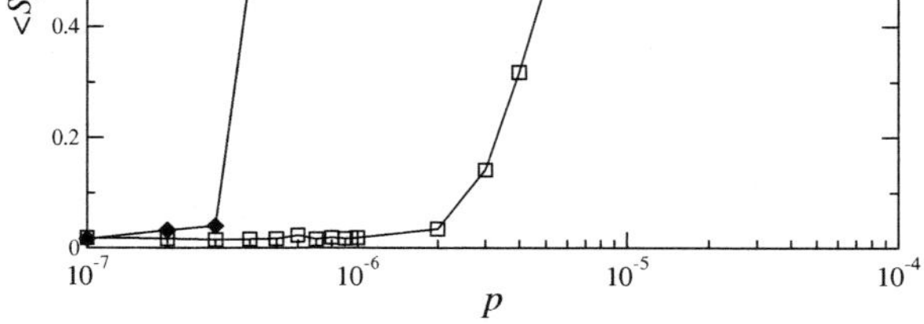

Fig. 8 Relative size of the largest domain vs the rewiring probability p for $N = 1024$, $F = 3$, $q = 20$ and observation times $\tau = 10^8$ (*squares*), $\tau = 10^{11}$ (*diamonds*) and $\tau = 10^{13}$ (*circles*). For a long enough observation time, $\langle S_{max} \rangle / N$ approaches to the value $\langle S_{max} \rangle / N \simeq 0.9$ independent on the value of p

dependence, we calculated S for fixed $q = 20$ in the connected phase I, and for three different observation times. As we observe in Fig. 8, for a fixed time, S/N is close to zero for very small values of p and it increases up to $S/N \simeq 0.9$ for larger values of p when the observation times are $\tau = 10^8$ and $\tau = 10^{11}$. However this transition in S seems to disappear when S is measured at longer times ($\tau = 10^{13}$). We expect that if we wait long enough the size of the largest component would reach the value $S/N \simeq 0.9$ for any value of p above zero. The last result means that there is a discontinuity in the critical value q_c at $p = 0$. For any $p > 0$, the behavior of S as a function of q is essentially the same in the long time limit.

3.3 Summary

The voter model for opinion formation exhibits a non-trivial dynamical behavior that depends on the spatial dimension of the interacting network of voters. In a finite network the system always reaches the consensus state due to fine size fluctuations. However, in infinite large networks, the consensus depends on the dimension. For $d \leq 2$ consensus is always reached in a characteristic time that depends on d. For $d > 2$, including random networks, the system reaches a stationary state where the two opinions coexist. This state is characterized by the average value of the stationary density of interfaces that monotonically increases with the average connectivity of the network.

The Axelrod model for the dissemination of culture is an incredibly rich non-equilibrium model that posses a quite interesting nonlinear dynamics and it is simple enough to be treated with traditional statistical physics tools. A phase transition from an ordered to a disordered phase is found when the degree of initial diversity in the population is increased above a critical value. When a co-evolutionary dynamics is added to the model, the order-disorder transition still happens but at a larger value q_c. The network remains connected below q_c, but it disintegrates in multiple small components above q_c (fragmentation). There is also a second transition at a much larger value q^* where a giant network component grows again as q is increased above q^* (recombination). We have also found that changing the rewiring rate does not affect the final result on co-evolution, it just adds an overall time scale to the system. The robustness of the co-evolving model was checked in [14] under the influence of an external noise. While in a fixed network the addition of a small amount of noise has the effect of ordering the system, in a co-evolving network the diversity is sustained even under the influence of a small external noise; i.e, the co-evolving Axelrod model is robust.

Acknowledgments We acknowledge financial support from MEC (Spain) through projects CONOCE2 (FIS2004-00953) and FISICOS (FIS2007-60327).

References

1. Holley R. Liggett T., Ann. Probab. **4,** 1975, 195.
2. T.M. Liggett, *Interacting Particle Systems* (Springer, New York 1985).
3. R. Axelrod, The dissemination of culture: A model with local convergence and global polarization, J. Conflict Res. **41,** 203–226 (1997).
4. I. Dornic, H. Chaté, J. Chavé, and H. Hinrichsen, Critical Coarsening without Surface Tension: The Universality Class of the Voter Model, Phys. Rev. Lett. **87,** 045701–045074 (2001).
5. C. Castellano, D. Vilone, and A. Vespignani, Incomplete ordering of the voter model on small-world networks, Europhy. Lett. **63,** 153–158 (2003).
6. K. Suchecki, V.M. Eguíluz, and M. San Miguel, Conservation laws for the voter model in complex networks, Europhy. Lett. **69,** 228–234 (2005).
7. V. Sood and S. Redner, Voter Model on Heterogeneous Graphs, Phys. Rev. Lett. **94,** 178701–178704 (2005).
8. K. Suchecki, V.M. Eguíluz, and M. San Miguel, Voter model dynamics in complex networks: Role of dimensionality, disorder and degree distribution, Phys. Rev. E (2005).
9. L. Frachebourg and P.L. Krapivsky, Exact results for kinetics of catalytic reactions, Phys. Rev. E **53,** R3009–3012 (1996).
10. D.J. Watts and S.H. Strogatz, Collective dynamics of 'small-world' networks, Nature **393,** 440–443 (1998).
11. P.L. Krapivsky, Kinetics of monomer-monomer surface catalytic reactions, Phys. Rev. A **45,** 1067–1072 (1992).
12. A.L. Barabási and R. Albert, Emergence of Scaling in Random Networks, Science **286,** 509–512 (1999).
13. F. Vazquez, J.C. Gonzalez-Avella, V. M. Eguiluz and M. San Miguel, Phys. Rev. E (2007), arXiv:0708.0776.
14. D. Centola, J.C. Gonzalez-Avella, V. M. Eguiluz , and M. San Miguel, J. Conflict Resolution (2007).

Enhancement of Signal Response in Complex Networks Induced by Topology and Noise

Juan A. Acebrón, Sergi Lozano and Alex Arenas

Abstract The effect of the topological structure of a coupled dynamical system in presence of noise on the signal response is investigated. In particular, we consider the response of a noisy overdamped bistable dynamical system driven by a periodic force, and linearly coupled through a complex network of interactions. We find that the interplay among the heterogeneity of the network and the noise plays a crucial role in the signal response of the dynamical system. This has been validated by extensive numerical simulations conducted in a variety of networks. Furthermore, we propose analytically tractable models based on simple topologies, which explain the observed behavior.

1 Introduction

The *stochastic resonance effect*, first conceived as a plausible mathematical explanation of the phenomenon of *glacial cycles*, has deserved an important part of the applied mathematics and physics literature of the last thirty years. It represents the interesting effect manifested in subthreshold nonlinear dynamics where a weak input signal can be amplified by the assistance of noise. Such a favorable result can be quantitatively explained by the matching of two time scales: the period of the input signal (the deterministic time scale) and the Kramers rate (i.e. the inverse of the average switch rate induced by the sole noise: the stochastic time scale) [1].

General statements about the necessary conditions for the emergence of stochastic resonance are usually related to three main aspects: (i) a dynamic non-linear

J.A. Acebrón (✉)
Departament d'Enginyeria Informatica i Matematiques, Universitat Rovira i Virgili, Tarragona, Catalonia, e-mail: juan.acebron@urv.cat

V. In et al. (eds.), *Applications of Nonlinear Dynamics*, Understanding Complex Systems,
DOI 10.1007/978-3-540-85632-0_16,

system endowing a potential with energetic activation barriers, usually bistable systems (ii) a small amplitude (usually periodic) external signal, and (iii) a source of noise inherent or coupled to the system. Neverthless, the response of the system can be enhanced by coupling it through several configurations, all-to-all, next neighbors [2], etc. This has been done mainly resorting to a linear coupling among oscillators.

In [3] we showed that this phenomenon is not restricted to the presence of stochastic fluctuations in the system, giving evidence that an equivalent amplification of a external signal can be obtained in a deterministic system of coupled bistable potentials in complex heterogeneous networks, producing what we termed topological resonance. Here, our goal is to show that the presence of noise can improve further the response of the system, by combining the topological resonance with the classical stochastic resonance phenomena.

The paper is organized as follows: in Sect. 2, we present the system model to be analyzed. The noiseless case is reviewed in Sect. 3. In Sect. 4, the role played by synchronization in the phenomenon is carefully analyzed. This is followed in Sect. 5 by an investigation of the effect of noise on the coupled dynamics. Finally, our results are summarized and discussed in Sect. 6.

2 Model Equations

The prototype system, where the stochastic resonance phenomenon has been carefully analyzed, is represented by the overdamped motion of a particle in a bistable potential subject to a periodic signal in presence of random fluctuations. In its simple form, it consists in the following dimensionless equation

$$\dot{x} = x - x^3 + A\sin(\omega t) + \eta(t) \tag{1}$$

where the bistable potential is $V(x) = -x^2/2 + x^4/4$, the external signal $A\sin(\omega t)$ and $\eta(t)$ a Gaussian white noise with $<\eta(t)>= 0$, and $<\eta(t)\eta(t')>= 2D\delta(t-t')$.

The deterministic system we will study corresponds to a network of elements obeying Eq. (1). The network is expressed by its adjacency matrix A_{ij} with entries 1 if i is connected to j, and 0 otherwise. For simplicity, from now on we will consider only undirected unweighted networks. Mathematically the system reads

$$\dot{x}_i = x_i - x_i^3 + A\sin(\omega t) + \lambda L_{ij}x_j + \eta_i(t) \qquad i = 1, ..., N, \tag{2}$$

where $L_{ij} = k_i\delta_{ij} - A_{ij}$ is the Laplacian matrix of the network, being $k_i = \sum_j A_{ij}$ the degree of node i. Here, $\eta_i(t)$ are again Gaussian white noises uncorrelated among them.

3 Model Without Noise

In this section, we review the role played by the heterogeneity in the dynamical behavior of the system, already presented in [3]. To this purpose, and to avoid mixing of nonlinear effects, the noise term has been neglected.

3.1 Numerical Results

Numerical experiments was performed over two classes of networks, considered representative of homogeneous and heterogeneous networks. For the former, an all-to-all connectivity is chosen, while for the latter the Barabasi-Albert (BA) network. Such a network model is probably the most studied growing model, which provides a scale-free (power-law) degree distribution for the degree of its nodes [4,5].

The mean gain G over different initial conditions was computed as function of the coupling λ, being the gain defined as $G = \max_i x_i/A$. The results obtained in [3] showed that the scale-free networks present a clear amplification of the external signal, for a significant range of values of λ that depends on the average degree, however the all-to-all connectivity does not amplify the signal.

3.2 Analytical Treatment

To understand the observed behavior, we analyzed a simple topology consisting on a star network. Such a topology is simple enough to be mathematically tractable, and simultaneously capable to capture the heterogeneity found in scale free-type networks. In fact, the resulting dynamical system can be seen as composed of two parts: the dynamic of the hub (the highly connected node in the network with $N-1$ links), x_H, and the remaining nodes, y_i, linked to the hub. Then, the dynamical system (2) can be written out as

$$\dot{x}_H = [1-\lambda(N-1)]x_H - x_H^3 + A\sin(\omega t) + \lambda \sum_{i=1}^{N-1} y_i, \tag{3}$$

$$\dot{y}_i = (1-\lambda)y_i - y_i^3 + A\sin(\omega t) + \lambda x_H, \qquad i = 1,\cdots,N-1. \tag{4}$$

If the coupling λ is sufficiently small, the dynamic of the nodes can be decoupled from that of the hub, obtaining the well known equation of an overdamped bistable oscillator. In absence of forcing, the system possesses two stable fixed points centered around ± 1, as a first approximation, which corresponds to the minimum of the potential energy function $V_i(x) = -y_i^2/2 + y_i^4/4$. When the amplitude of the forced signal is subthreshold, each node i oscillates around the minimum of its potential with the same frequency of the forcing signal. Then, Eq. (3) can be solved for the ith node, and asymptotically for long time yields,

$$y_i(t) \sim \xi_i - \frac{A}{\omega^2+4}\left[\omega\cos(\omega t) - 2\sin(\omega t)\right], \quad t \to \infty \tag{5}$$

with $\xi_i = \pm 1$ depending on the initial conditions. Inserting the solution above into the equation governing the dynamic of the hub, it can be rewritten as

$$\dot{x}_H = -V'_H(x_H) + A'\sin(\omega t) + B'\cos(\omega t) + \lambda \eta_i, \tag{6}$$

where $\eta_i = \sum_{i=1}^{N-1} \xi_i$, and $A' = A[1+2\lambda(N-1)]/(\omega^2+4)$, $B' = -2\lambda(N-1)/(\omega^2+4)$. Here $V_H(x,\eta) = -[1-\lambda(N-1)]x^2/2 + x^4/4 - \lambda x\eta$ represents the effective potential felt by the hub. Notice that the problem has been reduced to the motion of an overdamped oscillator in an effective potential driven by a reamplified forcing signal coming from the global sum of the remaining nodes.

From the equation above, two important facts can be observed. Firstly, the clear separation of scales governing both subsystems. In fact, a slight changes of coupling, decreases dramatically the potential barrier height for the hub, $h = [1-(N-1)\lambda]^2/4$, while remains almost constant for the remaining nodes. Moreover, when $\lambda = 1/(N-1)$ the barrier for the hub disappears, leading to a unique single fixed point $X_H = 0$. Secondly, the two possible solutions for the nodes, namely oscillations around ± 1, affect the dynamic of the hub as it were induced by a quenched disorder [6]. Indeed, in the limit $N \to \infty$, and by the central limit theorem, η_i behaves as a random variable governed by a gaussian probability distribution with variance $\sigma^2 = N$. This is because the initial conditions were randomly chosen. Note that both mechanisms may cooperate to allow the hub to surmount the potential barrier, and both are due mainly to the heterogeneity present in the network.

When $\lambda \neq 1/(N-1)$, Eq. (6) can be solved in the asymptotic long-time limit, carrying out the same calculation as before for the remaining nodes. This can be done in absence of forcing considering that now the hub relaxes to the equilibrium points, $x_H^{(0)}$. Such points are the minimum of the effective potential, and can be determined solving the cubic equation $V'_H(x_H^{(0)}) = -[1-\lambda(N-1)]x + x^3 - \lambda\eta$, being $V''_H(x_H^{(0)}) = -[1-\lambda(N-1)] + 3(x_H^{(0)})^2 > 0$. Therefore, as a first approximation, the resulting dynamics for the hub corresponds to oscillations around the equilibrium point with lowest potential value, being described by

$$\dot{\bar{x}}_H = -V''_H(x_H^{(0)})\bar{x}_H + A'\sin(\omega t) + B'\cos(\omega t), \tag{7}$$

where $\bar{x}_H = x_H - x_H(0)$. Hence, the solution is easily found, and yields

$$x_H(t,\eta) \sim x_H^{(0)} - \frac{1}{\omega^2 + a_H^2} \tag{8}$$
$$\{[B'\omega - A' a_H]\sin(\omega t) - [A'\omega + B' a_H]\cos(\omega t)\}, \quad t \to \infty$$

where $a_H = V_H''(x_H^{(0)})$. Note that $x_H(t,\eta)$ depends implicitly on the random variable η through the equilibrium point $x_H^{(0)}$. Knowing the long time evolution for the hub, its gain can be readily evaluated, and the result is

$$G(\eta) = \frac{1}{A}\frac{1}{a_H^2 + \omega^2}\sqrt{(B'\omega - A' a_H)^2 + (A'\omega + B' a_H)^2} \tag{9}$$

In practice, to cancel out the dependence on the initial conditions, averaging over them is required. However, this turns out to be equivalent to averaging over η_i. Therefore, the mean gain can be estimated, and is given by

$$< G >= \frac{1}{\sqrt{2\pi N}} \int_{-\infty}^{\infty} d\eta \, e^{-\frac{\eta^2}{2N}} G(\eta). \tag{10}$$

The influence on the size of the network, N can be analyzed resorting to the equation above. In particular, we are interested in the case of N large, and for $\lambda = 1/(N-1)$, when the mean gain attains its maximum value. Making use of the Laplace method, it is possible to find an asymptotic solution to Eq. (10), in the limit of infinitely many nodes, $N \to \infty$, which is

$$< G >= G(0) + \frac{1}{N} \frac{d^2 G}{d\eta^2}\bigg|_{\eta=0} \eta + O(N^{-2}), \tag{11}$$

Therefore, the maximum mean gain can be extracted from a network, turns out to be bounded from above by $G(0)$. This simple analysis was shown to match pretty well the results of the numerical experiments in [3].

3.3 Extension to Scale-Free Networks

Although a scale-free network is a much more intricate structure than the star network, it shares the finger print of heterogeneity namely the existence of hubs. A neat picture can be given considering that each highly connected node acts locally as it were the hub of a star network, with a degree k picked up from the degree distribution. Therefore, for a given coupling λ, we can find several star-like networks in different stages, depending on the degree of its local hubs. Recall that when $\lambda = 1/k$, the maximum signal amplification is attained for a hub with degree k. Furthermore, increasingly larger values of λ would activate local hubs with smaller degrees. In other words, the corresponding hub would provide the maximum gain of the network for such a coupling. Figure 1 shows such a predicted behavior in case of a

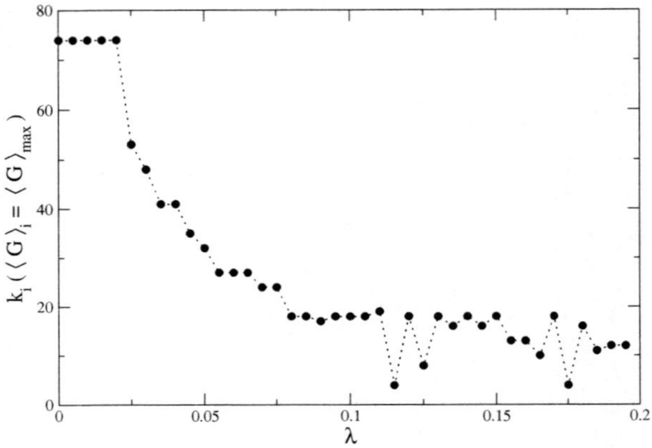

Fig. 1 Degree of the node in a BA network which provides the maximum mean gain $< G >$, as function of the coupling. Here $N = 500$, $A = 0.01$, and $\omega = 2\pi \times 10^{-1}$

BA network with average degree $< k >= 3$. Taking into account that the network possesses several hubs, it should exist a wide range values of λ for which the maximum gain is achieved. This contrasts with the results found for the star network, and anticipates what we already pointed out for scale-free networks in [3], showing that such a network consists on a much more robust topology where the phenomena can be found.

4 The Role of Synchronization

As opposed to other nonlinear coupled effects reported in literature [2], where synchronization seems to play a beneficial role in signal amplification, in our case it is responsible for deteriorating it. In [3], we showed that the mechanism described above remains valid until a full synchronization occurs. In the limit of large coupling, the nodes become tightly connected, synchronize and behave as a single node. By synchronization, here we mean spatial synchronization, measured it as the number of nodes occupying the two possible equilibrium points (the positive, n_+ and the negative one, n_-). Such a degree of synchronization, conveniently normalized by the number of nodes, was depicted in [3]. In particular, full synchronization was attained once a critical coupling was surpassed. Moreover, numerical simulations revealed that the path to synchronization is more pronounced for scale-free networks with lower average degree. Then, for this case a smaller range of values for which maximum gain is sustained was observed. Notably similar curves have been found in the prototype Kuramoto model [7], which describes synchronization phenomena in nonlinear coupled oscillators, on top of complex networks [8].

5 Model With Noise

Having found that the heterogeneity of the network can induce a signal amplification, a natural concern arises on the robustness of the phenomena when random fluctuations are taken into account. To this purpose, we performed some numerical experiments on the model equation (2), computing the average amplification $< G >$, as well as the the single-to-noise ratio (SNR) for several values of the noise and coupling. This has been done mainly for the star-like network configuration. We integrate numerically the stochastic differential equation (2) using the Euler-Maruyama scheme, and compute the power spectral density for the hub, which we know that provides the maximum signal amplification. The power spectrum was obtained averaging 128 noise realizations of 2^{20} time steps each. The SNR is defined here as the ratio of the signal power divided by the noise power in the signal bin. The signal power has been computed by substracting the noise background from the total power in the signal bin. The chosen amplitude for the input signal now is larger than in Fig. 1, to facilitate the interwell potential motion. In this way, the amount of noise neccesary to surmount the barrier potential is reasonably small, and thus the computational cost required to reduce the statistical error is reduced.

Figure 2 illustrates how the amplification and SNR curves changes as the noise increases. The amplification grows initially with increasing noise until reaching a maximum value, decreasing for larger values of the noise. A small amount on noise

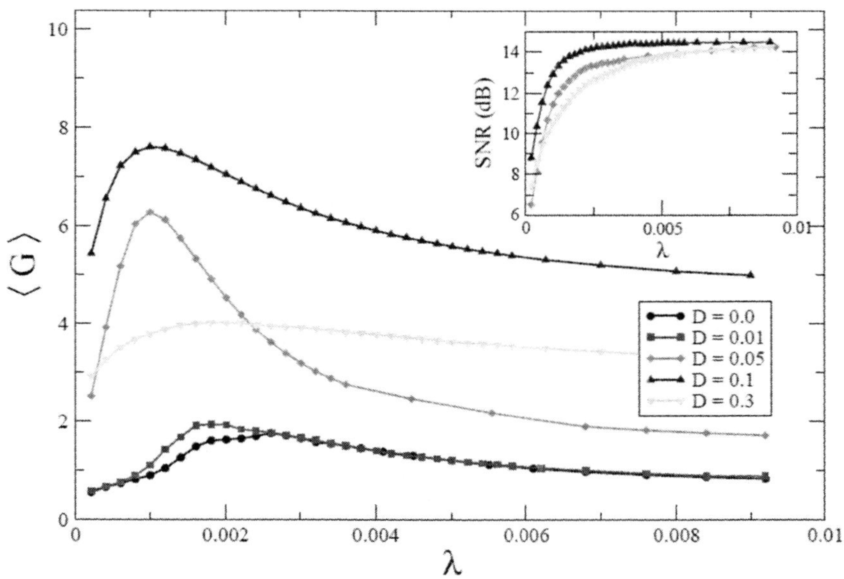

Fig. 2 Average amplification $\langle G \rangle$ as function of the coupling λ for four different values of the noise strength. Inset: The SNR output for the hub in the star-like network. Parameters are $A = 0.1$, $\omega = 2\pi \times 10^{-2}$

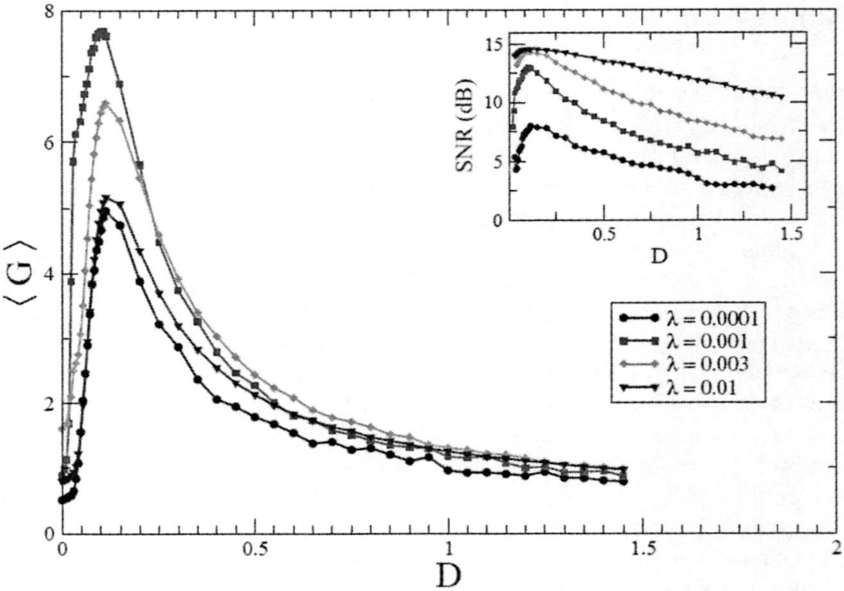

Fig. 3 Average amplification $\langle G \rangle$ as function of noise D for four different values of the coupling. Inset: The SNR output for the hub in the star-like network. Parameters are $A = 0.1$, $\omega = 2\pi \times 10^{-2}$

facilitates the hub to surmount the potential barrier, mainly because the reamplified signal coming from the leaves now increases. This can be explained resorting to the dynamics of the leaves in presence of noise. As a first approximation this corresponds to the motion of a noisy overdamped oscillator in a bistable potential, whose dynamics is well understood in the framework of the stochastic resonance [9]. There, the response of the system to a deterministic signal attains a maximum value for a nonzero value of the noise signal. Far above such a value the amplification response decreases when the noise increases. This affects the average amplification for the hub, in that it decreases correspondingly. In addition, Fig. 2 shows that the amplification curves peak for lower coupling values for the same reason explained above. As for the SNR, it grows monotonically with increasing coupling, and noise values.

In Fig. 3, the average amplification as well as the SNR as function of the noise is plotted for several values of the coupling strength. Note that the amplification achieves a maximum value for $\lambda = 0.001$, decreasing for larger values, while for the SNR it grows monotonically with the coupling when the noise is sufficiently large. Therefore, for a given value of the noise, increasing the coupling improves the signal response of the system in terms of SNR. The explanation rests on the fact that incre casing the coupling reduces the potential barrier, and thus facilitates the appearance of the characteristic SNR found in stochastic resonance phenomena.

6 Conclusions

We have extended the topological resonant-like effect found in [3] now in presence of external noise. Further, we have shown that an enhanced topological resonant-like effect can be induced by the optimization of the noise. The interplay between the cooperative interaction of nodes connected to the hub due to the topology, and the beneficial effects of the noise in surmounting the potential barrier, play a crucial role in the amplification of external signals. Moreover, an adjustment of the coupling can be used to improve further the system response, and may be important in the design of accurate remote sensing arrays.

We acknowledge support from the Spanish Ministry of Science and Technology, Grant FIS2006-13321-C02, and J.A.A. from the Ramon y Cajal programme. The assistance and usage of the resources provided by the BSC-CNS supercomputing facility is well appreciated.

References

1. Bulsara, A.R., Gammaitoni, L.: Tuning in to Noise. Phys. Today **49**, 39 (1996).
2. Lindner, J.F., Meadows, B.K., Ditto, W., Inchiosa, M.E., Bulsara, A.R.: Array Enhanced Stochastic Resonance and Spatiotemporal Synchronization. Phys. Rev. Lett. **75** 3 (1995).
3. Acebrón, J.A., Lozano, S., and Arenas, A.: Amplified signal response in scale-free networks by collaborative signaling. Phys. Rev. Lett. **99**(12), 128701 (2007).
4. Bollobas, B.: Random Graphs (2nd edition), Cambridge University Press, New York (2001).
5. Barabasi, A.L., Albert, R.: Emergence of scaling in random networks. Science **286**, 509–511 (1999).
6. Tessone, C.J., Mirasso, C.R., Toral, R., Gunton, J.D.: Diversity-induced resonance. Phys. Rev. Lett. **97** 194101 (2006)
7. Acebrón, J.A., Bonilla, L.L., Perez Vicente, C.J., Ritort, F., Spigler, R.: The Kuramoto model: A simple paradigm for synchronization phenomena. Rev. Mod. Phys. **77**, 137 (2005).
8. Gómez-Gardeñes, J., Moreno, Y., Arenas, A.: Paths to Synchronization on Complex Networks. Phys. Rev. Lett. **98**, 034101 (2007)
9. Gammaitoni, L., Hänggi, P., Jung, P., Marchesoni, F.:Stochastic resonance. Rev. Mod. Phys. **70** 223 (1998).

Critical Infrastructures, Scale-Free Networks, and the Hierarchical Cascade of Generalized Epidemics

Markus Loecher and Jim Kadtke

Abstract Characterizing the dynamics of large critical infrastructures is one of the most important national policy issues of recent years. Significant insight into these problems has been obtained by modeling the infrastructures as complex networks, using for example small-world and scale-free topological properties. Here, we investigate one aspect of this problem, namely how the flows of relevant quantities through these networks (e.g. information, electric power) are affected by particular heterogenous topological properties of the underlying network. We investigate some of these properties by employing known models of the spread of infection-like events in such networks, for example node failures or cyber-worm spreading. Specifically, we apply recent results on the dynamics of random walkers to the analysis of idealized epidemic outbreaks in scale-free networks. By replacing the node degree with the *random walk centrality* (RWC) we observe a refined hierarchical cascade leading to a greatly enhanced predictability for the order of infected nodes. We confirm our model results on data from real-world Internet maps at the autonomous system level. In addition we study the propagation of a stochastic disturbance as a function of a "small world" parameter. The present results are highly relevant for understanding information flow in scale-free networks and the development of dynamic and adaptive strategies that aim to mitigate network failures and attacks.

1 Introduction

One of the most important issues in homeland, national, and international security today is the development of operational and policy guidelines for the protection and recovery of large critical infrastructure networks. Because of their size and complexity, and the more recent addition of the internet as a common backbone, assess-

M. Loecher (✉)
Mao Networks, e-mail: loecher@mailaps.org

V. In et al. (eds.), *Applications of Nonlinear Dynamics*, Understanding Complex Systems, 211
DOI 10.1007/978-3-540-85632-0_17, © Springer-Verlag Berlin Heidelberg 2009

ing the behavior of these networks under stress has become increasingly difficult. While failure modes of these systems are important to understand, active attack of these networks has also become a realistic issue within the last several years. It will become critically important in the future that policy and defensive operations be informed by improved underlying dynamical models.

It has been proposed by several authors that recent results in complex network theory may provide simple and improved characterization of heterogeneous real-world networks. Particularly in the dynamics of the internet and cyberspace, network measures such as scale-free and small-world properties appear to be relevant. In fact, of the 17 critical infrastructure sectors, it appears that most approximately exhibit complex network properties [11]. This fact gives some promise that improved policies for network robustness and defense may be developed by taking such properties into account.

In this paper, we seek some insight into this general policy area by studying a limited technical problem. Namely, we aim to gain insight into how complex network topological properties affect the flow of relevant quantities through the network itelf, for example information flow through the internet or power flow through the electric grid. We attack this problem within the general theoretical framework of idealized disease spread, which can describe infection by a cumputer virus, or fault propagation in an engineered network. We draw on existing theoretical work to perform new numerical investigations of relevant behavior.

In this vein, studying the dynamics of the spread of "diseases" has a long history and is of utmost importance for the development of mitigation and containment strategies. More recently some of these results have been applied to technological networks, for example the spread of computer viruses in cyber-networks. Mathematical models usually assume a network of contacts among individuals along which the disease can be transmitted. The connectivity pattern of these networks has long been acknowledged as highly relevant to the time evolution of epidemic outbreaks as well as the final stationary state (see [1, 7, 8, 12, 18] and references therein).

Particularly interesting, and somewhat ubiquituous, heterogeneous connectivity patterns are found in scale-free (SF) networks. There, the number of links (degree) k attached to each node is characterized by a heavy-tailed distribution obeying a power-law of the form $P(k) \sim k^{-\gamma}$, with $2 \leq \gamma \leq 3$. For $\gamma < 2$, there is no meaningful "characteristic" degree (hence the term scale-free) but more importantly the probability of finding "hubs" or "superspreaders" is not negligible. These highly connected nodes are ultimately responsible for the rapid spreading of infections and the absence of an epidemic threshold [14, 15]. This insight is of great practical interest for, e.g. computer virus diffusion and the development of optimal strategies to protect technological networks. However, the detailed dynamics of the epidemic outbreaks has received far less attention than the stationary final states of affected networks.

In this paper, we refine and expand upon the recently observed [3] hierarchical dynamics of epidemic outbreaks in scale-free networks. While generally confirming the progressive cascade which infects hubs first and then percolates across successively smaller degree classes, we demonstrate significant improvements in the epidemic-path predictability. We illustrate the applicability of recent results in random walk theory to obtain a more precise temporal ordering of the nodes affected by the propagating infectious cascade. In particular, the *random walk centrality* (RWC) [13], which utilizes global topological information to rescale the node degree, serves as a greatly enhanced predictor for the most likely cascade originating from a single randomly chosen infection seed. We furthermore study the correlations of propagating signals in scale free networks as a function of a "small world" parameter.

2 Outline

This contribution is structured as follows. In Sect. 3, we outline the susceptible-infected (SI) model on SF graphs and then summarize and expand upon the recently observed [3] hierarchical dynamics of epidemic outbreaks in highly heteregoneous networks. In particular we demonstrate some of the shortcomings of using only local information (the node degree) to precisely quantify the temporal cascade. Section 4, provides convincing evidence for the superior performance of the RWC over the node degree in case such global connectivity information is available. We then show the relevance of our model insights for real Internet maps at the autonomous system (AS) level[1] and follow this analysis by our insights into signal propagation.

3 Hierarchical Propagation

Here, we focus on the standard susceptible-infected (SI) model [1] which is widely utilized in the study of computer virus infections. Each node of the network represents an individual and each link is a connection along which the infection can spread to other systems. Individuals exist only in two discrete states,"healthy" (susceptible) or "infected". If we denote the fraction of susceptible and infected nodes at time t by $s(t)$ and $i(t)$, respectively, then $s(t)+i(t)=1$. At each time step, a susceptible node is infected with probability λ if it is connected to one or more infected nodes. The densities of susceptible and infected individuals can be decomposed into the respective degree classes k: $s(t)=\sum_k s_k(t)P(k)$ and $i(t)=\sum_k i_k(t)P(k)$.

It can be shown [3] that the timescale of the exponential depletion of susceptible nodes is proportional to the degree k: $s_k(t)=1-i_k(t)=s_k^0 e^{-\lambda k \Phi(t)}$. Hence, regardless of the initial density s_k^0, the fraction of infected vertices belonging to degree class k' will rapidly surpass all densities $i_k(t)$ of lower degree classes, $k<k'$. The resulting hierarchical cascade was confirmed in numerical spreading experiments in BA networks [2, 3], where the initial seeds were taken to be a multitude of nodes selected at random.

We emphasize that this (global) statistical predisposition toward infecting hubs and large nodes very early in the process is clearly constrained by the local connectivities and thus by the distance to the initial seed. As the diameters of scale-free networks tend to grow at most logarithmically [5] with the number of nodes N, this "distance constraint" becomes less significant for a wide distribution of initial infection seeds. In order to fully understand the detailed spread of the epidemic, we randomly choose *one* node i_S as the initial seed and record both the degree $k(t)$ as well as the distance $d(t)$ from i_S for all newly infected nodes (NINs) at time t. Here, distance between two nodes is defined as the length of the shortest path between them. We have performed numerical simulations on a large variety of scale-free networks but for brevity will show results only for a real snapshot of the Internet at the autonomous system (AS) level as well as two recently introduced model networks [10, 19]. The AS level represents a coarse-grained description of the Internet, in which ASs are defined as independently administered domains which autonomously determine internal communications and routing policies. We are aware of the possibility of statistical bias in the AS topology due to sampling [4].

We summarize the results from extensive numerical simulations of SI outbreaks in the AS network in the following two Figures. Figure $1(a)$ shows the average degree of the NINs as a function of time t, defined as [2,3]:

$$\bar{k}_{NIN}(t_n) = \frac{\sum_k k[i_k(t_n) - i_k(t_{n-1})]}{i(t_n) - i(t_{n-1})}, \tag{1}$$

defined only if at least one new node was infected during timestep n, i.e. if $i(t_n) > i(t_{n-1})^2$. The observed initial plateau corresponds to an average degree of NINs for a low density of infected vertices given by $\langle k^2 \rangle / \langle k \rangle$ [3]. After this initial quick infection of the large-degree nodes, the epidemic successively spreads along smaller nodes and $\bar{k}_{NIN}(t)$ decreases smoothly with time. The lowest degree vertices tend to be reached last so that $\bar{k}_{NIN}(t \gg 1)$ converges toward the minimum degree of the network.

In Fig. $1(b)$ and (c) we plot the time evolution of the fraction of infected nodes within classes coarse-grained by degree (b) and distance to initial seed (c), respectively. These graphs provide further insights into the interplay of distance with the statistical inclination to infect hubs very quickly. Clearly, nodes closer to the infection seed get infected earlier, irrespective of their degree. We note in passing that similar observations regarding the distance to the seed have been published in [6]. At the same time the degree distribution $P_{NN}(k)$ of the nearest neighbors of a randomly chosen node is weighted by the number of links, $P_{NN}(k) = kP(k)/\langle k \rangle$. Therefore, the average degree of the nearest neighbors of a randomly chosen node is much larger than that of the node itself which explains the observed simultaneous cascades in Fig. $1(b)$ and (c).

Having established the existence of hierarchical propagation in a real representation of the Internet at the AS level begs the question of whether and how this added knowledge can be utilized to predict the most likely "path" along which an infection spreads. The observed hierarchical cascade implies that the degree of a node can serve as a statistical predictor of how early this node will get infected in the course of a spreading epidemic. The immediate questions to raise are (1) how good is such

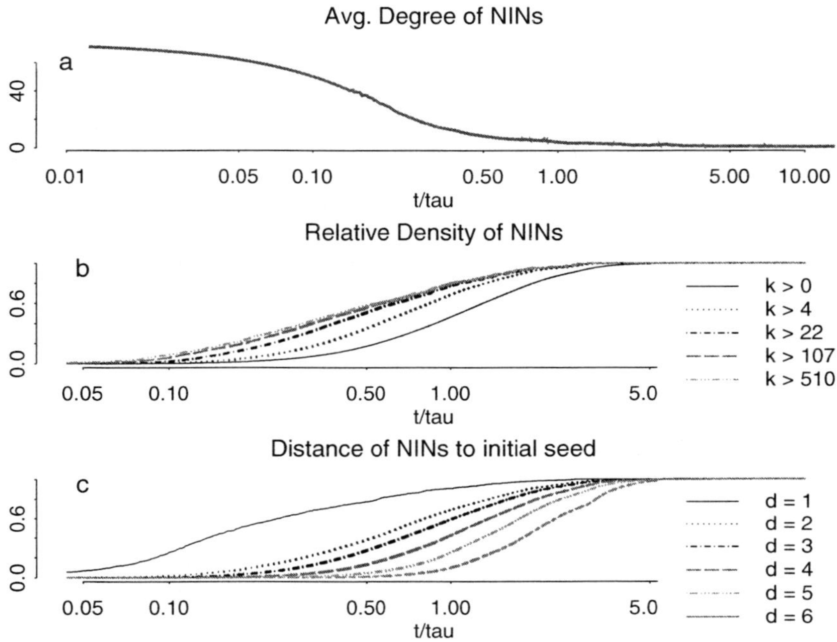

Fig. 1 (*color online*). (**a**) Time behavior of the average degree of the newly infected nodes (NINs) for SI outbreaks in the AS network; $N = 11461$, $\langle k \rangle = 5.7$, $\gamma = 2.2$ and $k_{max} = 2432$. Time is rescaled by the (exponential) growth rate $\tilde{\tau} = 1139$ iterations for $\lambda = 0.001$. (**b**) and (**c**) Cumulative fraction of infected nodes binned by degree and distance to initial seed, respectively, as a function of time. Note that an entry in the legend in (**b**) such as $k > 0$ should be interpreted in the context of the remaining bins, i.e. $4 \geq k > 0$

a prediction, and (2) are there even better approaches. Here we attempt to address these points in a quantitative fashion. We have found that a rescaled version of the node degree, which was derived in the context of random walk theory [13], serves as a much superior statistical predictor of the temporal order of infection. The computation of this predictor which we refer to as "random walk centrality" [13] (RWC) requires knowledge of the global network topology.

4 Random Walk Centrality

One peculiarity of heterogeneous networks is that a random walk from node i to j is not symmetric with the motion in the opposite direction. The asymmetry is characterized with the difference in the mean first passage times (MFPTÕs). It was revealed [13] that the difference is determined by a potential-like quantity: the RWC C of a node j is defined as the ratio of its normalized node degree $\tilde{K}_j = K_j/\mathcal{N}$ and the characteristic relaxation time $\tau_j = R^0_{jj}$:

$$C_j = \frac{\tilde{K}_j}{\tau_j}, \text{ with } \mathcal{N} = \sum_l K_l. \tag{2}$$

C_j is called the random walk centrality since it quantifies how central a node j is located regarding its potential to receive information randomly diffusing over the network [13]. To be more precise: Consider two nodes i and j with $C_i > C_j$. Assume that each of them launches a signal simultaneously, which is wandering over the network. Since the mean first passage time T between two nodes $i \neq j$ is simply $\langle T_{ij} \rangle - \langle T_{ji} \rangle = C_j^{-1} - C_i^{-1}$, one expects that the node with larger RWC will receive the signal emitted by its partner earlier. Hence, the RWC can be regarded as a measure for effectiveness in communication between nodes. The random walk centrality is a measure of the speed with which randomly walking messages reach a vertex from elsewhere in the network, a sort of random-walk version of closeness centrality. We note in passing that it is closely related to the *eigenvector centrality* a variant of which forms the essence of Google's *PageRank*. In a homogeneous network with translational symmetry, all nodes have the same value of the RWC. On the other hand, in a heterogeneous network the RWC has a distribution, which leads to the asymmetry in the random dynamic process.

We refer the reader to Ref. [13] for the exact definition of the relaxation time τ and here only remark that the 0th order moment matrix R^0, and thus τ, can be computed via Monte Carlo simulations or analytically by utilizing the full global connectivity matrix A_{ij} of the network, defined in[3].

For all networks that we have investigated, $i_C(t)$ as a function of the RWC becomes much smoother and nearly monotonous. To quantify this smoothness and its impact on predictability we follow Ref. [13] and measure for each node j the time T_j at which $i_j(t)$ exceeds a certain threshold i^*. We then compute the fraction f of node pairs (l,m) satisfying $T_l < T_m$ that violate the "expected" relations $K_l > K_m$ and $C_l > C_m$ respectively. For the AS network, this fraction improves from $f_K = 0.16$ to $f_C = 0.04$ and for the PFP network from $f_K = 0.14$ to $f_C = 0.01$, when rearranging the nodes according to the RWC.

An interesting deactivation growth model resulting in highly-clustered and scale-free networks has been proposed by Klemm and Eguíluz (KE) [10]. KE networks resemble regular chains with diameters that scale linearly with the network size N [16].

The original constructive algorithm is as follows: Each node of the network is assigned a binary state variable. A newly generated node is in the active state and keeps attaching links until eventually deactivated. Taking a completely connected network of m active nodes as an initial condition, each step of the time-discrete dynamics consists of the following three stages: (i) A new node joins the network by attaching a link to each of the m active nodes. (ii) The new node becomes active. (iii) One of the active nodes is deactivated. The probability that node i is chosen for deactivation is $p_i = ak_i^{-1}$ with normalization $a = \sum_j k_j^{-1}$. The mode generates networks with degree distribution $P(k) = 2m^2k^{-3} (k \geq m)$ and average connectivity $k = 2m$.

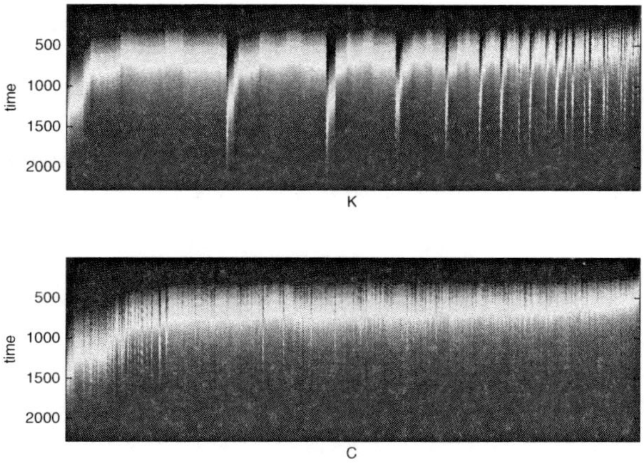

Fig. 2 (*color online*). Time evolution of the fraction of infected sites $i(t)$ as a function of ascending node degree K (*top*) and ascending RWC C (*bottom*) for the KE network with parameters $N = 1000$, $m = 6$

Although distance to the initial seed plays a much greater role in KE-networks, we nevertheless observe hierarchical propagation. We performed numerical simulations of SI outbreaks on KE networks with parameters $N = 1000$ and $m = 6$, the results of which are displayed in Fig. 2. One immediately notices the bands in the upper panel that correspond to node clusters which are infected much later than their degree would suggest. The RWC adjusts for the position of these nodes, leading to a much improved temporal prediction; we find the fraction of violators to decrease from $f_K = 0.27$ to $f_C = 0.1$.

5 Correlation Study

Co-occurrence of high clustering and short distances between nodes was originally termed as the "small world" phenomenon [17]. However, such networks which are typically created by randomly rewiring links of a regular lattice with probability $p \ll 1$, do not usually exhibit heavy tails in their degree distributions. Long-range connections can be introduced into the KE model by modifying stage (i) in the dynamical rules as follows [9]. For each of the m links of the new node it is decided randomly whether the link connects to the active node (as in the original model) or it connects to a random node. The latter case occurs with a probability μ. In this case the random node is chosen according to linear preferential attachment, i.e. the probability that node j obtains a link is proportional to the node's degree k_j. Varying μ in the interval $[0,1]$ allows the study of the cross-over between the BA model ($\mu = 1$) and the original high-clustering chain ($\mu = 0$). When increasing μ from zero to small finite values, the average shortest path length L drops rapidly and approaches the low value of the BAmodel. The clustering coefficient C remains practically constant in this same range.

In this paper, we investigate the propagation of disturbances in KE networks as a function of the small world parameter μ. "Disturbance" occurs in the context of stochastic background activity, which we attempt to generate according to simple but realistic rules. Our network traffic generator simulates "normal" packet traffic as follows:

- Each node can be a router or a host.
- Hosts create packages with probability p (0.15).
- Packet lifetime is drawn from user-defined distribution.
- Routers only route packages, do not create.
- Packages are randomly distributed along outgoing links.
- No finite queues, i.e. at each time step, all packets are routed until expiration.

For various values of μ we study the diffusion of perturbations by perturbing the ambient network state with a large flux burst. Our goal is to estimate correlation between nodes and dependencies on node degree, link distance, etc. Figure 3 shows the effect on the ten closest nodes of such a periodic addition of a rather large number of packets to one node in the original KE network ($\mu = 0$). We see that most nodes are almost unaffected. The situation in Fig. 4 is dramatically different. Here, there are many long range connections ($\mu = 0.15$) and the linear chain character of the KE network has been rendered into a "small world". As the distance distribution now is very narrow, we instead show the effect of the packet surplus on the ten largest nodes. We notice easily that not only are most large nodes strongly affected but even the packet flux at the "source" is modified and noisier. Given the periodicity of the disturbance, Pearson's linear correlation coefficient ρ is an adequate measure to quantify the observed correlations or lack thereof. Clearly, the two main factors determining the magnitude of ρ are degree and distance to the perturbed node. Figures 5 and 6 illustrate the complex interplay of degree and distance dependence for two values of μ. A highly heterogeneous pattern is visible in Fig. 5 which is due to the large range of distances in the $\mu = 0$ lattice. A very different picture emerges in Fig. 6 where the nonzero value for μ has introduced long range connections. As all pairwise distances now are distributed over a very narrow range we have the luxury of binning the correlations according to their distance. Figure 6 shows a clear, almost monotonic relationship between ρ and the degree if distance is held constant.

To conclude this exploratory analysis, we show in Figs. 7 and 8 the dependence of ρ on distance marginalized over degree for the two different values of μ. In the regular KE network with $\mu = 0$, correlations decay exponentially with distance. Distances cover a wide range, reflecting the chain-like topology of the KE structure. The small-world character established by a finite $\mu = 0.05$ leads to both an extremely small network diameter and a much slower correlation decay as is conveyed by Fig. 8.

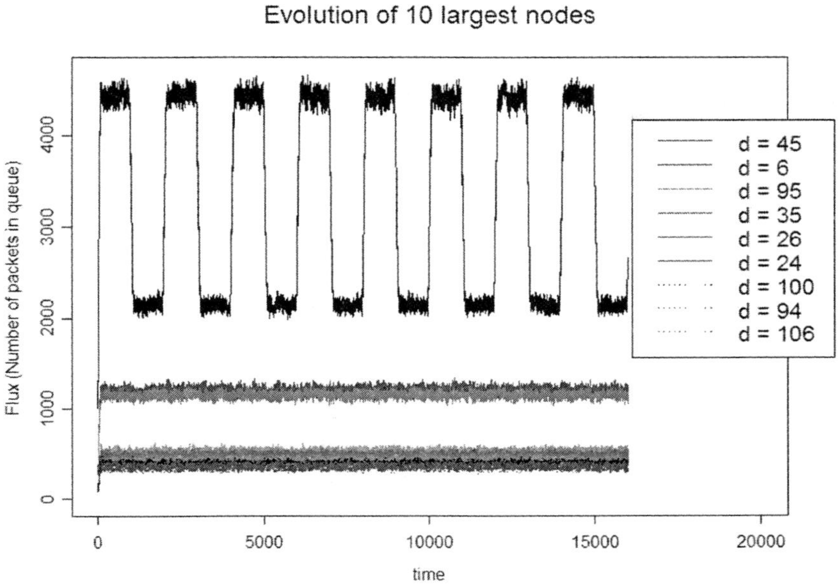

Fig. 3 (*color online*). Time evolution of packet flux at the ten closest nodes to the disturbance. For $\mu = 0.0$ the KE network is topologically similar to a linear chain and neighboring nodes tend to display very modest reactions to the large packet flux on node 0

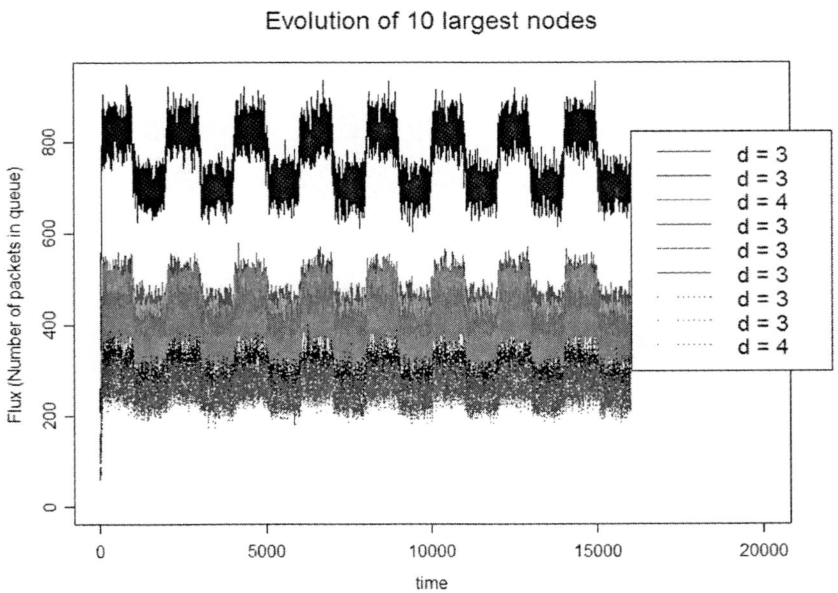

Fig. 4 (*color online*). Time evolution of packet flux at the ten largest nodes all close to the disturbance. F $\mu = 0.15$

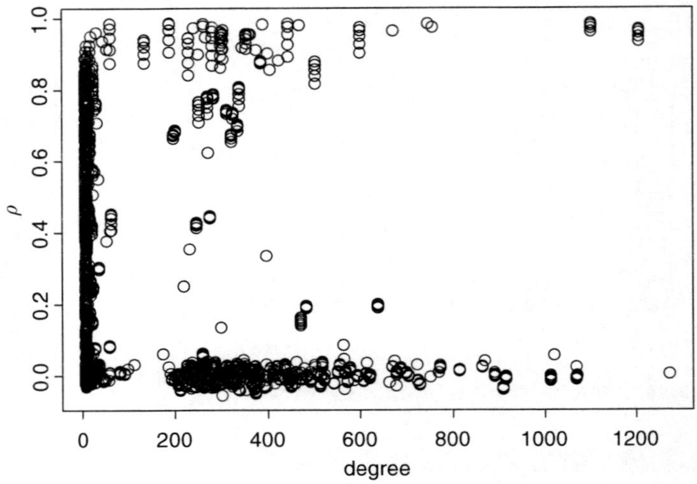

Fig. 5 $\mu = 0.0$: No clear relationship between ρ and the degree is discernible. Note that we marginalize over distance

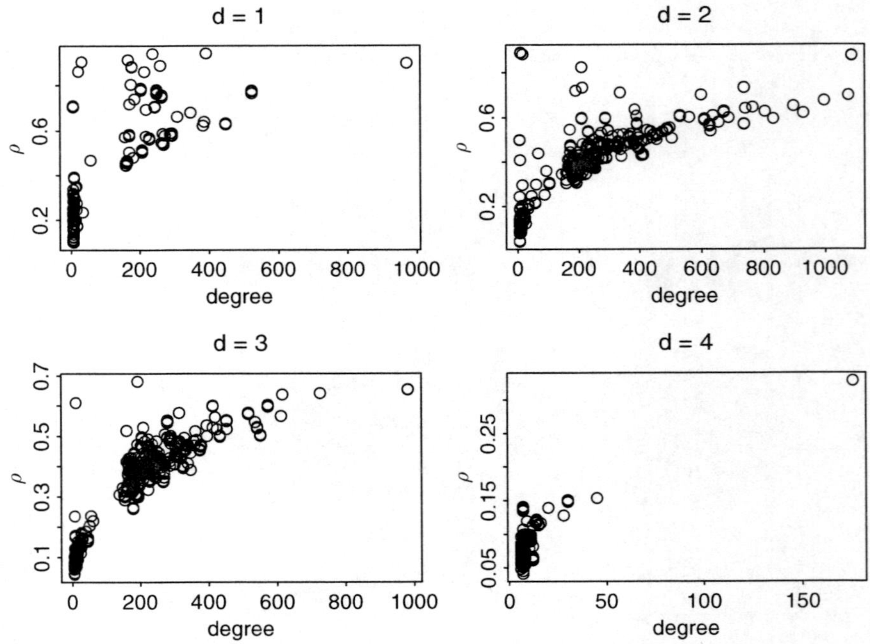

Fig. 6 For $\mu = 0.05$ distinct nearly monotonic patterns emerge when accounting for distance

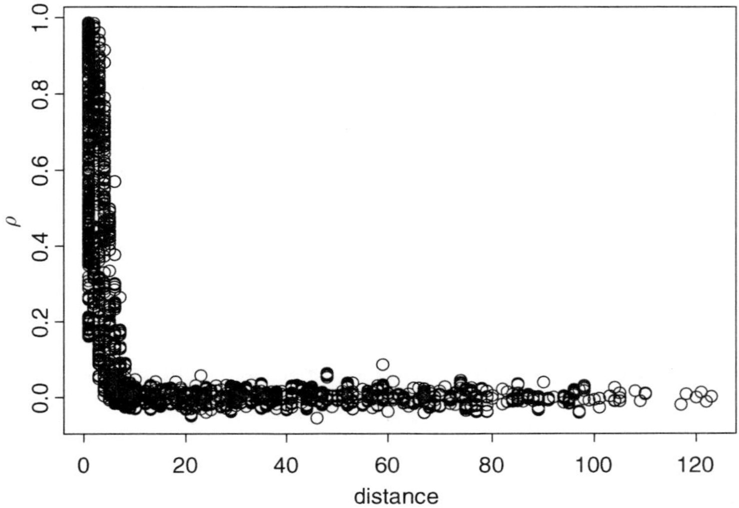

Fig. 7 Correlation ρ as a function of distance to the disturbance node for $\mu = 0.0$ for all nodes/degrees. We observe an exponential decay

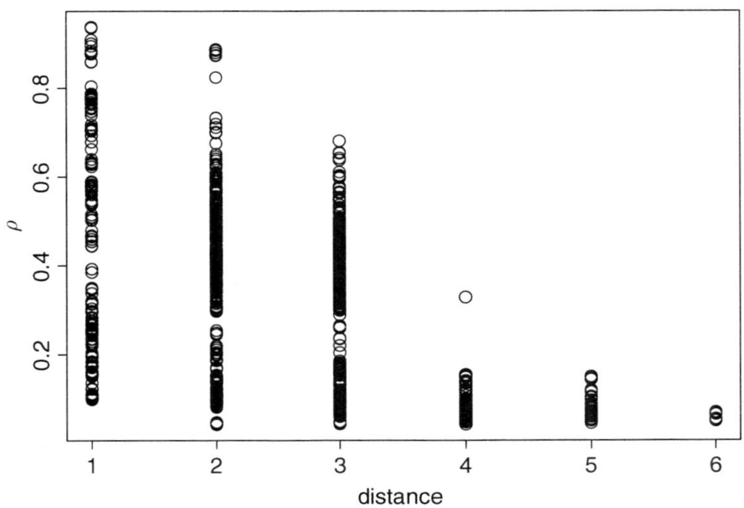

Fig. 8 Correlation ρ as a function of distance to the disturbance node for $\mu = 0.05$ for all nodes/degrees

6 Conclusion

In summary, we have demonstrated several results which are relevant from both a technical and policy standpoint.

From a technical standpoint, we have elaborated upon results concerning the likely statistical propagation paths of an infection-like phenomena. We show explicitely the complicated interplay between the hierachical cascade dynamics due to node degree, and the distance in links from the infection origin. We have demonstrated the improved statistical prediction capability of the random walk centrality over the simple node degree. We have also investigated the effect of distance from infection point using a flux perturbation scheme. What remains is to identify a unifying description of likely propogation dynamics which takes into account both the RWC and the distance in links. We have also demonstrated these results in data from a snapshot of the real-world internet as the AS level.

We have additionaly shown that there are fundamental differences in the τ fluctuations of the RWC between networks that otherwise show close similarities between other important topological features such as degree distribution, degree correlations, betweenness centrality, and characterstic path length. We therefore propose that this relaxation time distribution, or macroscopic summaries thereof, be used as an additional relevant topological property of complex networks.

From a policy standpoint, use of the above results on the RWC/distance interplay can provide improvements in tools for characterizing risk in real-world systems, such as infection or failure path probability maps, as well as related tools such as the calculation of improved vaccination or mitigation strategies. The use of the τ fluctuation distribution as an additional property to characterize networks may also allow better model selection for problems dealing with simulation of real-world networks. In general, because the application of such tools to real-world critical infrastructure systems is only beginning, a wide range of issues remains. We propose to attack some of these problems in future work.

Acknowledgments This work was supported by ARDA under Air Force Research Laboratory Contract No. F30602-03-C-0234. The views expressed are those of the authors and do not necessarily reflect the views of the supporting agency. The authors thank Martin Fong of Stanford Research Institute for help in calculating the large inverse matrix for the AS level analysis. M.L. wishes to thank Shi Zhou for providing the AS-Data, A. Barrat, M. Barthélemy and A. Vespignani for valuable discussions and Jae-Dong Noh for sharing the matrix formalism used in [13].

Notes

1. The Data Kit 0204 was collected as part of CAIDAs Skitterinitiative, http://www.caida.org. Support for Skitter is provided by DARPA, NSF, and CAIDA membership.
2. Averaging over a large number of temporal ensembles will reduce the times t_n for which $k_{NIN}(t_n)$ is not defined.
3. Private communications with Jae-Dong Noh; define $U_{ji} = A_{ij}/K_i$ and $V_{ji} = K_j/N$, then $R_{ij} = \langle i|[I-(U-V)]-1|j\rangle$.

References

1. R. M. Anderson and R. M. May. Infectious diseases in humans. Oxford: Oxford University Press, 1992.
2. M. Barthelemy, A. Barrat, R. Pastor-Satorras, and A. Vespignani. Dynamical patterns of epidemic outbreaks in complex heterogeneous networks. condmat/0410330, 2004. to appear in J. Theor. Bio 2005.
3. M. Barthelemy, A. Barrat, R. Pastor-Satorras, and A. Vespignani. Velocity and hierarchical spread of epidemic outbreaks in scale-free networks. Phys. Rev. Lett., 92:178701, 2004.
4. A. Clauset and C. Moore. Accuracy and scaling phenomena in internet mapping.Phys. Rev. Lett., 94(1):018701, 2005.
5. R. Cohen and S. Havlin. Scale-free networks are ultrasmall. Phys. Rev. Lett., 90(5):058701, 2003.
6. P. Crepey, F. P. Alvarez, and M. Barthelemy. Epidemic variability in complex networks. Phys. Rev. E, 73(4):046131, 2006.
7. S. Eubank, H. Guclu, A. Kumar, M. V. Marathe, A. Srinivasan, Z. Toroczkai, and N. Wang. Modelling disease outbreaks in realistic urban social networks. Nature, 429:180184, 2004.
8. H. W. Hethcote and J. A. Yorke. Gonorrhea: transmission and control. Lect. Notes Biomath., 56:1105, 1984.
9. K. Klemm and V. M. Eguluz. Growing scale-free networks with small-world behavior. Phys. Rev. E, 65:057102, 2002.
10. K. Klemm and V. M. Eguluz. Highly clustered scale-free networks. Phys. Rev. E, 65:036123, 2002.
11. T. G. Lewis. Critical Infrastructure Protection in Homeland Security: Defending a Networked Nation. Wiley-Interscience, 2006.
12. R. M. May and R. M. Anderson. The transmission dynamics of human immunodeficiency virus (hiv). Phil. Trans. R. Soc. Lond. B, 321:565, 1988.
13. J. D. Noh and H. Rieger. Random walks on complex networks. Phys. Rev. Lett., 92:118701, 2004.
14. R. Pastor-Satorras and A. Vespignani. Epidemic dynamics and endemic states in complex networks. Phys. Rev. E, 63:066117, 2001.
15. R. Pastor-Satorras and A. Vespignani. Epidemic spreading in scale-free networks. Phys. Rev. Lett., 86:3200, 2001.
16. A. Vazquez, M. Boguna, Y. Moreno, R. Pastor-Satorras, and A. Vespignani. Topology and correlations in structured scale-free networks. Phys. Rev. E, 67(4):046111, 2003.
17. D. J.Watts and S. H. Strogatz. Collective dynamics of small-world networks. Nature, 393:440, 1998.
18. J. A. Yorke, H. W. Hethcote, and A. Nold. Dynamics and control of the transmission of gonorrhea. Sex. Transm. Dis., 5:51, 1978.
19. S. Zhou and R. J. Mondragon. Accurately modeling the internet topology. Phys. Rev. E, 70(6):066108, 2004.

Noisy Nonlinear Detectors

A. Dari and L. Gammaitoni

1 Sensing in the presence of noise and nonlinearity

A sensor is a device specifically designed in order to measure a physical quantity [1]. Sensing is what a sensor does: an activity usually accomplished by means of a transducer, i.e. a device that converts the physical quantity to be measured into another physical quantity that is easier to measure. The measurement operation can result in an analog output, where the quantity to be measured is represented (most of the times) by the position of an indicator, or in a digital output, where the same quantity is represented by a number. In all the cases it is highly desirable that the sensor is designed so that the output behavior is as close as possible to the (unknown) behavior of the quantity to be measured.

There are at least two significant difficulties in designing a good sensor. The first one is related to the property called linearity: for proper operation it is usually required that the sensor output is a linear function of the measured input physical quantity, i.e. to a given variation of the sensed quantity corresponds a proportional variation of the output number. This property is clearly related to the transducer functioning that, most of the times, behaves in a linear way only in a small interval of values or under restricted conditions, thus limiting the sensor application range. As an example of linearity constraints, we might consider tha optical sensors [2, 3] usually employed for digital imaging like the popular CMOS Image Sensors. These sensor is cheaper than the traditional CCD sensor but, at the same time, shows a marked nonlinearity (distorsion) on its functioning. Another example can be taken from magnetic field sensors, like the magneto-resistive (MR) [4] and the super conductive quantum interference device (SQUID) [5], where both are inherently nonlinear.

There is no general and easy way around the nonlinear functioning of a transducer. In the last sixty years many attempts ([6–8]) have been made to find a lineariz-

A. Dari (✉)
NiPS Laboratory, Dipartimento di Fisica, Universitá di Perugia, Ialy

V. In et al. (eds.), *Applications of Nonlinear Dynamics*, Understanding Complex Systems, 225
DOI 10.1007/978-3-540-85632-0_18, © Springer-Verlag Berlin Heidelberg 2009

ing technique that might solve the problem. Present techniques are mainly focussed on algebraic manipulation of the digitalized output in order to transform it into a linear function of the input. The different approaches ranges from simple feedback and Jacobian linearization, to more sophisticated approaches like Carleman [9] linearization and artificial neural network applications [10]. Unfortunately, in order to accomplish results of good quality the knowledge of an accurate sensor mathematical model is required and this is not always available. In other cases, look-up tables or calibration curves are used to overcome the nonlinearity problem. This approach is usually faster but less accurate and, most importantly, it is very sensitive to the presence of noise.

Indeed the presence of noise, intended here as any spurious disturb that superimposes and/or distorts the original signal produced by the transducer, is the second significant difficulty that is inherent in any sensor design. Strictly speaking noise is not necessarily coming from the transducer action but it can be present since the beginning in the physical quantity under measurement. A classical example of this condition is the Johnson (thermal) noise that unavoidably affect any measurement of the resistance of any conductor at finite temperature. Noise can affect the functioning of a transducer in a number of different ways with the final results of limiting its sensitivity, its dynamic range and increasing the distortion.

1.1 A Simple Example

Although noise and nonlinearity are usually considered undesirable features of sensors, there are cases where they can be employed in a cooperative way in order to improve the response linearity, instead. The most simple example of this virtuous behavior is probably represented by the so called *Dithering* effect. Dithering is a technique employed in ADC (Analog-to-Digital Converters) devices invented long time ago [11–13] in order to mitigate the effect of amplitude quantization. Due to the intrinsic hard nonlinearity (thresholding) of ADCs the quantized signal is usually affected by distortion and lack of signal details smaller than the quantizing threshold. A mathematical representation of these errors is obtained by introducing the *quantization error* $\eta = x - y$, where x is the output of the ADC thresholding to the input y:

$$x = \begin{cases} 0 & if \quad y < b \\ 1 & if \quad y \geq b \end{cases}$$

where b is the threshold value. Usually η is treated as an additive noise whose statistical properties depend upon the statistical properties of the input y. In order to minimize such an error has been suggested to modify the input signal by adding an external signal (called *dither signal*) before quantization. A number of studies on the proper choice of dither signals have been performed since then. The main conclusions can be summarized as follows: (i) the addition of a proper dither signal can produce the independence and whitening of the error η resulting in both a reduction of the signal output distortion and an improvement on the system dynamic range; (ii) the best choice for the dither signal is a random dither uniformly distributed within an interval of amplitude equal to the quantization step. As it is apparent, in this case the presence of external noise is virtuously used to mitigate the device nonlinearity.

A somehow related [14] but much more complex phenomenon, where the presence of noise can improve the linear efficiency of a device is the Stochastic Resonance [15–17] effect. Here a bistable system is acted on by a sub-threshold periodic signal. The amount of the periodic component transferred to the system output can be maximized (closer to the linear response) by an optimal amount of input noise. The optimality condition is quantitatively determined by the matching of two time scales: the period of the sinusoid (the deterministic time scale) and the Kramers rate (i.e. the inverse of the average switch rate induced by the sole noise: the stochastic time scale). Thus the term "stochastic resonance".

1.2 The NANDs Scenario

In a number of cases the nonlinearity present in the functioning of the transducers manifest itself as an hysteretic input-output characteristics. Examples include magnetic field sensors, like the fluxgate sensor [18], wherein hysteresis in the **M–H** transfer characteristic is present in the ferromagnetic core [19], ferroelectric sensors [20], and mechanical sensors [21], e.g. acoustic transducers, made with piezoelectric materials. To deal with the bistable character of the sensor dynamics the detection of a target signal (dc or low frequency) is usually based on a spectral technique [18] wherein a known periodic bias signal is applied to the sensor so as to saturate it and drive it rapidly between its two stable states. Usually, the amplitude of the bias signal is taken to be quite large in order to render the response largely independent of the noise. The effect of a target dc signal is, then, to skew the potential, resulting in the appearance of features at the even harmonics of the bias frequency ω [22] in the system response. The spectral amplitude at 2ω is proportional to the bias frequency and the square of the target signal amplitude; hence, the spectral amplitude can be used to yield the target signal.

Recently, however, a new approach has been proposed [23], where the full nonlinear dynamics of the bistable sensor is exploited with the help of noise, in order to carry-on the measurement task. Specifically this new approach is based on the monitoring of the system output residence times intended as the time that the system spends in each of the two stable states before a transition from one state to the other occurs. Due to the presence of noise the residence time behaves as a random variable whose mean value is the same for both states in the case of symmetric system. The presence of an external signal to be detected and measured is assessed via the monitoring of the difference of the two residence times. In fact, without any external target signal, the residence times will be, on average, identical. The presence of an external target signal usually renders the potential asymmetric with a concommittant difference in the mean residence times which, to first order, is proportional to the asymmetry-producing target signal itself. This procedure has some advantages compared to the standard procedure. Among them, it can be implemented experimentally without complicated feedback electronics, with or without the presence of bias signals. In fact, the difference in residence times is quantifiable even in the *absence* of the periodic bias signal, with only noise driving the sensor between its steady states. Finally, the residence-times based technique works without the knowledge of the computationally demanding power spectral amplitude of the system output and it performs well in the presence of noise. In the following we will show how to compute the residence time difference and extract the information on the signal to be sensed.

2 Residence Time Based Detection Strategy: A Novel Approach

The control of the switch dynamics or, as is sometime called, the residence time approach to magnetic field detection has been already discussed within the framework of highly sensitive fluxgate magnetometers [24]. In this paper we focus our attention in the measurement strategy to be employed in order to sense the magnetic field.

The starting point is represented by the monitoring of the switch dynamics [25] in a symmetrically unperturbed bistable system. In order to fix our ideas we will consider in the following a two state discrete dynamical model. The Schmitt Trigger [26] (ST) model is often used as toy model for more complex bistable dynamics.

We consider a ST with static thresholds located at $\pm b$, an external signal $A(t)$ is fed into the system and the output $x(t)$ is considered (Fig. 2). As it is well known, $x(t)$ is a dicotomic signal assumed here of amplitude $x(t) = \pm x_m$: being $x(0) = x_m$, then the system rests in this state as long as $A(t) > -b$ with $t > 0$. As soon as $A(t) = -b$ the system output switch instantaneously into the lower state $x(t) = -x_m$ and remains so until $A(t) < b$. This simple functioning can be summarized in Fig. 1

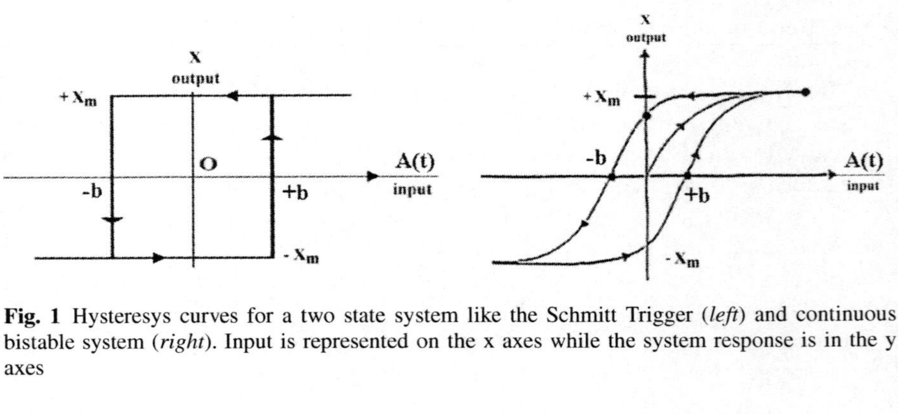

Fig. 1 Hysteresys curves for a two state system like the Schmitt Trigger (*left*) and continuous bistable system (*right*). Input is represented on the x axes while the system response is in the y axes

where the ST hysteresis cycle (left) is displayed together with an hysteresis cycle (right) typical of continuous dynamical system, for comparison.

For our purposes we consider the input signal $A(t)$ composed by a small DC target signal ε ($\varepsilon \ll b$) embedded in a (usually large) noise background represented here by the stochastic process $\zeta(t)$. As it is customary in this cases we will assume $\zeta(t)$ to be Gaussian band-limited (exponentially correlated) having zero mean, and correlation time τ_0. The noise $\zeta(t)$ originates from a white-noise driven Ornstein-Uhlenbeck (OU) process: $\dot{\zeta} = -\lambda\zeta + \sigma F(t)$, where $\lambda \equiv \tau_0^{-1}$, and $F(t)$ is white noise having zero mean and unit intensity σ^2. It is well-known that $\zeta(t)$ has correlation function $< \zeta(t)\zeta(s) >= \sigma_\zeta^2 \exp[-|t-s|/\tau_0]$ with variance $\sigma_\zeta^2 = \sigma^2\tau_0/2$.

In some cases it is useful to add an external periodic bias signal $B(t)$ that adds control to the functionality of the bistable sensor. As it will be apparent in the following the role of the bias signal is to help the input signal to cross the threshold at $\pm b$. Such a crossing is, in fact, necessary for the switch to happen and the output change to be recorded. On the other hand we note that in the presence of a Gaussian noise $\zeta(t)$ of the kind considered here, crossings will always happen at a due time. Thus the role of the bias signal $B(t)$ is merely confined to make such crossings to happen near a given time. We will show in the following however that the highest sensitivity for our bistable sensor is achieved when the bias signal is absent.

2.1 The Residence Time Approach

Having presented the main features of the dynamic system under study we are now ready to discuss the characteristics of our measurement strategy. This was already presented in Ref. [23, 25] and is here briefly summarized for convenience.

The use of residence time as an indicator of the system dynamics was initially proposed in [27] as a quantifier for the *Stochastic Resonance* [16] phenomenon. We define the residence time in the upper state as the time interval T_+ between the instant of a lower-to-upper state switch of the ST and the subsequent instant of the first upper-to-lower switch. Analogously it can be defined the residence time in the lower state as the time T_- interval between the instant of a upper-to-lower state switch of the ST and the subsequent instant of the first lower-to-upper switch. Being in the cases of interest the switch dynamic affected by the presence of the noise, T_+ and T_- behaves as random variable and the relevant quantities are their mean values known as mean residence times: $< T_+ >$ and $< T_- >$ respectively.

Without any external target signal, i.e. when $\varepsilon = 0$, the residence times will be, on average, identical. I.e. $< T_+ >=< T_- >$ and $< \Delta T >= | < T_+ > - < T_- > | = 0$. The presence of a target signal ($\varepsilon \neq 0$) renders the position of the switching thresholds $\pm b$ asymmetric with a concommittant difference in the mean residence times: $< \Delta T >= | < T_+ > - < T_- > | > 0$. Due to this feature it has been proposed to monitor the difference between the mean residence times of the two states of the system and to use this observable as a quantifier for detecting the presence of the target signal [23, 25].

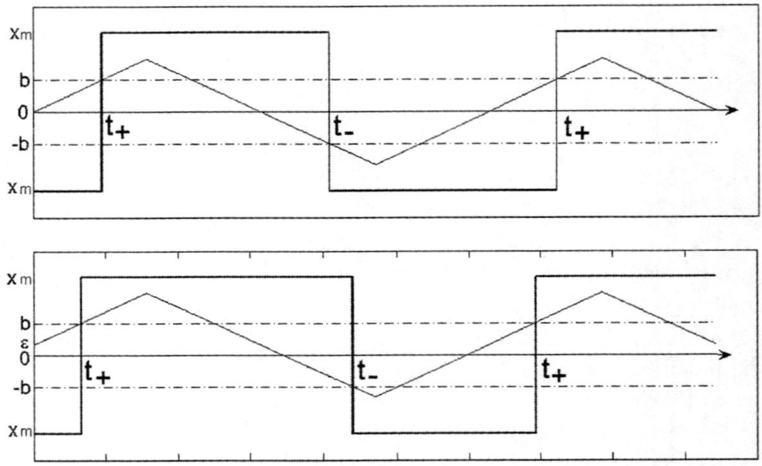

Fig. 2 Input and output time series for a standard Schmitt Trigger for the case where no DC signal is present and the residence times are equal (*upper*) and the case where there is a DC signal of amplitude ε (*lower*). In this second case it is apparent that the two residence times are not equal anymore

We note that the mean residence time difference, as described above, can be extended with little effort to dynamical bistable systems with more complex histeresys cycle like those in Fig. 2 (right). In this case it is necessary to introduce additional criteria to clearly discriminate when the switch event has effectively taken place. However this can be done unambiguously [27] and the mean residence time difference computed accordingly.

2.2 The Detection Strategy

The measurement of the mean residence time difference $<\Delta T>$ in a real bistable device can be a quite troublesome activity. In fact, to obtain the $<\Delta T>$ it is necessary to monitor the switches between the two states by keeping track of the chronology of each switch. Then the single T_+ and T_- have to be computed from the switching times and finally, upon accumulation of such quantities the mean residence times $<T_+>$, $<T_->$ are computed to generate $<\Delta T>$. In general, if a micro or nano sensor has to be built, incorporating all the machinery for computing $<\Delta T>$ can be a difficult task. Fortunately, as we are going to show, such a computation can be avoided without loosing the information on $<\Delta T>$. To show how this is possible we start considering the functioning of a ST in the presence of a triangular bias signal $B(t)$ of amplitude A, with a non-zero target signal and null noise.

$$B(\tau) = \begin{cases} 4A(\tau - n) & if \quad n \le \tau < n + \frac{1}{4} \\ 2A(1 - 2(\tau - n)) & if \quad n + \frac{1}{4} \le \tau < n + \frac{3}{4} \\ 4A(-1 + (\tau - n)) & if \quad n + \frac{3}{4} \le \tau < n + 1 \end{cases}$$

with $\tau = t/T_0$, where T_0 is the period of the triangular wave and $n = int(\tau)$, i.e. the integer part of τ. With such an input signal the ST output reads:

$$x(\tau) = \begin{cases} -x_m & if \quad n \le \tau < \tau_+ + n \\ +x_m & if \quad \tau_+ + n \le \tau < \tau_- + n \\ -x_m & if \quad \tau_- + n \le \tau < n + 1 \end{cases}$$

with $\tau_+ = t_+/T_0$ and $\tau_- = t_-/T_0$, where t_+ is the instant of the first upward crossing of the upper threshold (lower-to-upper state switch) and t_- is the instant of the first downward crossing of the lower threshold (upper-to-lower state switch).

For sake of simplicity let's consider the problem of sensing an unknown signal that is constant in time. The DC hidden signal of amplitude ε adds to the bias signal $B(t)$. The relevant crossing times in the absence of noise are readily computed under the condition $A(t) = B(t) + \varepsilon = \pm b$ as:

$$\tau_+ = \frac{b - \varepsilon}{4A} \tag{1}$$

$$\tau_- = \frac{b + \varepsilon}{4A} + \frac{1}{2} \tag{2}$$

As we have anticipated before, we are interested to compute the mean residence time difference $< \Delta T >= | < T_+ > - < T_- > |$. In the absence of noise, clearly $< \Delta T >= \Delta T = |T_+ - T_-|$. But $T_+ = t_- - t_+$ and $T_- = T_0 - T_+$. Thus

$$T_+ = T_0(\tau_- - \tau_+) = T_0 \left(\frac{A + \varepsilon}{2A} \right) \tag{3}$$

$$T_- = T_0(1 - (\tau_- - \tau_+)) = T_0 \left(\frac{A - \varepsilon}{2A} \right) \tag{4}$$

$$\Delta T = |2T_+ - T_0)| = |2(t_- - t_+) - T_0| = T_0 \frac{\varepsilon}{A} \tag{5}$$

or

$$\frac{\Delta T}{T_0} = \frac{\varepsilon}{A} \tag{6}$$

Equation (6) show that indeed the mean residence time difference carries the information about the hidden signal amplitude ε. Most importantly the amplitude of ΔT is linearly dependent from ε. A condition this one that, as we have seen before is quite desirable in a sensor.

In the presence of noise the deterministic values that we have computed here are still valid provided that two additional conditions are respected: (i) the noise intensity is small compared to the threshold: $\sigma \ll b$; (ii) the average value of the relevant quantities is considered [25, 28].

As we have pointed out before, the direct computation of ΔT as obtained through the measurement of t_- and t_+ is quite troublesome. In order to find a way around this problem we considered the following quantity:

$$I(t) = \int_0^t x(s)ds \tag{7}$$

In a single period T_0 we have:

$$I(1) = \int_0^1 x(s)ds = \int_0^{\tau_+} x(s)ds + \int_{\tau_+}^{\tau^-} x(s)ds + \int_{\tau_-}^1 x(s)ds = x_m[2(\tau_- - \tau_+) - 1] \tag{8}$$

By substituting the values for τ_+ and τ_- we obtain $I(1) = x_m \frac{\varepsilon}{A}$.

For a generic time $\tau > 0$ we have:

$$I(\tau) = nx_m\frac{\varepsilon}{A} + x_m \begin{cases} -\tau & if \quad n \le \tau < \tau_+ + n \\ (\tau - 2\tau_+) & if \quad \tau_+ + n \le \tau < \tau_- + n \\ 2(\tau_- - \tau_+) - \tau & if \quad \tau_- + n \le \tau < n+1 \end{cases}$$

Where, as we stated before, $n = int(\tau)$. As it is apparent the function $I(\tau)$ is composed by two distinct contributions: one step function, that increases its value every

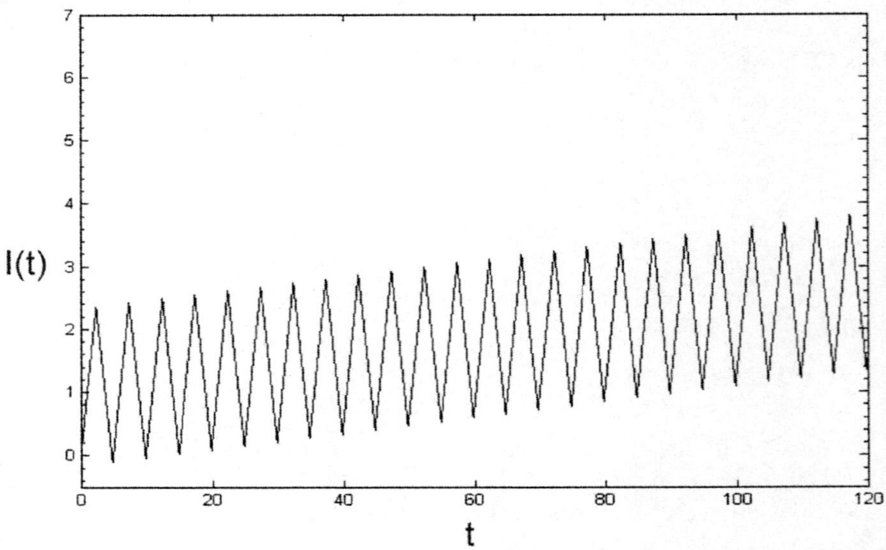

Fig. 3 Integral $I(t)$ of the system output. $I(t)$ is composed by two distinct contributions: one step function that increases its value every period of a quantity proportional to ε and a triangular function that superimposes onto the step function

period of a quantity proportional to ε and a triangular function that superimposes onto the step function (see Fig. 3).

It is interesting to note that if we sample $I(\tau)$ at discrete times $t_i = i + \tau_0$ with $i = 0, 1, \ldots$, we obtain values that lie on a straight line. For different τ_0 we have different lines but they are all parallels, i.e. they have all the same angular coefficient m:

$$m = x_m \frac{\varepsilon}{A} \tag{9}$$

Operatively, instead of measuring the computationally demanding ΔT we can use this result to obtain a measure of ε by monitoring the slope of a straight line obtained from the output integral, sampled at time interval equal to the period T_0 of the triangular bias. It is worth noticing that the slope computed in this way is equivalent to the average $< x >$ taken over any time interval equal to an entire multiple of the period T_0.

2.3 The Role of Noise

Before further analyzing this new approach to the measurement of ε, let's see what happens when the input signal is affected by noise. In Fig. 4 we show the behavior

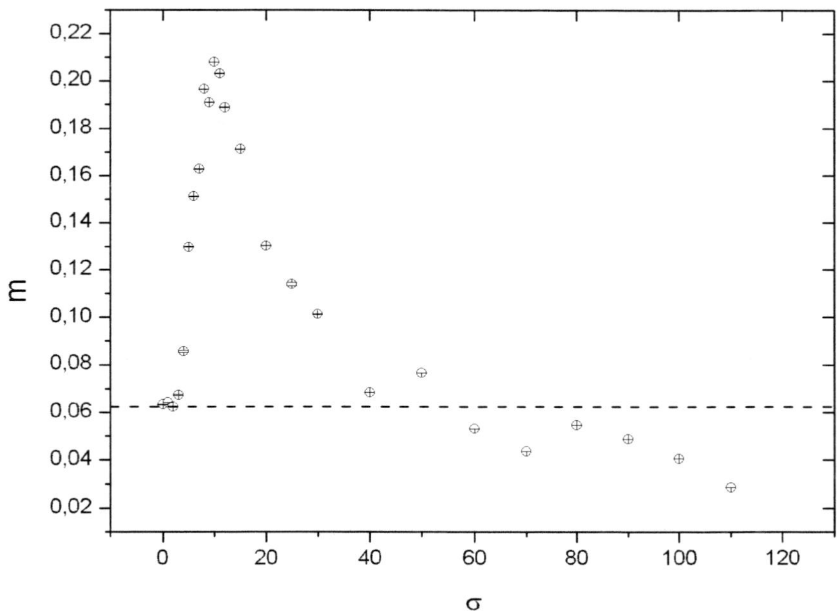

Fig. 4 Slope m versus the noise standard deviation σ, for a fixed value of the bias signal amplitude A and frequency $\nu_0 = 1/T_0$. The dashed line represents the theoretical prediction for the zero noise limit

of the slope m versus the noise standard deviation σ, for a fixed value of the bias signal amplitude A and frequency $\nu_0 = 1/T_0$.

The data are obtained by digital simulation of a ST. The horizontal line that intercepts the simulated data at $\sigma = 0$ represents the prediction in (9). On turning on the noise a non monotonic behavior for m sets in. A rapid grow up to a maximum value is followed by a slow decrease. We observe that here, the presence of a small amount of noise improves significantly the value of m (up to a factor 3.5), much on the line of what we have seen for the *Dithering* effect and the *Stochastic Resonance* phenomenon. Such an increase has been observed also in the direct measurement of the *DT* [23] and is due to a phenomenon called *resonant trapping* [29].

In a real measurement condition, one is usually faced with a situation where the hidden signal value ε and the noise intensity σ are fixed by external constraints while the signal bias amplitude and frequency can be chosen at will in order to maximize the sensor performances. In the following we will ignore the signal frequency, both because it has been treated elsewhere [25] and both because it is also related to the signal shape (see e.g. the case of the combined bias signal, square + triangular wave addressed in [25]).

In Fig. 5 we show the behavior of the slope m versus the bias signal amplitude A for fixed noise standard deviation σ and bias frequency $\nu_0 = 1/T_0$. The horizontal line represents the theoretical prediction for $A = 0$ obtained by computing the slope

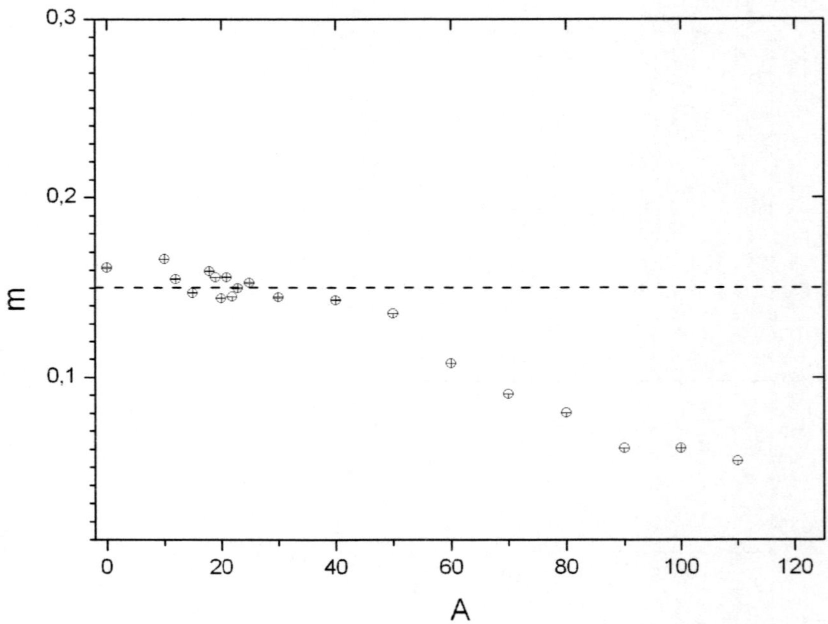

Fig. 5 slope m versus the amplitude A of the triangular forcing bias for a fixed value of the noise standard deviation $\sigma = 40$ and frequency $\nu_0 = 1/T_0$. The dashed line represents the theoretical prediction for the case where only noise is present (zero bias signal amplitude limit)

m with 8, where τ_+ and τ_- are estimated here in the limit where the threshold crossing mechanism is activated by the sole noise. In this limit τ_+ can be estimated as the *Mean First Passage Time* of the stochastic process $\xi(t)$, i.e. the average time that $\xi(t)$ takes to assume the value $\xi(t) = -b$ provided that $\xi(0) = b$. Analogously for τ_-. In the other limit (large A) the behavior of m can be easily reproduced by eq. 9 (dotted curve in the figure).

In conclusion, as it is apparent from Fig. 5, provided that threshold is low enough in order that the crossings can be induced by the sole noise, the best choice for the bias signal amplitude is $A = 0$, i.e. no bias signal. Quite a surprising result.

References

1. T. Grandke , W. H. Ko (eds), Sensors, A Comprehensive Survey, Volume 1, Fundamentals and General Aspects, Wiley, New York, (1989)
2. A. Rush, Nonlinear Sensors Impact Digital Imaging, Electron. Eng. December (1998)
3. W. Martin, High Dynamic CMOS Image Sensors, G.I.T. Imaging Microsc., pp. 2628 (2001)
4. J.R. Petta, T. Ladd, Weissman, M.B., Nonlinear AC response and noise of a giant magnetoresistive sensor, Magnetics, IEEE Transactions on , vol. 36, no.4, pp. 2057–2061, Jul (2000)
5. J. Beyer, D. Drung, F. Ludwig and Th. Schurig, Linearity of sensitive $YBa_2Cu_3O_{7-x}$ dc superconducting quantum interference device magnetometers, J. Appl. Phys. 86, 3382 (1999)
6. H. E. Singleton, Theory of nonlinear transducers, Technical report (MIT Research Laboratory of Electronics); 160 (1950) http://hdl.handle.net/1721.1/4896
7. D. Patranabis, S. Ghosh and C. Bakshi, Linearizing transducer characteristics, IEEE Trans. Instrum. Meas. 37 (1988) 6669.
8. R.P. Areny and J.G. Webster, Sensors and Signal Conditioning, second ed., Wiley, New York (2001)
9. K. Kowalski and W. Steeb, Nonlinear dynamical systems and Carleman linearization. World Scientific Publishing Co. Singapore, (1991)
10. N.J.M. Marques and B. Martin-del-Brio, Sensor linearization with neural networks, IEEE Trans. Ind. Electron. 48 (2001) 12881290.
11. W.R. Bennet, Bell Syst. Technol. J. 27, 446 (1948).
12. L. Schuchman, IEEE Trans. Commun. Technol. COM 12, 162 (1964).
13. B. Widrow, Trans. AIEE 79, 555 (1960).
14. L. Gammaitoni, Phys. Rev. E 52, 4691 – 4698 (1995).
15. R. Benzi, G. Parisi, A. Sutera and A. Vulpiani, Tellus 34, 10 (1982).
16. L. Gammaitoni, P. HŁnggi, P. Jung and F. Marchesoni, Rev. Mod. Phys. 70, 223 1998.
17. A.R. Bulsara and L. Gammaitoni, Phys. Today 49, 39 (1996).
18. W. Bornhofft and G. Trenkler in *Sensors, A Comprehensive Survey Vol. 5* eds. W. Gopel, J. Hesse and J. Zemel (VCH, New York 1989).
19. M. Rao, H. Krishnamurthy and R. Pandit Phys. Rev. **B42**, 856 (1990). F. Liorzou, B. Phelps, D. Atherton; IEE Trans. Magnetics **36**, 418 (2000).
20. I. Bunget and M. Prohescu *Physics of Solid Dielectrics* (Elsevier, NY 1984).
21. J. Fraden, *Handbook of Modern Sensors*, and references therein (AIP Press, NY 1997).
22. A.R. Bulsara, M.E. Inchiosa and L. Gammaitoni, Phys. Rev. E 55(4):4049–4056, 1997
23. L. Gammaitoni and A. Bulsara Phys. Rev. Lett. **88**, 230601 (2002).
24. B. Ando, S. Baglio, A.R. Bulsara and L. Gammaitoni, Circuits and Systems, 2003. ISCAS apos;03. Proceedings of the 2003 International Symposium on Volume 4, Issue , 25–28 May 2003 Page(s): IV-768 – IV-771 vol.4
25. A.R. Bulsara, C. Seberino, L. Gammaitoni, M.F. Karlsson, B. Lundqvist and J.W.C. Robinson. Phys. Rev. E. 67 16120, 2003.
26. J. Millman, *Microelectronics* (McGraw Hill, New York 1983).
27. L.Gammaitoni, F. Marchesoni, E. Menichella-Saetta and S.Santucci, Phys. Rev. Lett. 62, 349 (1989).
28. A. Nikitin, N.G. Stocks and A.R. Bulsara, Phys. Rev. E 68, 036133 (2003).
29. F. Apostolico, L. Gammaitoni, F. Marchesoni and S. Santucci, Phys. Rev. E 55, 36 (1997).

Cochlear Implant Coding with Stochastic Beamforming and Suprathreshold Stochastic Resonance

Nigel G. Stocks, Boris Shulgin, Stephen D. Holmes,
Alexander Nikitin and Robert P. Morse

Abstract Cochlear implants are prosthetic devices used to provide hearing to people who would otherwise be profoundly deaf. The deliberate addition of noise to the electrode signals can in principle increase the amount of information transmitted via an effect termed suprathreshold stochastic resonance (SSR). However, for SSR to occur the noise must be added in such a way so as to achieve, at least partial, independent stochastic activity across the nerve fibre population.

We are investigating the use of stochastic beamforming to achieve greater independence. The strategy involves presenting each electrode with a linear combination of independent Gaussian noise sources. Because the cochlea is filled with conductive salt solutions, the noise currents from the electrodes interact and the effective stimulus for each nerve fibre will therefore be a different weighted sum of the noise sources. We show that it is possible to get one independent point of excitation for each electrode. Also, we report for the first time, that stochastic beamforming lead to the observation of SSR in a cochlear implant.

Keywords Cochlear implant · Stochastic resonance · Beamforming · Noise

1 Introduction

Cochlear implants are a medical intervention routinely used to enable many adults and children with profound deafness to regain functional hearing [1, 2]. In many cases of profound deafness, the hair cells in the cochlea that normally convert sound into electrical signals have been severely damaged or completely destroyed. This can occur because of congenital factors, or happen in later in life through many causes such as disease or the use of some essential pharmaceuticals. Frequently, the

N.G. Stocks (✉)
School of Engineering, University of Warwick, Coventry, CV4 7AL, UK

V. In et al. (eds.), *Applications of Nonlinear Dynamics*, Understanding Complex Systems,
DOI 10.1007/978-3-540-85632-0_19,

conditions that damage the hair cells do not greatly affect the nerve of hearing – the cochlear nerve. This enables the hearing of many profoundly deaf people to be partially restored by direct electrical stimulation of the cochlear nerve by a set of about 16–22 electrodes that are surgically implanted into the cochlea [3]. This is achieved by a cochlear implant, which has three parts: the set of surgically implanted electrodes, an external speech processor that is used to encode a microphone signal into a set of appropriate electrode signals, and a radio link that transmits signals from the external speech processor to the internal electrodes.

When cochlear implants were first introduced, candidates were considered only if they got no benefit from conventional hearing aids, which simply amplify sounds. Cochlear implants, however, have now been shown to substantially improve the quality of life for people with some residual hearing [4, 5]. Even though speech processed through a cochlear implant sounds unnatural, speech comprehension in quiet can be very good because of the redundancy in speech [6]. But in poor listening conditions, the speech comprehension of cochlear implant users is severely impaired, even at noise levels that do not greatly perturb listeners with normal hearing. There is therefore a need to design better cochlear implant coding strategies to improve the naturalness of speech and speech comprehension. Expectations have also increased and many cochlear implant users would like the opportunity to enjoy music, which is less redundant than speech. To realize these expectations, cochlear implants must increase the amount of information that is represented by the spatio-temporal pattern of cochlear nerve discharges.

1.1 The Use of Noise in Cochlear Implants

The normal ear contains many sources of noise [7, 8], which may be an essential part of normal auditory coding [9–14]. We have previously proposed that Gaussian noise should be added to cochlear implant signals [10, 15, 16]. We have shown in physiological and computational studies that the addition of noise can theoretically result in nerve activity that more closely resembles that evoked by acoustic stimulation [10, 17], and, that the information transmitted by an array of nerve fibres can be enhanced by the addition of noise to a common input signal [18]. We have also shown that multiplicative noise, which is also present in the normal auditory system [7, 19], can theoretically enhance information transmission [20], and it may be that the information transmission is optimized by some combination of additive and multiplicative noise. Besides theoretical studies, tentative evidence that noise may be part of normal coding comes from psychophysical experiments with cochlear implant users. Noise is thought to increase frequency discrimination [21] and to enhance the sensitivity to amplitude modulation [22, 23]. Benham and Zeng [24] have also shown that noise can reduce the minimum detectable level of a harmonic in a complex tone, although it is not clear whether this effect is based on temporal or loudness cues.

The idea that noise can be useful is closely related to work on stochastic resonance, the phenomenon observed in some nonlinear systems where an increase in the input signal-to-noise ratio can lead to enhanced detection or information transmission. This effect, where noise can be beneficial, is now well established and has been demonstrated in many physiological experiments [25–27]. Our recent studies have extended this work to consider the effect of noise on the global information transmitted by an ensemble of neurons [15, 28]. Under these conditions, another form of stochastic resonance, termed suprathreshold stochastic resonance (SSR), can occur leading to a much improved information transfer.

Given the arguments for restoring the naturally occurring noise in the deafened ear and the results from SR, we believe the case for the deliberate use of noise in cochlear implants is substantial. Traditionally, however, noise is regarded as detrimental, and with good reason: if noise is added in an uncontrolled manner, it will almost certainly lead to worse performance [10,16,29]. Only if partially independent stochastic activity can be achieved in each nerve fibre can mechanisms like SSR be effective. One of our central goals is therefore to develop ways in which noise can be introduced to achieve statistical independence across nerve populations. This is the focus of this paper.

1.2 Stochastic Beamforming

As noted earlier, the direct stimulation of each electrode by independent Gaussian noise would not be expected to lead to independent noise at each nerve fibre. Greater independence could be achieved by using a larger number of electrodes that are closer to the nerve fibres. The number of electrodes, however, is limited by manufacturing constraints and the greater risk of further damage to the cochlea with longer or larger diameter electrode arrays [e.g. [30]]; animal experiments have shown that such damage may cause further degeneration of the cochlear nerve. Similarly, electrode arrays that are closer to the nerve fibres are more likely to damage the cochlea.

An alternative strategy is to present each electrode with a linear combination of independent Gaussian noise sources (Fig. 1). Because the normal and the deafened the cochlea is filled with conductive salt solutions, the current from an electrode spreads throughout the cochlea [31,32]. The noise currents from the electrodes therefore interact and the effective stimulus for each nerve fibre will be a different weighted sum of the noise sources. To some extent therefore, the effective stimulus for a nerve fibre will be independent of the effective stimulus of neighbouring fibres and all the noise sources. We refer to this approach as stochastic beamforming.

This strategy is compatible with improved electrode designs. The strategy has similarities with beamforming where a spatial filter operates on the output of an array of microphones to enhance directional selectivity. The approach also has similarities with the deconvolution method for reducing current spread proposed by Townshend and White [33]; in this deconvolution method, the matrix that maps electrode currents to excitation level at each point of neural excitation is determined and the inverted matrix can theoretically be used to get excitation at a single place. The main difference is that the source signals for the stochastic beamforming are

Gaussian processes rather than information-bearing signals and the effectiveness of the strategy will therefore be less sensitive to the matrix values.

For a particular cochlear implant subject, the electrode position and the amount of current spread are fixed. The fundamental objective is therefore to find the linear combination of noise sources that leads to the greatest independence between nerve discharges. More specifically, with reference to Fig. 1, let S be the structure matrix that represents linear current spread such that the observed signals X (at specified spatial locations) are a linear combination of the electrode sources E given by:

$$X = SE \tag{1}$$

The assumption that the spread of current is linear is supported by physiological measurements from the guinea pig cochlea [34] and the success of lumped resistive-models of current spread [35, 36].

In turn let the electrode sources, E, be composed of a linear combination of independent noise sources, N, which are governed by the matrix, Γ, such that

$$E = \Gamma N. \tag{2}$$

Hence the observed signals are given by

$$X = S\Gamma N. \tag{3}$$

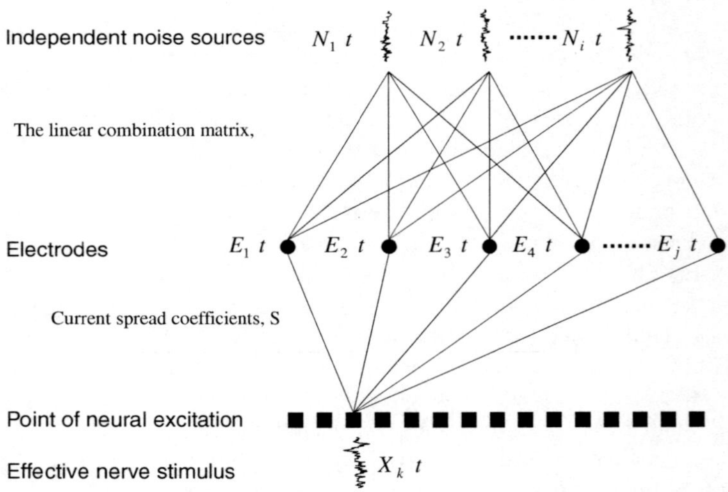

Fig. 1 The stochastic beamforming strategy. Each of the j electrodes is stimulated by a noise waveform, $E_j(t)$, that is a weighted sum of i independent Gaussian noise sources; The weightings are governed by the matrix Γ. Through current spread, each of the k fibres is stimulated by a different noise waveform $X_k(t)$, which is the weighted sum of the electrode currents; the weightings are given by the structure matrix S, which characterizes the distances between the fibres and the electrodes and the degree of current spread

We require that all elements of X are statistically independent. This requires that the covariance matrix of X is diagonal. From (3) the covariance matrix of X can be written

$$<XX^T>=<S\Gamma NN^T\Gamma^T S^T> . \tag{4}$$

Noting that $<NN^T>=\sigma^2 I$, (because the noise sources are independent) where σ^2 is the variance of the noise sources, this simplifies to:

$$<XX^T>=\sigma^2 S\Gamma\Gamma^T S^T \tag{5}$$

We define a matrix U such that

$$U=S\Gamma. \tag{6}$$

Equation (5) can then be re-written as

$$<XX^T>=\sigma^2 UU^T. \tag{7}$$

If follows that the covariance matrix will be diagonal if the rows of U are independent (orthogonal), in which case $UU^T=k^2 I$, and hence

$$<XX^T>=\sigma^2 k^2 I. \tag{8}$$

From Eq. (6) the required linear combination of independent noise sources is therefore given by

$$\Gamma=S^{-1}U \tag{9}$$

The analysis shows that it is possible to get one independent point of excitation (one null) for each electrode. The matrix U, however, is not unique and any set of orthogonal vectors can be used in its construction. The simplest matrix that diagnolises (7) is the identity matrix. i.e. $U=I$, and hence, trivially, $\Gamma=S^-1$. This enables further constraints to be placed on U, such as power consumption by the speech processor or the distribution of spontaneous rate across the neural array is maximally flat.

We have done a preliminary computational study on stochastic beamforming [16]. In this study, independent noise was presented to each electrode but the noise waveforms did not satisfy Eq. (9) and the strategy was therefore unlikely to have been optimal. Nonetheless, we showed that some independence between the noise exciting different regions of the cochlea was theoretically possible. Here we show that the trivial solution where the linear combination matrix is the inverse of the current spread matrix leads to much less correlated excitation compared with the direct additive strategy.

2 Current Spread Model

We evaluated the stochastic beamforming strategy using a simple model of electrical stimulation in which the attenuation of current with distance from an electrode was modelled by an exponential decay (with a space-constant of 3.6 dB/mm) [37, 38]. For simplicity, we modelled the spiral structure of the scala tympani as an uncoiled cylinder and the array of nerve fibres was modelled as being in a plane orthogonal to the electrodes. We took the length of the uncoiled cochlea to be 34 mm [39] and the maximum electrode insertion depth from the round window to be 25 mm [40]. Furthermore, to match the Clarion cochlear implant (Advanced Bionics Ltd), the 16 electrodes in the model were spaced 1 mm apart (from 10 to 25 mm from the round window). In typical cochlear implant patients, each electrode is about 0.5 to 1 mm away from the nearest afferent cell body [41, 42], which we have taken to be the region of initial excitation. In this study, we have been conservative and taken the distance from each electrode to the nearest afferent cell body (receptor) be 1 mm.

Two noise strategies were studied. In the first strategy, each electrode was directly stimulated by an independent noise source and the noise intensity at each electrode was equal. In the second strategy we used stochastic beamforming; given the geometry and current spread we calculated the current spread coefficients and then took the linear combination matrix to be the inverse of the current spread matrix, i.e. the simplest solution of Eq. (9). Again, the intensities of the independent noise sources were equal. For the two strategies, a cross-correlation coefficient was used to measure the degree of independence between the effective stimulus at a reference position (the receptor closest to the central electrode in the array, i.e. electrode 8) and the effective stimulus at more distant receptors. For each strategy, the outputs of the independent Gaussian noise sources were calculated at 10 μs intervals for 2 seconds and the effective stimulus for each receptor was calculated using Eq. (3).

With the direct additive noise strategy, the correlation between the effective stimulus at the central receptor and the effective stimulus at more distant receptors decreased gradually with distance (Fig. 2). For a receptor 1 mm (one electrode spacing) away from the central receptor, the cross-correlation coefficient for the effective stimuli was 0.96, and, even for a receptor 7 mm away, the cross-correlation coefficient for the effective stimuli was 0.30. In other words with direct additive noise, the correlation between the effective stimuli for two receptors was substantial even when the receptors were some distance apart. With stochastic beamforming, the correlation between the effective stimulus at the central receptor and the effective stimulus at more distant receptors decreased more rapidly than with direct additive strategy and was 0 when the receptor was an integer multiple of the electrode spacing away from the reference receptor. For a receptor more than 1 mm from the central receptor, the absolute cross-correlation coefficient for the effective stimuli was always less than 1.7, and, when the separation between the receptors was over 3 mm, the cross-correlation coefficient was negligible.

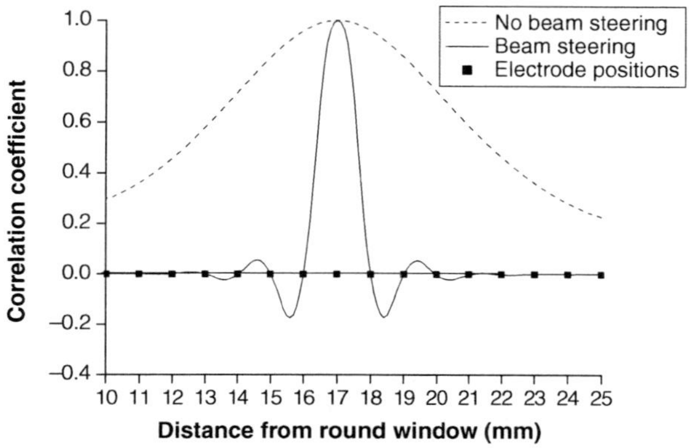

Fig. 2 Correlation between the noise at a reference fibre 17 mm from the round window and neighbouring fibres for two noise strategies: no stochastic beamforming or beamforming. With no beaming (*dashed line*) each electrode was simply stimulated by an independent noise source. With beamforming, (*solid line*) each electrode was stimulated by a linear sum of noise sources (Eq. (2)). The square symbols show the positions of the electrodes within the cochlea

3 Cochlear Implant Simulation

To evaluate the effectiveness of the beamforming technique a cochlear implant simulation was undertaken. A schematic of the simulation is shown in Fig. 3. We performed computer simulations using a coding strategy similar to the Simultaneous Analog Strategy (SAS) used in the Clarion cochlear implant (Advanced Bionics Ltd). In accord with the normal tonotopic arrangement of the cochlea, bandpass filters with low-frequency passbands are used to stimulate electrodes in the apex of the cochlea and those with a high-frequency passband are used to stimulate electrodes in the base. The SAS strategy is normally used with seven channels of bipolar stimulation however here we have implemented just four channels to enable each channel to have multiple electrodes (six electrodes per channel). The use of multiple electrodes is necessary for the implementation of the stochastic beamforming strategy. We have also used monopolar stimulation because bipolar stimulation can evoke nerve discharges during both phases of stimulation [43], whereas nerve fibres in the normal ear respond to only one phase [44]. The temporal responses of cochlear nerve fibres to bipolar electrical stimulation are therefore unlike those to acoustic stimulation in the normal ear.

To make the study relevant to previous studies of speech comprehension, the simulated cochlear implant signals were derived from real speech from the BKB sentence list [45]. Sentences were concatenated to give a total signal duration of twelve seconds. The signal processing chain consisted of a pre-emphasis filter before the bank of bandpass filters (Fig. 3), which was identical to that used for SAS; the filter was a first-order highpass filter (cutoff 1500 Hz) cascaded with a third order lowpass filter (cutoff 7800 Hz). The bandpass filters were 3rd order butterworth filters with bands 250–500, 500–875, 875–1150, 1150–1450 Hz. The output of the band-

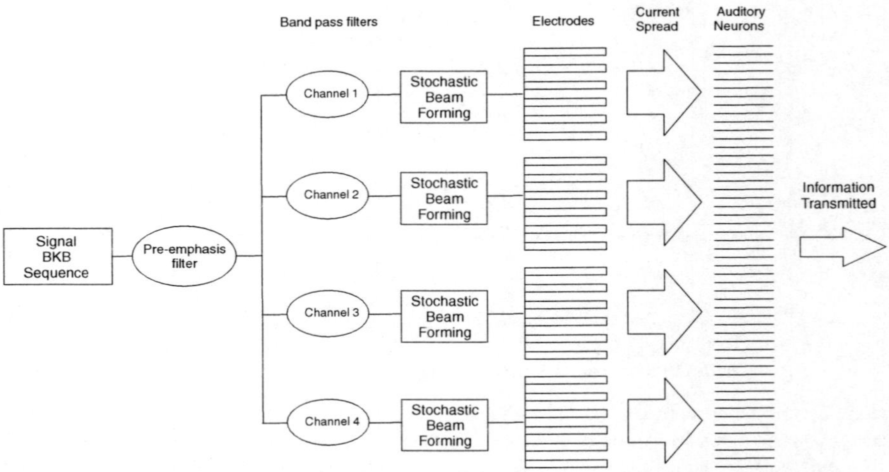

Fig. 3 Schematic of the simulation. After pre-emphasis the signal is split into four frequency bands, the resulting signal in each band was then used to stimulates six electrodes – the beamforming strategy was applied to each set of electrodes. The output of the model was taken to be the summed response of the individual fibres – shifted to compensate for the delays in the channels. 58 nerve fibres were used in the simulations

pass filters were used to stimulate six electrodes; each electrode was also stimulated with a linear mix of Gaussian noise sources as discussed in the previous section. This enabled each channel to implement stochastic beamforming and hence produce six nulls in correlation at the points of neural stimulation. The cochlear nerve fibres were modelled using a leaky-integrate and fire neuron with dynamic threshold. Details of the nerve fibre model and the parameters used can be found in [21]. The output of the simulation was taken to be the summed response of all the nerve fibres – however the spike trains were shifted to take into account the processing delays in the channels. For the purposes of the this study 58 fibres were used.

To quantify the performance of the implant we estimate the information rate (lower bound), R, between the signal $s(t)$ and the summed output of the nerve fibres $r(t)$ using standard information-theoretic methods [46],

$$R = -\frac{1}{2\pi}\int_0^\infty \log_2(1-|\gamma(\omega)|^2)d\omega, \tag{10}$$

where the coherence function, $\gamma(\omega)$, is defined as

$$\gamma(\omega) = \frac{\hat{S}_{sr}(\omega)}{\sqrt{|\hat{S}_{ss}(\omega)|^2}\sqrt{|\hat{S}_{rr}(\omega)|^2}} \tag{11}$$

and $\hat{S}_{sr}(\omega)$ is the cross-spectral density between the signal and the summed spike train response, and $\hat{S}_{ss}(\omega)$, $\hat{S}_{rr}(\omega)$ are the power spectra of the input signal and summed spike train response respectively.

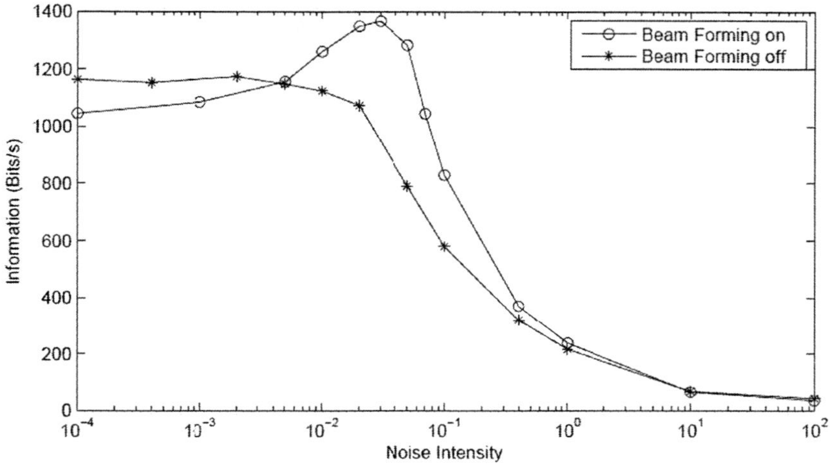

Fig. 4 Plot of information rate against noise intensity. Open circles are with the beamforming switched on, stars are with it switched off

The results of the simulation are show in Fig. 4. Two curves are shown, one with beamforming switched on (open circles) and the other with it switched off (stars). Clearly, with the beamforming off, increasing noise leads to a decrease in the information rate, whilst with the beamforming on an SSR type effect is observed. It should be noted that the difference in information at small noise levels (i.e. higher information with beamforming off) is not significant. Implementing the beamforming leads to a change in gain in the channels due to the introduction of the beamforming matrix - thus effectively changing the signal levels on each channel. This has yet to be corrected for in our simulations.

4 Discussion and Conclusions

The simulations have demonstrated that, in principle, stochastic beamforming can be used to obtain less correlated stimulation of cochlear nerve fibres compared with directly adding independent noise to each electrode. Furthermore, with the beamforming on, a stochastic resonance type effect was observed in the transmitted information. This gives some confidence that the decorrelation of the noise sources at the point of neural stimulation has resulted in SSR.

The practical effectiveness of the strategy will depend on details of the current spread matrix. Because of anatomical difference between patients and various placements of the electrode array, the matrix will differ between patients. We can use one of two approaches: we can try to obtain the current-spread matrix for individual subjects and require that we get exact nulls, or we can relax the requirement for complete independence and choose the linear combination parameters to satisfy a more lenient criterion for a range of current spreads that reflects patient variability; for example, the linear combination parameters can be chosen to minimize the maximum covariance of the effective receptor stimuli over different current-spread matrices, minimize the minimum covariance of effective receptor stimuli over different current-spread matrices, or minimize the variance of the covariance values. We regard each of these options as a different stochastic beamforming strategy.

In the simulations, however, the linear combination matrix was based on exact knowledge of the current spread matrix. In practice this matrix will have to be obtained from psychophysical experiments or measurements of the evoked compound action potential to electrical stimulation - a measurement that all modern cochlear implants can make. These measurements will not be precise and we have yet to determine the sensitivity of the method to variability in the current spread measurements. However, we have established that current spread is not an obstacle to the implementation SSR, and hence potential enhanced performance, in cochlear implants.

Finally, we note that an alternative method has been proposed [47] to induce independent stochastic neural activity across the population of nerve fibres. This involves the use of a high rate pulsatile conditioner applied at multiple electrodes. We wish to suggest that there may exist a third method for achieving independent stochastic neural activity – the use of pharmacological methods. In particular, it has been noted that nerve fibres can display enhanced stochasticity when exposed to neurotrophins [48]. As methods for the delivery of pharmacological agents in cochlear implants become more advanced, it is conceivable that neurotrophins, or other pharmacological products, may well become viable as a method for introducing stochasticity into cochlear implants. This would enable SSR to occur without the use of electrically applied noise or conditioners.

Acknowledgments This work was funded by the EPSRC (grant GR/R35650/01 and EP/D05/1894/1(P)).

References

1. Clark, G.M., Cochlear Implants: Fundamentals and Applications. 2003, New York: Springer-Verlag.
2. Allum, D.J., Cochlear implant rehabilitation in children and adults. 1996, London: Whurr.
3. Rosen, S., Cochlear implants, in Scott-Brown's Otolarngology (6th edition) Volume 2, D. Stephens, Editor. 1996, Butterworth Heinemann: Oxford.
4. Fraysse, B., et al., Cochlear implants for adults obtaining marginal benefits from acoustic amplification – A European study. American Journal of Otology, 1998. 19: p. 591–597.
5. Summerfield, A.Q. and D.H. Marshall, Cochlear implantation in the UK 1990–1994: report by the MRC Institute of Hearing Research on the evaluation of the national cochlear implant program. 1995, HMSO.
6. Dorman, M.F., Speech perception by adults, in Cochlear implants: audiological foundations, R.S. Tyler, Editor. 1993, Singular Publishing Group: London. p. 145–190.
7. Evans, E.F., Cochlear nerve and cochlear nucleus, in Handbook of sensory physiology, W.D. Keidel and W.D. Neff, Editors. 1975, Springer: Berlin. p. 1–108.

8. DeVries, H., Brownian motion and the transmission of energy in the cochlea. Journal of the Acoustical society of America, 1952. 24: p. 527–533.
9. Longtin, A., A.R. Bulsara, and F. Moss, Time interval sequences in bistable systems and the noise-induced transmission of information by sensory neurons. Physical Review Letters, 1991. 67: p. 656–659.
10. Morse, R.P. and E.F. Evans, Enhancement of vowel coding for cochlear implants by addition of noise. Nature Medicine, 1996. 2: p. 928–932.
11. Henry, K.R., Noise improves transfer of near-threshold, phase-locked activity of the cochlear nerve: evidence for stochastic resonance? Journal of Comparative Physiology A, 1999. 184: p. 577–584.
12. Greenberg, S., Possible role of low and medium spontaneous rate fibres in the coding of waveform periodicity, in Auditory frequency selectivity, B. Moore and R. Patterson, Editors. 1986. p. 241–248.
13. Jaramillo, F. and K. Wiesenfeld, Mechanoelectrical transduction assisted by Brownian motion: a role for noise in the auditory system. Nature Neuroscience, 1998. 1: p. 384–388.
14. Lowenstein, Peripheral mechanisms of equalibrium. British Medical Bulletin, 1956. 12: p. 114–118.
15. Stocks, N.G., D. Allingham, and R.P. Morse, The application of suprathreshold stochastic resonance to cochlear implant coding. Noise and Fluctuation Letters, 2002. 2: p. L169–L181.
16. Morse, R.P. and G.F. Meyer, The practical use of noise to improve speech coding by analogue cochlear implants. Chaos, Solitons and Fractals, 2000. 11: p. 1885–1894.
17. Morse, R.P. and E.F. Evans, Additive noise can enhance temporal coding in a computational model of analogue cochlear implant stimulation. Hearing Research, 1999. 133: p. 107–119.
18. Allingham, D., et al. Noise enhanced information transmission in a model of multichannel cochlear implantation. in SPIE Conference on Noise and Fluctuations. 2004. Gran Canaria.
19. Meddis, R., Simulation of the mechanical to neural transduction in the auditory receptor. Journal of the Acoustical society of America, 1986. 79: p. 702–711.
20. Morse, R.P. and N.G. Stocks. Enhanced cochlear implant coding using multiplicative noise. in SPIE Conference on Fluctuations and Noise. 2005. Austin, USA.
21. Zeng, F.-G., Q.-J. Fu, and R.P. Morse, Human hearing enhanced by noise. Brain Research, 2000. 869: p. 251–255.
22. Chatterjee, M. and M.E. Robert, Noise enhances modulation sensitivity in cochlear implant listeners: Stochastic resonance in a prosthetic sensory system? Jaro, 2001. 2(2): p. 159–171.
23. Chatterjee, M. and S.I. Oba, Noise improves modulation detection by cochlear implant listeners at moderate carrier levels. Journal of the Acoustical Society of America, 2005. 118(2): p. 993–1002.
24. Benham, S.E. and F.-G. Zeng, Noise improves suprathreshold discrimination in cochlear implant listeners. Hearing Research, 2003. 186: p. 91–93.
25. Douglass, J., et al., Noise enhancement of information transfer in crayfish mechanoreceptors by stochastic resonance. Nature, 1993. 365: p. 337–340.
26. Levin, J.E. and J.P. Miller, Broadband neural encoding in the cricket cercal sensory system enhanced by stochastic resonance. Nature, 1996. 380: p. 165–168.
27. Russell, D.F., L.A. Wilkens, and F. Moss, Use of behavioural stochastic resonance by paddle fish for feeding. Nature, 1999. 402: p. 291–294.
28. Stocks, N.G., Suprathreshold stochastic resonance in multilevel threshold systems. Physical Review Letters, 2000. 84: p. 2310–2313.
29. Wilson, B.S., The future of cochlear implants. British Journal of Audiology, 1997. 31: p. 205–225.
30. Leake-Jones, P.A. and S.J. Rebscher, Cochlear pathology with chronically implanted scala tympani electrodes, in Cochlear Prostheses: An International Symposium, C.W. Parkins and S.W. Anderson, Editors. 1983, New York Academy of Sciences: New York. p. 203–223.
31. Simmons, F.B. and T.J. Glattke, Comparison of electrical and acoustical stimulation of the cat ear. Annals of Otology Rhinology Laryngology, 1972. 81: p. 731–738.
32. Merzenich, M.M., D.N. Schindler, and M.W. White, Feasibility of multichannel scala tympani stimulation. Laryngoscope, 1974. 84: p. 1887–1893.

33. Townshend, B. and R.L. White, Reduction of electrical interaction in auditory prostheses. IEEE Transactions on Biomedical Engineering, 1987. 34: p. 891–897.
34. Clopton, B.M. and F.A. Spelman, Neural mechanisms relevant to the design of an auditory prosthesis: location and electrical characteristics. Annals of Otology Rhinology Laryngology, 1982. 91: p. 9–14.
35. Suesserman, M.F. and F.A. Spelman, Lumped-parameter model for in-vivo electrical stimulation. IEEE Transactions on Biomedical Engineering, 1993. 40: p. 237–245.
36. Zarowski, A., F.J. Vanpoucke, and S. Peeters. A lumped electrical model for electrode contacts and cochlear current spread: structure, identification from EFI and in vivo results. in 2003 Conference on Implantable Auditory Prostheses. 2003. Asilomar, USA.
37. Wilson, B., et al., Speech processors for auditory prostheses: 7th Quarterly NIH Report N01-DC-2-2401. 1994, Research Triangle Institute.
38. Bruce, I.C., et al., A stochastic model of the electrically stimulated auditory nerve: Single-pulse response. IEEE Transactions on Biomedical Engineering, 1999. 46(6): p. 617–629.
39. Nadol, J.B., Jr., Comparative anatomy of the cochlea and auditory nerve in mammals. Hearing Research, 1988. 34: p. 253–266.
40. Loizou, P.C., Mimicking the human ear. IEEE Signal Processing Magazine, 1998. 15(5): p. 101–130.
41. Shepherd, R.K., S. Hatsushika, and G.M. Clark, Electrical stimulation of the auditory nerve: the effect of electrode position on neural excitation. Hearing Research, 1993. 66: p. 108–120.
42. Finley, C.C., B.S. Wilson, and M.W. White, Models of neural responsiveness to electrical stimulation, in Cochlear implants: models of the electrically stimulated ear, J.J. Miller and F.A. Spelman, Editors. 1990, Springer: New York.
43. van den Honert, C. and P.H. Stypulkowski, "Temporal response patterns of single auditory nerve fibres elicited by periodic electrical simuli". Hear. Res., 29, 207–222, 1987.
44. Kiang, N.Y.S., Discharge patterns of single fibres in the cat's auditory cortex, Cambridge: MIT Press 1965.
45. Bench, J., A. Kowal, and J. Bamford, "The BKB (Bench-Kowal-Bamford) sentence lists for partially hearing children". Brit. J. Audiol., 13, 108–112, 1979.
46. Stein, R.B., A.S. French, and A.V. Holden, Biophys. J. **12** 295–322 (1972).
47. Rubinstein, J.T., et al., "Pseudospontaneous activity: stochastic independence of auditory nerve fibres with electrical stimulation". Hearing Research 127, 108–118, 1999.
48. Oestreicher, E., M. Knipper, A. Arnold, H. P. Zenner, D. Felix Neurotrophin 3 potentiates glutamatergic responses of IHC afferents in the cochlea in vivo European Journal of Neuroscience 12 (5), 1584–1590, 2000.

Applying Stochastic Signal Quantization Theory to the Robust Digitization of Noisy Analog Signals

Mark D. McDonnell

Abstract Suprathreshold stochastic resonance is a variant of stochastic resonance that has been shown to occur in parallel arrays of independently noisy, but otherwise identical, binary threshold devices. It can be described as a form of stochastic signal quantization that utilizes independently random noise sources to digitize an analog signal. This paper outlines a generalization of this effect, and discusses several sensor applications in which it can occur, such as analog-to-digital converter circuits, distributed sensor networks and the reduction of in-band noise via coherent integration in radar, sonar and sodar systems. All of these scenarios can be modeled using a framework called a 'stochastic pooling network.'

1 Introduction and Background

Two challenges are ubiquitous for many forms of signal and information processing tasks. These are (i) robustness to the effects of random noise, and (ii) extraction of only the 'information' which is relevant for the task. The latter challenge can require the use of lossy compression, that is, the situation where 'information' is intentionally discarded and can never be recovered. This is often desirable, because the lost information is considered to be either redundant or irrelevant. Robustness to noise is often achieved using a network or array of sensors.

It is less obvious that a network approach could lead to lossy compression. In this paper we consider a signal processing model in which noise reduction and lossy

M.D. McDonnell (✉)
Institute for Telecommunications Research, University of South Australia,
Mawson Lakes, Australia, e-mail: Mark.McDonnell@unisa.edu.au

V. In et al. (eds.), *Applications of Nonlinear Dynamics*, Understanding Complex Systems, 249
DOI 10.1007/978-3-540-85632-0_20, © Springer-Verlag Berlin Heidelberg 2009

compression can be simultaneously achieved. This model is quite general, and is now sometimes known as a *stochastic pooling network*. A special case of such a network previously has been shown to exhibit *suprathreshold stochastic resonance* [1,2].

Stochastic pooling networks have some very general properties closely related to their capabilities for noise reduction and compression, and are therefore very suitable for modeling a wide range of systems, including biological neurons [3]. However, the focus of this paper is on a different situation, that of digital data acquisition in artificial sensors by an Analog-to-Digital Converter (ADC).

The paper is organized into four parts. First, the remainder of the introduction describes the scenario where noise reduction by averaging – 'digital averaging' is performed on digitized signals, and how this can be modeled by a stochastic pooling network. Section 2 contains a review of previous work on Suprathreshold Stochastic Resonance (SSR). This will aid discussion in Sect. 3, which contains analysis of 'digital averaging'. It is demonstrated that SSR occurs in the more general pooling network that results. Finally, Sect. 4 outlines some open questions and speculations for future directions for this research.

1.1 Averaging of Noisy Digital Signals

Digitization of analog signals consists of two processes—sampling in time, and quantization of amplitude. Quantization is always a lossy process, and some distortion will result as the original analog signal can never be perfectly recovered [4].

Continuing advances in digital signal processing technology have led to a trend for receivers in communications networks and sensors to become as digitized as possible and for ADCs to be 'as close to the antenna as possible.' One example of a simple signal processing task affected by such a shift of the ADC location in a receiver, is that of random noise reduction.

Although frequency domain filtering—whether analog or digital—can reduce out of band noise, this has no effect on in-band noise or interference. Under certain conditions, such noise can be reduced by *averaging*—also known in this context as *integration*. This fact is exploited in many signal processing applications including radar and sonar systems.

For example, in a radar system the signals resulting from analog in-phase and quadrature (I & Q) demodulation are often averaged over a number of radar pulses prior to a matched-filter detection stage. This is known as *coherent* or *pre-detection* integration, with the resulting performance gain known as the integration gain. Assuming infinite precision, coherently averaging N independently noisy observations of the same signal is known to reduce the noise by a factor of $\sqrt{N}$.

The process of noise reduction via averaging assumes that

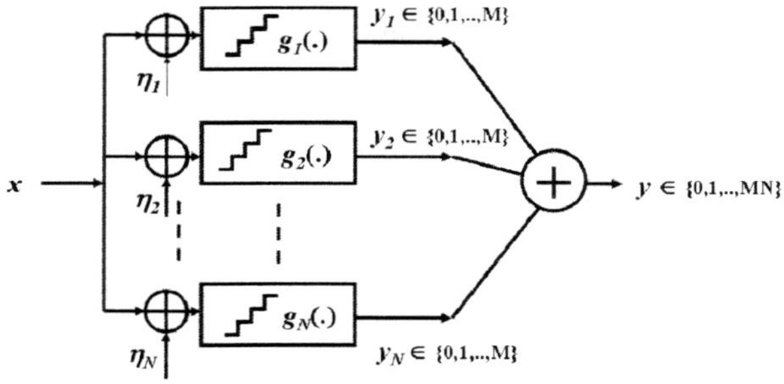

Fig. 1 Stochastic Pooling Network Model. The network consists of N nodes, each of which operates on independently noisy versions of the same signal, x. The output of the i–th node is $y_i = g_i(x + \eta_i)$. The nonlinear function, $g_i(\cdot)$ is lossy, and converts the signal into a discrete representation. Due to this fact, each y_i is assumed to be a nonnegative integer. The overall network sums the output from each node

1. the in-band noise—usually assumed to be random noise—is independent for all the noisy signals that are averaged;
2. more than one independently noisy measurement is available, e.g., a number of radar pulses rapidly transmitted in the same direction.

The performance gain obtained by averaging is reduced if the noise has a nonstationary distribution or is positively correlated from measurement to measurement[1]. However, here we assume that the noise is random and independent and identically distributed (*iid*) in each measurement. We make this assumption because we are only interested in comparing averaging of analog signals with averaging of digitized signals.

This situation can be modeled by the stochastic pooling network shown in Fig. 1, which indicates how N independently noisy versions of the same signal, x, are quantized, and then averaged. Note that conceptually we do not need to differ between adding and averaging in this situation, since the output of the network is discrete.

Before analyzing this 'digital averaging' situation, the remainder of this introductory section describes pooling networks more generally, and outlines their relationship to suprathreshold stochastic resonance.

1.2 Noise-Enhanced Signal Quantization: Suprathreshold Stochastic Resonance and Stochastic Pooling Networks

Suprathreshold Stochastic Resonance (SSR) [1, 2, 5–7] is a form of noise enhanced signal processing—i.e., Stochastic Resonance (SR) [8]—that occurs in a parallel array of identical, but independently noisy, nonlinear threshold devices, including neuron models [9–11]. We refer to the *original* [1] parallel array as the *SSR model*.

[1] Negative correlation can in fact lead to improved averaging performance [9].

Like other forms of SR, the performance of the SSR model is improved by an optimal level of noise. However, unlike conventional SR, this occurs not just for weak input signals, or subthreshold input signals, but for input signals with arbitrary magnitude.

The SSR model was originally intended to model neural populations, however with the relaxation of some of its specifications, it also serves as a good model for other scenarios. The name 'stochastic pooling networks' has recently been used to describe such generalizations [3, 12, 13].

An N-element stochastic pooling network, as illustrated in Fig. 1, has three key features:

Independently noisy nodes: in the absence of noise, each element in the network operates on *independently* noisy versions of the *same* signal;

Lossy compression in each node: each element in the network performs a *lossy* operation on its inputs, so that each output has only a finite number of states (or less states than its input, if the input is discretely valued); hence, the operation of each element is inherently nonlinear;

Node outputs are pooled; nodes are unlabeled: the outputs of each element in the network are 'pooled' to form an overall network output. This means that no labeling of the output of any individual network element is required, unlike, for example, the case in a flash ADC.

It is emphasized that each element in a stochastic pooling network is *nonlinear*, and lossy, since it maps more than one value of the input to the same output value. It may be as simple as a one-bit quantizer, or be a more complex entity like a multi-bit ADC or a full digital 'sensor' like a DIMUS sonar hydrophone array, or fully dynamical systems like sensory rate-coding neurons. The essential features are the ones measured: those of noise reduction and compression, and these, with suitable definitions, become independent of system dynamics.

Note that each element does not necessarily perform the same processing in the absence of noise. This means that the SSR model is a specific case of a stochastic pooling network in which each element is an identical one-bit quantizer (in the absence of noise), and the outputs of each quantizer are summed.

Although the SSR model and the generalization discussed in this paper have some similarities to dithering in ADCs, the crucial difference is that conventional dithered signals provide only one independent noise source per sample. Hence dithering can be considered to be equivalent to an $N = 1$ pooling network.

There are several possible ways of analyzing the performance of a pooling network. The mutual information [14] between X and y is discussed in this paper, as the main aim here is to discuss a signal processing scenario that forms a simple generalization of the basic SSR model, and the original work on SSR used information theory [1].

Figure 1 shows a schematic diagram of a stochastic pooling network model consisting of N parallel 'sensors'. Each sensor receives the same sample, x of a random input signal, X. This random signal is assumed to consist of a sequence of independent samples drawn from a distribution with stationary probability density function (PDF), $f_X(x)$. The i–th device in the model is subject to continuously valued *iid*—independent and identically distributed—additive noise, η_i ($i = 1,..,N$), drawn from a probability distribution with PDF $f_\eta(\eta)$. Each noise signal is required to also be independent of the signal, X.

Each sensor nonlinearly processes its input, $x+\eta_i$, with the lossy function $g_i(\cdot)$, to give an output signal, $y_i = g_i(x+\eta_i)$, which can take only a finite number of states, M. Without loss of generality, we label each possible value of y_i with the nonnegative integers, so that $y_i \in \{0,1,..,M\}$. In this paper it is assumed that the outputs from each element, y_i, are pooled simply by *summing*, to give an overall output signal, $y = \sum_{i=1}^{N} y_i$, so that $y \in \{0,1,..,MN\}$.

The key to analyzing pooling networks is a way of finding the conditional output distribution, $P_{y|x}$. Once this is found, the output distribution, $P_y(n)$, easily follows, since

$$P_y(n) = \int_{-\infty}^{\infty} P_{y|x}(n|x) f_X(x) dx \quad n \in 0,\ldots,N. \tag{1}$$

It is always assumed that the signal PDF, $f_X(x)$, is known.

The mutual information between the input signal, x, and the output, y, of the network is that of a semi-continuous channel, and can be written as [1]

$$\begin{aligned} I(x,y) &= H(y) - H(y|x) \\ &= -\sum_{n=0}^{N} P_y(n)\log_2 P_y(n) - \left(-\int_{-\infty}^{\infty} P(x)\sum_{n=0}^{N} P(n|x)\log_2 P(n|x)dx\right), \end{aligned} \tag{2}$$

where $H(y)$ is the entropy of y and $H(y|x)$ is the average conditional entropy of y given X [14]. The units of mutual information is *bits per sample*.

2 Pooling Networks Consisting of Binary Elements

In an N-element SSR model, as initially described by Stocks [1], the nonlinear processing performed in each element is a simple one-bit quantization, of the form

$$g_i(s) = \begin{cases} 1 & s \geq \theta \\ 0 & \text{otherwise}, \end{cases} \tag{3}$$

for all $i = 1,..,N$. For the i–th element, s is the sum of the common signal and the i–th independent noise signal, $s = x+\eta_i$. Letting the output of the i–th element be y_i, then y_i is unity if $x+\eta_i$ is greater than the threshold value, θ and zero otherwise.

As mentioned, the threshold value, θ, is the same for *all* elements. This means that in the absence of noise, all elements give the same output value for the common input signal, x, and the overall network output is either 0 or N. However, even for the smallest noise signals, this is no longer the case, and the network output y becomes a discrete random variable. If the noise has infinite support, for example Gaussian, then y can theoretically take any value between 0 and N for all possible values of x.

This means that y is a $\log_2(N+1)$ bit *quantization* of x, rather than 1 bit. This facts leads to the occurrence of SSR in the model: performance is improved for even the smallest amount of noise, as well as for signals which are not subthreshold. In fact, performance is optimal for a threshold equal to the mean value of the signal [5].

How well y represents the input signal depends on the conditional distribution, $P_{y|x}(y=n|x)$, as well as the signal distribution. Since the conditional output distribution of all elements in the network is binary, and identical, $y|x$ is binomially distributed

$$P_{y|x}(n|x) = \binom{N}{n} P_{1|x}^{n}(1-P_{1|x})^{(N-n)}, \tag{4}$$

where

$$P_{1|x} = 1 - F_{\eta}(\theta - x), \tag{5}$$

following the notation of [1], is the probability that any given element is in state 1, given x, and $F_{\eta}(\cdot)$ is the cumulative distribution function (CDF) of the noise.

For any specified threshold value, and signal and noise distributions, it is straightforward given Eqs. (4) and (5) to numerically calculate measures of how well y represents x. In particular, mutual information [1, 5, 10, 15], mean square error [16], correlation coefficient [2] and Fisher information [11, 17, 18] have been used.

Here, however, we focus on mutual information, so as a benchmark for comparison with subsequent results, Fig. 2 shows the mutual information between the input and output signals of the SSR model, for the examples of (a) Gaussian signal and noise, and (b) uniform signal and Gaussian noise. Other signal and noise distributions show very similar qualitative behavior [7, 16]. The mean of both the signal and the noise are zero, as is the value of the thresholds, θ. The x-axis shows increasing 'noise-intensity',

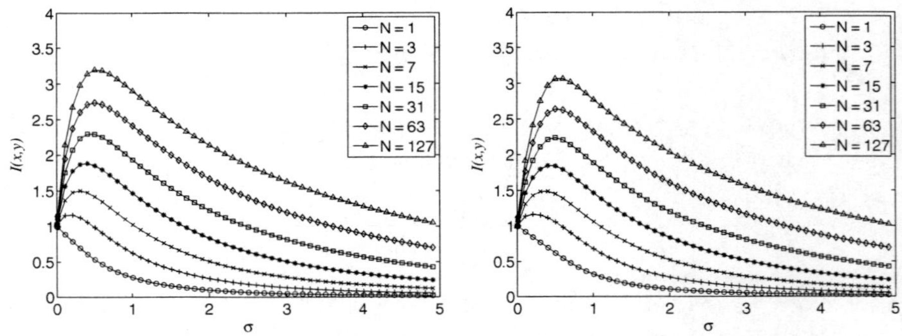

(a) Gaussian signal and noise (b) Uniform signal and Gaussian noise

Fig. 2 Mutual information plotted against increasing noise intensity, σ, for a stochastic pooling network consisting of N identical binary quantizers, as first studied by Stocks [1]. Suprathreshold stochastic resonance can be seen to occurs – the optimal noise intensity occurs for a nonzero vlaue of σ. Increasing values of N, and two examples are shown: (**a**) Gaussian signal, with Gaussian noise, and (**b**) uniform signal, with Gaussian noise

$$\sigma = \frac{\mathrm{E}[\eta^2]}{\mathrm{E}[x^2]}, \tag{6}$$

converted to decibels. Hence, σ is the inverse of the signal-to-noise ratio (SNR) at the input to each element (*not* converted to decibels), and $\sigma = 0$ corresponds to the absence of noise, or infinite input SNR.

Notice that in the absence of noise the mutual information is exactly one bit per sample. However as the noise intensity increases the mutual information increases, until it reaches a maximum, and then starts to decrease again as σ increases further. This is the signature of stochastic resonance. In this model this phenomenon is referred to as *suprathreshold* stochastic resonance, since significant noise-enhancement does not require the signal to be substantially subthreshold [1].

2.1 Extending the SSR Model: Non-Identical Thresholds

As recognized by Stocks in his seminal paper [1], a simple change to the SSR model leads, in the absence of noise, to a model of a flash analog-to-digital converter (ADC) [19]. This is an ADC that consists of N parallel comparators, which due to a chain of resistors each have a different threshold voltage.

The conventional way to write the overall transfer function between analog input signal x, and discretely valued output state y for an ideal $\log_2(N+1)$ bit ADC, or scalar quantizer is

$$y(x) = \begin{cases} 0 & x < \theta_1 \\ 1 & \theta_1 \leq x < \theta_2 \\ \vdots & \vdots \\ N-1 & \theta_{N-1} \leq x \leq \theta_N \\ N & x \geq \theta_N \end{cases} \tag{7}$$

Note that there is no uncertainty about which about state occurs for a given value of x; the mapping is deterministic.

If one extends the SSR model to a stochastic pooling network consisting of binary threshold elements with *non-identical* threshold values, then each non-linear operation can be written as

$$g_i(s) = \begin{cases} 1 & s \geq \theta_i \\ 0 & \text{otherwise}, \end{cases} \tag{8}$$

where $i = 1,..,N$, and $\theta_1 < \theta_2 < ... < \theta_N$.

At this point it is instructive to note that the overall output, y, cannot be expressed in the manner of Eq. (7), as the noise at the input to each element is *independent*, and this can allow any value of y for any value of x (provided the noise has a long tailed PDF). The end result is a *random quantization* of the input, as indicated by expressing the transfer function as $y(x) = \frac{1}{2}\sum_{i=1}^{N} \text{sign}[x + \eta_i - \theta_i] + \frac{N}{2}$.

It is very difficult to find a compact analytical expression for the transition probabilities, $P_{y|x}(y = n|x)$, however they can be found numerically using an efficient recursive algorithm, assuming each threshold value, and the noise distribution is known [7, 16].

Given this method for finding $P_{y|x}$, it is straightforward to calculate the mutual information for arbitrary threshold values, using Eq. (2). This also means that it is possible to numerically optimize the set of threshold values, $\{\theta_1, \theta_2, .., \theta_N\}$, for maximum mutual information, using standard numerical optimization [6, 7]. The result of such an optimization contains some surprises, as described previously [6, 7]. For example, for the case of $N = 2$, the two optimal threshold values are widely spaced for small σ, but as σ increases, the two values get closer together, and eventually both have the same value.

This behavior also occurs for larger N, however there is more than one value of σ where the number of unique threshold values decreases. As σ decreases, these points appear to be bifurcation points. The same qualitative behavior persists as N increases. However, in all cases that have been studied, provided σ is sufficiently large, the SSR model situation of all threshold identical is optimal.

For larger σ many local optima exist, for example, when $N = 15$ there is a value of σ where three clusters of 5 identical thresholds each is near optimal and provides mutual information very close to that of the global optimum. Indeed, clustering of thresholds to identical values has been found to provide good, near optimal solutions, provided the independent noise on each threshold device is not negligible [20]. The situation described here for an $N = 15$ pooling network is equivalent, *with one exception*, to the situation where the outputs of $M = 5$ *independent* pooling networks, consisting of $N = 3$ threshold nodes, are added together. In the situation discussed until now, noise is independent for every threshold element. In the pooling network model, as shown in Fig. 1, since the noise is only independent at the input to each node, it would result in *identical* noise signals being present on each threshold within a node. This new scenario of non-independent noise is now discussed in Sect. 3.

3 Stochastic Pooling Networks Consisting of Multi-bit ADCs

In this section, the assumption of a stochastic pooling network with binary elements is discarded. Instead, multi-bit ADCs (or quantizers) are allowed. Just as the basic SSR model consists of N identical threshold devices, here we assume that the network consists of N identical – although independently noisy – elements, but that each element is a quantizer with M thresholds, that is

$$g_i(s) = \begin{cases} 0 & s < \theta_1 \\ 1 & \theta_1 \leq s < \theta_2 \\ \vdots & \vdots \\ M-1 & \theta_{M-1} \leq s \leq \theta_M \\ M & s \geq \theta_M \end{cases}, \tag{9}$$

for all i, where $\theta_1 < \theta_2 < ... < \theta_M$.

Since each quantizer is identical, the set of threshold values is the same for all i. This means that in the absence of noise, the outputs of each element in the network are identical, and the overall network output is always N times the individual output, and only $M+1$ states are attainable. However, just like the SSR case, when independent noise is present in each network element, the output of each element becomes a random variable, and the overall network output can take any of $N(M+1)$ states. This means that the presence of noise allows the possibility of a $\log_2(N) + \log_2(M+1)$ bit quantizer, and SSR can be expected to occur.

However, in this situation, the thresholds within each ADC are not subject to *iid* noise, but identical noise. This is a big difference to the situation of Sect. 2, in which the noise signals could be considered either as predominantly internal to an ADC, say thermal noise, which is *iid* on each threshold, or predominantly external, but *iid* across N single bit ADCs. Here however, it is assumed that the noise is indeed external background noise and there is no *iid* noise on each threshold device.

In order to calculate the mutual information for this situation, a method is needed to calculate the transition probabilities, $P_{y|x}(y = n|X = x)$. The technique used in Sect. 2 [16] cannot be used, since not every threshold in the whole network has *iid* noise. However, a generalization of the method can be used, and the remainder of this section presents some results that illustrate some important points about this model.

Recall that N is the number of independently noisy ADCs in a pooling network, and each device is identical, with M thresholds. Figure 3 shows the mutual information for four different values of M, for a Gaussian signal and Gaussian noise. The $M = 1$ case in Fig. 3(a) is the SSR model case, and shows the same data as Fig. 2(a) but with the x-axis rescaled to show the noise intensity as SNR, in dB. Figure 3(b), (c), (d) show the mutual information for the cases of $M = 2$. $M = 3$ and $M = 7$ respectively.

The following observations can be noted for all values of M.

1. As the input SNR becomes large (this is equivalent to $\sigma \to 0$), the mutual information decreases to $\log_2(M+1)$. The reason for this is that each ADC is quantizing the same signal, with the same thresholds – only $M+1$ output states are available.
2. The mutual information can be seen to increase well above $\log_2(M+1)$ for some sufficiently small input SNR, meaning that suprathreshold stochastic resonance occurs.

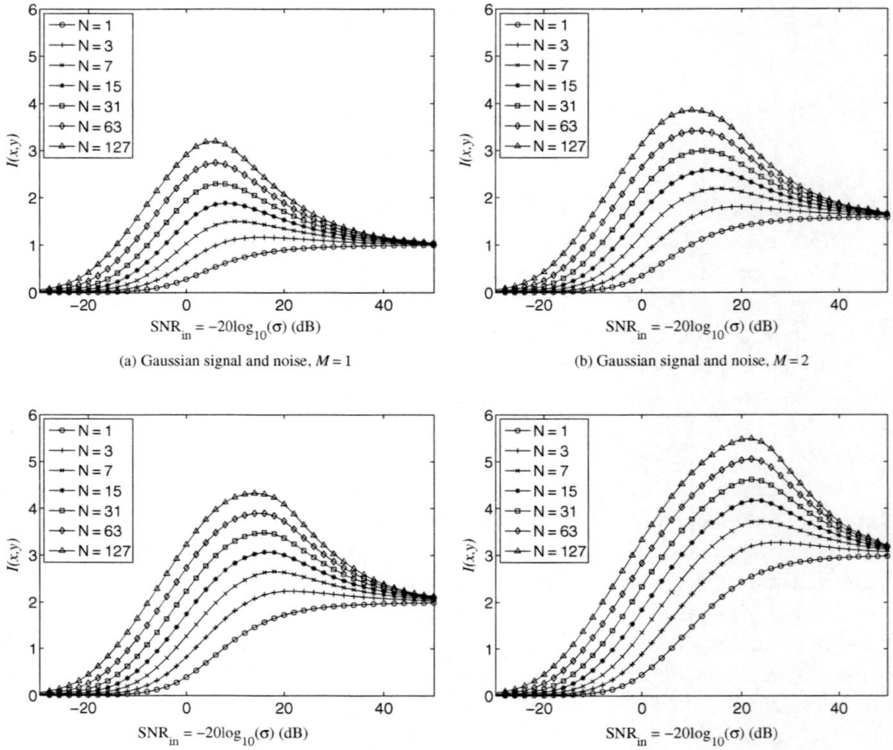

(a) Gaussian signal and noise, $M = 1$

(b) Gaussian signal and noise, $M = 2$

(c) Gaussian signal and noise, $M = 3$

(b) Gaussian signal and noise, $M = 7$

Fig. 3 Mutual information in pooling networks consisting of ADCs that quantize with M threshold values. Four values of M are shown for selected values of N between 1 and 127

3. The optimal input SNR appears to converge to the same value as N increases, however as M increases, this optimal input SNR also increases. In terms of σ, this means that σ decreases.
4. As the input SNR decreases beyond the SSR optima, the mutual information decreases, approximately linearly with decreasing input SNR, until the input SNR is so small that it slowly decays to zero.

Thus, a generalization of the basic SSR model version of a stochastic pooling network, to the case of N independently noisy, but otherwise identical, M state ADCs, shows the same behavior as the basic SSR model. This demonstrates that SSR is not a phenomenon restricted to single-bit (two state) devices, but is more general, and can be expected to occur in other types of pooling networks.

Increasing M also has the affect of increasing the mutual information for all input SNRs. However, as shown in Fig. 4(a) – which shows the mutual information for $N = 127$ and $M = 1, 2, 3$, and 4 – the extra mutual information achievable by increasing M decreases as the input SNR decreases. For input SNRs less than about 0 dB, there is negligible difference in the mutual information for each value of M.

This indicates that for sufficiently small input SNRs, it would suffice to use a single bit ADC, rather than more bits. For slightly larger values of input SNR, it would suffice to use, say, a two bit ADC, rather than $M > 2$. These observations agree well with ad-hoc rules of thumb used by engineers to select the number of bits to use in an ADC, that is, the maximum number of bits is estimated to be that which allows the voltage difference between adjacent threshold levels to be no smaller than the expected dynamic range of the noise.

While Fig. 4(a) illustrates that the mutual information can be increased for sufficiently large input SNRs by increasing M, while N remains constant, this is achieved by increasing the total number of output states available, i.e. $MN+1$. On the other hand, if we fix the number of output states to a constant value, and vary M and N to achieve this value, the mutual information behaves somewhat differently. This means the total maximum number of output bits is $\log_2(MN+1)$ bits per sample.

The mutual information for this case for varying values of M is illustrated in Fig. 4(b) for the case of $MN \simeq 511$. The values of M used are $1,3,7..,511$, so that the number of bits available in each of the N ADCs varies from 1 to 9 bits, as indicated by the legend. Since M and N are integers, exactly 512 output states is not achievable for all selected values of M, and the nearest integer value of N to make $MN \simeq 511$ has been used. This difficulty does not affect the qualitative behavior of the mutual information.

As the noise intensity decrease to zero (i.e. large input SNR), the mutual information increases with increasing M, and is equal to the number of bits in a single ADC. However, as the input SNR decreases, it is clear that the mutual information can be made larger by using less bits per ADC, and utilizing more than $N = 1$ element in the network, that is, by making use of the SSR effect. For sufficiently small input SNR, more mutual information is achieved by N single bit ADCs than any other case. Unlike the case shown in Fig. 4(a), using SSR can be seen to have an advantage even for quite large input SNRs.

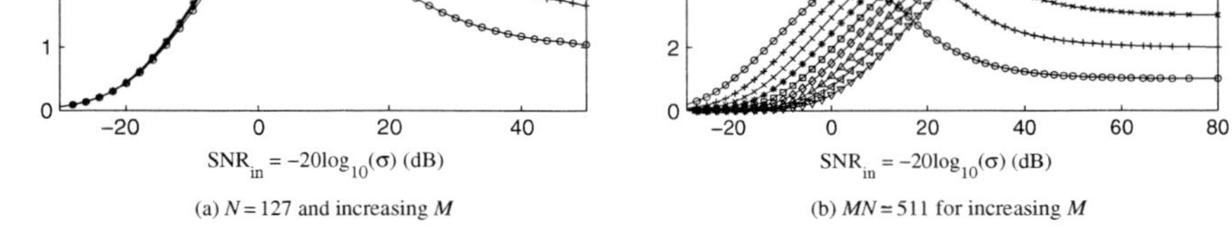

Fig. 4 The effect of (a) increasing the number of bits in each of $N = 127$ ADCs, and (b) modifying the ratio of M to N while MN remains approximately constant

4 Conclusions and Future Directions

While the results shown above illustrate that SSR can be generalized from the case of N single bit thresholds, it has to be pointed out that it is possible to achieve mutual information that decreases monotonically with decreasing input SNR, if one *optimizes* the thresholds for a given N and M. However, this is extremely difficult to achieve, and would be prohibitive in practise, in some cases. Furthermore, given the initial motivating context of averaging signals after digitization, in practical situations, the ADC would not be able to be modified between measurements. The results presented here indicate that fully understanding the SSR effect could lead to improvements in the design of digital receiver systems.

It is also of interest to determine the maximum achievable mutual information – i.e., channel capacity – for given signal and noise distributions for all values of input SNR. Clearly for large SNRs, this is limited from above by $\log_2(NM+1)$. For small SNRs, I conjecture that a pooling network approaches within a small constant of the capacity that results from analog averaging the outputs from N additive noise channels, i.e. $0.5\log_2(1+\frac{N}{\sigma^2})$. In other words, digital averaging approaches the same performance achievable for analog averaging. This is supported by numerical simulations. Current work is aimed at mathematically proving this, and finding a general formula that holds for all input SNRs, with the large and small input SNR formulas as limiting cases.

Acknowledgments The author gratefully acknowledges funding support from the Australian Research Council (ARC), and the DSTO in Australia, and discussions with Nigel Stocks and Pierre-Oliver Amblard.

References

1. N.G. Stocks, Physical Review Letters **84**(11), 2310 (2000)
2. M.D. McDonnell, D. Abbott, C.E.M. Pearce, Fluctuation and Noise Letters **2**(3), L205 (2002)
3. P.O. Amblard, S. Zozor, N.G. Stocks, M.D. McDonnell, in *Proc. SPIE Noise and Fluctuations in Biological, Biophysical, and Biomedical Systems*, vol. 6602, ed. by S.M. Bezrukov (2007), vol. 6602, p. 66020S
4. R.M. Gray, D.L. Neuhoff, IEEE Transactions on Information Theory **44**(6), 2325 (1998)
5. N.G. Stocks, Physical Review E **63**(4), 041114 (2001)
6. M.D. McDonnell, N.G. Stocks, C.E.M. Pearce, D. Abbott, Physics Letters A **352**, 183 (2006)
7. M.D. McDonnell, N.G. Stocks, C.E.M. Pearce, D. Abbott, *Stochastic Resonance: From Suprathreshold Stochastic Resonance to Stochastic Signal Quantisation* (Cambridge University Press, 2008 (In Press))
8. L. Gammaitoni, P. Hänggi, P. Jung, F. Marchesoni, Reviews of Modern Physics **70**(1), 223–287 (1998)
9. S. Durrant, J. Feng, Suprathreshold stochastic resonance in neural processing tuned by correlation (2006). To be submitted to Physical Review E
10. N.G. Stocks, R. Mannella, Physical Review E **64**, 030902(R) (2001)
11. T. Hoch, G. Wenning, K. Obermayer, Physical Review E **68**(3), 011911 (2003)
12. P.O. Amblard, S. Zozor, O.J.J. Michel, in *Proc. 2006 IEEE International Conference on Acoustics, Speech, and Signal Processing*, vol. 3 (2006), vol. 3, pp. 716–719

13. S. Zozor, P. Amblard, C. Duchêne, Fluctuation and Noise Letters **7**, L39 (2007)
14. T.M. Cover, J.A. Thomas, *Elements of Information Theory* (Wiley, New York, 1991)
15. N.G. Stocks, Physics Letters A **279**(5-6), 308 (2001)
16. M.D. McDonnell, D. Abbott, C.E.M. Pearce, Microelectronics Journal **33**(12), 1079 (2002)
17. D. Rousseau, F. Duan, F. Chapeau-Blondeau, Physical Review E **68**, 031107 (2003)
18. M.D. McDonnell, N.G. Stocks, D. Abbott, Physical Review E **75**, 061105 (2007)
19. V.K. Madisetti, D.B. Williams, *The Digital Signal Processing Handbook* (CRC Press, 1999)
20. M.D. McDonnell, N.G. Stocks, C.E.M. Pearce, D. Abbott, in *Proc. SPIE Smart Structures, Devices, and Systems II*, vol. 5649, ed. by S.F. Al-Sarawi (2005), vol. 5649, pp. 75–84

Resonance Curves of Multidimensional Chaotic Systems

Glenn Foster, Alfred W. Hübler and Karin Dahmen

Abstract We study resonance curves of nonlinear dynamical systems with chaotic forcing functions. We use the calculus of variations to determine the forcing function that induces the largest response. We compute the resonant forcing for a set of model systems and determine the response of the dynamical system to each forcing function. We show that the response is largest if the model system matches the dynamical system. We find that the signal to noise ratio is particularly large if one of the Lyapunov exponents is large.

1 Introduction

The limiting response of damped nonlinear oscillators to sinusoidal forcing functions can usually be characterized with frequency response curves and Arnold tongues. The response is typically largest and synchronized, *-i.e.* the periodic forcing function is in resonance, if the deriving frequency is within a frequency interval centered around the frequency of the unperturbed oscillator. Therefore the frequency response curve is often called a resonance curve. Resonance curves can be derived with secular perturbation theory and have a large range of important applications, including synchronization [1], stochastic resonance [2, 3], and nonlinear transport phenomena [4]. Probably the most important application is resonance spectroscopy, *-i.e.* the identification of dynamical systems based on peaks in resonance curves. However only undamped and weakly damped linear oscillators, have sharp resonance peaks, whereas the quality of the resonance peak is small for systems with large damping or large nonlinearity [5,6]. In 1987 Reiser et al. [7] showed that forcing functions which are proportional to the time reflected unperturbed dynamics has zero reaction power and produces a large response even for damped oscillators

G. Foster (✉)
Department of Physics, University of Illinois at Urbana-Champaign,
Urbana, IL 61801, USA, e-mail: gfoster@uiuc.edu

V. In et al. (eds.), *Applications of Nonlinear Dynamics*, Understanding Complex Systems,
DOI 10.1007/978-3-540-85632-0_21, © Springer-Verlag Berlin Heidelberg 2009

with a large nonlinearity. And in the same year Kautz [8] used optimal control theory [9] to derive minimal energy forcing functions to determine the most problem escape paths and activation energies of damped nonlinear oscillators. Beale [10] and Grassberger [11] generalized Kautz's approach to systems which do not necessarily have an energy function. Grassberger shows that the initial exponential growth rate of optimal forcing functions is equal and opposite to the negative Lyapunov exponent, if there is only one negative Lyapunov exponent. Chang et al. [12] use minimal escape paths of damped nonlinear oscillators to compute resonance curves. Plapp et al. [13] show that the minimization of the energy and the minimization of the reaction power lead to similar solutions, Krempl et al. [14] apply the methodology to quantum systems, and later Wargitsch et al. [15] show that the solutions are the same if the duration of the pulse is optimized. Graham et al. [16] introduce generalized potentials for dynamical systems and Beri et al. [17] simplify the boundary problem associated with Grassberger's equations with topological methods and solve it numerically. Only recently Foster el al. [18] published analytical solutions to the boundary value problem for systems with multiple or no negative Lyapunov exponents and show that the scalar product of the optimal forcing function and the separation of nearby trajectories is a conserved quantity.

In this paper, we use the analytical solution by Foster et al. to determine resonance curves of chaotic systems for system identification. We show that minimal escape path resonance curves have a large signal-to-noise ratio. We consider a dynamical systems, and a set of models where one of these models has the same mapping function as the dynamical system. Typically, the set of models is parameterized with a model parameter a_m. For each model we determine optimal forcing functions, apply the optimal forcing function to the dynamical system, and determine the response. The response versus the model parameter is called the resonance curve. This is a generalization of the traditional meaning of resonance curves. For instance, if the dynamical system is a weakly damped, linear oscillator, and the models are set weakly damped of linear oscillators where the frequency is the model parameter, then resonant forcing functions are sine-functions with a time dependent amplitude and the resonance curve is very similar to a standard frequency response curve.

2 Resonant Forcing and Resonance Curves

We consider the iterated map dynamics where the mapping function f contains an unknown parameter a:

$$\mathbf{x}_{n+1} = \mathbf{f}(\mathbf{x}_n, a) + \mathbf{F}_n + \mathbf{r}_n \tag{1}$$

where $\mathbf{x}_n \in R^d$ denotes the state of the d-dimensional system at time step $n = 0, 1, \cdots, N-1$, and f is differentiable in $\mathbf{x}$ within a given region of interest. $\mathbf{F}_n \in R^d$ is a small forcing function at time step n. $\mathbf{r}_n = (r_{n,1}, r_{n,2}, \cdots, r_{n,d})$ is very small, additive, band-limited, white noise where each component at each time step is a random number with variance $< (r_{n,i})^2 >= r^2/(Nd)$ without correlations, i.e.

$< r_{n,i} r_{n,j} >= 0$ for $i \neq j$ where $i, j = 1, 2, \ldots, d$. The response is predicted with a model:

$$\mathbf{X}_{n+1} = \mathbf{f}(\mathbf{X}_n, a_m) + \mathbf{F}_n \tag{2}$$

where $\mathbf{X}_n$ is the state of the model system at time n. The model parameter a_m is within a given range $a_{min} \leq a_m \leq a_{max}$ which is assumed to contain the correct value, i.e. $a_{min} \leq a \leq a_{max}$. If the model is exact ($a = a_m$, $\mathbf{x}_0 = \mathbf{X}_0$, $\mathbf{r}_n = 0$), the difference between the observed response, $R^2(\mathbf{F}) = |\mathbf{x}_N - \mathbf{y}_N|^2$, and the predicted response $R_m^2 = |\mathbf{X}_N(a_m) - \mathbf{Y}_N(a_m)|^2$ is zero, where y_n is the unperturbed system dynamics, Y_n is the unperturbed model dynamics, and $\mathbf{F} = \{\mathbf{F_0}, \mathbf{F_1}, \cdots, \mathbf{F_{N-1}}\}$ is a forcing function. The fact that $D^2(\mathbf{F}, a_m) = R^2 - R_m^2$ is zero if the model is correct, can be used for system identification. System identification is unique if $D^2(\mathbf{F}, a_m)$ has only one root. The number of roots depends on the forcing function and the noise. In order to achieve a large signal to noise ratio we study the final response of the system to a set of forcing functions, which contains the resonant forcing function. The resonant forcing function is the forcing function which produces the largest response among all forcing functions with the same magnitude $F^2 = \sum_{n=0}^{N-1} |\mathbf{F}_n)|^2$. In the case of a laser field F^2 is a measure for the total energy of the radiation pulse [19, 20]. F is a metric for the forcing function $\mathbf{F}$. We consider a set of forcing functions S, where all forcing functions have the same magnitude F, and each forcing function maximizes the final response of a particular model with model parameter a_m. Then the forcing function depends on the model parameter $\mathbf{F} = \mathbf{F}(\mathbf{a_m})$. If the model is correct the forcing function maximizes the response of the dynamical system among all forcing functions with the same magnitude. Consequently $\mathbf{F}(a)$ maximizes the response of the system among all forcing functions within the set S. The function $R(\mathbf{F}(a_m))$ is called a resonance curve. The resonance curve has an absolute maximum for $a_m = a$, i.e. $R(\mathbf{F}(a)) \geq R(\mathbf{F}(a_m))$ for all a_m in the range $a_{min} \leq a_m \leq a_{max}$.

We use the calculus of variations with Lagrange function

$$L = R_m^2/2 + \sum_{n=0}^{N-1} \mathbf{k}_n \left(\mathbf{x}_{n+1} - \mathbf{f}\left(\mathbf{x}_n\right) - \mathbf{F}_n\right) + \mu \left(\mathbf{F}_n\right)^2/2 \tag{3}$$

to determine the forcing function with yields the largest final response R_m^2 where $\mathbf{Y}_0 = \mathbf{X}_0$. $\mathbf{k}^{(n)}$ and μ are Lagrange multipliers. The stationary points of the Lagrange function provide necessary conditions for the maximum response. Elimination of the Lagrange multipliers $\mathbf{k}^{(n)}$ gives the following set of equations for the resonant forcing function

$$\left(J_{n+1}(b)\right)^T \mathbf{F}_{n+1} = \mathbf{F}_n \tag{4}$$

and

$$\mathbf{x}_N - \mathbf{y}_N = -\mu \mathbf{F}_{N-1} \tag{5}$$

where n = 0,1, ..., N−1, and $J_n(b) = (\partial f_i(b)/\partial X_j)|_{\mathbf{X}_n}$ is the Jacobi matrix evaluated at $\mathbf{X}_n$. The dynamics of a small displacement $\mathbf{d}_n = \mathbf{X}_n - \tilde{\mathbf{X}}_n$ between two neighboring trajectories at $\mathbf{X}_n$ and $\tilde{\mathbf{X}}_n$ is $\mathbf{d}^{(n+1)} = J^{(n)} \mathbf{d}^{(n)}$. Hence the scalar product of the resonant forcing and the displacement is a conserved quantity

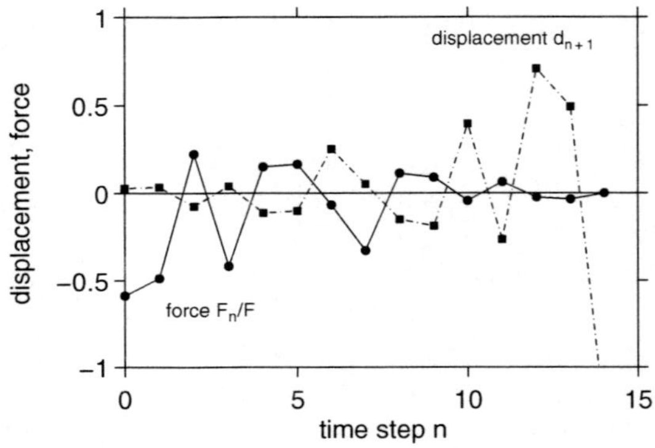

Fig. 1 The resonant forcing F_n (*circles*) and the displacement of two neighboring trajectories d_{n+1} (*squares*) versus time step n for a chaotic logistic map dynamics. This plot illustrates that the resonant forcing complements the displacement of neighboring trajectories of the unperturbed system, i.e. $F_n d_{n+1} = constant$. When the magnitude of the displacement is large, then the magnitude of the resonant force is small, and if the displacement is positive, the resonant force is negative

$$P = \mathbf{F}_n \cdot \mathbf{d}_{n+1} \tag{6}$$

for $n = 0, 1, \cdots, N-1$. We used no approximations to derive Eq. (6). The resonant forcing function complements the displacement dynamics of the model. If the system is one-dimensional, then the resonant forcing is proportional to the inverse of the displacement at each time step, i.e. $F_n = P/d_{n+1}$. Thus if the displacement dynamics is periodic, then the resonant forcing has the same periodicity and if the displacement dynamics is chaotic, then the resonant forcing has the same type of aperiodicity. Figure 1 shows that the displacement and the optimal forcing function are complementary for a chaotic logistic map dynamics, $x_{n+1} = ax_n(1-x_n) + F_n$, where $a = 3.61$, $N = 15$, $F = 0.0001$ and $x_0 = 0.897$.

3 Resonance Curves for Small Forcing Functions

Next we assume that the forcing function is small and expand the Jacobi matrix about the unperturbed dynamics to lowest order, i.e. $J_n \approx (\partial f_i/\partial x_j)|_{\mathbf{y}_n}$. To lowest order, the difference between the trajectory of the driven system and the unperturbed system reads: $\mathbf{x}_N - \mathbf{y}_N = \mathbf{F}_{N-1} + \sum_{n=1}^{N-1} \left(\prod_{i=1}^{n} J_{N-i}\right) \mathbf{F}_{N-1-n}$. With Eqs. (4) and (5) we obtain

$$M\mathbf{F}_{N-1} = -\mu \mathbf{F}_{N-1} \tag{7}$$

where $M = I + \sum_{n=1}^{N-1} M_n$ and $M_n = \left(\prod_{i=1}^{n} J_{N-i}\right)\left(\prod_{i=1}^{n} J_{N-i}\right)^T$. I is the identity matrix. M is a symmetric matrix with up to d orthogonal eigenvectors $\mathbf{e}_i$, where $M\mathbf{e}_i = \mu_i \mathbf{e}_i$, $i = 1, 2, \ldots, d$ and $\mathbf{e}_i^2 = 1$. The corresponding eigenvalues μ_i are positive.

The eigenvectors of matrix M are the solutions of Eq. (7) $\mathbf{F}_{N-1} = \pm F_{N-1}\mathbf{e}_i$, where $F_{N-1} = |\mathbf{F}_{N-1}|$ and $\mu = -\mu_i$. Equations (4) and (7) yield $F^2 = \mu_i F_{N-1}^2$ and $R^2 = (\mathbf{x}_N - \mathbf{y}_N)^2 = \mu_i^2 (F_{N-1})^2 = \mu_i F^2$. Hence the final forcing which parallels the eigenvector with the largest eigenvalue of M, $\hat{\mu} = \max\{\mu_i\}$ produces the largest response, and the largest response is

$$R^2 = \hat{\mu} F^2 \tag{8}$$

and with Eq. (7) we obtain

$$\hat{\mathbf{F}}_{N-1} = \pm \frac{F}{\sqrt{\hat{\mu}}} \hat{\mathbf{e}} \tag{9}$$

where $\hat{\mathbf{e}}$ is the eigenvector that corresponds to the largest eigenvalue of M, and for $n = 0, 1, \cdots, N-2$ the time dependence of the resonant forcing function is $\mathbf{F}_n = \pm \frac{F}{\sqrt{\hat{\mu}}} \left(\prod_{i=1}^{N-1-n} (J_{n+i})^T \right) \hat{\mathbf{e}}$. Figure 2 shows resonance curves of a chaotic Henon map dynamics as a function of the map parameter a_m and b_m. The Henon map is $x_{1,n+1} = 1 - a(x_{1,n})^2 + x_{2,n} + F_{1,n}$ and $x_{2,n+1} = b x_{1,n} + F_{2,n}$, where $\mathbf{x}_n = (x_{1,n}, x_{2,n})$ is the state and $\mathbf{F}_n = (F_{1,n}, F_{2,n})$ is the forcing function. a and b are parameters. The magnitude of the forcing function is $F = 0.0001$, the noise level is $r = 0$, and the number of time steps is $N = 3$. The numerical values of the peak location of the resonance curve is in good agreement with the system parameters $a = 1.08$, $b = 0.3$.

For systems with only one variable $x_{n+1} = f(x_n, a) + F_n + r_n$, the eigenvalue of M is (see Eq. (7))

$$\hat{\mu} = 1 + \sum_{n=1}^{N-2} \prod_{i=1}^{n} \left(\left. \frac{\partial f}{\partial x} \right|_{y(N-i)} \right)^2 \tag{10}$$

From Eq. (4) we obtain for the resonant forcing function

$$F^{(n)} = \prod_{i=1}^{N-n-1} \left. \frac{\partial f}{\partial x} \right|_{y(n+i)} F^{(N-1)} \tag{11}$$

where $F^{(N-1)} = \pm F/\sqrt{\hat{\mu}}$. The response to the resonant forcing function is $R = \sqrt{\hat{\mu}} F$. Figure 3 shows the resonant forcing function (Eq. (11)) and the displacement

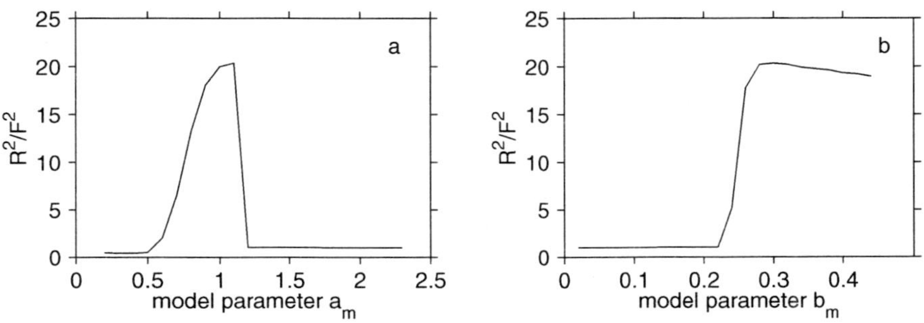

Fig. 2 The resonance curve of a chaotic Henon map versus model parameter a_m, where $b_m = b$ (a) and versus model parameter b_m where $a_m = a$. The parameters are $a = 1.08$, $b = 0.3$, $N = 3$ and $F = 0.0001$

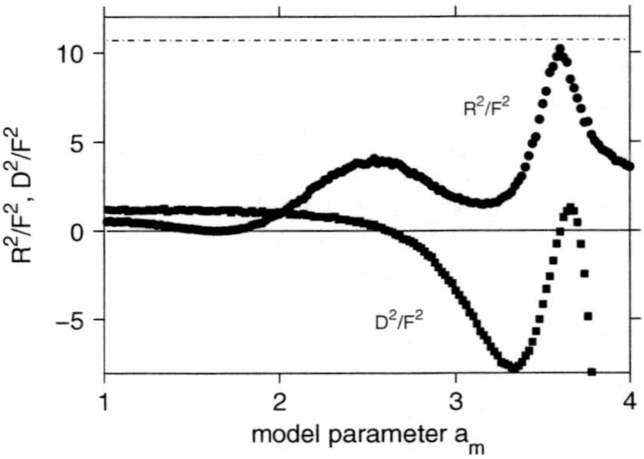

Fig. 3 The resonance curve for a chaotic logistic map (*circles*) and deviation D between the response model and the response of the dynamical system for a set of sinusoidal forcing functions (*squares*) versus the model parameter a_m. The number of time steps is $N = 4$, the noise level is $r = 0.0005$, and the magnitude of the forcing function is $F = 0.001$. The dashed line indicates the theoretical result for the maximum of the resonance curve given by Eq. (8). The maximum of the resonance curve, as well as one of the three roots of D are close to the parameter value of the dynamical system, $a = 3.6$

dynamics for a chaotic logistic map dynamics $x^{(n+1)} = 3.61x^{(n)}(1-x^{(n)}) + F^{(n)}$, for $n = 0, 1, \cdots, 14$, with the initial condition $y^{(0)} = x^{(0)} = 0.34$. The magnitude of the forcing function $F = 0.0001$. With Eq. (10) we compute $\hat{\mu} = 1500$. We find that the predicted response $R = 0.00387$ is close to the numerical value $R = 0.00378$.

The matrix $M^{(N)}$ describes how the magnitude of a displacement grows $|\mathbf{d}^{(N)}|^2 = \left(\mathbf{d}^{(0)}\right)^T M^{(N)} \mathbf{d}^{(0)}$. If $\mu_i^{(n)}$ are the eigenvalues of $M^{(n)}$ then the Lyapunov exponents are the limits $\lambda_i = \lim_{n\to\infty} \frac{1}{2n} \ln \mu_i^{(n)}$. The set of Lyapunov exponents will be the same for almost all starting points on an ergodic attractor.

For some chaotic systems the matrices $M^{(n)}$ have approximately the same eigenvectors and eigenvalues. For instance if the Jaccobian is constant, i.e. $J^{(n)} = J^{(0)}$ for $n = 1, 2, \cdots, N-1$, then $M^{(n)} = (J^{(0)})^n((J^{(0)})^T)^n = (J^{(0)}(J^{(0)})^T)^n = (M^{(1)})^n$ since $J^{(0)}(J^{(0)})^T = (J^{(0)})^T J^{(0)}$. This is the case for coupled Bernoulli map dynamics (see Eq. (15)). If the matrices $M^{(n)}$ have approximately the same eigenvectors and eigenvalues, then the eigenvalues obey the following relation $\mu_i^{(n)} \approx (\mu_i^{(1)})^n \approx e^{2n\lambda_i}$.

If an initial displacement is parallel to the eigenvector of M that corresponds to the largest Lyapunov exponent $\hat{\lambda} = max\{\lambda_i, i = 1, 2, \cdots, d\}$, then it has the largest growth rate, i.e. $d^{(n)} = e^{n\hat{\lambda}} d^{(0)}$. The final value of the optimal forcing function $\mathbf{F}^{(N-1)}$ is parallel to the eigenvector of M that corresponds to the largest Lyapunov exponent and earlier values obey the dynamics:

$$|\mathbf{F}^{(n)}|^2 = \left(\mathbf{F}^{(N-1)}\right)^T M^{(N-n)} \mathbf{F}^{(N-1)} = \mu_i^{(N-n)} (F^{(N-1)})^2 = \frac{1}{\mu_i^n} (F^{(0)})^2 \tag{12}$$

Hence the the growth rate of the magnitude of the optimal forcing function is equal to the opposite of the largest Lyapunov exponent:

$$F^{(n)} \approx e^{-\hat{\lambda} n} F^{(0)} \tag{13}$$

Since $M = I + \sum_{n=1}^{N-1} M^{(n)}$ we estimate $\hat{\mu} \approx \sum_{n=0}{}^{N-1} \hat{\mu}_1^n = \frac{1-\mu_1^N}{1-\mu_1} \approx \frac{1-e^{2\hat{\lambda}N}}{1-e^{2\hat{\lambda}}}$. Then the response can be approximated by

$$R^2 \approx \frac{1-e^{2\hat{\lambda}N}}{1-e^{2\hat{\lambda}}} F^2 \tag{14}$$

An example for a mapping function with a constant Jaccobian is a system of two coupled Bernoulli maps:

$$\begin{pmatrix} x_{1,n+1} \\ x_{2,n+1} \end{pmatrix} = \begin{pmatrix} mod(ax_{1,n} + kx_{2,n} + F_{1,n}) \\ mod(bx_{2,n} + kx_{1,n} + F_{2,n}) \end{pmatrix} \tag{15}$$

where the function $mod(x) = x - \lfloor x \rfloor$ returns the decimal part of x. a and b are the growth rates and k is the coupling constant. We assume that $a > b \geq 0$. For the corresponding model dynamics with the parameters a_m, b_m, and k_m, the eigenvalues of M_1 are $\hat{\mu}_1(a_m, b_m, k_m) = 0.5(a_m^2 + b_m^2 + 2k_m^2 + (a_m + b_m)\sqrt{(a_m - b_m)^2 + 4k_m^2})$ and the eigenvectors

$$\mathbf{e} = \frac{(a_m - b_m + \sqrt{(a_m - b_m)^2 + 4k_m^2}, 2k_m)}{\sqrt{\left(\left(a_m - b_m \pm \sqrt{(a_m - b_m)^2 + 4k_m^2}\right)^2 + 4k_m^2\right)}} \tag{16}$$

Since the Jacobian $J_n(a_m, b_m, k_m)$ is symmetric and constant, the eigenvectors of the M_1 are eigenvectors of M_n, and the Lyapunov exponents are $\lambda_{1/2} = \frac{1}{2} \ln \mu_{1/2}^{(1)}$ and the largest Lyapunov exponent is $\hat{\lambda} = \frac{1}{2} \ln(\hat{\mu})$. If $\hat{\mu}(a,b,k) > 1$ the unperturbed dynamics is chaotic. Hence $\mathbf{F}^{(n)} = \pm (J)^{N-1-n} \frac{F}{\sqrt{\hat{\mu}}} \hat{\mathbf{e}}$. The peak value of the resonance curve is given by Eq. (17).

If $k = 0$ then $\hat{\mu}_J = a_1$, $\hat{\mu} = \frac{a_1^{2N}-1}{a_1^2-1}$, $\hat{\mathbf{e}} = (1,0)$, $\mathbf{F}^{(n)} = \pm(a_1^{N-n-1} F/\sqrt{\hat{\mu}}, 0) = (F^{(0)}/a_1^n, 0)$, where $F^{(0)} = \pm a_1^{N-1} F/\sqrt{\hat{\mu}}$ and

$$R^2 = F^2 \frac{(a^N a_m^N - 1)^2 (a_m^2 - 1)}{(a_m^{2N} - 1)(a a_m - 1)^2} \tag{17}$$

The resonance curve does not depend on b_m. Hence the resonance curve can not be used to determine the parameter of the less chaotic dynamics b_m. For $k = 0$ the systems contains two decoupled Bernoulli maps, where $\lambda_i = \ln |a_i|$, $i = 1,2$ is the Lyapunov exponent of each map. The resonant forcing function is in the direction of the map with the larger Lyapunov exponent. Hence if both maps have a positive Lyapunov exponent and therefore both are chaotic, then the resonant forcing function forces only the map which is more chaotic. There is no forcing of the less chaotic map. For a periodic forcing $F_{1,n} = (-1)^n F/\sqrt{N}$ and $F_{2,n} = 0$ the difference between the system response R^2 and the model response R_m^2 is

$$D^2 = \left(\left(\frac{1-(-a)^N}{a+1} \right)^2 - \left(\frac{1-(-a_m)^N}{a_m+1} \right)^2 \right) \frac{F^2}{N} \tag{18}$$

Figure 4 shows the numerical and theoretical resonance curve for a Bernoulli map dynamics and the deviation D^2. The response to the sinusoidal forcing function and D are much smaller than the response to the optimal forcing function.

Finally we compute the response to random forcing $\mathbf{r}_n = (r_{1,n}, r_{2,n}, \cdots, r_{d,n})$ where each component of the forcing function at each time step is a random number with variance $< (r_{i,n})^2 >= F^2/(Nd)$ and without correlations $< r_{i,n} r_{j,n} >= 0$ for $i \neq j$. Then the expectation value of the response is

$$R_r^2 = \left(\frac{1}{d} \sum_{i=1}^{d} \frac{1-e^{2\lambda_i N}}{1-e^{2\lambda_i}} \right) \frac{r^2}{N} \tag{19}$$

From Eqs. (14) and (19) we conclude that the response for the optimal forcing is large compared to response from random forcing, if the largest Lyapunov exponent is much larger than the other Lyapunov exponents. Figure 5 shows the signal-to-noise ratio, i.e. the ratio response for optimal forcing and random forcing, R^2/R_r^2, as a function of the largest Lyapunov exponent for a chaotic Bernoulli map dynamics. The signal-to-noise ratio is particularly large, if the largest Lyapunov exponent is much larger than the other Lyapunov exponents.

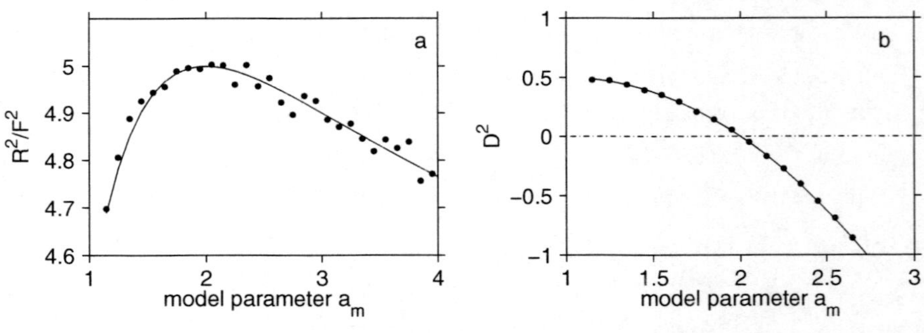

Fig. 4 The resonance curve (**a**) and the difference between the model response and the system response versus the model parameter (**b**) for a chaotic Bernoulli map with parameter $a = 2$. The continuous lines are the theoretical values given by Eqs. (18) and (19). The resonance curve has an absolute maximum if the model parameter matches the system parameter. In contrast, the difference of the response is zero if the model parameter matches the system parameter

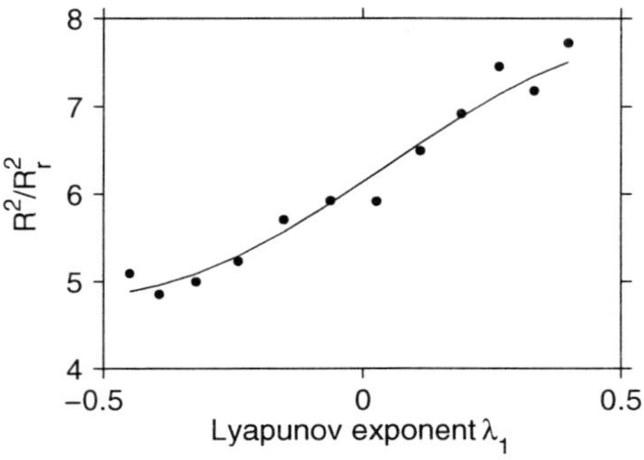

Fig. 5 Signal-to-noise ratio R^2/r_r^2 versus the largest Lyapunov exponent of a coupled Bernoulli map, where $N = 4$, $a_2 = 0.5$, $k = .2$, and $F = r = 0.0001$. The continuous line is the theoretical value given by Eqs. (14) and (19). The bullets are the expectation values determined from 1000 simulations. This figure illustrates that the signal to noise ratio is particularly large if one Lyapunov exponent is significantly larger than the others

4 Summary

We compute resonance curves of nonlinear dynamical systems with chaotic forcing functions (see Figs. 2, 3, and 4). We use the calculus of variations to determine the forcing function that induces the largest response (Eqs. (4) and (5)). We find that the product of resonant forcing and the displacement of nearby trajectories is a conserved quantity (Eq. (6)), i.e. when the displacement dynamics is irregular, the resonant forcing function is irregular too (see Fig. 1). Figure 2 illustrates that the response is largest if the model system matches the dynamical system. Figure 5 shows that the signal to noise ratio is particularly large if one of the Lyapunov exponents is large. Traditional resonance curves show the limiting response to very long sinusoidal perturbations. However in many experimental situations forcing functions have to be rather short, because experimental constraints, such as the coherence length and the energy content, impose limits on the duration of a laser pulse. In this paper, resonance curves show the final response to short, but most efficient forcing functions. For damped linear oscillators, these forcing functions are sine-functions where the amplitude grows exponentially at a rate which is equal and opposite to its negative Lyapunov exponent. For quasi-periodic systems, optimal forcing functions are quasi-periodic and have a constant amplitude [21]. For chaotic one dimensional map dynamics the optimal forcing function is aperiodic and its amplitude decreases exponentially at a rate which is equal and to its positive Lyapunov exponent. For some systems the Lyapunov exponent does not describe the evolution of small perturbation [22]. Even in this case we can conclude that for one-dimensional mapping functions, the magnitude optimal forcing function decreases on average if the magnitude of the separation of nearby trajectories increases on average and vice versa, since the product of the optimal forcing function and the separation of nearby trajectories is a conserved quantity (see Eq. 6).

Acknowledgments This material is based upon work supported by the National Science Foundation Grant No. NSF PHY 01-40179 and NSF DMS 03-25939 ITR. A.H. thanks the Santa Fe Institute for support.

References

1. I. Siddiqi, R. Vijay, F. Pierre, C. M. Wilson, L. Frunzio, M. Metcalfe, C. Rigettiand, R. J. Schoelkopf, M. H. Devoret, D. Vion, et al., Phys. Rev. Lett. **94**, 027005 (2005)
2. R. L. Badzey and P. Mohanty, Nature **437**, 962 (2005)
3. C. Wargitsch and A. W. Hubler, Il Nuovo Cimiento **17D**, 969 (1990)
4. S. Wimberger, R. Mannella, O. Morsch, and E. Arimondo, Phys. Rev. Lett. **94**, 130404 (2005)
5. T. Eisenhammer, A. W. Hubler, T. Geisel, and E. Luscher, Phys. Rev. A **41**, 3332 (1990)
6. L. E. Arsenault and A. W. Hubler, Phys. Rev. E **51**, 3561 (1995)
7. G. Reiser, A. W. Hubler, and E. Luscher, Z. Naturforschung **42a**, 803 (1987)
8. R. L. Kautz, Phys. Lett. A **125**, 315 (1987)
9. W. H. Fleming, Appl. Math. Optim. **4**, 329 (1977)
10. P. D. Beale, Phys. Rev. A **40**, 3998 (1989)
11. P. Grassberger, J. Phys. A: Math. Gen. **22**, 3283 (1989)
12. K. Chang, A. Kodogeorgiou, A. W. Hubler, and E. A. Jackson, Physicsa D **51**, 99 (1991)
13. B. B. Plapp and A. W. Hubler, Phys. Rev. Lett. **65**, 2302 (1990)
14. S. Krempl, T. Eisenhammer, A. Hubler, G. Mayer-Kress, and P. W. Milonni, Phys. Rev. Lett. **69**, 430 (1992)
15. C. Wargitsch and A. Hubler, Phys. Rev. E **51**, 1508 (1995)
16. R. Graham, A. Hamm, and T. Tel, Phys. Rev. Lett **66**, 3089 (1991)
17. S. Beri, R. Mannella, D. G. Luchinsky, A. N. Silchenko, and P. V. E. McClintock, Phys. Rev. E **72**, 036131 (2005)
18. G. Foster, A. W. Hubler, and K. Dahmen, Phys. Rev. E **75**, 0036212 (2007)
19. V. N. Smelyanskiy and M. I. Dykman, Phys. Rev. E **55**, 2516 (1997)
20. I. A. Khovanov, D. G. Luchinsky, R. Mannella, and P. V. E. McClintock, Phys. Rev. Lett. **85**, 2100 (2000)
21. J. Xu and A. W. Hubler, Phys. Rev. B **67**, 184202 (2003)
22. G. Paladin, M. Serva, and A. Vulpiani, Phys. Rev. Lett. **74**, 66 (1995)

Learning of Digital Spiking Neuron and its Application Potentials

Hiroyuki Torikai

Abstract The digital spiking neuron consists of shift registers and behaves like a simplified neuron model. By adjusting wirings among the registers, the neuron can generate spike-trains with various characteristics. In this paper some analysis results on spike-train properties are outlined. Also a learning algorithm for the neuron is introduced and its application potentials are discussed.

1 Introduction

Simplified spiking neuron models have been studied from fundamental and application viewpoints [1–7]. These neuron models can be regarded as *analog nonlinear dynamical systems that can behave like spiking neurons*. On the other hand, we have claimed that it is also interesting to investigate a *digital dynamical system (like cellular automaton and shift register generator [8–12]) that can behave like a spiking neuron* [13–18]. We then proposed a *digital spiking neuron* (DSN) which consists of digital state cells and operates on a discrete time [13]. The dynamics of the DSN is inspired by analog integrate-and-fire neuron models [1–3] as explained in Fig. 1. The important parameter of the DSN is pattern of wirings among the digital state cells because the DSN can realize rich dynamics and various characteristics of the spike-train by adjusting the wiring pattern.

In this paper, first, some analysis results [14] on the spike-train properties are outlined. Next, a learning algorithm [14] for the neuron is introduced and its application potentials are discussed. As an understandable application example, ranging based on *ultra-wide band* (UWB) impulse radio signal [19–22] is considered. It is show that the learning algorithm can search wiring patterns that realize spike-trains whose characteristics are suited to the ranging application. Another application of

H. Torikai (✉)
Graduate School of Engineering Science, Osaka University, Osaka, Japan
e-mail: torikai@sys.es.osaka-u.ac.jp

V. In et al. (eds.), *Applications of Nonlinear Dynamics*, Understanding Complex Systems, DOI 10.1007/978-3-540-85632-0_22,

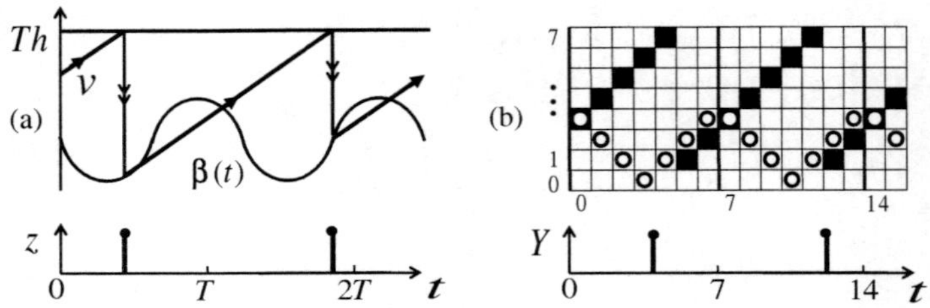

Fig. 1 (**a**) Basic dynamics of integrate-and-fire neuron models in [1–3]. A potential v repeats so-called *integrate-and-fire dynamics* between a firing threshold Th and a reset level $\beta(t)$. The reset level is modulated by a periodic signal $\beta(t)$ that can realize rich bifurcation phenomena of the spike-train z. (**b**) Basic dynamics of the digital spiking neuron in [13]. The black box corresponds to the potential v and the white circle corresponds to the periodic reset level $\beta(t)$. The black box repeats shift-and-reset dynamics that can be regarded as a digital version of the integrate-and-fire dynamics. Y is an output spike-train

the DSN (approximation of various analog neuron models by the DSN toward mimicking dynamics of biological neurons) can be found in [18].

The DSN and its learning algorithm are suited to reconfigurable hardware implementation (e.g., FPGA) because their dynamics are based on digital operations and change of wirings. On the other hand it is troublesome to implement a learning algorithm for the analog neuron models in [1–3] if the algorithm utilizes change of analog parameter values.

2 Digital Spiking Neuron

In this section the *digital spiking neuron* (DSN) [13, 14] is introduced and its basic dynamics is explained by using Fig. 2. The DSN operates on a discrete time

$$t = 0, 1, 2, \cdots .$$

As shown in Fig. 2(a), the DSN has M pieces of p-cells that are indexed by $i \in \{0, 1, \cdots, M-1\}$, where $M \geq 2$. Each p-cell has a digital state $p_i(t) \in \{0, 1\} \equiv \boldsymbol{B}$, where "$\equiv$" denotes "*definition*" hereafter. The p-cells are ring-coupled and are governed by

$$p_{i+1(\text{mod } M)}(t+1) = p_i(t). \tag{1}$$

In this paper initial states of the p-cells are fixed as follows: $p_0(0) = 1$ and $p_k(0) = 0$ for all $k \neq 0$. Then the p-cells oscillate periodically with period M. As shown in Fig. 2(a), the DSN has one-way reconfigurable wirings from p-cells to x-cells. In the dotted box, the left terminals accept the following state vector of the p-cells:

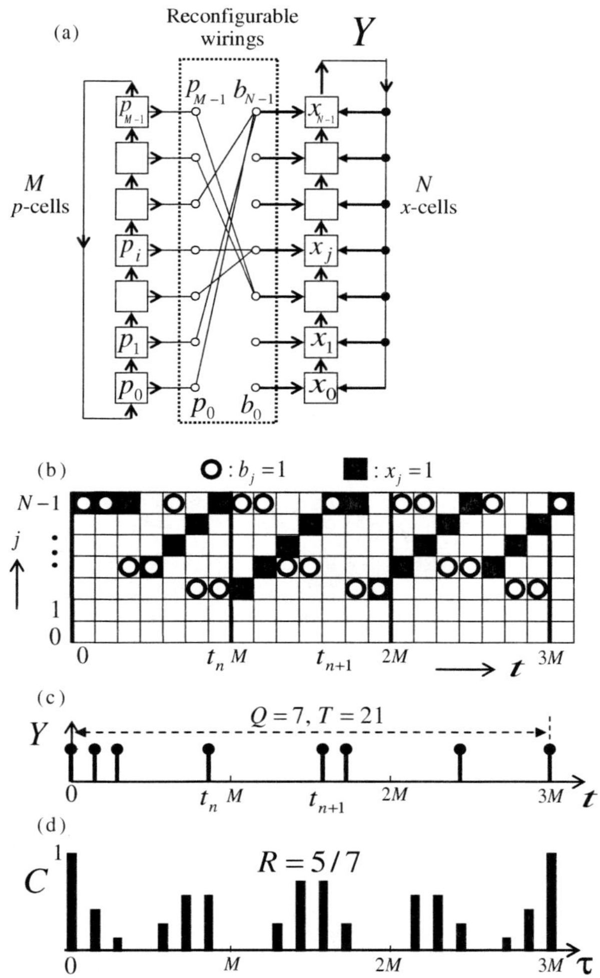

Fig. 2 (**a**) Digital spiking neuron. $M = N = 7$. The wiring matrix $\boldsymbol{A}$ is given by Eq. (2). (**b**) Shift-and-reset dynamics. (**c**) Output spike-train $Y(t)$. (**d**) Auto-correlation function $C(\tau)$ of $Y(t)$

$$(p_0(t), \cdots, p_{M-1}(t))^t \equiv \boldsymbol{P}(t) \in \boldsymbol{B}^M.$$

The right terminals output the following *base signal*

$$(b_0(t), \cdots, b_{N-1}(t))^t \equiv \boldsymbol{b}(t) \in \boldsymbol{B}^N$$

where $N \geq 2$. Each left terminal p_i has one wiring, and each right terminal b_j can accept any number of wirings. In order to describe pattern of the wirings, let us introduce an $N \times M$ *wiring matrix* $\boldsymbol{A}$ whose elements are defined by

$$a(j,i) = \begin{cases} 1 & \text{if terminal } p_i \text{ is wired to terminal } b_j, \\ 0 & \text{otherwise} \end{cases}$$

where $i = 0,1,\cdots,M-1$ and $j = 0,1,\cdots,N-1$. The DSN in Fig. 2(a) has the following wiring matrix:

$$\boldsymbol{A} = \begin{pmatrix} 0\,0\,0\,0\,0\,0\,0 \\ 0\,0\,0\,0\,0\,0\,0 \\ 0\,0\,0\,0\,0\,1\,1 \\ 0\,0\,1\,1\,0\,0\,0 \\ 0\,0\,0\,0\,0\,0\,0 \\ 0\,0\,0\,0\,0\,0\,0 \\ 1\,1\,0\,0\,1\,0\,0 \end{pmatrix}. \tag{2}$$

The wiring matrix $\boldsymbol{A}$ has one "1" in each column. Using the wiring matrix $\boldsymbol{A}$, the base signal $\boldsymbol{b}(t)$ is given by

$$\boldsymbol{b}(t) = \boldsymbol{A}\boldsymbol{P}(t). \tag{3}$$

In Fig. 2(b), white circles illustrate a base signal $\boldsymbol{b}(t)$ which corresponds to the wiring pattern in Fig. 2(a). Here let us consider the x-cell. Each j-th x-cell has a digital state $x_j(t) \in \boldsymbol{B}$. Also, as shown in Fig. 2(a), the x-cell has three digital inputs $\{b_j, x_{N-1}, x_{j-1}\}$, where $x_{-1} \equiv 0$. The x-cell operates as follows: $x_j(t+1) = x_{j-1}(t)$ if $x_{N-1}(t) = 0$, and $x_j(t+1) = b_j(t)$ if $x_{N-1}(t) = 1$. Let

$$(x_0(t),\cdots,x_{N-1}(t))^t \equiv \boldsymbol{X}(t) \in \boldsymbol{B}^N$$

be a state vector of the x-cells, and let

$$S((x_0,\cdots,x_{N-1})^t) = (0,x_0,\cdots,x_{N-2})^t$$

be a shift operator. Then the dynamics of the x-cells is described by

$$\boldsymbol{X}(t+1) = \begin{cases} S(\boldsymbol{X}(t)) & \text{if } x_{N-1}(t) = 0, \\ \boldsymbol{b}(t) & \text{if } x_{N-1}(t) = 1. \end{cases} \tag{4}$$

Basic dynamics of the x-cells is shown in Fig. 2(b). In this figure a black box indicates that the corresponding x-cell has state $x_j(t) = 1$. If the black box is below the highest position which is indexed by $j = N-1$, the black box is shifted upward. If the black box reaches the highest position at $t = t_n$ (i.e., $x_{N-1}(t_n) = 1$), the black box at $t = t_n + 1$ is reset to the position of the white circle at $t = t_n$. At this reset moment, the DSN outputs a spike $Y(t_n) = x_{N-1}(t_n) = 1$. Repeating such *shift-and-reset* dynamics, the DSN generates the following spike-train:

$$Y(t) \equiv x_{N-1}(t), \;\; t = 0,1,2,\cdots. \tag{5}$$

As a result, the dynamics of the DSN is described by Eqs. (1), (3), (4) and (5). Also the DSN is characterized by the following parameters:

$$\# \; p\text{-cells} = M, \quad \# \; x\text{-cells} = N,$$
$$\text{elements } a(j,i) \text{ of wiring matrix } \boldsymbol{A}$$

where "#" denotes "*the number of*" hereafter. For given M and N, the total number of the wiring matrixes is given by

$$\# \text{ wiring matrixes} = N^M. \tag{6}$$

Depending on the parameters $(M,N,\boldsymbol{A})$, the DSN can exhibit rich phenomena including the following ones [13].

- The DSN generates a periodic spike-train $Y(t)$ in a steady state. The periodic spike-train $Y(t)$ can have various patterns of spike positions.
- The DSN may generate different spike-trains for different initial states. Such a phenomenon is so-called co-existence for initial state.
- The DSN may exhibit a complicated transient phenomenon to co-existing periodic spike-trains.

In this paper we focus on generation of spike-trains that have long periods and complicated patterns of spike positions.

3 Wiring Matrix and Spike-Train

In this section we consider a fundamental problem: how a change of the wiring matrix $\boldsymbol{A}$ affects to basic characteristics of the spike-train $Y(t)$?

First, let us consider basic characterization of the spike-train $Y(t)$. As shown in Fig. 2(c), let the n-th spike position be denoted by t_n, where $n = 1,2,3,\cdots$. Then a spike-train $Y(t)$ can be characterized as follows.

Definition 1. A spike-train $Y(t)$ is said to be a *periodic spike-train* if there exists a positive integer Q such that $t_n = t_{n+Q} \pmod{M}$ for all $n \geq 1$. The minimum integer Q such that $t_n = t_{n+Q} \pmod{M}$ is said to be *spike number* of the periodic spike-train $Y(t)$. An integer $T \equiv t_{Q+1} - t_1$ is said to be *period* of $Y(t)$. A sequence $\boldsymbol{S} \equiv (t_1,t_2,\cdots,t_Q)$ is said to be *periodic spike position sequence* of $Y(t)$.

In the case of Fig. 2(c), the spike-train $Y(t)$ is characterized by spike number $Q = 7$, period $T = 21$ and periodic spike position sequence $\boldsymbol{S} = (0,1,2,6,11,12,17)$.

Next, let us consider change of the wiring matrix $\boldsymbol{A}$. In order to simplify the consideration, let us focus on the following condition:

$$M = N. \tag{7}$$

That is, the numbers of p-cells and x-cells are identical. Now the DSN is characterized by the parameters $(M, \boldsymbol{A})$. Under the condition in Eq. (7), the spike number Q and the period T of a spike-train $Y(t)$ satisfy the following properties:

$$1 \leq Q \leq M, \quad M \leq T \leq M^2. \tag{8}$$

In Fig. 2, we can confirm that the DSN satisfies the condition in Eq. (7) and the spike-train $Y(t)$ satisfies the properties in Eq. (8). Let the wiring pattern of the DSN in Fig. 2(a) be changed into the wiring pattern of the DSN in Fig. 3(a). Such change of wirings is referred to as *re-wiring*. The re-wired DSN in Fig. 3(a) has the following re-wired matrix:

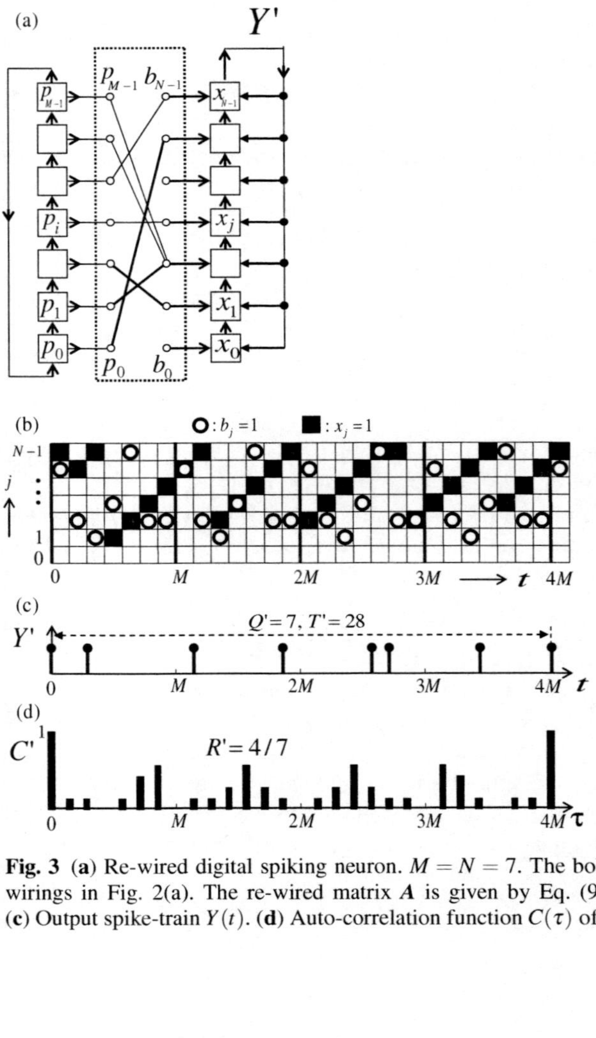

Fig. 3 (**a**) Re-wired digital spiking neuron. $M = N = 7$. The bold wirings are different from the wirings in Fig. 2(a). The re-wired matrix $\boldsymbol{A}$ is given by Eq. (9). (**b**) Shift-and-reset dynamics. (**c**) Output spike-train $Y(t)$. (**d**) Auto-correlation function $C(\tau)$ of $Y(t)$

$$\boldsymbol{A}' = \begin{pmatrix} 0\,0\,0\,0\,0\,0\,0 \\ 0\,0\,1\,0\,0\,0\,0 \\ 0\,1\,0\,0\,0\,1\,1 \\ 0\,0\,0\,1\,0\,0\,0 \\ 0\,0\,0\,0\,0\,0\,0 \\ 1\,0\,0\,0\,0\,0\,0 \\ 0\,0\,0\,0\,1\,0\,0 \end{pmatrix} \tag{9}$$

where the prime "$\prime$" implies "*re-wired version*" hereafter. As shown in Fig. 3(c), the re-wired DSN generates a spike-train $Y'(t)$ which is characterized by spike number $Q' = 7$, period $T' = 28$ and periodic spike position sequence $\boldsymbol{S}' = (0,2,8,13,18,19,24)$. In Fig. 2(c) and Fig. 3(c), we can confirm the following facts.

- The spike number does not change (i.e., $Q' = Q$) under the re-wiring.
- The periodic spike position sequence changes (i.e., $\boldsymbol{S}' \neq \boldsymbol{S}$) under the re-wiring.

These facts can be generalized as follows. Let two integers (r,s) be referred to as *re-wiring positions*, where $0 \leq r < s \leq M-1$. Let us create a re-wired matrix by the following rule.

Re-wiring rule: For given re-wiring positions (r,s), the wiring matrix $\boldsymbol{A}$ (whose element is $a(j,i)$) is changed into the re-wired matrix $\boldsymbol{A}'$ (whose element is $a'(j,i)$) as follows:

$$\begin{aligned} a''(j,r) &= a(j+s-r+M\ (\mathrm{mod}\ M), s), \\ a''(j,s) &= a(j+r-s+M\ (\mathrm{mod}\ M), r), \\ &\text{for all } j = 0,1,\cdots,M-1, \end{aligned} \tag{10}$$

$$\begin{aligned} &a'(M-r+i\ (\mathrm{mod}\ M), i) \\ &\qquad = a''(M-s+i\ (\mathrm{mod}\ M), i), \\ &a'(M-s+i\ (\mathrm{mod}\ M), i) \\ &\qquad = a''(M-r+i\ (\mathrm{mod}\ M), i), \\ &\quad \text{for all } i = 0,1,\cdots,M-1 \end{aligned} \tag{11}$$

where $a''(j,i)$ is element of an $M \times M$ matrix $\boldsymbol{A}''$ which is used as an intermediate in this re-wiring.

Then we can give the following theorem.

Re-wiring theorem: Assume that a DSN with a wiring matrix $\boldsymbol{A}$ generates a periodic spike-train $Y(t)$ with spike number $Q = M$ and periodic spike position sequence $\boldsymbol{S}$. Then, for arbitrary re-wiring positions (r,s), the DSN with the re-wired matrix $\boldsymbol{A}'$ generates a periodic spike-train $Y'(t)$ with spike number $Q' = Q = M$ and periodic spike position sequence $\boldsymbol{S}' \neq \boldsymbol{S}$.

A proof of this theorem can be found in [14].

4 Learning by Re-Wirings

The main scope of this section is to show significances of the re-wiring theorem and to explore learning potential of the DSN by using an understandable application example: ranging based on *ultra-wide band* (UWB) impulse radio signals.

Let us define an auto-correlation function $C(\tau)$ of the spike-train $Y(t)$ as follows:

$$C(\tau) \equiv \tfrac{1}{Q}\Sigma_{t=0}^{T-1} Y(t)Y(t+\tau). \tag{12}$$

Figure 2(d) shows the auto-correlation function $C(\tau)$ of the spike-train $Y(t)$ in Fig. 2(c). The auto-correlation function $C(\tau)$ has the following basic properties: $C(0) = 1$, $0 \leq C(\tau) \leq 1$ and $C(\tau+T) = C(\tau)$. Let us define the 2nd peak R of the auto-correlation function $C(\tau)$ as follows:

$$R \equiv \max_{1 \leq \tau \leq T-1} \{C(\tau)\}.$$

The auto-correlation function $C(\tau)$ in Fig. 2(d) has the 2nd peak $R = 5/7$. Roughly speaking, the 2nd peak R of auto-correlation characterizes complexity of the periodic spike position sequence $\boldsymbol{S}$, i.e., R tends to be smaller as $\boldsymbol{S}$ becomes more complicated. Now let us consider application of the DSN to UWB-based ranging. Referring to [21], we consider the following points as simplified problem settings.

- (P1) Longer period T can realize wider measurement range.
- (P2) Larger spike number Q is preferable because larger spike density Q/T can realize higher resistivity to noise.
- (P3) Lower 2nd peak R of auto-correlation can realize higher resistivity to noise.

In the case of Fig. 2, the spike-train $Y(t)$ is characterized by $(T,Q,R) = (21, 7, 5/7)$. In the case of Fig. 3, the spike-train $Y'(t)$ is characterized by $(T',Q',R') = (28,7,4/7)$. Recall that Eq. (8) says that the possible maximum value of the spike number Q is $M = 7$. Hence the spike-trains $Y(t)$ in Fig. 2(c) and $Y'(t)$ in Fig. 3(c) are optimal from the viewpoint of the point (P2). The change of $(T,R) = (21,5/7)$ to $(T',R') = (28,4/7)$ suggests that the re-wiring of $\boldsymbol{A}$ may improve characteristics of the spike-train $Y(t)$ from the viewpoints of the points (P1) and (P3). Based on this suggestion, we propose the following basic learning algorithm whose flow-chart is shown in Fig. 4.

Learning algorithm

Step 1: Initialization
Initialize the wiring matrix $\boldsymbol{A}$ as

$$a(j,i) := \begin{cases} 1 \text{ for } j = M-1, \\ 0 \text{ otherwise} \end{cases} \tag{13}$$
$$\text{for all } i = 0,\cdots, M-1 \text{ and } j = 0,\cdots, M-1$$

where ":=" denotes "*substitution of the right side into the left side*" hereafter. Initialize a counter k for iteration number as $k := 0$.

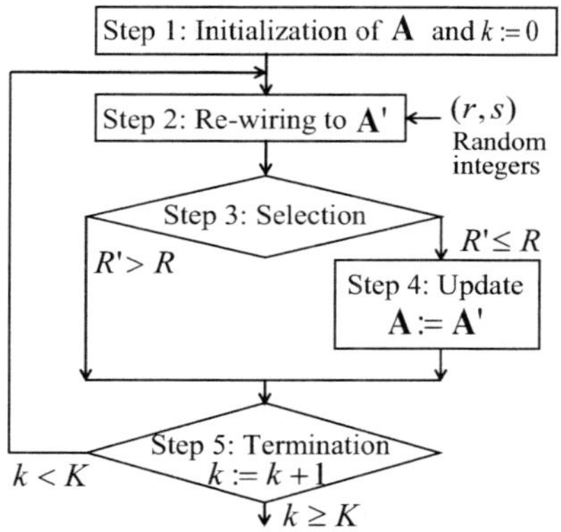

Fig. 4 Learning algorithm based on re-wiring

Step 2: Re-wiring
Generate two random integers (r,s) such that $0 \le r < s \le M-1$. Apply the re-wiring rule with re-wiring positions (r,s) to $\boldsymbol{A}$, and obtain a re-wired matrix $\boldsymbol{A}'$.

Step 3: Selection
Compare the 2nd peaks R and R' of auto-correlations that correspond to the wiring matrix $\boldsymbol{A}$ and the re-wired matrix $\boldsymbol{A}'$, respectively. If $R' \le R$ then go to Step 4. If $R' > R$ then go to Step 5.

Step 4: Update
Update the wiring matrix as $\boldsymbol{A} := \boldsymbol{A}'$. Go to Step 5.

Step 5. Termination
Let K be a given maximum iteration number. Increment the counter k by one. If $k < K$ then go to Step 2. If $k \ge K$ then terminate the algorithm.

Let us explain how the learning algorithm works based on an example of the learning in Fig. 5. In this figure the numbers of p-cells and x-cells are set to $M = 20$. Figure 5(a) shows the spike-train $Y(t)$ of the DSN just after the initialization ($k = 0$). The spike-train $Y(t)$ is characterized by period $T = 20$, spike number $Q = 20$ and 2nd peak $R = 1$ of auto-correlation. After $k = 2$ learning iterations, the DSN generates the spike-train $Y(t)$ in Fig. 5(b) that is characterized by $(T,Q,R) = (80,20,14/20)$. After $k = 10$ learning iterations, the DSN generates the spike-train $Y(t)$ in Fig. 5(c) that is characterized by $(T,Q,R) = (220,20,5/20)$. We can see that the spike number Q always takes the possible maximum value $M = 20$ which is optimal from the viewpoint of the point (P2). We can also see that, as the learning proceeds, longer period T and smaller 2nd peak R of auto-correlation can be obtained. This suggests that the learning algorithm can search better wiring matrixes $\boldsymbol{A}$ from the viewpoints of the points (P1) and (P3).

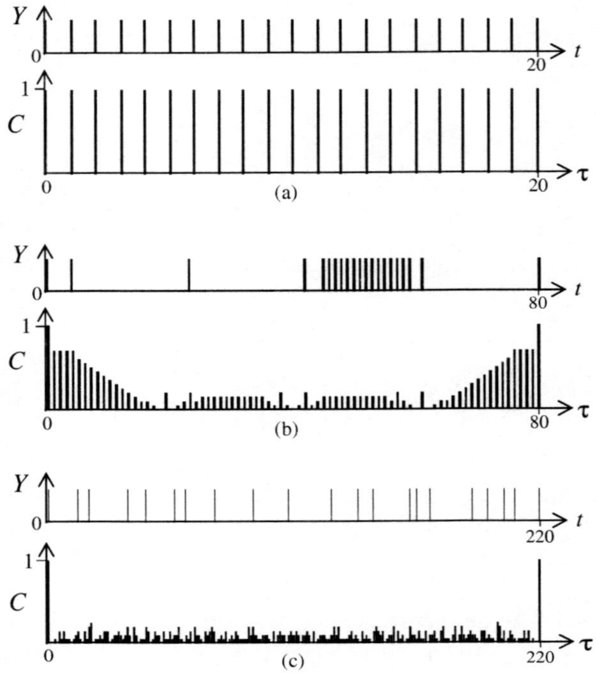

Fig. 5 Example of the learning $M = 20$. (**a**) After initialization. $(T,Q,R) = (20,20,20/20)$. (**b**) After $k = 2$ learning iterations. $(T,Q,R) = (80,20,14/20)$. (**c**) After $k = 10$ learning iterations. $(T,Q,R) = (220,20,5/20)$

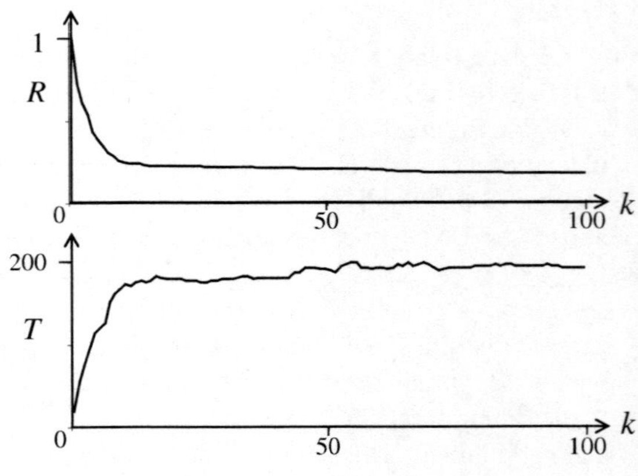

Fig. 6 Learning characteristics for iteration number k. $M = 20$. R is the 2nd peak of auto-correlation function of $Y(t)$. T is period of $Y(t)$. The graphs are averages for 40 learning trials

Figure 6 shows characteristics of the 2nd peak R of auto-correlation and the period T for the learning iteration k. Let us consider basic relations between such learning dynamics and the points (P1) and (P3). (The point (P2) is considered in a later discussion part.) The 2nd peak R decreases because the learning algorithm utilizes the point (P3) as a kind of cost function, i.e., smaller 2nd peak R is better. On the other hand, the period T increases although the learning algorithm does not utilize the point (P1) which says longer period T is better. Roughly speaking, the reasons for the increase of T are: (i) the period T and the 2nd peak R have approximately negative correlation because the 2nd peak R tends to be smaller as the period T becomes longer; and (ii) R is decreased by the learning algorithm.

As shown in Fig. 6, the learning seems to converge after about 1.0×10^2 iterations. On the other hand, the total number of the wiring matrixes $\boldsymbol{A}$ is $M^M = 20^{20} \simeq 1.0 \times 10^{26}$ (see Eq. (6)). That is, convergence of the learning is quite faster (10^2 iterations) compared to the whole search space (10^{26} candidates of the wiring matrixes $\boldsymbol{A}$). Let us consider basic relations between such learning dynamics and the re-wiring theorem as follows.

Discussion: Roles of re-wiring theorem in learning

For given M, let $\mathscr{Q}$ be a set of all the wiring matrixes $\boldsymbol{A}$, i.e., $\mathscr{Q}$ is the whole parameter space. Let $\mathscr{Q}_M \subset \mathscr{Q}$ be a set of all the wiring matrixes $\boldsymbol{A}$ that realize the possible maximum spike number $Q = M$. Note that $\mathscr{Q}_M$ is the set of optimal wiring matrixes $\boldsymbol{A}$ from the viewpoint of the point (P2). After the initialization in Eq. (13), the wiring matrix $\boldsymbol{A}$ belongs to the set $\mathscr{Q}_M$. The re-wiring theorem guarantees that, during the learning iterations, any re-wired matrix $\boldsymbol{A}'$ belongs to the set $\mathscr{Q}_M$. Then we can state the followings.

- The re-wiring theorem can guarantee the possible maximum spike number $Q = M$ that is optimal from the viewpoint of the point (P2). Hence the learning algorithm does not have to execute any calculation about the point (P2).
- The number of elements in the whole parameter space $\mathscr{Q}$ is M^M. The number of elements in the parameter set $\mathscr{Q}_M$ is $(M-1)!$ which is quite smaller than M^M for large M. By using the re-wiring theorem, we can focus on $\mathscr{Q}_M$ as a search space. For example, in the case of Figs. 5 and 6, we can focus on the search space $\mathscr{Q}_M$ with $M! = (20-1)! \simeq 1.2 \times 10^{17}$ elements although the whole parameter space $\mathscr{Q}$ has $M^M = 20^{20} \simeq 1.0 \times 10^{26}$ elements. This focusing of the search space on $\mathscr{Q}_M$ by the re-wiring theorem is one reason for the fast convergence of the learning algorithm.

We note that outside of the parameter set $\mathscr{Q}_M$, the DSN may exhibit complicated phenomena, e.g., transient to co-existing multiple periodic spike-trains (multiple periodic attractors). Detailed analysis for change of the wiring matrix $\boldsymbol{A}$ (e.g., change of $\boldsymbol{A} \in \mathscr{Q}_M$ into $\boldsymbol{A}' \notin \mathscr{Q}_M$ by not obeying the re-wiring rule) is a challenging future problem. We also note that this paper firstly considers learning of the DSN. In this early stage, more detailed analysis of the learning dynamics and detailed comparison of the ranging performances to conventional methods (e.g., [22]) are premature and should be included in future problems.

5 Conclusions

The DSN can realize various dynamics of the spike-train by adjusting the wiring matrix. We have considered the theorem that can clarify basic relations between change of the wiring matrix and change of characteristics of the spike-train. Also we have considered the learning algorithm and have shown that the algorithm can search wiring matrixes that can realize spike-trains whose characteristics are suited to the ranging application. Then we have shown that the presented theorem can realize efficiency of the learning algorithm. Future problems include the following points. (a) Detailed analysis of the presented learning dynamics, and development of a learning algorithm based on biological evidences such as [23]. (b) FPGA implementation of the learning algorithm. We note that some preliminary results on FPGA implementation of the DSN can be found in our other papers such as [15]. (c) Development of a PCNN of DSN and analysis of its performances such as power consumption and computational abilities .

References

1. H. Torikai and T. Saito, Synchronization phenomena in pulse-coupled networks driven by spike-train inputs, IEEE Trans. Neural Networks, vol. 15, no. 2, pp. 337–347, 2004.
2. G. Lee and N. H. Farhat, The bifurcating neuron network 2, Neural networks, vol.15, pp. 69–84, 2002.
3. R. Perez and L .Glass, Bistability, period doubling bifurcations and chaos in a periodically forced oscillator, Phys. Lett. vol.90A, no.9, pp. 441–443, 1982.
4. E. M. Izhikevich, Dynamical systems in neuroscience, MIT Press, 2006.
5. J. J. Hopfield and A. V. M. Herz, Rapid local synchronization of action potentials: Toward computation with coupled integrate-and-fire neurons, Proc. Natl. Acad. Sci. USA, vol. 92, pp. 6655–6662, 1995.
6. S. R. Campbell, D. Wang and C. Jayaprakash, Synchrony and desynchrony in integrate-and-fire oscillators, Neural computation, vol. 11, pp. 1595–1619, 1999.
7. R. Eckhorn, Neural mechanisms of scene segmentation: recordings from the visual cortex suggest basic circuits for linking field models, IEEE Trans. Neural Networks, vol.10, no.3, pp. 464–479, 1999.
8. S. Wolfram, Universality and complexity in cellular automata, Pysica D, vol. 10, pp. 1–35, 1984.
9. L. O. Chua, S. Yoon and R. Dogaru, A nonlinear dynamics perspective of Wolfram's new kind of science. Part I: threshold of complexity, Int. J. Bif. and Chaos, vol. 12, no.12, pp. 2655–2766, 2002.
10. T. Tamura, J. Kuroiwa and S. Nara, Errorless reproduction of given pattern dynamics by means of cellular automata, Physical Review E 68, 036707, 2003.
11. S. C. Kim and B. G Lee, A theory on sequence spaces and shift register generators, IEEE Trans. Comm. vol.44, no.5, pp. 609–618, 1996.
12. D. E. Knuth, The art of computer programming, vol. 2, 3rd edn., Addison Wesley, 1998.
13. H. Torikai, H. Hamanaka and T. Saito, Reconfigurable Digital Spiking Neuron and its Pulse-Coupled Network: Basic Characteristics and Potential Applications, IEEE Trans. CAS-II, vol. 53, no.8, pp. 734–738, 2006.
14. H. Torikai, Basic Characteristics and Learning Potential of a Digital Spiking Neuron, IEICE Trans. Fundamentals, 2007 (to appear).

15. H. Torikai, Y. Shimizu and T. Saito, Various spike-trains from a digital spiking neuron: analysis of inter-spike intervals and their modulation, Proc. of IJCNN, 2006, pp. 7591–7598.
16. H. Torikai, Y. Shimizu and T. Saito, A digital spiking neuron and its simple learning, Proc. of NOLTA pp. 263–266, 2006.
17. T. Kabe, H. Torikai and T. Saito, Synchronization via multiplex spike-trains in digital pulse-coupled networks, Lecture Note on Computational Science, 4234, Springer, Neural Information Processing, III, pp. 1141–1149 (2006). *The Best Student Paper Award of ICONIP2006.*
18. H. Torikai, A. Funew and T. Saito, Approximation of Spike-trains by Digital Spiking Neuron, Proc. of IJCNN, paper #1708, 2007.
19. G. M. Maggio, N. Rulkov and L. Reggiani, Pseudo-chaotic time hopping for UWB impulse radio, IEEE Trans. CAS-I, vol. 48, no. 12, pp. 1424–1435, 2001.
20. I. Oppermann, L. Stoica, A. Rabbachin, Z. Shelby and J. Haapola, UWB wireless sensor networks: UWEN - a practical example, IEEE Communications Magazine, vol. 42, no.12, pp. S27-S32, 2004.
21. M. G. Benedetto and G. Giancola, Understanding Ultra Wide Band Radio Fundamentals, Prentice Hall, 2004.
22. W. Chu and C. J. Colbourn, Sequence designs for ultra-wideband impulse radio with optimal correlation properties, IEEE Trans. Information Theory, vol. 50, no.10, pp. 2402–2407, 2004.
23. G. Bi and M. Poo, Synaptic modifications in cultured hippocampal neurons: Dependence on spike timing, synaptic strength, and postsynaptic cell type, J. Neurosci., vol. 18, pp. 10464–10472, 1998.

Dynamics in Manipulation and Actuation of Nano-Particles

Takashi Hikihara

Abstract In mico and nano technology, in situ actuation of small particles is an interesting problem. The micro-cantilever fabricated in mechanical vibrators or oscillators has a potential to effect particles through interaction force. The concept of manipulation is including the capture, release, and transport of particles under the practical circumstance: in air and liquid. These technologies are inevitable for the recent development of biotechnology. The systematic approach can reach to the achievement of actuation of particles. In this paper, the theoretical consideration for the manipulation of particles is discussed from the standpoint of nonlinear dynamics. In particular, the possibility of particle manipulation is formulated based on Hamiltonian in the classical mechanics. The control directly links to the treatment of the resonance and the dissipation of the system. Moreover, the particle actuation is also proposed with the use of intrinsic localized mode (ILM) in MEMS.

1 Introduction

Nonlinear dynamics is substantial in nano and micro technology and physics. However, the nonlinearity has not been positively accepted in the investigation of new science and technology. This paper discusses the nonlinear dynamics in manipulation of single molecules bonding to material surface by van der Waals force towards actuations.

Recently, the manipulation of single atoms and molecules has been experimentally achieved on surfaces with use of scanning tunneling microscope (STM) and atomic force microscopy (AFM) in lateral processes as well as vertical processes [1–5]. The lateral processes can carry atoms and molecules parallel to the surface

T. Hikihara (✉)
Department of Electrical Engineering, Kyoto University, Katsura, Nishikyo, Kyoto 615-8510, Japan; Photonics and Electronics Science and Engineering Center, Kyoto University, Katsura, Nishikyo, Kyoto 615-8510, JAPAN, e-mail: hikihara@kuee.kyoto-u.ac.jp

V. In et al. (eds.), *Applications of Nonlinear Dynamics*, Understanding Complex Systems,
DOI 10.1007/978-3-540-85632-0_23,

and the vertical process can capture an adatom by vibratory excitation of the target-substrate bond. These processes have been also studied theoretically [6–8] and numerically [9]. Quite recently, the dynamical approaches are discussed to control the behavior of molecules [10, 19]. The modeling of single atom manipulation was based on the van der Waals molecular vibrational predissociation in T-shaped interconnection with Morse potential [11–15]. The research let us confront to the probability of fragmentation of atoms in chaotic dynamics. Moreover, the theoretical consideration of dissipation is formulated by [16, 17].

In the following section, we introduce a periodically distributed potential structure. The dynamical behavior of nano-particle is discussed in the potential under external forcing. Then the possibility of particles actuation is possibly achieved by the design of vector flow in the spatially periodic potential. Moreover, it is shown that the nonlinear wave, named intrinsic localized mode (ILM), has a potential to excite particles by MEMS array for the transport of particles.

2 Manipulation of Particles: Capture and Release

In this section, the dynamics of nano-particle manipulation is discussed. We have already investigated the capture of nano-particle by AFM probe [9]. Here, the modeling and dynamics are reviewed.

The probe of atomic force microscopy can directly capture and release nano-particles in its potential well. Here the setup is considered as shown in Fig. 1. The interaction force between a probe and a particle depends on Lennard-Jones potential :

$$V_{\mathrm{LJ}}(x_{\mathrm{p}}) = 4\varepsilon \left[\left(\frac{\sigma}{x_{\mathrm{p}}} \right)^{12} - \left(\frac{\sigma}{x_{\mathrm{p}}} \right)^{6} \right], \tag{1}$$

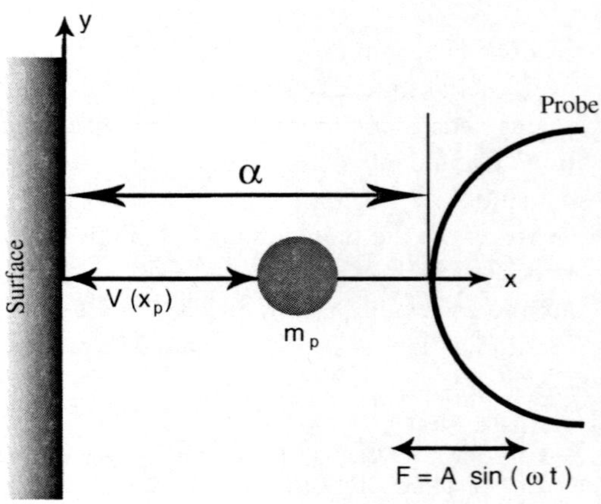

Fig. 1 Setup of particle manipulation between probes and material surface. Probe is the tip of STM or AFM

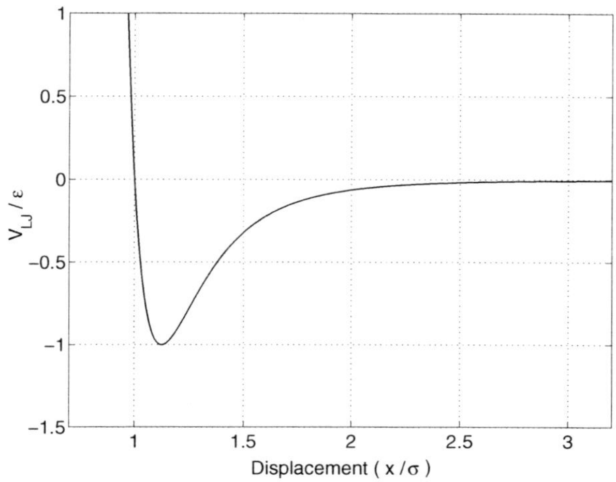

Fig. 2 Lennard-Jones potential between surface and particle

where ε denotes the depth of the potential well and σ the collision diameter. It is the distance at which the potential is zero in each potential field. The distance at which $V_{\mathrm{LJ}} = -\varepsilon$ and the interparticle force becomes zero is $2^{1/6}\sigma$. The shape of LJ potentials is shown in Fig. 2.

The potential governs the dynamics of particles over the material surface. The interaction force by AFM probe can be approximated by a sinusoidal force depending on the change of the distance between probe and particle. For the insight of the dynamics, Hamiltonian based approach seems substantial.

The Hamiltonian can be written in the form:

$$H = \frac{(m_{\mathrm{p}}\dot{x}_{\mathrm{p}})^2}{2m_{\mathrm{p}}} + V^{\mathrm{s}}(x_{\mathrm{a}}) + V^{\mathrm{a}}(\alpha - x_{\mathrm{p}}), \tag{2}$$

where the upper suffix s denotes the Lennard-Jones potential from the surface and the upper suffix a from the AFM probe. α depicts the equilibrium position of the probe. In this setup, α is sinusoidally excited. Between the probe and the surface, the potential shape becomes double well. Under the external excitation, particles behave depending on the change of potential shape between the surface and the probe. We have already showed the possibility of the suspension and capture of single particle by vibratory probe [9]. At the resonant frequency, the particle moves out of the well. If we need to control the behavior of particle under excitation, the dissipation must be introduced to stabilize it. However, the dissipation in nano-systems is hard to understand and formalize [21]. As we have the results by [16], the formalization and application in control owe to us.

3 Actuation of Particles: System Model and Generation of Flow

In this section, we discuss the possibility of particle transport by micro-mechanical system. The capture and release are the most fundamental effects on particles. The dynamic suspension is also possible in the potential well [9]. Therefore, the transport of single particle is necessary to buildup a nano-factory.

The transport of particle implies the continuous capture and release of particles in the spatially distributed structure. In this paper, we discuss the structure as shown in Fig. 3. In analyzing area (gray zone), there are two fixed masses and one vibrating mass. They have the same radius R and mass M. The vibration of one mass can induces the flow of particles in the channel. We assume the spatial periodicity on the system to obtain the continuous flow.

The Hamiltonian of the system without excitation given by

$$H_{\rm p} = \frac{(m_{\rm p}\dot{x}_{\rm p})^2}{2m_{\rm p}}\frac{(m_{\rm p}\dot{y}_{\rm p})^2}{2m_{\rm p}} + V(x_{\rm p}, y_{\rm p}). \tag{3}$$

Where the potential becomes the sum of the neighboring mass,

$$V(x_{\rm p}, y_{\rm p}) = \sum_i \{V(r_{2{\rm i}-1} - R) + V(r_{2{\rm i}} - R)\}. \tag{4}$$

Here we treat the neighboring mass in Velet neighbors. The potential has spatial periodicity on x at $2L$:

$$V(x_{\rm p} + 2L, y_{\rm p}) = V(x_{\rm p}, y_{\rm p}).$$

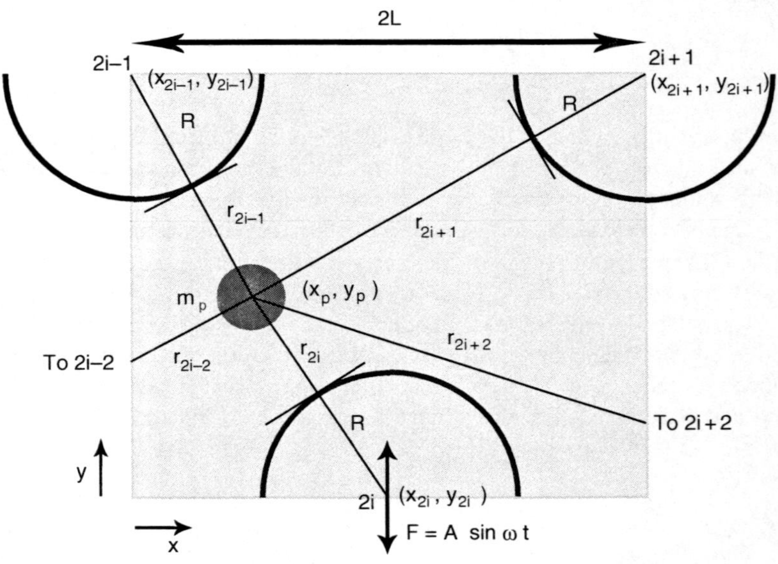

Fig. 3 Schematic transport band along separatricies in 2D

The distance between a particle and a mass is given by

$$r_j = \sqrt{(x_p - x_j)^2 + (y_p - y_j)^2}. \tag{5}$$

In the setup, the mass # 2i is excited in y-direction sinusoidally. That is,

$$y_{2i} = y_{2i}^0 + A \sin \omega t,$$

where y_{2i}^0 denotes the initial position of mass # 2i without excitation.

When the normalized coordinate system :

$$\begin{aligned} q_1 &= \frac{x_p}{\sigma}, \; q_2 = \frac{y_p}{\sigma}, \\ p_1 &= \frac{\dot{x}_p}{\sigma}, \; p_2 = \frac{\dot{y}_p}{\sigma}, \end{aligned} \tag{6}$$

is introduced, the model of particle dynamics is given by

$$\begin{aligned} \dot{q}_1 &= p_1, \qquad \dot{q}_2 = p_2, \\ \dot{p}_1 &= -\frac{\partial H}{\partial q_1}, \; \dot{p}_2 = -\frac{\partial H}{\partial q_2}. \end{aligned} \tag{7}$$

In this setup as shown in Fig. 3, the differential equation must have a forcing input caused by the oscillating mass at # 2i. When every mass at # 2i is excited, there is a simplicity depending on the periodical structure to analyze the dynamics in $R \times S$.

Here we discuss the dynamical vector flow of Eq. (7) under the external forcing. The vector flow in a segment is schematically drawn in Fig. 4. Depending on the

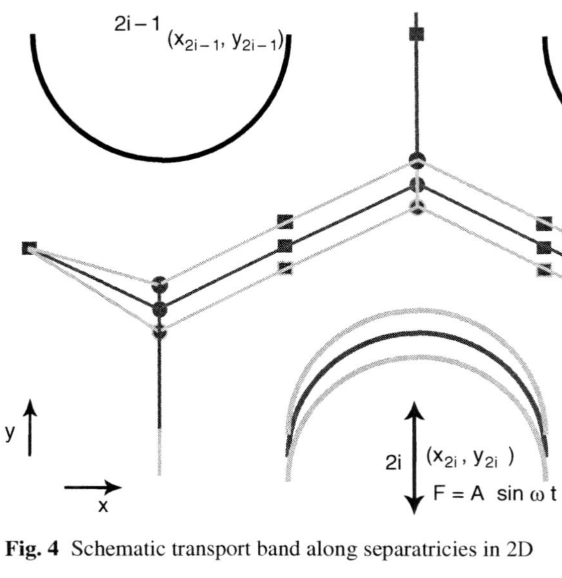

Fig. 4 Schematic transport band along separatricies in 2D

external periodic forcing, the separatrix roughly moves up and down. If we can capture particles during the upper and lower limits of separatrix, there appears a vector flow between masses. The lower than the separatrix, the force directs to the mass # 2i and the upper the mass # (2i−1) or (2i+1). Therefore, the periodical y-directional forcing on # 2i can change the direction of force acting on the particle. The mechanism is similar to the principle of a linear ion trap. If the particles is kept in the channel, the flow of particles appears. The parametric excitation can keep the particle flow without feedback loop along the separatrices.

4 Application of ILM in Coupled Cantilever Array

In order to transport particles, we consider the application of cantilever array as an actuator device. The MEMS technology makes it possible to fabricate the cantilever array in micro scale. The array is the nonlinear oscillators coupled nonlinearly. These types of MEMS devices are strongly interested.

In the nonlinear coupled oscillators, it is weel-known that there is a temporally periodic and spatially localized wave mode. Sievers and Takeno showed the existence theoretically [22]. Recently, Sato et al. [23] have shown the traveling phenomena of ILM in MEMS array experimentally. We proposed the possible application of ILM into a nano-actuator [24]. The proposed cantilever array is schematically shown in Fig. 5.

In the array, the torsional vibration appears as in Fig. 6. The dynamics is described by coupled Duffing equations with fixed boundary condition:

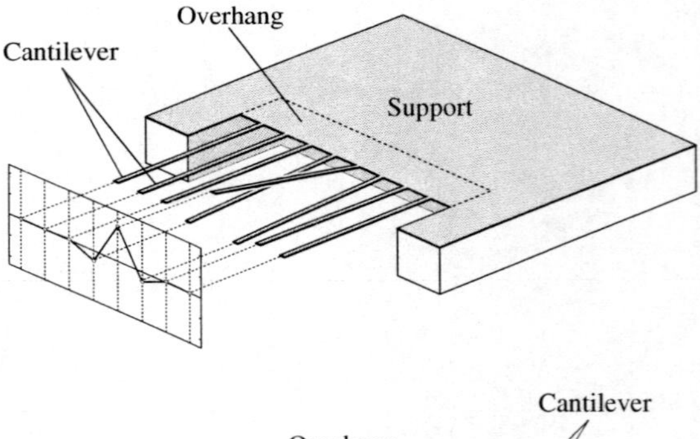

Fig. 5 Cantilever array. Left figure shows ILM, which has the localized amplitude distribution of periodic vibration

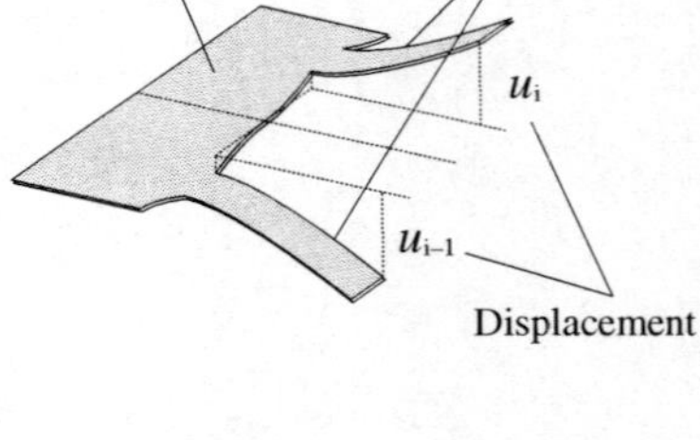

Fig. 6 Cantilever model. Coupling by overhang has nonlinear spring characteristics

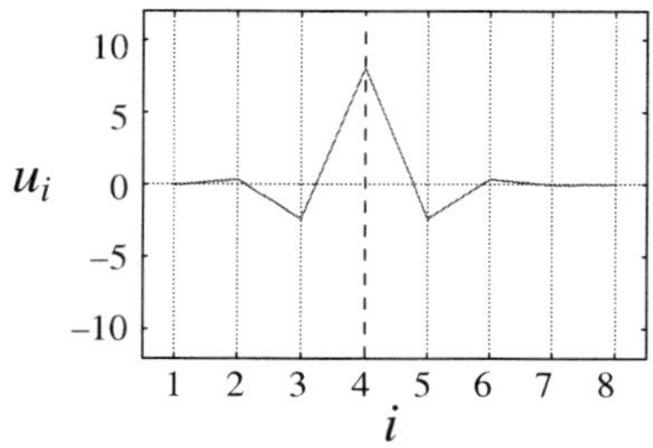

(a) Sievers-Takeno mode

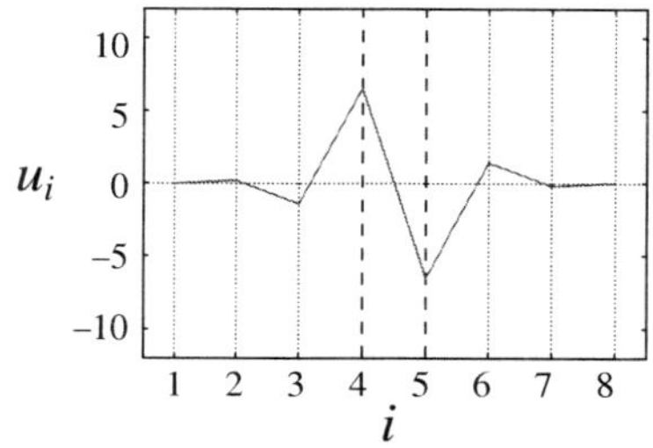

(b) Page mode

Fig. 7 Mode of intrinsic localized mode. Sievers-Takeno mode is symmetric and Page mode is asymmetric. Both have stable and unstable mode depending on coupling parameters

$$\begin{aligned} \frac{\mathrm{d}u_i}{\mathrm{d}t} &= v_i, \\ \frac{\mathrm{d}v_i}{\mathrm{d}t} &= -\alpha_1 u_1 - \beta_1 u_i^3 \\ &\quad -\alpha_2(2u_i - u_{i-1} - u_{i+1}) \\ &\quad -\beta_2(u_i - u_{i-1})^3 - \beta_2(u_i - u_{i+1})^3. \end{aligned} \tag{8}$$

Where u_i denotes the displacement of cantilever #i and v_i the velocity of the displacement. The coefficients α_1 and α_2 denote the on-site and coupling linear spring constant. β_1 and β_2 also depict the on-site and coupling nonlinear spring constant. The interaction force dues to the coupling of overhang (see Fig. 6). Numerical simulation shows the existence of the intrinsic localized mode as in Fig. 7. Depending on the nonlinearity, the several stable and unstable modes coexists at the same parameter [24].

The spatial transition of ILM is achieved by controlling a coupling parameter [24]. The appropriate shift of vibrating cantilever has a potential to carry particles along the separatrix of Lennard-Jones potential force field as discussed above. The MEMS array with nonlinear localized wave is expected to be a key device of the transport of nano-particles.

5 Summary

In this paper, we discussed the nonlinear dynamics for the manipulation of nano-particles. At first, the possibility of particle manipulation was theoretically considered based on Hamiltonian. Secondly, the actuation concept of particles was proposed. Third, it is explained that the cantilever array can be an actuator devices by using an intrinsic localized mode (ILM) and its moving control.

The control directly links to the treatment of resonant excitation and the dissipation of the system. Then our future investigation will be focused on the dissipation control in the nano-dynaqmical systems.

Acknowledgments The author would like to show his sincere thanks to Mr. Byungsoo Kim, the graduate student of the Colorado State University, for the wonderful discussion during his research visit to Kyoto University in 2007. He also acknowledge Mr. Masayuki Kimura, the graduate student of Kyoto University, for his collaboraton on the application of ILM in nano-actuators. Finally, he shows his cordial appreciation to Prof. M. Sato, Kanazawa University, for his continuous support to our research. This research work is partially supported by Global Center of Excellent Program by JSPS.

References

1. D. M. Eigler and E. K. Schweizer, "Positioning single atoms with a scanning tunnelling microscope," *Nature*, vol. 344, pp.524–526, 1990.
2. L. Bartels, G. Meyer, and K.-H. Rieder, "Basic Steps of Lateral Manipulation of Single Atoms and Diatomic Clusters with a Scanning Tunneling Microscope Tip," *Phy. Rev. Lett.*, 79, pp.697–700, 1997.
3. J. A. Stroscio and R. J. Celotta, "Controlling the Dynamics of a Single Atom in Lateral Atom Manipulation," *Science*, 306, pp.242–247, 2004.
4. N. Oyabu et al., "Single Atomic Contact Adhesion and Dissipation in Dynamic Force Microscopy," *Phy. Rev. Lett.*, 96, 106101, 2006.
5. Y. Sugimoto et al., "Mechanism for Room-Temperature Single-Aton Latteral Manipulations on Semiconductors using Dynamic Force Microscopy," *Phy. Rev. Lett.*, 98, 106104, 2007.
6. X. Bouju, C. Joachim, and C. Girard, "Single-atom motion during a lateral STM manipulation," *Phy. Rev. B*, R7845, 1999.
7. N. Lorente and M. Persson, "Theoretical aspects of tunneling-current-induced bond excitation and breaking at surfaces," *The Royal Soc. Chem.*, Faraday Disscuss., 117, pp.277–290, 2001.
8. A. Kuhnle, et al., "Understanding atom movement during lateral manipulation with the STM tip using a simple simulation method," *Surf. Sci*, 499, pp.15–23, 2002.
9. T. Hikihara and K. Yamasue, "A Numerical Study on Suspension of Molecules by Microcantilever Probe", The Fifth International Symposium on Linear Drives for Industry Applications (LDIA2005), Awaji, Hyogo, September 25–28 (2005).
10. T. Hikihara, Dynamics of Single Atoms and Molecules Manipulated at Material Surface, International Symposium on Nonlinear Theory and its Application (NOLTA) 2007, Vancouver, Canada, Sept. 16–19, 2007.
11. J. A. Beswick, G. Delgado-Barrio, and J. Jortner, "Vibrational predissociation lifetimes of the van der Waals molecule HeI_2," *J. Chem. Phys.*, 70(8), pp.3895–3901, 1979.
12. R. J. Wolf and W. L. Hase, "Quasiperiodic trajectories for a multidimensional anharmonic classical Hamiltonian excited above the unimolecular threshold," *J. Chem. Phys.*, 73(8), pp.3779–3789, 2006.
13. R. M. Hedges and W. P. Reinhardt, "Classical and quantum dynamics of long lived doubly excited vibrational states of triatomic molecules," *J. Chem. Phys.*, 78(6), pp.3964–3975, 1983.
14. S. K. Gray, S. A. Rice, and D. W. Noid, "The classical mechanics of vibrational predissociation: A model based study of phase space structure and its influence on fragmentation rates," *J. Chem. Phys.*, 84(7), pp.3745–3752, 1986.
15. M. Toda, "Crisis in Chaotic Scattering of a Highly Excited van der Waals Complex," *Phy. Rev. Lett.*, 74(14), pp.2670–2673, 1995.
16. V. Putkaradze, Dissipative kinetic equations and their applications to manipulations of particles using AFM tips, International Symposium on Nonlinear Theory and its Application (NOLTA) 2007, Vancouver, Canada, Sept. 16–19, 2007.
17. B. Kim, T. Hikihara and V. Putkaradze, Numerical Study of Atom Interchange on Material Surface under Periodic Force, International Symposium on Nonlinear Theory and its Application (NOLTA) 2007, Vancouver, Canada, Sept. 16–19, 2007.

18. M. O. Allen, "Introduction to Molecular Dynamics Simulation," in *Computational Soft Matter: From Synthetic Polymers to Proteins, Lecture Notes, NIC Series*, 23, ISBN 3-00-012641-4, pp. 1–28, 2004.
19. D. Vainchtein and I. Mexiċ, "Capture into Resonance: A Method for Efficient Control," *Phy. Rev. Lett.*, 93, 084301, 2004.
20. B. V. Chirikov, "A Universal Instability of Many-Dimensional Oscillator Systems," *Phy. Rep.*, 52, pp. 263–379, 1979.
21. V. Caciuc, H. Hölscher, adn S. Blügel, "*Ab initio* investigation of noncontact atomic force microscopy tip-surface instability in InAs(110) surface," *Phy. Rev. B*, 72, 035423, 2005.
22. A. J. Sievers and S. Takeno, Intrinsic Localized Modes in Anharmonic Crystals," *Phys. Rev. Lett.* 61, 970, 1988.
23. M. Sato, B.E. Hubbard, A.J. Sievers, et al., Observation of Locked Intrinsic Localized Vibrational Modes in a Micromechanical Oscillator Array, *Phys. Rev. Lett.* 90, 044102, 2003.
24. M. Kimura and T. Hikihara, A Numerical Study on Invariant Manifold Related to Transition of Intrinsic Locarized Mode in Coupled Cantilever Array, International Symposium on Nonlinear Theory and its Application (NOLTA) 2007, Vancouver, Canada, Sept. 16–19, 2007.

Nonlinear Buckling Instabilities of Free-Standing Mesoscopic Beams

S.M. Carr, W.E. Lawrence and M.N. Wybourne

Abstract We review our observations of a rather striking change in the static equilibrium shape of free-standing mesoscopic buckled beams as a function of the length-to-width aspect ratio. For low values of the aspect ratio the beams exhibit the classical mechanical bifurcation known as Euler buckling, while at higher values the shapes become surprisingly complex. The data are consistent with a nonlinear Ginsburg-Landau elastic field theory that incorporates asymmetry in higher-order buckling modes. Results of mechanical and electromechanical actuation of the beams are discussed with respect to the theory. The observed effects may have an impact on the emerging technology associated with bistable nanomechanical systems.

1 Introduction

The classical and quantum aspects of mesoscopic and nanoscale mechanical structures are currently of interest, due to the potential unexplored physics and the possible applications in nanotechnology. The nonlinearity and bistability exhibited by both static and dynamic mesomechanical systems may be utilized in various contexts, including noise-enabled precision measurements [1], stochastic resonance [2, 3], non-equilibrium dynamical transitions [4–6], quantum information [7], and mechanical memory [8, 9].

A particularly intriguing aspect of mesoscopic mechanical systems is that when cooled such that the thermal energy is less than the quantum of energy associated with one or more collective vibrational modes, the systems may operate in a regime at the interface between quantum and classical physics [10]. Realization of this

S.M. Carr (✉)
Quantum Processes and Metrology Group, Physics Laboratory, National Institute of Standards and Technology (NIST), Gaithersburg, MD 20899 USA

V. In et al. (eds.), *Applications of Nonlinear Dynamics*, Understanding Complex Systems, DOI 10.1007/978-3-540-85632-0_24,

regime is an outstanding experimental challenge, a challenge which provided part of the motivation for the work described in this paper.

Here we review our work on the buckling of free-standing mesoscopic beams. We begin in Sect. 2 by describing the utilization of residual strain in the top-down fabrication process to enable mesoscopic buckling. In Sect. 3 we present our observations of the observed shapes of the buckled beams. The static equilibrium buckled shape of the beams unexpectedly evolved from the expected fundamental mode to a superposition of buckling modes as the length-to-width aspect ratio was increased. Section 4 contains our results of mechanical actuation of the beams. There we also outline a nonlinear elastic field theory that we have employed in order to provide insight into the nature of the observed shapes. We compare the theory with the experimental results of mechanical actuation of a beam buckled predominantly in the fundamental mode. We discuss in Sect. 5 our experimental findings on electromechanical actuation of several beams, where we show that it was possible to induce electromechanical transitions between buckling modes. Finally, in Sect. 6 we present our conclusions.

2 Residual Strain and Mesoscopic Buckling

Mesoscopic doubly-clamped beams were fabricated using e-beam lithography and plasma etching. The starting material was a silicon wafer coated with a 500 nm thermal oxide. We show the steps of the fabrication process schematically in Fig. 1(a): (i) electron-beam lithography is used to expose a pattern in a resist which is then developed, (ii) metal is thermally evaporated over the entire surface, (iii) the undeveloped resist and covering metal layer are removed in a lift-off step, (iv) reactive ion etching (RIE) was used to transfer the pattern to the SiO_2 in one step, leaving free-standing SiO_2 beams with supports fixed to the silicon substrate. The undercut depth measured using atomic force microscopy (AFM) is consistent with the RIE etching parameters. For some of the beams in the array shown in Fig. 1(b) the length-to-width aspect ratio, (L/d), is large enough for the fundamental static buckling displacement to be visible in the image. Micrographs of an individual beam that has buckled after release from the silicon substrate are shown in Fig. 1(c, d).

Euler buckling is normally described in terms of applied compression; one end of a beam is held fixed and the other end is moved towards the fixed end, eventually leading to a buckled state. For the mesoscopic beams we study, the residual compressive strain develops during the growth and room-temperature relaxation of the thermal oxide, and is presumably fixed for beams fabricated from the same wafer. After release from the Si substrate, the beams can buckle due to this residual compressive strain at the Si/SiO_2 interface. Therefore, the static buckling displacement will depend on the value of the residual strain, and on the length L and width d of the beams through the aspect ratio (L/d).

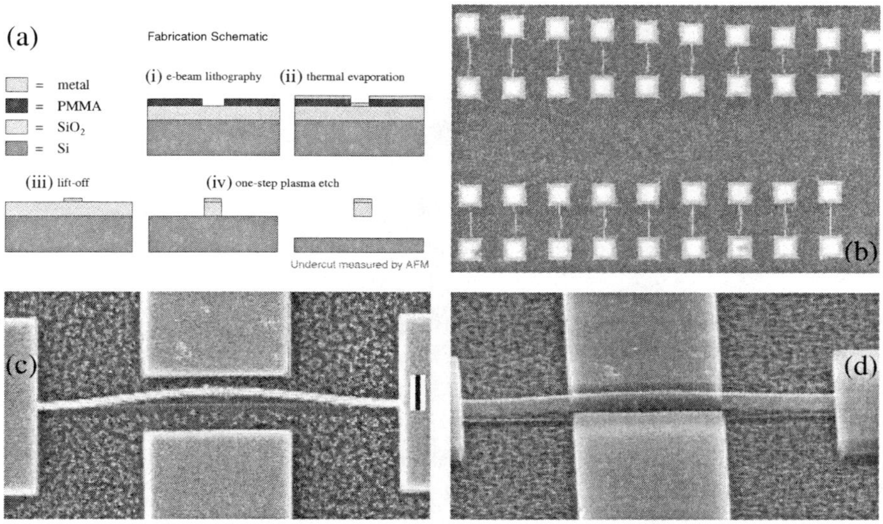

Fig. 1 (**a**) A schematic of the top-down process used to fabricate the free-standing doubly-clamped beams. The steps (i)–(iv) are discussed in the text. (**b**) A top-view scanning electron microscope (SEM) micrograph showing an array of free-standing SiO_2 beams. (**c**) A top-view SEM micrograph of a single beam buckled in the fundamental mode. The rectangular structures near the center of the beam can be used as gates for electromechanical actuation. The scale bar represents 1 μm. (**d**) A tilt-view SEM micrograph of the beam in (**c**), showing the undercut

3 Observed Shapes

We fabricated and analyzed more than 100 free-standing SiO_2 beams. Beam dimensions were thickness w of 500 nm, widths of 235 nm $\lesssim d \lesssim$ 320 nm, and lengths of 2 μm $< L <$ 60 μm, thus the aspect ratios of the beams studied covered the range 5 $\lesssim (L/d) \lesssim 255$. For all the beams studied $(w/d) \approx 2$ so that buckling occurred in the d direction. Previous micromechanical studies have found the resulting buckled shape to take the form expected from the lowest-order solution to the Euler equation for clamped-clamped boundary conditions [11], similar to the shape shown in Fig. 2(a).

For fixed strain it is expected that as the aspect ratio is increased the midpoint displacement will increase, but the functional form of the buckled shape will remain essentially unchanged. Contrary to this expectation, we found [12] a dramatic change in the static equilibrium shape as the aspect ratio was increased, as shown in Fig. 2.

To explore the nature of the buckling spectrum, each beam was imaged and digitally recorded using a scanning electron microscope (SEM). Image processing of the SEM micrographs was used to extract the shape $y(x)$ of each beam. Several examples of the shapes extracted using this method are shown in Fig. 2.

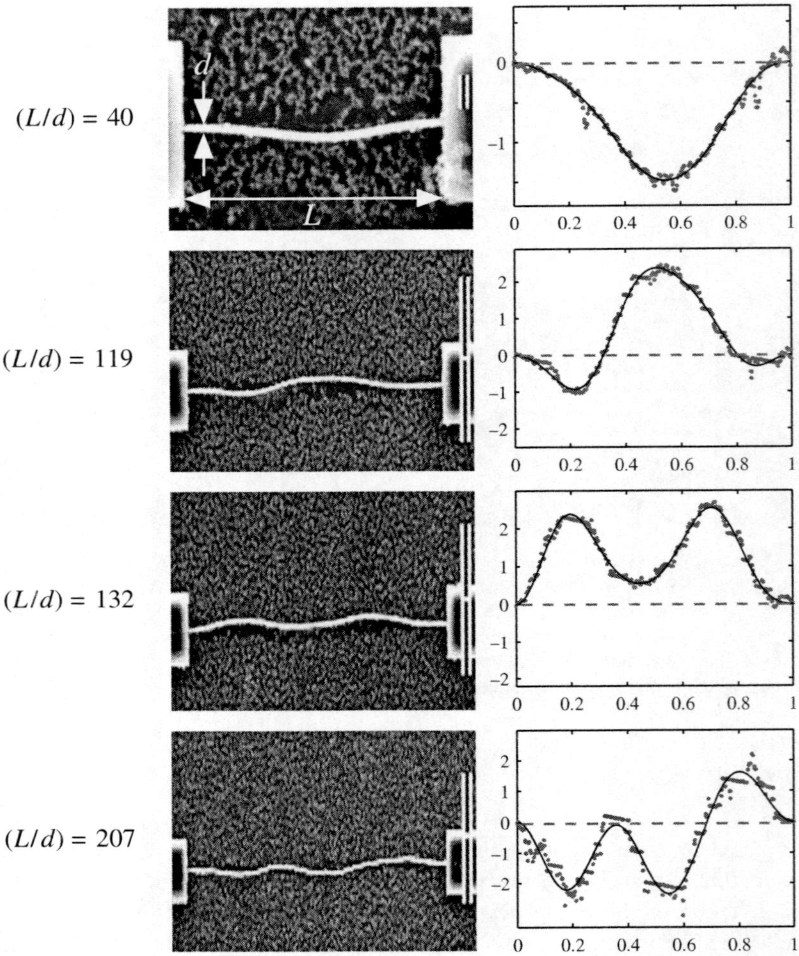

Fig. 2 Observed static equilibrium buckling shapes of several free-standing SiO_2 beams [12]. Each row represents a beam with the aspect ratio (L/d) indicated in the *left column*. The *middle column* contains SEM micrographs of the beams. The length L and width d are defined in the top micrograph. The *scale bar* represents 1 μm in the top micrograph and 10 μm in the other three micrographs. The *right column* contains the shapes of the beams extracted from the micrographs through an image processing algorithm. The *dot markers* are the extracted data and the *black curve* is a fit to the points using Eq. (2). For the plots in the *right column* the x-axis is (x/L) and the y-axis is $(y(x)/d)$. The *dashed grey line* represents the centerline $(y(x)) = 0$

4 Mechanical Actuation and Bistability

We investigated the elasticity and bistability of the mesoscopic buckled beams by using a mechanical probe tip to apply a local transverse force. A mechanical probe attached to a linear micropositioner was brought into contact with the beam at the position indicated with an arrow in Fig. 3(a). The beam was first pushed in the direction of the arrow towards the centerline (dashed black line), but not beyond the centerline, then the probe was retracted and brought out of contact with the beam. Repeating these steps many times, it was found that the beam always returned to the initial equilibrium profile (dashed white line), from which we conclude that the beam responds elastically about the initial equilibrium profile. Next, the beam was pushed just beyond the centerline, which resulted in the beam assuming the new local equilibrium profile shown in Fig. 3(a). When the beam was pushed back toward the centerline, it spontaneously snapped back into the initial profile. This mechanical switching between the two states was repeated many times, with the beam always settling into either the initial equilibrium profile or the local metastable equilibrium profile, depending on the direction and amount of pushing.

Similar mechanical actuation was performed on a beam buckled into the shape shown in Fig. 3(b). For this beam, the initial equilibrium profile (dashed white line) is a shape with two antinodes on the same side of the centerline (dashed black line). A mechanical probe was brought into contact with the beam at the position indicated with an arrow. Similar procedures were followed as in the previous case and repeated many times with similar results. We conclude that the dashed white line indicates stable equilibrium and the beams as imaged (solid white line)

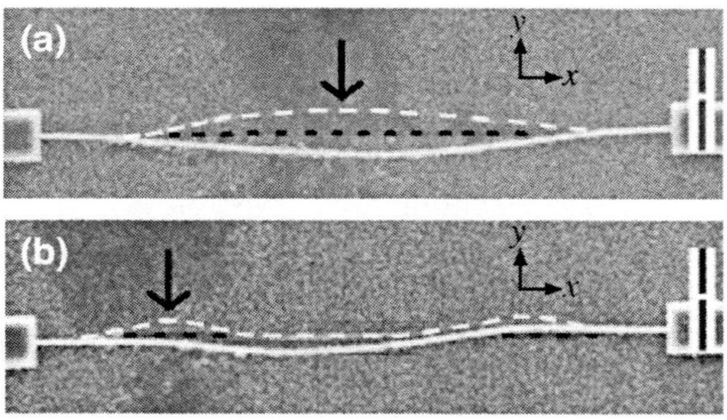

Fig. 3 SEM micrographs indicating the in-plane mechanical bistability of the buckled beams [13]. The *dashed black lines* are guides to the eye to indicate the centerline, which is the (*straight*) shape of the beams before the plasma etching. After the undercut was completed during the plasma etching, the beams buckled into the shapes indicated by the *dashed white lines*. A mechanical probe was brought into contact with the antinodes indicated with *arrows* in (**a**) and (**b**) and the antinode was slowly pushed until a state was found on the opposite side of the *centerline*. The beams were then imaged in this metastable state. (**a**) A beam buckled predominantly in the fundamental mode. (**b**) A beam buckled into a shape with two antinodes on the same side of the *centerline*. In (**a**) and (**b**) the scale bar represents 10 μm

indicate metastable equilibrium. Thus, even for beams with complex shapes, we find an elastic response about the equilibrium profile and mechanical bistability.

To provide insight into the nature of the observed shapes, consistent with the results of mechanical actuation discussed above, we have explored a phenomenological description based on a nonlinear elastic field theory [14–17]. In this description the field is the shape of the beam, $y(x)$, that minimizes the potential energy functional

$$V[y(x)] = \frac{1}{2}Fb^2 \int (y'' - \tilde{y}'')^2 \, dx + \frac{1}{2}F\varepsilon \int (y')^2 \, dx + \frac{F}{8L}\left(\int (y')^2 \, dx \right)^2, \quad (1)$$

where F is the linear elastic modulus and $b = d/\sqrt{12}$ is the bending moment. The first term accounts for the energy of bending, while the second and third terms account for stretching (compression), where $\varepsilon = (L - L_o)/L_o$ is the longitudinal strain with L_o the equilibrium length of the beam. The nonlinear third term accounts for the additional strain due to the actual shape of the beam [11].

The bending potential energy in Eq. (1) is a quadratic functional of the departure of the exact shape $y(x)$ from the unstrained equilibrium shape $\tilde{y}(x)$. The unstrained equilibrium shape $\tilde{y}(x)$ is the shape that a beam would take if released at its endpoints. This function, which is unknown *a priori*, characterizes the asymmetry present in the fabrication process.

It may be instructive to consider Eq. (1) in the limit $\tilde{y}(x) = 0$ and without nonlinearity. Then the associated Euler-Lagrange equation of static equilibrium is $-Fb^2 y'''' + F\varepsilon y'' = 0$. This expression may be familiar from the theory of elastic stability [11], first described by Euler, and consequently often referred to as Euler buckling. In this limit it is possible to describe the static eigenfuctions (shapes) and eigenvalues (critical strains) at the onset of elastic instability. Some form of nonlinearity is essential in order for the system to be stable beyond the Euler buckling instability. The particular form of nonlinearity we employ in Eq. (1) accounts for the stretching of the beam due to the finite transverse static displacement, under the condition of fixed endpoint separation.

We express the shape as a superposition of eigenfunctions,

$$y(x) = \sqrt{L/8} \sum_{n}^{N} Y_n \, g_n(x), \quad (2)$$

where the Y_n are the physical peak-to-peak static displacements in mode n. A basis set appropriate for clamped boundary conditions is

$$g_n(x) = \sqrt{2/L} \left(1 + (-1)^{(n-1)/2} \cos k_n x \right), \qquad (\text{odd } n), \quad (3)$$

$$g_n(x) = \sqrt{2/L} \, C_n \left(\gamma_n k_n x + (-1)^{(n-2)/2} \sin k_n x \right), \qquad (\text{even } n), \quad (4)$$

with the wavenumbers k_n and constants C_n and γ_n determined from linear Euler buckling theory [14], as noted above. We have used the decomposition Eq. (2) with an upper limit $N = 6$ to fit the observed shapes extracted from the image processing algorithm. Several examples of such fits are shown as the black curves in Fig. 2. The fits allowed us to extract the mode coefficients Y_n, obtain the mechanical buckling spectrum Y_n vs. (L/d), and compare the experimental data with theory [12–14].

Substituting the superposition Eq. (2) into Eq. (1), we find that the resulting potential energy functional has the form of a general multi-mode Ginzburg-Landau potential,

$$V\{Y_n\} = \sum_n \left(\frac{1}{2} \alpha_n Y_n^2 - \tilde{\alpha}_n \tilde{Y}_n Y_n \right) + \frac{1}{4} \sum_{n,m} \beta_{nm} Y_n^2 Y_m^2, \tag{5}$$

where the decomposition Eq. (2) has also been used for the unstrained equilibrium shape $\tilde{y}(x)$ with eigenfunctions $g_n(x)$ and mode coefficients $\tilde{Y}_n$. The parameters appearing in Eq. (5) are

$$\tilde{\alpha}_n = FLb^2 k_n^4/8, \qquad \alpha_n = \tilde{\alpha}_n (1 - \varepsilon/\varepsilon_n), \qquad \beta_{nm} = \sqrt{\tilde{\alpha}_n \tilde{\alpha}_m}/8b^2, \tag{6}$$

with the critical strains $\varepsilon_n = -b^2 k_n^2$. We have used the multi-mode Ginzburg-Landau potential Eq. (5) to explore the static equilibrium ground states of the system [12, 14], defined as the stable minimum of $V\{Y_n\}$ in the presence of nonlinearity and asymmetry.

For a beam buckled predominantly in the fundamental ($n = 1$) mode, Eq. (5) reduces to

$$V\{Y_1\} = \frac{1}{2} \alpha_1 Y_1^2 - \tilde{\alpha}_1 \tilde{Y}_1 Y_1 + \frac{1}{4} \beta_{11} Y_1^4. \tag{7}$$

First consider the case $\tilde{Y}_1 = 0$. Then the effective potential $V\{Y_1\}$ is a symmetric double-well with extrema defined by $(dV\{Y_1\}/dY_1)|_{Y_1 = Y_{eq}} = Y_{eq}(\beta_{11} Y_{eq}^2 - |\alpha_1|) = 0$. Thus the minima of this potential are given by $Y_{eq}^2 = |\alpha_1|/\beta_{11}$ or $Y_{eq} = \pm\sqrt{|\alpha_1|/\beta_{11}}$. In Fig. 4 we plot Eq. (7) with $\tilde{Y}_1 = 0$, the symmetric double-well, as a dashed grey line.

We can use the theoretical framework described above to investigate the effective potential(s) that are consistent with the results of the mechanical actuation shown in Fig. 3(a). Recall that in Fig. 3(a) the dashed white line is the ground-state equilibrium shape, the beam as imaged is the metastable shape, and the dashed black line is the centerline. Since the beam in Fig. 3(a) is buckled predominantly in the fundamental ($n = 1$) mode, we can estimate the difference between the ground-state and metastable state static equilibrium displacements (Y_{eq}) directly from the micrograph. We found this difference to be $\lesssim 5\%$. In Fig. 4 we show an asymmetric double-well potential that is consistent with the observed $\lesssim 5\%$ difference. In Fig. 4 the ground-state (left well) corresponds with the dashed white line in

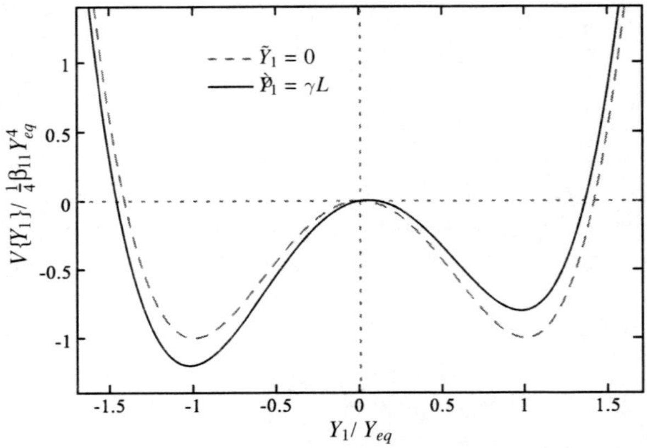

Fig. 4 Effective potential energy curve for the fundamental ($n = 1$) buckling mode. The *solid-line curve* is an effective potential for the fundamental mode that is consistent with the observed asymmetric bistability of the sample shown in Fig. 3(a). For comparison, the symmetric double-well potential ($\tilde{Y}_1 = 0$) is shown as a *dashed grey line*

Fig. 3(a) and the metastable state (right well) corresponds with the beam as imaged in Fig. 3(a).

We have previously considered certain models for the asymmetry. Perhaps the simplest model is $\tilde{Y}_1$ = constant (independent of Y_1), which corresponds to a continuous buckling transition rather than the critical buckling transition from Euler buckling theory [18]. One such model is $\tilde{Y}_1 = \gamma L$, where γ is the static displacement per unit length. This is shown in Fig. 4 as the solid blue line where $\gamma = 0.012$. We have also considered displacement-dependent asymmetry, found to describe beams fabricated with longer etching times [13, 14], with the form $\tilde{Y}_1 = \xi_1 Y_1$ where ξ_1 is a dimensionless constant. For the constant asymmetry model described by $\tilde{Y}_1 = \gamma L$ (Fig. 4), the resulting asymmetric bistability is consistent with observations for beams buckled predominantly in the fundamental mode.

5 Electromechanical Transitions

As shown in Fig. 1(c) and (d), it is relatively straightforward to incorporate gates in the fabrication process that can be used to explore electromechanical effects. These gates, like the beam supports, consist of SiO_2 in a rectangular structure that is fixed to the silicon substrate and covered with a thin metal layer. The conducting metal layer allows charge and electric potential to be transmitted to the vicinity of the beam. The resulting local electrical fields can generate electromechanical forces. The relative positioning accuracy afforded by e-beam lithography allows

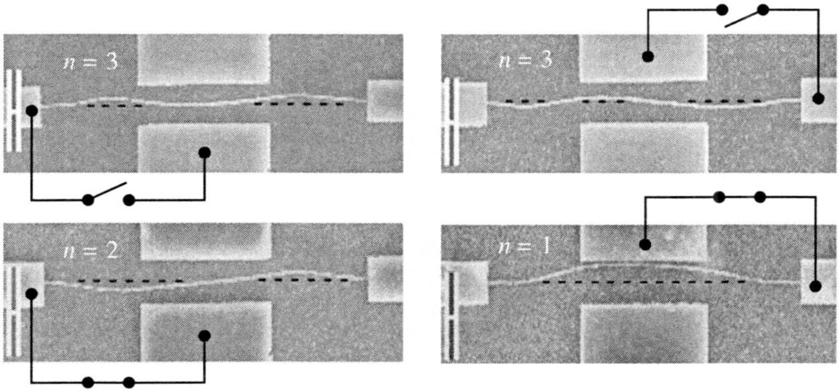

Fig. 5 Electromechanical transitions between buckling modes induced by utilizing the rectangular gates [13]. The predominant mode for each shape is indicated. In the *top row* are SEM micrographs of two beams that have been imaged after plasma etching; hence these shapes are the static equilibrium ground states. The initial electrical isolation of the supports and the gates is represented schematically as an open-circuit. In the *bottom row* are the observed shapes after electrically shorting the indicated supports and gates. The *dashed black lines* are guides to the eye to indicate the *centerline* between the *left* and *right* supports. The scale bars represent 10 μm

small physical separations; for example, in Fig. 1(c) the separation between the edge of the upper gate and the nearest edge of the buckled beam is approximately 200 nm.

In the top row of Fig. 5 we show the observed static equilibrium shapes of two different beams after plasma etching. During plasma etching the beams and electrodes are electrically floating. After the plasma process is terminated, the beams and electrodes remain at some floating electric potential. We found that it was possible to induce a shape change simply by providing a conducting path between the beam and an electrode. The bottom row of Fig. 5 shows the shapes after the beams were electrically connected to one of the electrodes. For the beam in the left column the dominant mode initially (open-circuit) is $n = 3$, whereas after the connection (short-circuit) the dominant mode is $n = 2$. Therefore, we refer to this observation as an $n = 3 \rightarrow n = 2$ electromechanical transition. For the beam in the right column the dominant mode initially is $n = 3$, whereas after the connection the dominant mode is $n = 1$, so in this case we observe an $n = 3 \rightarrow n = 1$ electromechanical transition.

We note that a quantity potentially relevant to the observed electromechanical transitions is the in-plane capacitance between beam and gate. From an elementary electrostatics calculation, we estimate this capacitance to be on the order of 100 aF. For perspective, a capacitance of 100 aF means that an electric potential difference of 1 mV requires a coupled charge of only a single electron. The continuum elasticity model presented in Sect. 4 could be extended to include electrostatics and explore theoretically the electromechanical effects of charge redistribution.

6 Conclusions

We have reviewed our work on the observation and actuation of mesoscopic buckled beams. The static equilibrium buckled shape of the beams unexpectedly evolved from the expected fundamental mode to a superposition of buckling modes as the aspect ratio was increased. Through mechanical actuation, it was demonstrated that the beams respond elastically about the static equilibrium profile and exhibit asymmetric static bistability. We also demonstrated electromechanical transitions between buckling modes by using gates fabricated in close proximity to the buckled beams.

The asymmetric static bistability exhibited by the beams may be compared with recent results demonstrating dynamic bistability of mesoscopic mechanical systems. For the results presented here, the residual strain, the length-to-width aspect ratio, and the inevitable asymmetry in the fabrication process, together define the static potential energy landscape for a particular beam. Strongly driven mesomechanical systems have been utilized to explore the rich nonlinear dynamics associated with transitions between dynamical states [4–6], where a non-equilibrium effective potential governs the dynamics. We hope that further experimental and theoretical work will elucidate the connections between these systems.

References

1. J.S. Aldridge and A.N. Cleland, Phys. Rev. Lett. **94**, 156403 (2005).
2. A.R. Bulsara, Nature (Lond) **437**, 962 (2005).
3. R.L. Badzey and P. Mohanty, Nature (Lond) **437**, 995 (2005).
4. I. Kozinsky, H.W.Ch. Postma, O. Kogan, A. Husain, and M.L. Roukes, Phys. Rev. Lett. **99**, 207201 (2007).
5. C. Stambaugh and H.B. Chan, Phys. Rev. Lett. **97**, 110602 (2006).
6. M.I. Dykman, I.B. Schwartz, and M. Shapiro, Phys. Rev. E **72**, 021102 (2005).
7. S. Savel'ev, A.L. Rakhmanov, X.D. Hu, A. Kasumov, and F. Nori, Phys. Rev. B **75**, 165417 (2007);
8. R.L. Badzey, G. Zolfagharkhani, A. Gaidarzhy, and P. Mohanty, Appl. Phys. Lett. **85**, 3587 (2004).
9. T. Rueckes, K. Kim, E. Joselevich, G.Y. Tseng, C.L. Cheung, and C.M. Lieber, Science **289**, 94 (2000).
10. K.C. Schwab and M.L. Roukes, Phys. Today **58**, 36 (2005).
11. L. D. Landau and E. M. Lifshitz, *Theory of Elasticity* (Pergamon, Oxford, 1986).
12. S.M. Carr, W.E. Lawrence, and M.N. Wybourne, Eur. Phys. Lett. **69**, 952 (2005).
13. S.M. Carr, W.E. Lawrence, and M.N. Wybourne, IEEE Trans. Nanotech. **4**, 655 (2005).
14. W.E. Lawrence, M.N. Wybourne, and S.M. Carr, New J. Phys. **8**, 223 (2006).
15. S.M. Carr, W.E. Lawrence, and M.N. Wybourne, Phys. Rev. B **64**, 220101(R) 2001.
16. S.M. Carr, W.E. Lawrence, and M.N. Wybourne, Physica B **316-317**, 464 (2002).
17. W.E. Lawrence, Physica B **316-317**, 448 (2002).
18. S.M. Carr and M.N. Wybourne, Appl. Phys. Lett. **82**, 709 (2003).

Developments in Parrondo's Paradox

Derek Abbott

Abstract Parrondo's paradox is the well-known counterintuitive situation where individually losing strategies or deleterious effects can combine to win. In 1996, Parrondo's games were devised illustrating this effect for the first time in a simple coin tossing scenario. It turns out that, by analogy, Parrondo's original games are a discrete-time, discrete-space version of a flashing Brownian ratchet—this was later formally proven via discretization of the Fokker-Planck equation. Over the past ten years, a number of authors have pointed to the generality of Parrondian behavior, and many examples ranging from physics to population genetics have been reported. In its most general form, Parrondo's paradox can occur where there is a nonlinear interaction of random behavior with an asymmetry, and can be mathematically understood in terms of a convex linear combination. Many effects, where randomness plays a constructive role, such as stochastic resonance, volatility pumping, the Brazil nut paradox etc., can all be viewed as being in the class of Parrondian phenomena. We will briefly review Parrondo's paradox, its recent developments, and its connection to related phenomena. In particular, we will review in detail a new form of Parrondo's paradox: the Allison mixture—this is where random sequences with zero autocorrelation can be randomly mixed, paradoxically producing a sequence with non-zero autocorrelation. The equations for the autocorrelation have been previously analytically derived, but, for the first time, we will now give a complete physical picture that explains this phenomenon.

D. Abbott (✉)
Centre for Biomedical Engineering (CBME) and School of Electrical & Electronic Engineering, The University of Adelaide, Adelaide, SA 5005, Australia
e-mail: dabbott@eleceng.adelaide.edu.au

V. In et al. (eds.), *Applications of Nonlinear Dynamics*, Understanding Complex Systems, DOI 10.1007/978-3-540-85632-0_25, © Springer-Verlag Berlin Heidelberg 2009

1 Introduction

Parrondo's paradox is where losing situations combine in order to win, and is exemplified by simple coin tossing games [1] that readily yield to physical and mathematical exploration [2, 3]. The Parrondian paradigm is one of 'survival of the weakest' and is a counterintuitive nonlinear effect. Parrondo's original games comprise simple coin tossing and are thus not game theory in the von Neumann sense [4] where players make decisions—however, in their original form they can be thought of as game theory in the Blackwell sense [5], and more recently Behrends has extended Parrondo's original games to include player strategy [6, 7] thus bringing them into the von Neumann realm. Consequently, in the following review we will use the term *game-theoretic* in its most inclusive sense—in the new field of quantum game theory, it is interesting to note that the phrase 'game theory' is also used broadly. In general, the emerging interest in game theory in the field of physics [8] uses the term in its broadest sense.

This Chapter is constructed as follows. Firstly, we we briefly go through how Parrondo's original games are constructed and how they work. We then review some recent developments in the study of Parrondo effects in a number of diverse fields and also review some interesting closely related phenomena. In particular, we discuss *volatility pumping* on the stock market, in its simplest form, and point out its similarities as a ratcheting effect.

Finally we conclude with a discussion on the thermodynamics of chance and then exploit thermodynamic analogies to develop a physical picture to explain a new intriguing Parrondo effect: the *Allison mixture* [9]. Here, the Allison mixture is the counterintuitive situation where the random shuffling of random sequences begins to 'erase' their randomness. In other words, two sequences that are incompressible can be randomly interleaved resulting in a sequence that has some compressibility.

1.1 Parrondo's Original Games

The original Parrondo games [1, 11] were devised in 1996 [12], as a pedagogical analogy of a flashing Brownian ratchet [13]. Since then they have stimulated research in diverse areas from economics [14], through to physical quantum systems [15–17], and population genetics [18–20]. For a more complete review, see [21].

Part of the original appeal of Parrondo's games is that they clearly illustrated effect of 'losing to win' for the first time, with a toy model involving simple coin tossing games. Other related phenomena existed prior to Parrondo's games [13, 22–28], but Parrondo was the first to show the effect in a clear game-theoretic form. His work was a landmark discovery because the simple analytical solution to his model enabled many workers to grasp the theory behind the phenomenon of 'losing to win.' Another significant event was when Parrondo's original games were first shown be formally related to the Fokker-Planck equation [29], then independently confirmed [30], and rigorously systemized [31]. This is significant as it opens up a

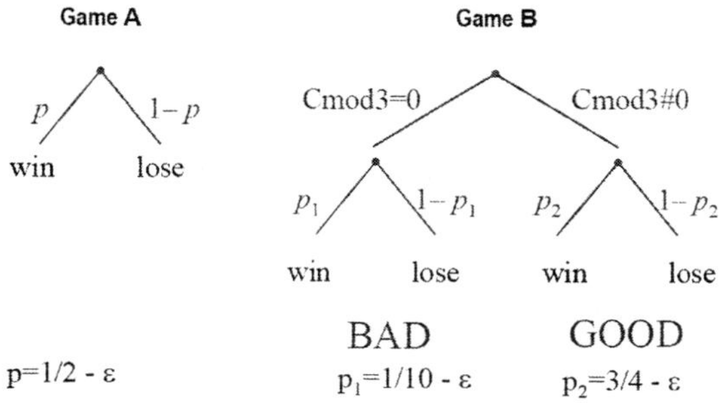

Fig. 1 The construction of Parrondo's original games. Game A is a simple coin toss that simulates the U_{flat} state of the flashing ratchet. The coin's bias is ε, which simulates the gradient of the flashing ratchet. Note that Game A is a losing game. Game B is composed of two coins. The 'good' coin is favorable and simulates the ratchet tooth's long slope and the 'bad' coin simulates the shorter slope of the ratchet tooth. For simplicity, your capital C goes up or down by \$1 every time you win or lose. You toss the bad coin if your capital is a multiple of three, otherwise you toss the good coin—this modulo arithmetic simulates the periodicity of the ratchet profile. The parameters of Game B are such that it is a losing game overall. When we switch periodically or randomly between the two losing games, surprisingly, we win

formal link between thermodynamics and games of chance (see Sect. 3). Parrondo's games were originally inspired by the flashing Brownian ratchet [13, 21], and via the Fokker-Planck equation they are intrinsically related.

The key idea of Parrondo's games is that you can have two or more sets of games that are individually losing—however, if you periodically or randomly switch between the losing strategies, there are conditions under which it is possible to counter-intuitively win. The games are constructed as indicated in Fig. 1 to cleverly simulate the action of the flashing ratchet that has been expounded elsewhere [1,11]. Game A simulates the flat potential and Game B simulates the sawtooth potential. As we can see, in Fig. 2, Game A and Game B are indeed losing games when played in isolation. Now, when we switch between the two games either periodically or randomly our winnings surprisingly increase.

It has been shown using Discrete Time Markov Chain (DMTC) analysis [32] that the games are governed by very simple inequalities—Game A is losing provided,

$$\frac{1-p}{p} > 1 \tag{1}$$

and Game B is losing provided,

$$\frac{(1-p_1)(1-p_2)^2}{p_1 p_2^2} > 1 \tag{2}$$

and the random combination of Game A and Game B wins provided,

$$\frac{(1-q_1)(1-q_2)^2}{q_1 q_2^2} < 1 \tag{3}$$

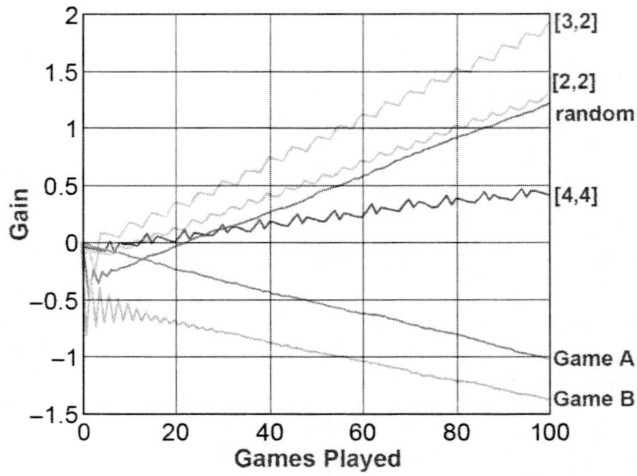

Fig. 2 The output of Parrondo's original games. The graph shows the amount of money gained versus the number of times we play. The parameters in this computer simulation are those listed in Fig. 1 with $\varepsilon = 0.005$. As expected, we see that Game A and Game B are individually losing. If we play two rounds of A followed by two rounds of B, indicated as $[2,2]$, we find that we remarkably win. In fact any periodic combination wins and the graph shows [3,2] and $[4,4]$ as examples. The curve marked 'random' is that case where we decide to play A or B on the flip of an unbiased coin. The results are averaged over 50,000 trials to produce smooth curves, however, the trends are still observable for individual trials

where p, p_1, and p_2 are defined in Fig. 1 and $q_1 = \gamma p + (1-\gamma)p_1$ and $q_2 = \gamma p + (1-\gamma)p_2$. Here, γ is the probability that Game A is selected and $1-\gamma$ is the probability of playing Game B. There are many ways to form a physical picture of why Parrondo's games work as they do—the picture becomes clearer once it is realized that Game A is coupled to Game B via the capital dependent rule. The first physical picture, due to Parrondo, is to simply to view the games as a discrete analogy to the flashing ratchet. An alternative picture is to see that Game B has a state dependence on capital that is forcing it to lose, and that Game A is acting as a source of noise that is breaking up that state dependence—this has been dubbed the Boston Interpretation, as it grew out of discussions at H.E. Stanley's group at Boston University [33]. In fact, it has been shown that as the amount of Game A 'noise' is gradually increased, by increasing γ, we find that the winnings follow a stochastic resonance-like curve [34]. Perhaps the most powerful interpretation is that of Moraal [35], who was the first to show that the games work due to a convex linear combination. This was later, and independently, reported by Costa *et al.* [36], which is recommended as a clearer exposition for the new reader. The realization that a convex linear combination is at the heart of the games is a significant one, as it then more readily links Parrondo's paradox to control and optimization problems. Another key point in understanding why the games work is that the Inequalities 2 and 3 are nonlinear—Parrondo's paradox is essentially a nonlinear phenomenon. The fundamental reason why the governing inequalities are nonlinear is due to the state-dependence in Game B—in mathematical terms this is equivalent to saying that Game B is not a martingale. Finally, it should be noted that there are possibly an infinite number of ways to construct different nonlinear games that exhibit Parrondian effects. The open question is to search for the interesting cases that map onto physical and biological systems, and to investigate those that display the largest regions of parameter space for the effect to occur. Progress has been made with developing differently constructed Parrondo games [37–39], but so far it is still early days and there is still much to explore in this regard.

2 Developments in Parrondo's Paradox and Related Phenomena

From an engineering viewpoint it is known that mixing unstable systems can result in stability, and the connection to Parrondo's paradox has been pointed out [40]—further investigation in this area may be of relevance to optimization problems. In the area of neural networks it is known that a network can perform better at *network generalization* if noise is added to the training data set [41]—this evokes the idea of losing in order to win and connections to Parrondo effects have yet to be investigated. Parrondo effects in spin systems [35] and quantum game theory [15–17], have been reported. Increasing our understanding of how to control decoherence is one of the motivating factors behind these developments in quantum game theory.

It has been shown [42] that Parrondo's original games can be rather elegantly described in terms Onsager rate equations [43,44]—this suggests the possibility for future work in searching for chemical reactions that display Parrondo-like effects. Parrondo effects have also inspired work in the study of negative mobility phenomena [45], reliability theory [46], noise induced synchronization [47], spatial patterns via switching [48], and in controlling chaos [49,50].

The area of biology is still ripe with open questions for the study of Parrondo-like effects. There are many examples in biology of 'losing to win'—for example *sickle cell anemia* is deleterious and yet it can protect the host from contracting malaria. Conjectures have been mooted for the application of Parrondo's paradox to biogenesis [10], the dynamics of gene transcription in GCN4 protein [51], and the dynamics of transcription errors in DNA [51]. Parrondo's paradox has been studied in various interesting scenarios involving population genetics [18–20,52].

In conventional sociobiology the standard dogma is that when chosing an optimal mate we are attracted to beauty, as those features we see in beauty are in fact indicators of a healthy mate—therefore to efficiently propagate our genes we seek attractive mates [53]. If this was really the complete picture, one might expect ugliness to have been selected out by now. The present theory does not seem to account for the common fact that two ugly parents can often produce an attractive child. Perhaps there is a Parrondian payoff in being a 'loser' at the dating game, where survival of the weakest can come into play. Arizmendi [54] has recently proposed a Parrondian dating model that fosters survival of the 'loser.' Satinover and Sornette [55,56] propose a Parrondian model where they show that short term optimization can turn a positive expected gain into a negative one, thus showing in some circumstances it can pay to be the uncompetitive 'loser' who sticks to the minority.

In terms of the stockmarket, Boman [57, 58] has used a Parrondian game framework as a toy model for studying the dynamics of insider information. A number of Parrondo-like toy models for switching between poor performing investments are well-known. For example, Maslov and Zhang demonstrate a model where switching between volatile assets and non-performing cash reserves produces an increase in the gains [23] in a fashion not too dissimilar from Luenberger's *volatility pumping* method [27]. There are also closely related models such as the *excess growth* model of Fernholz and Shay [59] and Tom Cover's *universal portfolio* [60]. There are open questions yet to be explored in systemizing these effects and drawing out the exact connections with the large body of literature on Parrondo's paradox.

Here, we focus on Luenberger's volatility pumping as it is a simple toy model that rather nicely illustrates the principles of 'winning' with poorer stocks in a clear way. Figure 3 simulates two stocks: one stock is stable but is mediocre and in the long run neither wins or loses significantly, the other stock has some growth but is volatile. A very simple toy model of volatility is used here, namely, that we randomly halve or double the stock value from day to day. Leunberger's method is to then sell both

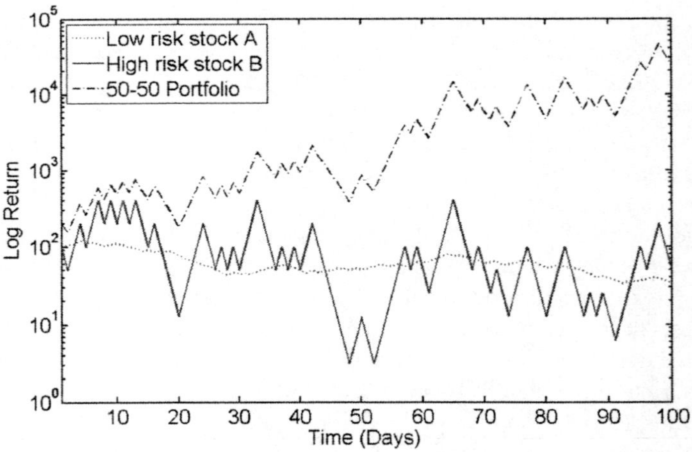

Fig. 3 Volatility pumping a low-risk stock with a high-risk stock. The *dotted curve* simulates a mediocre low-risk stock that in the long run neither wins nor loses. The *solid curve* represents a volatile stock that gives a 25% expected return, though is high-risk—a simple toy model of volatility is implemented here, where the stock simply halves or doubles, at random, its previous value at each time-step. The chained curve is found by selling both stocks at the end of each time step, adding the total cash to get $\$T$, then repurchasing them at the beginning of each time-step at a 50:50 split—that is, we purchase $\$T/2$ worth of the high-risk stock and $\$T/2$ worth of the low-risk-stock. This is process called *portfolio rebalancing*. Surprisingly, the chained curve grows exponentially, even though the two stocks individually do not perform as well. Both stocks start at Day 1 priced at \$100, and thus the combined portfolio (*chained curved*) starts at \$200. The vertical axis is the return in dollars plotted on a logarithmic scale. The return on the rebalanced portfolio is so large that we would not be able to see the individual curves, without the logarithmic plot

stocks each day and rebuy them implementing *portfolio rebalancing*. The rebalance operation is to take your total cash $\$T$, and buy $\$T/2$ of one stock and $\$T/2$ of the other—thus we maintain a 50:50 portfolio. This operation is repeated each day and remarkably it produces the top curve in Fig. 3 with exponential growth. Now, it should be noted this is a stripped down toy model to illustrate the key idea that rebalancing creates growth. Of course, in reality one would not buy and sell everyday as transaction costs would be prohibitive. Also the halving and doubling operation is an artificial construct that is cleverly designed so that we never hit rock bottom—in reality, hitting the $y = 0$ axis is always the problem. However, the beauty of any toy model is it enables us to explore the pertinent features of an interesting effect. The open questions are why does volatility pumping work and how should the portfolio rebalancing strategy be optimized for best performance in a real scenario? Whilst both these questions are still being actively debated [61], from the point of view of ratchet science, volatility pumping must clearly be the result of an asymmetry that rectifies fluctuations in the market. This is the principle behind every Brownian ratchet and volatility pumping is no exception. The action of maintaining the 50:50 portfolio split guarantees that we are always buying low and selling high, which is the source the ratchet asymmetry.

3 Thermodynamics of Games of Chance

The fact the Parrondo's original games can be exactly derived via discretization of the Fokker-Planck equation [29–31] is of fundamental interest because it can then serve as a useful toy model for investigating the discrete-continuous interface. As Parrondo's games are deeply rooted in the thermodynamics of Brownian ratchets, via the Fokker-Planck equation, this has enabled the use of thermodynamic analogies when understanding the operation of the games. Various authors [11, 51, 62] have suggested simple analogies between Parrondo's games and physical Brownian motors. Amengual et al. [63] have taken this a step further and have proposed how to characterize Parrondo's games in terms of a thermodynamic engine efficiency.

This raises the open question of whether a generalized thermodynamic picture can be applied to arbitrary games of chance. To this end, in about 2001, I performed the following simple thought experiment. I imagined two players flipping a simple unbiased coin. Player 1 wins \$1, given heads, and Player 2 wins \$1, given tails. The coin is unbiased, which means that if I take a video of the game, and run the movie backwards we would not be able to tell the difference. Therefore the system displays *time-reversibility*, which is what we expect of physical systems that are in thermal equilibrium. Now, let us imagine we have a biased coin such that, say, Player 1 progressively wins. The situation is no longer time-reversible, because the forward and backward movies are now clearly different—in the forward movie Player 1 appears to win, but in the reverse movie Player 2 appears to win. This time-irreversible situation corresponds to physical systems that are out of equilibrium. Thus the situation of *detailed balance* when the coin is unbiased is analogous to thermodynamic equilibrium, and a biased coin is rather like a system out of equilibrium. Initially, this thought experiment appeared trivial and not particularly useful, however, in the next section we put it to good use illustrating a significant link between random manipulation of integer sequences and thermodynamics.

For those readers who are new to the concept that time-reversibility relates to thermal equilibrium, I recommend to consider the simple mental picture of Feynman's ratchet and pawl [64]. The two ends of the system are at different temperatures and the ratchet wheel rotates in one direction. Now if you make the two ends the same temperature, there is no net rotation of the ratchet wheel. Now take a movie of each of these two cases, and run the movie backwards. What do you see?

- Case 1: When the temperatures are different, $T_1 \neq T_2$. Here the backwards movie looks different to the forwards movie. Because the wheel is turning in a particular direction, the backwards movie would have the wheel turning the opposite way. As the movies are different, this is the acid test that tells us the process is irreversible.
- Case 2: When the temperatures are same, $T_1 = T_2$. Here we have thermal equilibrium, and so there is no net energy coming into the system and therefore the wheel does not rotate in a net direction. It just randomly jiggles back and forth, and no work is done. We run the movie backwards and now we cannot tell the difference—the backwards movie just looks like random jiggles, as does the forwards movie. Thus this case is reversible.

Thus, in summary, in thermal equilibrium we have reversibility, when out of thermal equilibrium we have irreversibility. This is a well-known principle in thermodynamics, but we have introduced the specific ratchet example as a nice physical picture for visualizing this concept.

4 A New Parrondian Effect: The Allison Mixture

A new form of Parrondo's paradox, namely the *Allison mixture*, has recently been reported [9]. Let us imagine two sequences of random numbers; we shall call them Sequence 1 and Sequence 2—they are totally random in that they are independent and have zero autocorrelation. For simplicity, we can consider these sequences to be random strings of 1s and 0s—however, note that the effect I am about to describe is not limited to binary sequences but is in fact general. If we now randomly scramble these two sequences to generate a third new sequence, naively we would expect this resulting sequence to also be completely random. It turns out that this is not always the case: counterintuitively the final sequence can have a finite autocorrelation ρ even though the ρ's of originating sequences are zero—this is what we call an *Allison mixture.*

Let us now be a little bit more precise about how we actually scramble the two sequences, so we can then write out an analytical expression for ρ to show that it can be in fact non-zero. We start at an arbitrary nth position of Sequence 1. We either move to position $n+1$ of Sequence 1 with probability $1-\alpha_1$ or skip to position $n+1$ of Sequence 2 with probability α_1. Whenever we find ourselves in Sequence 2, we hop to the next location on Sequence 1 with probability α_2 or advance one step within Sequence 2 with probability $1-\alpha_2$. We continue hopping back and forth between the two sequences in this manner and each digit that we sequentially land on is called out to form the new sequence. In this way a third new sequence is generated from the original two sequences by random hopping, using separate transition probabilities α_1 and α_2 to keep everything perfectly general. For the sake of further generality, let the means of the two originating sequences be μ_1 and μ_2.

It has been shown that the autocorrelation ρ for the generated sequence is [9],

$$\rho = \frac{1}{\sigma^2}\frac{\alpha_2}{\alpha_1+\alpha_2}(\mu_1-\mu_2)^2(1-\alpha_1-\alpha_2) \tag{4}$$

where σ^2 is the variance of the final sequence. The full expression for the variance has been previously reported [9], but is not given here as it is not relevant for the following physical discussion. Our naive expectation is that a random mixture of random sequences should always result in $\rho = 0$—however, Eq. (4) reveals that ρ is only zero provided $\mu_1 = \mu_2$ or $\alpha_1+\alpha_2 = 1$. If we break both these conditions, then we can legally produce a sequence with a non-zero ρ. The mathematics dictates to us that this must be the case, but the question is why? And what is the physical picture and basis for what is going on?

The previous section on the thermodynamics of chance—Sect. 3—contains many of the necessary clues to unravel the physical picture. Firstly, let us address physically why to get $\rho \neq 0$ we must have $\mu_1 \neq \mu_2$—the implication is that the means of the sequences are analogous to the temperature of a physical process. Loosely speaking, temperature is some measure that is proportional to the average of all the jiggling within a solid object. In the case of the random sequences, μ_1 and μ_2 are the averages of all the jiggling or varying numbers and play the same role as temperature. Thus when $\mu_1 \neq \mu_2$, we have an irreversible situation—the sequences are irreversibly mixed and we therefore get autocorrelation in the final sequences, because there is information loss. Recall that the originating sequences are random, and thus are incompressible in the Chaitin-Kolmogorov sense and contain maximal information in the Shannon sense. Thus by subjecting them to an irreversible process we know from thermodynamics that we must *lose* information, and thus redundancy must have crept into the final sequence leading to $\rho \neq 0$. Now, in the special case, when $\mu_1 = \mu_2$, we have reversible mixing, because this is analogous to thermal equilibrium where $T_1 = T_2$. If the process is reversible then there is no information loss, no redundancy is added, and therefore $\rho = 0$.

However, this is only part of the picture, as Eq. (4) also predicts that to obtain a case where $\rho \neq 0$, we must also observe the $\alpha_1+\alpha_2 \neq 1$ condition. So what is the physical reason why $\alpha_1+\alpha_2 \neq 1$ is required to get non-zero ρ in the final sequence? To unravel the mystery we draw the state diagrams to illustrate the mechanism. Figure 4 illustrates the case when $\alpha_1+\alpha_2 \neq 1$—the caption explains why switching between the two states leads to *memory persistence* that causes

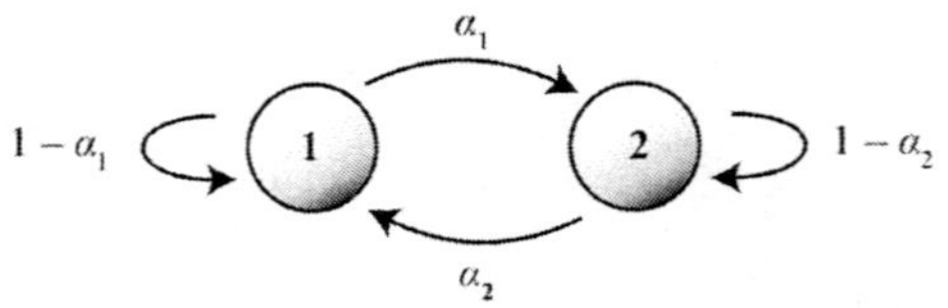

Fig. 4 State diagram for unbalanced switching process, when $\alpha_1 + \alpha_2 \neq 1$, giving rise to persistence. *Circle* 1 represents the state of landing in random Sequence 1 and *Circle* 2 represents the state we are in when we land on random Sequence 2. We jump between these two sequences to generate a new sequence. The apparent paradox of this Allison mixture is that the resulting sequence has non-zero autocorrelation even when the originating sequences have zero autocorrelation. Here, α_1 and α_2 are the transition probabilities of jumping between the two sequences. Notice, for example, as $\alpha_1 \to 0$ the probability of a self-transition to stay in Sequence 1 is high—so, if we are already in Sequence 1 we are likely to stay there. This can be thought of as a type of 'memory' of the system, which causes the new sequence to have non-zero autocorrelation. Note: this is *not* a form of memory in the sense that requires storage of a previous state—as there is no clear terminology, in the literature, for our probabilistic type of memory effect, we hereby call it *memory persistence*

correlation, and hence redundancy, in the final sequence (or anticorrelation in the case of antipersistence). Now for the case when $\alpha_1 + \alpha_2 = 1$, Fig. 5 illustrates we get detailed balance between the probability of entering a state and the probability of staying in a state. (Note that staying within a state is also called a *self-transition*). The detailed balance implies there is no memory persistence and hence $\rho = 0$. An alternative valid explanation is to use the argument of Sect. 3 that explains why detailed balance implies a reversible process. So essentially we must have $\alpha_1 + \alpha_2 \neq 1$ to ensure irreversibility, which is a necessary condition for obtaining $\rho \neq 0$.

So now we begin to see the connection between an Allison mixture and Parrondian effects that require an asymmetry to interact with random behavior. Figure 5 is the symmetric case where we have detailed balance, and Fig. 4 is the asymmetric case where detailed balance is broken. Symmetry breaking is the essence of all Parrondian and Brownian ratchet phenomena. The $\mu_1 \neq \mu_2$ condition is analogous to the $T_1 \neq T_2$ condition that is required for Feynman's ratchet to operate. The $\alpha_1 + \alpha_2 \neq 1$ condition is analogous to the ratcheting mechanism in Feynman's ratchet as these both are the sources of asymmetry. A number of open questions

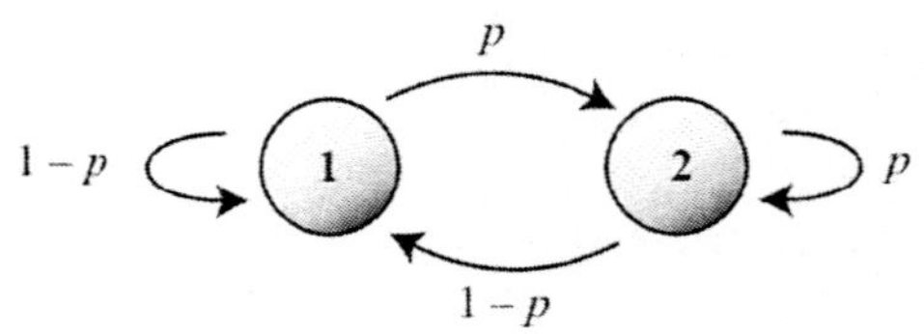

Fig. 5 State diagram for balanced switching process, when $\alpha_1 + \alpha_2 = 1$, resulting in no persistence. For simplicity we have inserted $p = \alpha_1 = 1 - \alpha_2$, which clearly reveals that the probability of entering a state exactly balances the probability of self-transition in that state. This detailed balance implies there is no memory persistence effect. Hence, the new generated sequence also has zero autocorrelation

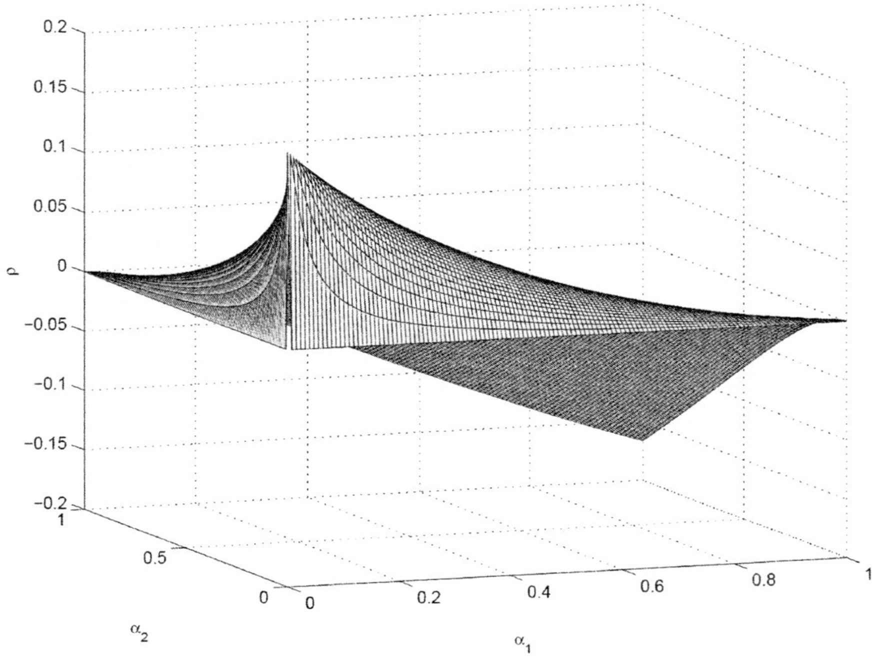

Fig. 6 Plot of the autocorrelation ρ of the generated sequence as a function of α_1 and α_2. In this specific example, the two originating binary sequences have $\mu_1 = 0.2$ and $\mu_2 = 0.6$. Pearce has named the peak of this graph the *pinnacle* [66]. The plot shows that the system displays some mathematically curious features. Surprisingly, $\rho = 0$ does not occur at a unique point in the parameter space. Another intriguing feature is that as $\alpha_1 \to 0$ and $\alpha_2 \to 0$, whether ρ ends up at zero or at Pearce's pinnacle depends on the direction that you approach the limits

remain concerning the mathematical features of this switched system that can be thought of as a two-state discrete-time hidden Markov model—for example, Fig. 6 illustrates that as $\alpha_1 \to 0$ and $\alpha_2 \to 0$ the direction of the limits intriguingly affect the final value of ρ. Another open question is that of a possible application for Allison mixtures—this remains to be seen, but possible areas of promise might be in encryption and in optimizing file compression. Another open question to ask is if there are any links between Allison mixtures and biological evolution or genetics? Could it be that the redundancy that appears in sequences of non-coding (or 'junk') DNA are the result of something along the lines of Allison mixing (i.e. ratcheted random mixing)? In the case of coding DNA, random mutations are a biased process—for example frame shift mutations in DNA are more likely to occur in sequences with runs of a single base and some single base mutations are more probable. This together with the process of selection, which again is random but with biases, results in order that is created in a set of DNA sequences. These sequences encapsulate in an ordered way information about the regularities of the organism in its environmental context.

5 Conclusion

There are two key take home messages that the study of Parrondo effects reveal: (i) the process of switching is a nonlinearity and therefore switching can radically alter the overall system behavior, and (ii) the interaction between noise and an asymmetry can give rise to directed motion even against a gradient, provided we are out of equilibrium.

Physicists have traditionally sought symmetry in Nature—a new challenge for future research is to now search for asymmetries and observe how they interact with noise or random behavior. On a more philosophical note, we might pose the question "Should we consider noise or randomness as a special form of order?" As well as our discussion on Parrondo effects, there are other examples that point to this: (i) a random walk displays self-similarity, (ii) randomly switched processes can produce fractals, (iii) according to Shannon, noise packs in maximal information, and (iv) as Chaitin points out, even the integers have noisy properties [65]. After all, noise is the most ordered way to avoid redundancy. There are many situations where noise appears to give rise to order and the challenge is to identify the general mathematical principles behind this.

Acknowledgments A special thanks is due to Adi R. Bulsara who prompted me to write this Chapter on the occasion of the celebration of his Feschrift, for his 55th birthday, held in Kauai, Hawaii, 2007. A warm thanks is due to Withawat Withayachumnankul who assisted with the preparation of the diagrams and Mark D. McDonnell who assisted with the LaTeX formatting. I would also like to thank Andrew G. Allison, Charles E. M. Pearce, Matthew J. Berryman, Paul C. W. Davies, Charles R. Doering, Adrian P. Flitney, Peter Hänggi, and Juan M. R. Parrondo, for a number of helpful and stimulating discussions on the topic.

References

1. Harmer, G. P. and Abbott, D., Losing strategies can win by Parrondo's paradox, *Nature* **402** 864 (1999).
2. Arena, P., Fazzino, S., Fortuna, L., and Maniscalco, P., Game theory and non-linear dynamics: the Parrondo Paradox case study, *Chaos, Solitons & Fractals* **17**(2–3) 545–555 (2003).
3. Behrends, E., The mathematical background of Parrondo's paradox, *Proc. SPIE Noise in Complex Systems and Stochastic Dynamics II*, Maspalomas, Spain, Ed: Zoltan Gingl, **5471** 510–517 (2004).
4. von Neumann, J. and Morgenstern, O., *Theory of Games and Economic Behavior,* Princeton University Press, New York, (1954).
5. Blackwell, D. and Girshick, M. A., *Theory of Games and Statistical Decisions,* John Wiley & Sons, New York (1954).
6. Behrends, E., Parrondo's paradox: a priori and adaptive strategies, Preprint: A-02-09, www.math.fu-berlin.de (2002).
7. Groeber, P., On Parrondo's games as generalized by Behrends, *Lecture Notes in Control and Information Sciences*, **341** 223–230 (2006).
8. Abbott, D., Davies, P. C. W., and Shalizi, C. R., Order from disorder: the role of noise in creative processes: A special issue on game theory and evolutionary processes–overview, *Fluctuation and Noise Letters,* **2** C1–C12 (2002).

9. Allison, A., Pearce, C. E. M., and Abbott, D., Finding keywords amongst noise: Automatic text classification without parsing, *Proc. SPIE Noise and Stochastics in Complex Systems and Finance,* Florence, Italy, Eds: János Kertész, Stefan Bornholdt, and Rosario N. Mantegna **6601** 660113 (2007).
10. Davies, P. C. W., Physics and life: The Abdus Salam Memorial Lecture, *Sixth Trieste Conference on Chemical Evolution, Trieste*, Italy, Eds: J. Chela-Flores, T. Tobias, and F. Raulin, Kluwer Academic Publishers 13–20 (2001).
11. Harmer, G. P. and Abbott, D., Parrondo's paradox, *Statistical Science* **14** 206–213 (1999).
12. Parrondo, J. M. R., How to cheat a bad mathematician, in *EEC HC&M Network on Complexity and Chaos* (#ERBCHRX-CT940546), ISI, Torino, Italy (1996), Unpublished.
13. Adjari, A. and Prost, J., Drift induced by a periodic potential of low symmetry: Pulsed dielectrophoresis, *C. R. Acad. Science Paris, Série II,* **315** 1635–1639 (1993).
14. Johnson, N. F., Jeffries, P., and Hui, P. M., *Financial Market Complexity*, Oxford University Press, Oxford (2003).
15. Lee, C. F., Johnson, N. F., Rodriguez, F., and Quiroga, L., Quantum coherence, correlated noise and Parrondo games, *Fluctuation and Noise Letters* **2**(4) L293–L297 (2002).
16. Flitney, A. P. and Abbott, D., Quantum Parrondo games, *Physica A* **314**(1–4) 35–42 (2002).
17. Meyer, D. A. and Blumer, H., Quantum Parrondo games: biased and unbiased, *Fluctuation and Noise Letters* **2**(4) L257–L262 (2002).
18. Wolf, D. M., Vazirani, V. V., and Arkin, A. P., Diversity in times of adversity: Probabilistic strategies in microbial survival games, *Journal of Theoretical Biology* **234** 227–253 (2005).
19. Reed, F. A., Two-locus epistasis with sexually antagonistic selection: A genetic Parrondo's paradox, *Genetics,* **176,** 1923–1929 (2007).
20. Masuda, N. and Konno, N., Subcritical behavior in the alternating supercritical Domany-Kinzel dynamics, *European Physical Journal B* **40** 313–319 (2004).
21. Harmer, G. P. and Abbott, D., A review of Parrondo's paradox, *Fluctuation and Noise Letters, 2*(2) R71–R107 (2002).
22. Pinsky, R. and Scheutzow, M., Some remarks and examples concerning the transient and recurrence of random diffusions, *Annales de l'Institut Henri Poincaré—Probabilités et Statistiques* **28** 519–536 (1992).
23. Maslov, S. and Zhang, Y., Optimal investment strategy for risky assets, *International Journal of Theoretical and Applied Finance* **1** 377–387 (1998).
24. Westerhoff, H. V., Tsong, T. Y., Chock, P. B., Chen Y., and Astumian, R. D., How enzymes can capture and transmit free energy contained in an oscillating electric field, *Proceedings of the National Academy of Science* **83** 4734–4738 (1986).
25. Key, E. S., Computable examples of the maximal Lyapunov exponent, *Probability Theory and Related Fields* **75** 97–107 (1987).
26. Abbott, D., Overview: Unsolved problems of noise and fluctuations, *Chaos*, **11** 526–538 (2001).
27. Luenberger, D. G., *Investment Science,* Oxford University Press, Oxford (1997).
28. Rosato, A., Strandburg, K. J., Prinz F., and Swendsen, R. H., Why the Brazil nuts are on top: Size segregation of particulate matter by shaking, *Physical Review Letters* **58** 1038–1040 (1987).
29. Allison, A. and Abbott, D., The physical basis for Parrondo's games, *Fluctuation and Noise Letters*, **2**(4) L327–L341 (2002).
30. Toral, R., Amengual, P., and Mangioni, S., Parrondo's games as a discrete ratchet, *Physica A*, **327**(1–2) 105–110 (2003).
31. Amengual, P., Allison, A., Toral, R., and Abbott, D., Discrete-time ratchets, the Fokker-Planck equation and Parrondo's paradox, *Proceedings of the Royal Society London A*, **460**(2048), 2269–2284 (2004).
32. Harmer, G. P., Abbott, D., and Taylor, P. G., The paradox of Parrondo's games, *Proceedings of the Royal Society London A* **456** 247–259 (2000).
33. Key, E. S., Kłosek, M. M., Abbott, D., On Parrondo's paradox: how to construct unfair games by composing fair games, *ANZIAM J.* **47**, 495–511 (2006).

34. Allison, A. and Abbott, D., Stochastic resonance in a Brownian ratchet, *Fluctuation and Noise Letters* **1**(4) L239–L244 (2001).
35. Moraal, H., Counterintuitive behaviour in games based on spin models, *Journal of Physics A*, **33** L203–L206 (2000).
36. Costa, A., Fackrell, M., and Taylor, P. G., Two issues surrounding Parrondo's paradox, *Advances in Dynamic Games: Applications to Economics, Finance, Optimization, and Stochastic Control,* Eds: Andrzej S. Nowak and Krzysztof Szajowski, **7** 599–609 (2005).
37. Parrondo, J. M. R., Harmer, G. P., and Abbott, D., New paradoxical games based on Brownian ratchets, *Physical Review Letters* **85** 5226–5229 (2000).
38. Kay, R. J. and Johnson, N. F., Winning combinations of history-dependent games, *Physical Review E 67* 056128 (2003).
39. Toral, R., Cooperative Parrondo's games, *Fluctuation and Noise Letters* **1** L7–L12 (2001).
40. Allison, A. and Abbott, D., Control systems with stochastic feedback, *Chaos* **11** 715–724 (2001).
41. Bishop, C. M., *Neural Networks for Pattern Recognition,* Oxford Press Oxford Chapter 9, 346–349 (1996).
42. Van den Broeck, C., Reimann P., Kawai, R., and Hänggi, P., Coupled Brownian motors, *Lecture Notes in Physics: Statistical Mechanics of Biocomplexity,* Eds: D. Reguera, M. Rubi, and J. M. G. Vilar, **527** Springer-Verlag: Berlin, Heidelberg, New York, 93–111 (1999).
43. Onsager, L., Reciprocal relations in irreversible processes I, *Physical Review* **37** 405–426 (1931).
44. Onsager, L., Reciprocal relations in irreversible processes II, *Physical Review* **38** 2265 (1931).
45. Cleuren, B. and Van den Broeck C., Random walks with absolute negative mobility, *Physical Review E,* **64** 030101 (2002).
46. Di Crescenzo, A., A Parrondo paradox in reliability theory, *The Mathematical Scientist* **32**(1) 17–22 arXiv:math/0602308v2 (2007).
47. Kocarev, L. and Tasev, Z., Lyanpunov exponents, noise-induced synchronization, and Parrondo's paradox, *Physical Review E* **65** 046215 (2002).
48. Buceta, J., Lindenberg, K., and Parrondo, J. M. R., Pattern formation induced by nonequilibrium global alternation of dynamics, *Physical Review E* **66** 036216 (2002).
49. Almeida, J., Peralta-Salas, D., and Romera, M., Can two chaotic systems give rise to order? *Physica D* **200** 124–132 (2005).
50. Boyarsky, A., Góra, P., and Shafiqul Islam, Md., Randomly chosen chaotic maps can give rise to nearly ordered behavior, *Physica D 210* 284–294 (2005).
51. Harmer, G. P., Abbott, D., Taylor, P. G., and Parrondo, J. M. R., Parrondo's games and Brownian ratchets, *Chaos* **11** 705–714 (2001).
52. Atkinson, D. and Peijnenburg, J., Acting rationally with irrational strategies: Applications of the Parrondo effect, *Reasoning, Rationality, Probability,* Eds: Maria Carla Galavotti, Roberto Scazzieri, and Patrick Suppes, CSLI Publications, Stanford (2007).
53. Diamond, J. M., *Why Sex is Fun?: The Evolution of Human Sexuality,* Harper Collins, New York (1997).
54. Arizmendi, C. M., Paradoxical way for losers in a dating game, *Proceedings of the AIP Nonequilibrium Statistical Mechaniucs and Nonliear Physics: XV Conference on Nonequilibrium Statistical Mechanics and Nonlinear Physics*, Mar del Plata, Argentina, 4–8 December, 2006, Eds: Orazio Descalzi, Osvaldo A. Rosso, and Hilda A. Larrondo, **913**, 20–25 arXiv:physics/0703189v1 (2007).
55. Satinover, J. B. and Sornette, D., 'Illusion of control' in time-horizon minority and Parrondo games, *The European Physical Journal B* **60**(3) 369–384 (2007).
56. Satinover, J. B. and Sornette, D., Illusion of control in a Brownian game, *Physica A* **386**(1) 339–344 (2007).
57. Boman, M., Johansson, S. J., and Lyback, D., Parrondo strategies for artificial traders, in *Intelligent Agent Technology: Research and Development,* Eds: Ning Zhong, Jiming Liu, Setsuo Ohsuga, Jeffrey Bradshaw, World Scientific, 150–159 arXiv:cs.ce/0204051 (2001).

58. Wah-Sui Almberg, W-S. and Boman, M., An active agent portfolio management algorithm, *Artificial Intelligence and Computer Science*, Ed: Susan Shannon, Nova Science Publishers, Inc. Hauppauge NY Chapter 4, 123–134 (2005).
59. Fernholz, R. and Shay, B., Stochastic portfolio theory and stock market equilibrium, *Journal of Finance,* **37** 615–624 (1982).
60. Cover, T. M. and Ordentlich, E., Universal portfolios with side information, *IEEE Transactions on Information Theory* **42**(2), 348–363 (1996).
61. Dempster, M. A. H. and Evstigneev, I. G., Volatility-induced financial growth, *Quantitative Finance* **7**(2) 151–160 (2007).
62. Heath, D., Kinderlehrer, D., and Kowalczyk, M., Discrete and continuous ratchets: From coin toss to molecular motor, *Discrete and Continuous Dynamical Systems—Series B*, **2** 153–167 (2002).
63. Amengual, P., Toral R., Allison, A., and Abbott, D., Efficiency of discrete-time ratchets, arXiv:cond-mat/0410173 (2004).
64. Feynman, R. P., Leighton, R. B., and Sands, M., *The Feynman Lectures on Physics,* **1** 46.1–46.9 Addison-Wesley, Reading, MA (1963).
65. Chaitin, G. J., *The Unknowable,* Springer-Verlag Berlin (1999).
66. Pearce, C. E. M., Allison, A., and Abbott, D., Perturbing singular systems and the correlating of uncorrelated random sequences, *Proceedings of the AIP International Conference on Numerical Analysis and Applied Mathematics,* Corfu, Greece, Eds: Theodore E. Simos, George Psihoyios, and Ch. Tsitouras, **936** 699 (2007).

Magnetophysiology of Brain Slices Using an HTS SQUID Magnetometer System

Per Magnelind, Dag Winkler, Eric Hanse and Edward Tarte

Abstract A high-transition-temperature superconducting quantum interference device (HTS SQUID) system has been developed and employed in neuromagnetic measurements of evoked fields from *in vitro* brain slices. Transverse hippocampal slices from rat were prepared and measured. The evoked neuronal activity produced neuromagnetic fields of ∼5 pT.

1 Introduction

Magnetophysiology is the neurophysiological study of the magnetic fields generated by the ion currents inside the neurons. Previously, these studies have been performed in systems equipped with low-transition-temperature (low-T_c) Superconducting QUantum Interference Devices (SQUIDs) [3, 6, 7, 9]. The separation between the sensor and the neural slice in these systems has been larger than 2 mm, which limits the strength of the recorded magnetic field. We have previously reported [8] that a high-T_c superconducting (HTS) SQUID magnetometer might yield a better signal-to-noise ratio (SNR) as it (more easily) can be moved closer to a thin window separating the vacuum chamber and the slice chamber.

In this letter we report on HTS SQUID magnetometer recordings of evoked magnetic fields (EMFs) from transverse hippocampal slices.

P. Magnelind (✉)
Department of Microtechnology and nanoscience, Chalmers University of Technology, SE-412 96 Göteborg, Sweden, e-mail: per.magnelind@mc2.chalmers.se. Presently with Los Alamos National Laboratory, P-21 MS-D454, Los Alamos, NM 87545, USA

V. In et al. (eds.), *Applications of Nonlinear Dynamics*, Understanding Complex Systems, DOI 10.1007/978-3-540-85632-0_26, © Springer-Verlag Berlin Heidelberg 2009

2 Materials and Methods

The SQUID system has been described in detail in refs. [4, 5]. However, a few changes has been made to the system. We have fabricated and mounted a directly-coupled magnetometer (see description below). The SQUID electronics used in the measurements reported here was a Magnicon SEL-1 operated with a bias-reversal frequency of 2kHz.

2.1 Directly-Coupled Magnetometer

A directly coupled magnetometer with a square pick-up loop with side length of $d_p = 2$ mm and an inner side length of $a_p = 0.5$ mm, see Fig. 1, was used in the measurements. The washer has structures of a linewidth of 5 μm and holes of $5 \times 40\ \mu\text{m}^2$ to reduce the flux-trapping in the structures, when exposed to the ambient fields.

The pick-up loop was coupled to a hairpin-shaped SQUID. The lengths and widths of the hairpin SQUID have been optimized using the 3D-MLSI program [2]. The outcome of the optimization was a SQUID consisting of a 4 μm wide strip that surrounds a 4 μm wide and 40 μm long slit. The junctions were 2 μm wide. The

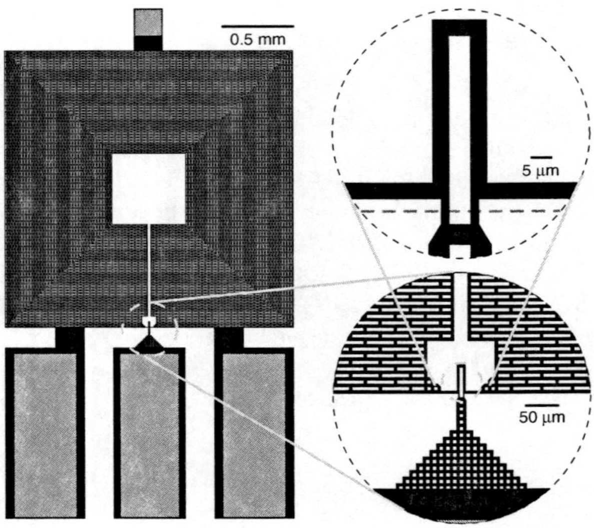

Fig. 1 Design of a directly coupled magnetometer. The outer size of the pick-up washer is 2 mm and the hole is 0.5 mm. The *left* insert shows the SQUID area and the holes of the washer. The size of the holes is 40×5 μm and the width of the superconducting strips is 5 μm. The *right* insert shows the area with the hairpin SQUID with 2 μm wide Josephson junctions. The length and the width of the strips are 40 μm and 4 μm, respectively. The strips are separated by 5 μm. The *dashed line* indicates the grain boundary

width of the strips, which inject the magnetic field-induced current from the pick-up loop into the SQUID, was 4 μm.

The magnetometer was patterned from a 200 nm thick $YBa_2Cu_3O_{7-\delta}$ film on a $SrTiO_3$ bicrystal substrate with a symmetric 24° misorientation. The film was covered by a thin Au layer *in situ*. E-beam lithography was used to create a lift-off stencil, through which additional Au was deposited forming contact pads. An amorphous carbon (a-C) layer was then deposited and Cr was deposited through another lift-off stencil for the SQUID structure. The unmasked a-C was removed by a low power O_2 plasma and the remaining a-C/Cr mask was used for the Ar-ion milling. Then, the remaining a-C was removed using O_2 plasma. Finally, a short ion milling was used to remove the *in situ* Au.

2.2 Experimental Methods

Transverse hippocampal tissue slices (see schematic drawing in Fig. 2) have been prepared according to the rules of The Swedish Animal Welfare Agency. Two Wistar rats of age 16 and 17 post-natal days (denoted p16 and p17, respectively) have been used in the experiments. The preparation technique for making the 400 μm thick slices is described in, e.g., [11].

The slices were transported to the measurement lab. During the ~20 min transport the slices were kept in a Ringer solution bubbled by a 95% O_2–5% CO_2 gas mixture. The slice compartments were floating in the solution. The transport system was insulated in order to keep the temperature at ~20 °C.

On arrival to the measurement lab the slices rested for at least an hour before a single slice was transferred to the recording chamber.

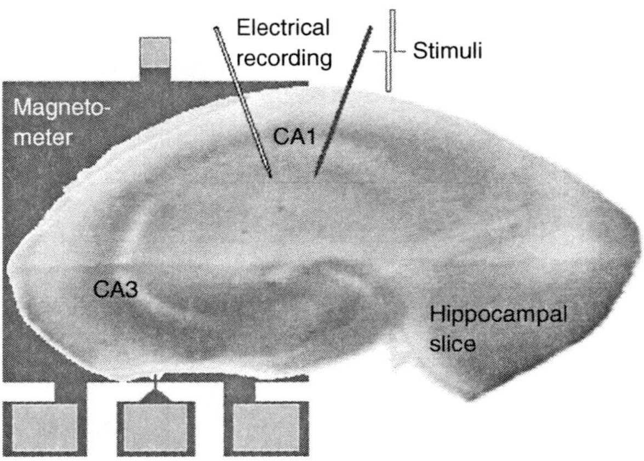

Fig. 2 Transverse hippocampal slice and the experimental setup with the 2 mm magnetometer and the stimulation and recording electrodes. The sizes of the magnetometer and the slice are approximately in correct relation. CA1 and CA3 denotes different parts of the hippocampal formation

A peristaltic pump provided a constant flow (~2 ml/min) of bubbled, room-temperature Ringer solution (see [11] for concentrations). Added to the perfusion Ringer was 0.1 mM picrotoxin (PTX) to block $GABA_A$ receptors and thus to inhibit inhibitory responses.

After a single slice was transferred to the recording chamber, positioned relative to the SQUID sensor, and stabilized by three nylon wires suspended on a Pt grid, it was allowed to settle down for at least 20 min.

The neurons in the CA1 region were activated by stimulating the Schaffer collaterals (see Fig. 2). Stimuli consisted of biphasic 0.1–0.2 ms negative/positive current pulses (20–100 μA) delivered through a tungsten wire electrode. The stimulation frequency was 0.2 Hz.

A glass micropipette filled with 1 M NaCl was inserted as indicated in Fig. 2, i.e. in the stratum radiatum. The recorded potential was amplified 1000 times and low-pass filtered at 3 kHz.

The output of the SQUID electronics was amplified up to 100 times with an EG&G 5113 amplifier and low-pass filtered at 1 kHz to remove the bias-reversal signal.

An NI-6013 data acquisition (DAQ) card was employed to collect the electrical and magnetic signals to the computer. The traces were acquired by a program written in Igor Pro (WaveMetrics, USA) and analyzed off-line in additional Igor Pro programs. The sampling rate was 10 kHz.

3 Results

The recorded EMFs from transverse hippocampal slices are shown in Fig. 3(a–c) and a recording of a excitatory post-synaptic potential (EPSP) is shown in Fig. 3(d). The EMFs are digitally low-pass filtered at 1 kHz and averaged from 500 sweeps (Fig. 3(a,b)) and 100 sweeps (Fig. 3(c)). The SQUID output voltage was amplified 10 and 100 times before the DAQ card for the EMFs in Fig. 3(a–c), respectively. The EPSP is averaged from 10 sweeps.

The magnetometer's noise measured directly after the EMFs in Fig. 3(a) is plotted in Fig. 4. During the noise measurement the slice was in the recording chamber but no stimuli were applied. The SQUID was biased ($I_{\mathrm{bias}} = 237$ μA) at the point with maximum voltage modulation ($\Delta V = 10$ μV). The effective area ($A_{\mathrm{eff}} = 0.071$ mm^2) of the magnetometer was determined by using a Helmholtz coil.

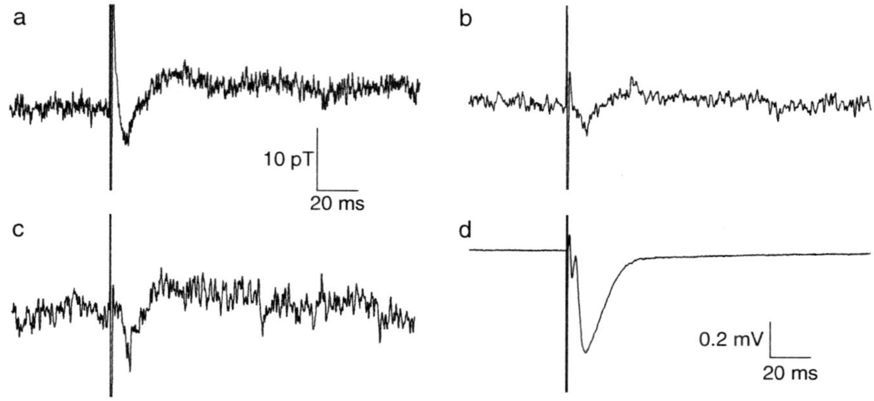

Fig. 3 Recorded evoked magnetic fields (EMFs) and excitatory post-synaptic potentials (EPSPs). The large, sharp signals in each trace are the stimulation artefacts. EMFs (average of 500 sweeps) of slices from a p16 rat (**a**) and a p17 rat (**b**). (**c**) EMF (average from 100 sweeps; p17 rat) measured with the electrophysiological amplifier turned off to exclude cross-talk in the recording chamber. (**d**) EPSP from 10 averaged sweeps from the p17 rat. The high frequency noise of the EMFs is mainly caused by the bias-reversal operation of the SQUID electronics at a frequency of 2 kHz

4 Discussion

To exclude influence of the electrical recording the amplifier was turned off. The characteristics of the EMFs in Fig. 3(b) and (c) are similar. Thus, the EPSP recording seems to have no, or a negligable influence, on the magnetic signal.

We have experienced low level cross-talk between the channels of the DAQ card. Therefore, the EMFs presented here were recorded with the other input short-circuited. The main reason for the cross-talk is the very large difference in signal strength between the two signals. An improvement in the amplification of the SQUID signal will enable simultaneous recordings.

The maximum magnetic field we obtained from the averaged sweeps was $\sim$5 pT with a noise of $\sim$3 pT_{pp}, resulting in an SNR of $\sim$2. The magnetic field noise in the white noise region was $S_B^{1/2} = 0.75\ \text{pT/Hz}^{1/2}$ giving a 42 pT noise floor for the 3 kHz bandwidth used. By improving the filtering of the bias-reversal signal a significant part of the noise would be excluded.

S.-I. Kyuhou and Y. Okada [3] have performed similar experiments on transverse hippocampal slices in a system with 3 mm separation. They obtained maximum fields of 3.5 pT for magnetic fields originating from population spikes of $\sim$2 mV amplitude. In one experiment an EMF of $\sim$3 pT was obtained for an EPSP of $\sim$0.4 mV, which is comparable to our case. In [7] they showed EMFs of $\sim$5 pT at 2 mm separation. In both studies they used transverse slices where the CA1 region was dissected out while we used intact slices, which to some part could explain why the magnetic fields measured at separations of $\geq$2 mm and $\leq$500 μm were of comparable sizes.

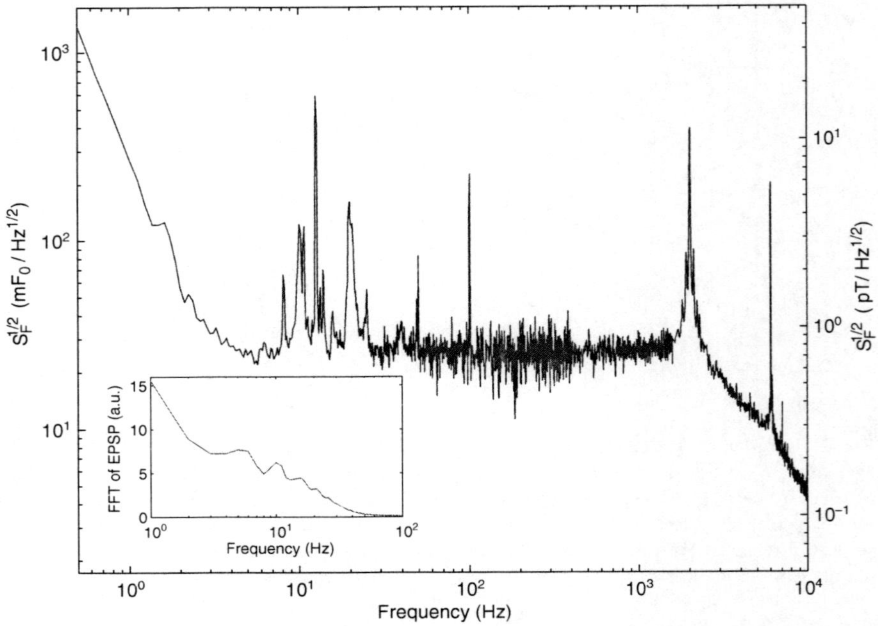

Fig. 4 Flux and field noise of the directly coupled magnetometer measured with a hippocampal slice in the recording chamber. No stimuli were applied during the noise measurement. The inset shows the FFT of the averaged EPSP in Fig. 3d. The magnetic low frequency noise around 10 Hz can be seen in the electrical signal as well and might be attributed to theta oscillations in the slice [10]

The magnetic field of the recorded signals and the SNR are limited by three factors: (i) the present separation between the sensor and the sample, (ii) the temperature of the slice beeing lower than in the rat, and (iii) the noise of the magnetometer, which is not fully optimized.

The separation between the magnetometer and the slice is presently estimated (by using the stereomicroscope) to be $<500\ \mu m$. By making a new window holder we expect to be able to bring the magnetometer less than 100 μm away from the bottom of the slice.

In the rat the temperature is $\geq 35\ ^\circ C$ and the lower temperature we use results in slower and smaller signals compared to slices kept at a physiological temperature [1]. The temperature of the slice chamber is a few degrees below room-temperature, due to the cold nitrogen boil-off which is cooling the interior of the magnetic shields. By mounting a non-metallic heater around the slice chamber the signal is expected to increase slightly.

By further optimizing the magnetometer fabrication a lower magnetic field noise would be achieved. With lower noise a larger bandwidth could be used. However, the neuronal signals are properly resolved with a bandwidth of a few kHz. The bandwidth of the SQUID electronics was intentionally reduced to 3 kHz using a 15 turn feedback coil and 30 kΩ feedback resistor, which results in a mutual inductance of 0.02 Φ_0/μA.

Low frequency $1/f$ noise in the magnetometer was present below 5 Hz. The origin of this noise needs to be investigated to bring the corner frequency down below 1 Hz.

The peaks in the noise spectrum (Fig. 4) around 10 Hz could originate from oscillations in the hippocampal slice. There are well described oscillations in the hippocampus which range between 5–12 Hz [10].

5 Conclusion

In conclusion, evoked magnetic fields of transverse hippocampal slices of rat have been recorded using a high-T_c SQUID magnetometer. Minor improvements of the system should result in superior signal-to-noise ratio compared to conventional low-T_c SQUID systems previously used for magnetophysiology.

Acknowledgments Thomas Schurig, Dietmar Drung, and Sylke Bechstein helped us with replacement SQUID electronics and other electronics issues. We thank Jakob Blomgren for help with some SQUID issues. Staffan Pehrson machined parts to and helped in the development of the system.

This work was supported by the Swedish Foundation for Strategic Research through its OXIDE program.

References

1. Aihara, H., Okada, Y., Tamaki, N.: The effects of cooling and rewarming on the neuronal activity of pyramidal neurons in guinea pig hippocampal slices. Brain Research **893**(1–2), 36–45 (2001)
2. Khapaev, M.M., Kidiyarova-Shevchenko, A.Y., Magnelind, P., Kupriyanov, M.Y.: 3D-MLSI: software package for inductance calculation in multilayer superconducting integrated circuits. IEEE Transactions on Applied Superconductivity **11**(1), 1090–1093 (2001)
3. Kyuhou, S., Okada, Y.C.: Detection of magnetic evoked fields associated with synchronous population activities in the transverse CA1 slice of the guinea pig. Journal of Neurophysiology **70**(6), 2665–2668 (1993)
4. Magnelind, P.: High-T_c SQUIDs for magnetophysiology – development of a magnetometer system and measurements of evoked fields from hippocampal neurons *in vitro*. Ph.D. thesis, Chalmers University of Technology, Göteborg, Sweden (2006)
5. Magnelind, P.E., Ivanov, Z., Tzalenchuk, A.Y., Tarte, E.J., Lõhmus, A.: A high-T_c SQUID based magnetophysiology system. IEEE transactions on applied superconductivity **13**(2), 368–370 (2003)
6. Murakami, S., Hirose, A., Okada, Y.C.: Contribution of ionic currents to magnetoencephalography (MEG) and electroencephalography (EEG) signals generated by guinea-pig CA3 slices. Journal of Physiology **553**(3), 975–985 (2003)
7. Okada, Y.C., Wu, J., Kyuhou, S.: Genesis of MEG signals in a mammalian CNS structure. Electroencephalography and Clinical Neurophysiology **103**(4), 474–485 (1997)

8. Tarte, E.J., Magnelind, P.E., Tzalenchuk, A.Y., Lõhmus, A., Ansell, D.A., Blamire, M.G., Ivanov, Z.G., Dyball, R.E.: High T_c SQUID systems for magnetophysiology. Physica C **368**(1–4), 50–54 (2002)
9. Tesche, C., Krusin-Elbaum, L., Knowles, W.D.: Simultaneous measurement of magnetic and electric responses of in vitro hippocampal slices. Brain Research **462**(1), 190–193 (1988)
10. Vertes, R.P.: Hippocampal theta rhythm: A tag for short-term memory. Hippocampus **15**(7), 923–935 (2005)
11. Wasling, P., Hanse, E., Gustafsson, B.: Developmental Changes in Release Properties of the CA3-CA1 Glutamate Synapse in Rat Hippocampus. Journal of Neurophysiol **92**(5), 2714–2724 (2004)

Dynamical Hysteresis Neural Networks for Graph Coloring Problem

Kenya Jin'no

Abstract We analyze syncronization phenomena in a dynamical hysteresis neural network. The dynamical hysteresis neural network accounts a phase differece in each neuron to be an information. Based on this result, we propose an application for solving 2-colorable graph coloring problems. A system that can treat dynamical information, is attracted to great attention. The reason why such system receives great attention, is that dynamical information processing functions can be found in biological neural networks. Especially, we think that a synchronization phenomenon plays an important role for signal processing in the brain. Our system consists of relaxation oscillators in which contain hysteresis elements. The relaxation oscillator is regarded as a multi-vibrator, namely, the system takes bistable, monostable, and astable state. By using stable and monostable state, we proposed a combinatorial optimization solver that its stable equilibrium point is regarded as an information. On the other hand, various kinds of attractors exist in the hysteresis neural network. We consider that periodic and aperiodic attractors have rich information. For exploiting such attractors, we proposed a novel dynamical associative memory whose information is represented by phase difference. This behavior relates synchronization phenomena. In this article, we will analyze synchronization phenomena in a coupled hysteresis neuron. And, we will propose an application which can dye 2-colorable graph.

1 Itroduction

Recently, a system that can treat dynamical information, is attracted to great attention [1–5]. The reason why such system receives great attention, is that dynamical information processing functions can be found in biological neural networks.

K. Jin'no (✉)
Kanto Gakuin University, 1-50-1, Mutsuura-Higashi, Kanazawa-ku, Yokohama, 236-8501, Japan, e-mail: kjinno@kanto-gakuin.ac.jp

V. In et al. (eds.), *Applications of Nonlinear Dynamics*, Understanding Complex Systems, 331
DOI 10.1007/978-3-540-85632-0_27, © Springer-Verlag Berlin Heidelberg 2009

Especially, we think that a synchronization phenomenon plays an important role for signal processing in the brain. In our previous works, we proposed and analyzed a hysteresis neural network [6, 7]. Our system consists of relaxation oscillators in which contain hysteresis elements. The relaxation oscillator is regarded as a multi-vibrator, namely, the system takes bistable, monostable, and astable state. By using stable and monostable state, we proposed a combinatorial optimization solver that its stable equilibrium point is regarded as an information [6]. On the other hand, various kinds of attractors exist in the hysteresis neural network. We consider that periodic and aperiodic attractors have rich information. For exploiting such attractors, we proposed a novel dynamical associative memory whose information is represented by phase difference [7]. This behavior relates synchronization phenomena. In this article, we will analyze synchronization phenomena in a coupled hysteresis neuron. And, we will propose an application which can dye 2-colorable graph.

2 Hysteresis Neuron

The objective hysteresis neuron [8, 9] is described as

$$\begin{cases} \dfrac{dx(t)}{dt} = -x(t) + e(y(t)), \\ \\ y(t) = h(x(t)) = \begin{cases} +1 & \text{for } x > -1, \\ -1 & \text{for } x < +1, \end{cases} \end{cases} \tag{1}$$

where $x(t) \in \Re$ is a state variable, $y(t) \in \{+1, -1\}$ is an output, and $e(y(t)) \in \Re$ denotes a feedback parameter. $h(x(t))$ is a bipolar piecewise linear hysteresis as shown in Fig. 1.

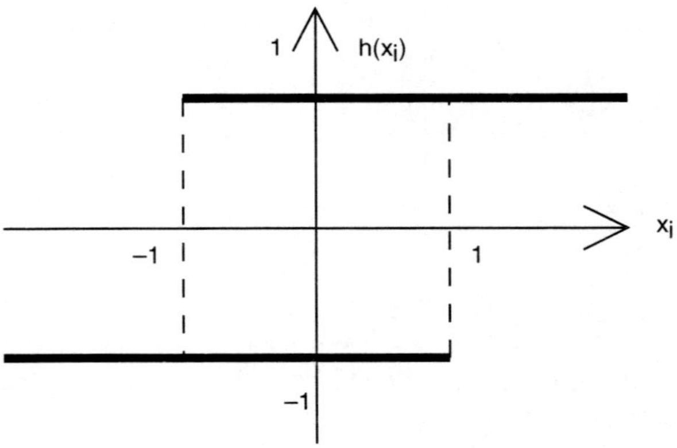

Fig. 1 A normalized bipolar hysteresis. The output is switched from $+1$ to -1 if x_i reaches the left threshold "-1", and vice versa

The state variable of (1) varies toward an equilibrium point that denotes $e(y(t))$. Since the equilibrium point depends on its output, the equilibrium point becomes a constant while the output does not change. If the equilibrium point does not exist on the hysteresis branch, the trajectory reaches the threshold of the hysteresis before it converges to the equilibrium point. When the trajectory hits the threshold of the hysteresis, the output changes its sign, and the corresponding equilibrium point changes too. If the equilibrium point $e(y(t))$ satisfies the following condition, the trajectory can not reach the equilibrium point [9].

$$e(y(t))y(t) \leq -1 \tag{2}$$

In this case, the system must exhibit an oscillatory state [9]. The oscillation period is given as

$$T = \log\left\{\frac{1-e(1)}{-1-e(1)}\right\} + \log\left\{\frac{-1-e(-1)}{1-e(-1)}\right\}. \tag{3}$$

Since the equilibrium point depends on the self feedback, the oscillation frequency can be controlled by the feedback parameter. Figure 2 shows examples of waveforms of the state variable in a hysteresis neuron. The different point between two waveforms is only the self feedback parameter: The feedback parameter is given as $e(y(t)) = -2y(t)$ in Fig. 2(a), and it is given as $e(y(t)) = -4y(t)$ in Fig. 2(b). The oscillation frequencies of Fig. 2(a) and (b) are $F_a = 1/2\log 3$ and $F_b = 1/2\log\frac{5}{3}$, respectively.

Next, we consider the case where some hysteresis neurons are coupled as

$$\tau_i \frac{dx_i(t)}{dt} = -x_i(t) + \sum_{j=1}^{N} w_{ij} y_j(t), \tag{4}$$

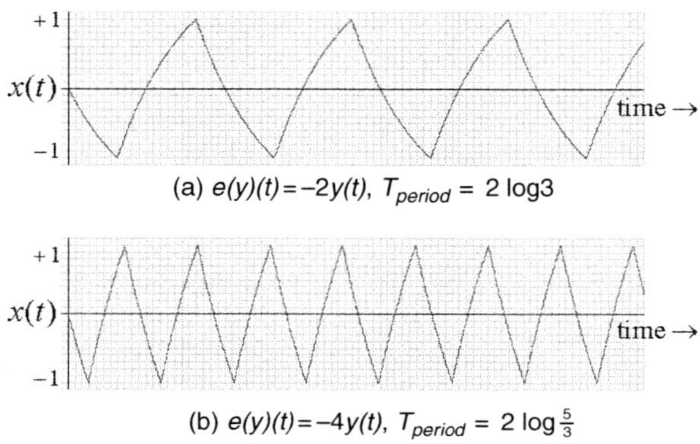

Fig. 2 These waveforms are from a hyteresis neuron. The upper waveform is observed in the case where the feedback parameter is given as $e(y(t)) = -2y(t)$. The oscillation period is $T_{period} = 2\log 3$. The lower waveform is observed in the case where the feedback parameter is given as $e(y(t)) = -4y(t)$. The oscillation period is $T_{period} = 2\log\frac{5}{3}$

where $\tau_i > 0$ is a time constant, N is the number of hysteresis neurons, $w_{ij} \in \Re(i \neq j)$ is a coupling coefficient, and w_{ii} is a self feedback parameter. For simplicity, we assume all time constants have the identical value, hereafter.

The equilibrium point $e_i(y(t))$ is given from Eq. (4) by

$$e_i(y(t)) = \sum_{j=1}^{N} w_{ij} y_j(t). \tag{5}$$

If the following condition is satisfied, the system exhibits oscillatory state [7].

$$e_i(y(t))y_i(t) \leq -1, \exists i. \tag{6}$$

Especially, if the following condition is satisfied, all neurons keep oscillatory state.

$$e_i(y(t))y_i(t) \leq -1, \forall i. \tag{7}$$

Then, the system has no stable output state completely.

We suppose that the connection coefficients satisfy the following relation:

$$\begin{cases} |w_{ii}| > \sum\limits_{j=1, j\neq i}^{N} |w_{ij}|, \\ w_{ii} + \sum\limits_{j=1, j\neq i}^{N} |w_{ij}| < -1, \forall i. \end{cases} \tag{8}$$

Namely, the connection coefficient matrix has a diagonally dominant. If the system adopts such assumption (8), the condition (7) is satisfied. Therefore, this system must exhibit oscillatory state. Figure 3 shows typical waveforms from two neurons system. In this figure, these waveforms correspond to the state variables, x_1 and x_2, respectively. Both results are observed into the following configurations of the initial values.

$$\begin{aligned} (x_1(0), y_1(0)) &= (+1, +1), \\ (x_2(0), y_2(0)) &= (0, +1). \end{aligned} \tag{9}$$

In the case of Fig. 3(a), the configuration of parameters is

$$\begin{pmatrix} w_{11} & w_{12} \\ w_{21} & w_{22} \end{pmatrix} = \begin{pmatrix} -4 & 1 \\ 1 & -4 \end{pmatrix}. \tag{10}$$

The cross-connection coefficient has a positive value. Namely, the coefficient can be regarded as an excitatory connection. In this case, the system exhibits an in-phase synchronization. On the other hand, the parameters configuration of Fig. 3(b) is

$$\begin{pmatrix} w_{11} & w_{12} \\ w_{21} & w_{22} \end{pmatrix} = \begin{pmatrix} -4 & -1 \\ -1 & -4 \end{pmatrix}. \tag{11}$$

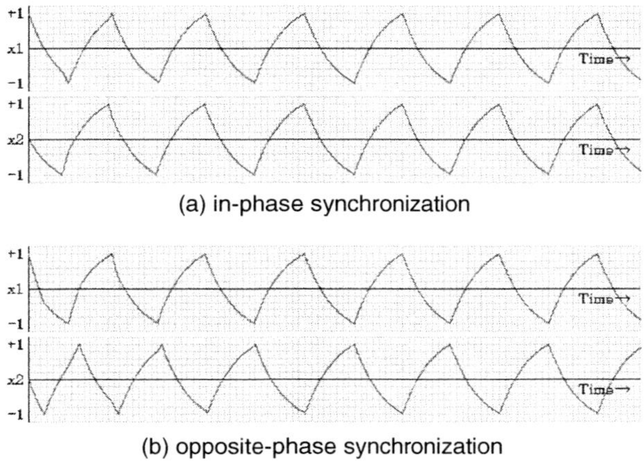

(a) in-phase synchronization

(b) opposite-phase synchronization

Fig. 3 The waveform from two coupled hysteresis neurons. The waveform of (**a**) is observed in the case where the system has an excitatory connection, and (**b**) is observed in the case where the system has an inhibitory connection

The cross-connection coefficient has a negative value, that is an inhibitory connection. In this case, the system exhibits an opposite-phase synchronization. For what has been discussed above, the fundamental frequency is affected by the self feedback parameter. The coupling parameter plays a role as frequency modulation. As the self feedback parameter W_{ii} is -4, its fundamental frequency F_f is $F_f = 1/2\log\frac{5}{3}$ representing in Fig. 2(b). The synchronized oscillation frequency F_s is $F_s = 1/2\log 2$ which is affected by the interaction with another neuron.

3 Analysis

We define a return map for analyzing these synchronization phenomena. Since this system always satisfies (7), a range of the state variable is $[-1,+1]$. Let Φ be a one dimensional domain such as

$$\Phi \equiv \{x_2(T) | |x_2(T)| \leq 1, x_1(T) = 1\}. \tag{12}$$

We can define a return map from Φ to itself:

$$\mathscr{F} : \Phi \mapsto \Phi \tag{13}$$

This mapping can be easily calculate by using our proposed algorithm [6]. Figure 4 shows the return map $\mathscr{F}$. The parameter configuration of Figure 4(a) obeys Eq. (10), and (b) obeys Eq. (11). These return maps contain two fixed points that are indicated as "R" and "A", respectively. The point "R" denotes a repellent fixed point, and "A" denotes an attractive fixed point. All initial values converge to an

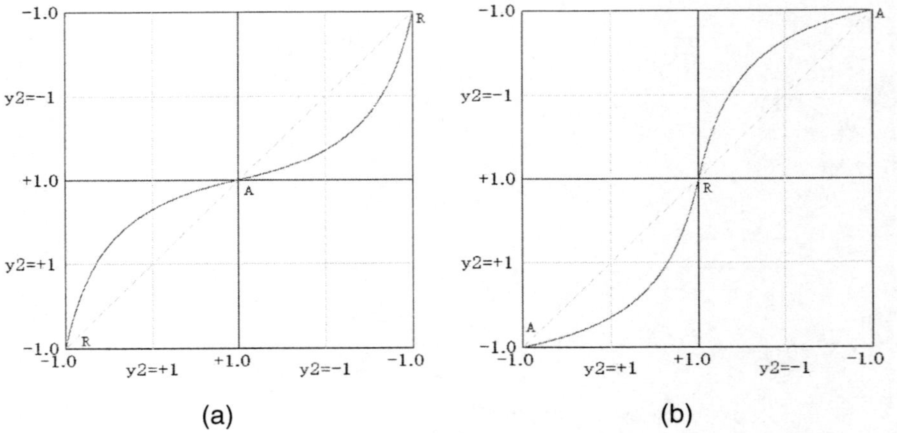

Fig. 4 Return map $\mathscr{F}$ of the two neurons system. (**a**) and (**b**) correspond to the case of Fig 3(**a**) and (**b**), respectively. "A" denotes an attractor, and "R" denotes a repeller

attractive fixed point. In the case of Fig. 4(a), the attractive fixed point is $x_2 = 1$. Based on the definition of this return map, this attractor represents an in-phase synchronization. When the parameters set as (10), all initial values converge to an in-phase synchronized attractor such as Fig. 3(a).

On the other hand, in the case of Fig. 4(b), the attractive fixed point is $x_2 = -1.0$. Based on the definition of this return map, this attractor represents an opposite-phase synchronization such as Fig. 3(b).

It follows from what has been said thus far that the excitatory connection leads to in-phase synchronization, and the inhibitory connection leads to opposite-phase phase synchronization.

4 Graph Coloring Problems

In this section, we investigate a dynamical hysteresis neural network for solving graph coloring problems. The description of graph coloring problems is that a vertex coloring of an undirected graph is an assignment of a label to each node. It is required that a minimum coloring of a graph is a coloring that uses as few different labels as possible. Dr. Wu studied the possibility of coloring graphs by means of synchronized coupled relaxation oscillators [11]. However, this system has some problems that there is a case which the system has a stable equilibrium point, and so on. Also, it is difficult to set the parameters. On the other hand, the behavior of our hysteresis neuron can be easily controlled by its equilibrium point. If the connected neurons have an inhibitory connection, they do not share the same phase. By using this ability, we apply an oscillatory hysteresis neural network for solving

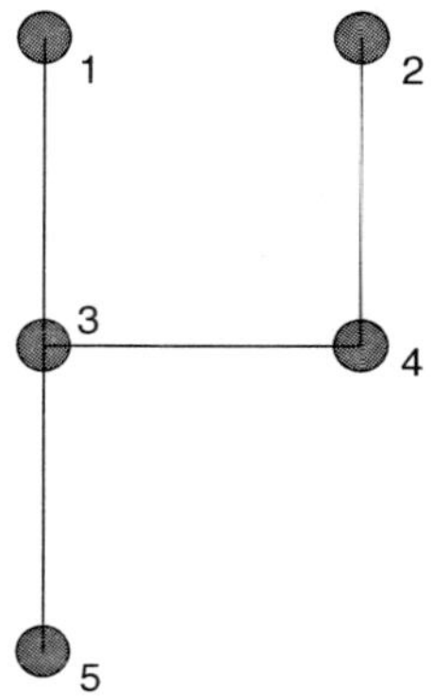

Fig. 5 Example of 2-colorable graph with 5 vertices [11]

graph coloring problems. Each hysteresis neuron is assigned to each vertex. Figure 5 shows a simplest example of 2-colorable graph with 5 vertices which is introduced in Ref. [11]. From Fig. 5, an adjacent matrix A of the graph is given as

$$A \equiv \begin{pmatrix} 0\,0\,1\,0\,0 \\ 0\,0\,0\,1\,0 \\ 1\,0\,0\,1\,1 \\ 0\,1\,1\,0\,0 \\ 0\,0\,1\,0\,0 \end{pmatrix}. \tag{14}$$

where $a_{ij} \in \{0,1\}$ is a component element of an adjacent matrix A. $a_{ij} = 1$ represents connected between i-th vertex and j-th vertex, and $a_{ij} = 0$ represents disconnected edge.

Based on this adjacent matrix, we determine a connection matrix for the oscillatory hysteresis neural network.

$$\begin{cases} w_{ij} = -\alpha \dfrac{a_{ij}}{\sum_{k=1}^{N} a_{ik}}, \text{ for } i \neq j, \\ w_{ii} = -\alpha - \varepsilon \end{cases} \tag{15}$$

where α and ε are positive constants.

For example, an connection matrix W for solving a graph such as Fig. 5, is given as

$$W \equiv \begin{pmatrix} -5 & 0 & -1.2 & 0 & 0 \\ 0 & -5 & 0 & -1.2 & 0 \\ -0.4 & 0 & -5 & -0.4 & -0.4 \\ 0 & -0.6 & -0.6 & -5 & 0 \\ 0 & 0 & -1.2 & 0 & -5 \end{pmatrix}. \tag{16}$$

where $\alpha = 1.2$ and $\varepsilon = 3.8$. In this case, the system exhibits a periodic attractor as shown in Fig. 6. Figure 6 shows waveforms of the state of each neuron, which

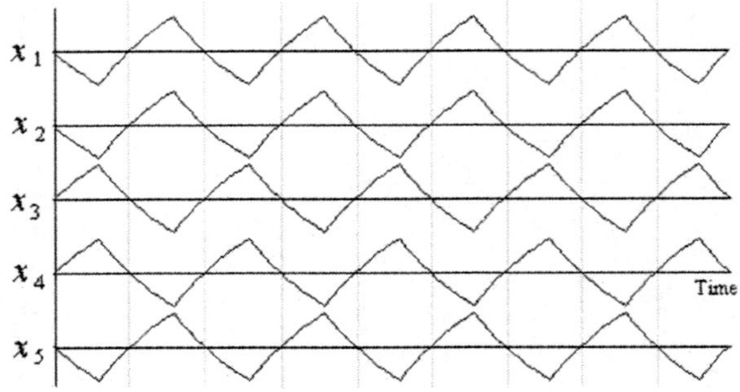

Fig. 6 The oscillatory state corresponds to the solution of 2-colorable graph with 5 vertices. The horizontal axis corresponds to the time evolution, and the vertical axis corresponds to the neuron

corresponds to each vertex. These waveforms can be classified into two clusters by using its phase. One cluster consists of x_1, x_2, and x_5, these exhibit in-phase synchronization. Another cluster contains 3rd and 4th neuron, therefore, these vertices are assigned the same color. Namely, a phase difference in each neuron corresponds to a solution of graph coloring problems. We show another example of 2-colorable graph with 16 vertices as shown in Fig. 7. Figure 8 shows a result which indicates a solution of Fig. 7.

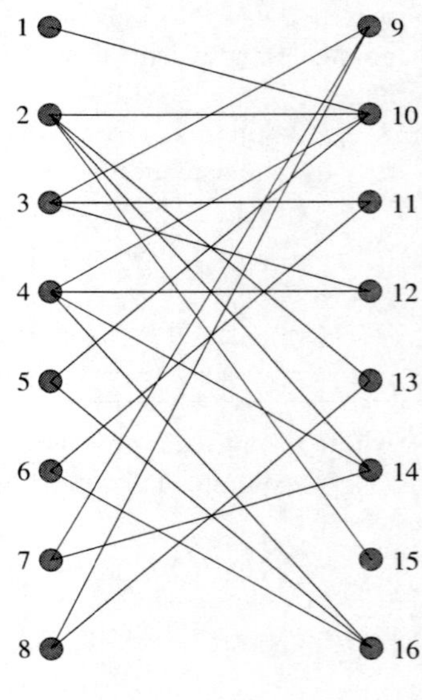

Fig. 7 Example of 2-colorable graph with 16 vertices [11]

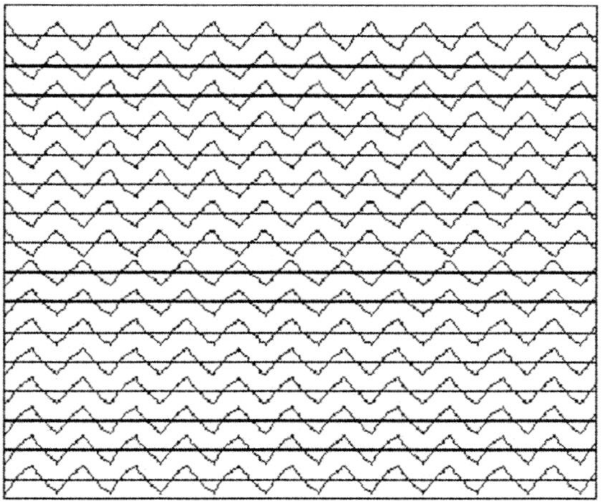

Fig. 8 Waveforms of the inner states of 16 neurons. These waveform can be classified into two kinds which corresponds to coloring graph of Fig. 7

5 Conclusions

In this article, we analyzed the synchronization phenomena in an oscillatory hysteresis neural network. We clarified that this system exhibits an in-phase synchronization when the cross-connection coefficient denotes an excitatory connection. Also, the system exhibits an opposite-phase synchronization when the cross-connection coefficient denotes an inhibitory connection. Based on these results, we proposed an application which can dye 2-colorable graphs. For solving k-colorable graph by proposed dynamical hysteresis neural network, is one of our future problems.

References

1. D.L. Wang, "Emergent Synchrony in Locally Coupled Neural Oscillators," *IEEE Trans. Neural Networks*, vol. 6, no. 4, pp. 941–948, 1995.
2. S. Campbell, and D.L. Wang, "Synchronization and Desynchronization in a Network of Locally Coupled Wilson-Cowan Oscillators," *IEEE Trans. Neural Networks*, vol. 7, no. 3, pp. 541–554, 1996.
3. E.M. Izhikevich, "Weakly Pulse-Coupled Oscillators, FM Interactions, Synchronization, and Oscillatory Associative Memory," *IEEE Trans. Neural Networks*, vol. 10, no. 3, pp. 508–526, 1999.
4. M.B.H. Rhouma, and H. Frigui, "Self-Organization of Pulse-Coupled Oscillators with Application to Clustering," *IEEE Trans. Pattern Anal. and Mach. Intell.*, vol. 23, no. 2, pp. 180–195, 2001.
5. K. Jin'no, and T. Saito, "Analysis and Synthesis of Continuous-Time Hysteretic Neural Networks," *IEICE Trans.*, vol. J75-A, no. 3, pp. 552–556, 1992. (*in Japanese*)

6. T. Nakaguchi, K. Jin'no, and M. Tanaka, "Hysteresis Neural Networks for N-Queens Problems," *IEICE Trans. Fundam.* vol. E82-A, no. 9, pp. 1851–1859, 1999.
7. K. Jin'no, "Oscillatory Hysteresis Associative Memory," in *Proc. Int'l Symp. Circuits Syst., Scottsdale, AZ, USA*, May 2002.
8. M.E. Zaghloul, J.L. Meador and R.W. Newcomb, *Silicon implementation of pulse coded neural networks,* Kluwer Academic Publishers, Norwell, MA, 1994.
9. T. Saito, "Chaos from a Forced Neural-Type Oscillator," *Trans. IEICE*, vol. E73, no. 6, pp. 836–841, 1990.
10. K. Jin'no, T. Nakamura and T. Saito, "Analysis of Bifurcation Phenomena in a 3 Cells Hysteresis Neural Network," *IEEE Trans. Circuits Syst. I*, vol. 46, no. 7, pp. 851–857, 1999.
11. C.W. Wu, "Graph Coloring via Synchronization of Coupled Oscillators," *IEEE Trans. Circuits Syst. I*, vol. 45, no. 9, pp. 974–978, 1998.

Semiconductor Laser Dynamics for Novel Applications

Jia-Ming Liu

Abstract The nonlinear dynamics of semiconductor lasers have been thoroughly studied in recent years. These complex dynamical characteristics, which range from narrow-linewidth oscillation to broadband chaos, open up the opportunity for many novel applications, such as chaotic communications, chaotic lidar, chaotic radar, photonic microwave generation, dual-frequency precision radar/lidar, all-optical AM-to-FM conversion, and single-sideband radio-over-fiber transmission. The utilization of the rich nonlinear dynamics of single-mode semiconductor lasers for these interesting applications is discussed in this paper.

1 Introduction

The nonlinear dynamics of semiconductor lasers generate diverse optical and microwave waveforms [1, 2] that can have many interesting novel applications. In the past ten years, most of the attention for the applications has been focused on the chaotic states for synchronized secure communications and cryptography [3,4]. Nevertheless, many useful applications do not require synchronization, and each dynamical state has its specific characteristics that can be explored for unique applications. For example, the novel concepts of chaotic lidar and chaotic radar have been demonstrated by utilizing the broadband characteristics of the chaotic states without synchronization [5, 6]. Even the simplest state of stable locking under strong optical injection exhibits desirable high-frequency modulation characteristics, such as bandwidth enhancement, chirp reduction, and noise reduction for optical communications applications [7–9]. All other dynamical states have their own useful applications as well. By taking specific advantage of the unique features of each dynamical state, these applications are all unique and advantageous in certain ways.

J.-M. Liu (✉)
Electrical Engineering Department, University of California, Los Angeles, CA 90095, USA
e-mail: liu@ee.ucla.edu

V. In et al. (eds.), *Applications of Nonlinear Dynamics*, Understanding Complex Systems, 341
DOI 10.1007/978-3-540-85632-0_28, © Springer-Verlag Berlin Heidelberg 2009

Because there are many different perturbation methods, each generating many different dynamical states even just with a single-mode laser, there are basically so many possible novel applications waiting to be explored and developed. This paper focuses only on the chaotic and period-one states of an optically injected single-mode semiconductor laser to show the diversity of such novel applications.

2 Nonlinear Dynamics of Semiconductor Lasers

A laser is inherently a nonlinear device because it has a threshold above which the population inversion in its gain medium is clamped. For any laser, its gain medium provides the basic nonlinearity; additional nonlinearity can come from other elements in the laser cavity. For a laser to exhibit complex nonlinear dynamics, however, the mere existence of nonlinearity is necessary but not sufficient. A semiconductor laser is a class B laser. A single-mode class B laser does not have sufficient degrees of freedom to enter a complex dynamical state such as chaos. Thus, it requires a suitable external perturbation to provide additional degrees of freedom for a single-mode semiconductor laser to have complex dynamics. While many different perturbation schemes are possible, the most well-studied are optical injection, optical feedback, and optoelectronic feedback. The same single-mode semiconductor laser exhibits very different nonlinear dynamical characteristics under the three different perturbation schemes. Under optical injection, it has high-frequency dynamics that follow a period-doubling route to chaos under optical injection; its dynamics can be controlled by the injection strength and the injection frequency detuning. Under optical feedback, it has high-frequency dynamics that follow either a period-doubling route or a quasi-periodicity to chaos but it also has low frequency fluctuation dynamics when operated near laser threshold; its dynamics can be controlled by the feedback strength and the feedback delay time. Under optoelectronic feedback, it has high-frequency pulsing dynamics that follows a quasi-periodicity route to chaotic pulsing.

The nonlinear dynamics of semiconductor lasers have a few salient features that can be exploited for many novel, useful applications. First, the large number of different dynamical states, each having its characteristic waveform and spectrum, allows the generation of a wide variety of diverse waveforms using a single semiconductor laser in different perturbation schemes [1]. Such waveforms include a narrow-band microwave frequency that is continuously tunable over a broad frequency range, precisely locked harmonic and subharmonic frequencies, broad-band chaotic waveforms, regularly pulsing waveforms, quasi-periodic pulsing waveforms, and chaotic pulsing waveforms, to name just a few. As an example, Fig. 1 shows the experimentally recorded spectra, temporal characteristics, and temporal auto-correlation characteristics of representative broad-band chaotic oscillation (CO), period-one oscillation (P), chaotic pulsing (CP), and quasi-periodic (Q) waveforms. Second, even in a fixed setting, with a given semiconductor laser in a given perturbation scheme, many different waveforms can be generated by varying

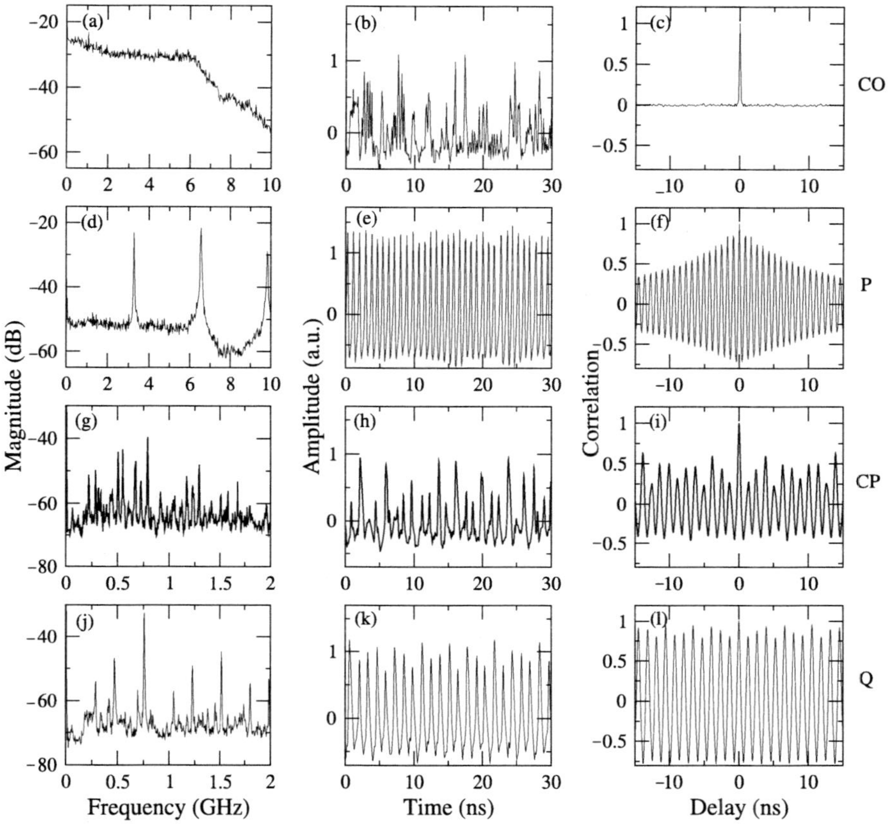

Fig. 1 Experimentally recorded spectra, temporal characteristics, and temporal auto-correlation characteristics of representative broad-band chaotic oscillation (CO), period-one oscillation (P), chaotic pulsing (CP), and quasi-periodic (Q) waveforms

the operating condition to vary the dynamical state through the bifurcation route. Fast switching between different waveforms, for example between a narrow-band single frequency waveform and a broad-band chaotic waveform, can be easily accomplished by switching between two targeted operating conditions. In most situations, it simply involves a change of the magnitude of a current injected to a laser. Third, all of these waveforms are microwave waveforms on an optical carrier that is the emission of the semiconductor laser. Thus, any optical wavelength, ranging from the infrared to the ultraviolet, that is available from semiconductor lasers can be selected to suit purpose of an application by selecting a semiconductor laser emitting at the desired wavelength. Fourth, these waveforms can be amplified, transmitted, detected, and handled as optical signals in certain applications. Alternatively, by first removing the optical carrier, they can be amplifies, radiated, detected, and processed as microwave waveforms in other applications.

All of these dynamical states can be synchronized between two properly matched and coupled single-mode semiconductor lasers, resulting in synchronized, identical waveforms emitted by separate lasers. Some applications, such as synchronized chaotic communications, required synchronized lasers, but most applications do not involve the use of two synchronized lasers. See Ref. [10] for a detailed review of the dynamics and synchronization of single-mode semiconductor lasers.

3 Applications of Chaotic Dynamics

A chaotic waveform has the characteristics of noise-like but deterministic temporal variations, a broad spectral bandwidth, and a narrow correlation peak with a small peak-sidelobe ratio (PSL) defined as the ratio of the maximum sidelobe to the peak [1]. These features can be seen in the characteristics of chaotic oscillation and chaotic pulsing shown in Fig. 1. A single-mode semiconductor laser can generate chaotic oscillation waveforms under optical injection or optical feedback; it can also generate chaotic pulsing waveforms under optoelectronic feedback. The fundamental difference between chaos and noise is that chaos is deterministic. Therefore, two properly matched and coupled chaotic oscillators can be synchronized. Two synchronized semiconductor lasers generate identical waveforms in synchronism. For a particular application, it is only necessary to exploit some, but not all, of these special characteristics of chaos. Three unique applications, namely chaotic optical communications [11, 12], chaotic radar (CRADAR) [5], and chaotic lidar (CLIDAR) [6], of the chaotic dynamics of semiconductor lasers are reviewed in the following.

3.1 Chaotic Optical Communications

Chaotic communications based on the transmission of messages encoded on chaotic waveforms have been a subject of great current interest, attracting increasingly intensive research activities in the past ten years [3, 4]. Potential benefits of chaotic communications include efficient use of the bandwidth of a communication channel, utilization of the intrinsic nonlinearities in communication devices, large-signal modulation for efficient use of the carrier power, reduced number of components in a system, and security of communications through chaotic encryption. Chaotic optical communications use fast chaotic optical waveforms to transmit messages at high bit rates over optical carriers [11, 12].

The example illustrated here uses the chaotic pulses generated by a semiconductor laser with optoelectronic feedback as the encrypting chaotic waveform and additive chaos modulation [11] as the message encoding method. As is shown in Fig. 2, the transmitter laser, which has an optoelectronic feedback loop, is

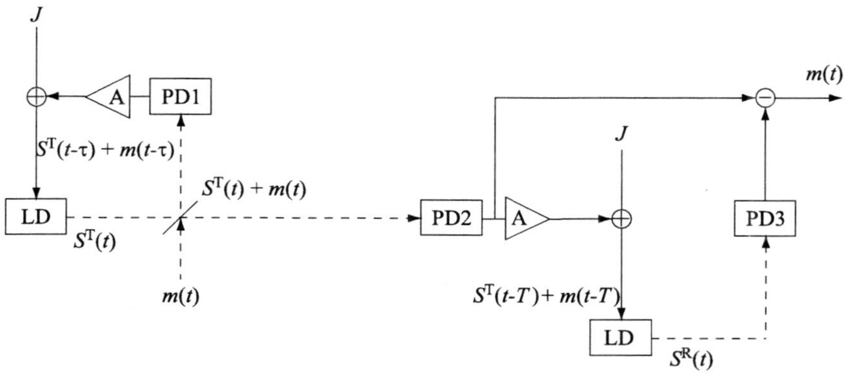

Fig. 2 Schematic setup of the chaotic communication system with chaotically pulsing semiconductor lasers for encoding and decoding messages through additive chaos modulation. LD: Laser Diode; PD: Photodetector; A: Amplifier

operated in a chaotic pulsing state at a properly adjusted the feedback delay timeτ. The receiver laser operates in an open-loop configuration and is driven by the signal from the transmitter. Message $m(t)$ is encoded by means of incoherent addition onto the output of the transmitter laser as $S^{\mathrm{T}}(t)+m(t)$. When $m(t)$ is encoded onto $S^{\mathrm{T}}(t)$ and the combined signal is sent to the receiver laser, it is also fed back to the transmitter laser. Therefore, the transmitter laser is driven by $S^{\mathrm{T}}(t-\tau)+m(t-\tau)$ because of the delay time in the feedback loop, and the receiver laser is driven by $S^{\mathrm{T}}(t-T)+m(t-T)$ because of the time delay in transmission, where T is the transmission time. With parameters matched between the transmitter and the receiver lasers, the receiver is synchronized to the transmitter with a time shift $T-\tau$. Therefore, the receiver laser output is $S^{\mathrm{R}}(t)=S^{\mathrm{T}}(t-T+\tau)$, and the message is recovered as $S^{\mathrm{T}}(t-T+\tau)+m(t-T+\tau)-S^{\mathrm{R}}(t)=m(t-T+\tau)$.

Figure 3 shows the recovery of a pseudorandom sequence at a bit rate of 2.5 Gb/s that is encoded and transmitted by the transmitter laser [13]. With the transmitter and the receiver steadily synchronized, the message is successfully received and decoded by the receiver laser. The top trace is the received signal, measured at PD2 in Fig. 2 and shifted by a time τ. The second trace, measured at PD3, is the local receiver laser output, which synchronizes with and duplicates the chaotic pulse output of the transmitter laser. The time shift between the transmitter and the receiver has been matched. Shown in the third trace, signal recovery is achieved by subtracting the receiver laser output (the second trace) from the received signal (the top trace). A decision threshold can be set at the position of the dashed line. The recovered signal shows very good quality of decoding compared with the original signal, which is also shown in the bottom trace as a reference.

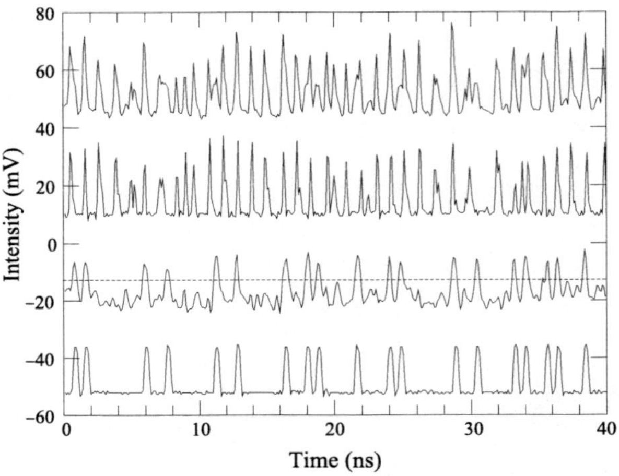

Fig. 3 Transmission of a pseudorandom NRZ bit sequence at a bit rate of 2.5 Gb/s in a chaotic communication system using semiconductor lasers with optoelectronic feedback. Time series of received signal (*top*), receiver laser output (*second*), recovered signal (*third*), and encoded signal (*bottom*)

3.2 Chaotic Lidar

Chaotic lidar (CLIDAR) [6] uses the broadband and single-peak correlation characteristics of a chaotic waveform for optical detection, ranging, and imaging purposes. The CLIDAR demonstrated in our laboratory, shown in Fig. 4, uses the nonlinear dynamics of a semiconductor laser to produce a chaotic waveform carried on an optical beam. The waveform has a very broad bandwidth that is easily generated and easily amplified by an optical amplifier. The system is capable of accurately measuring the range to a remote target. Combining the ranging capability with an imaging technique in the lateral plane, the system can be used as a three-dimensional imaging lidar.

As is shown in Fig. 4, the target and reference optical signals can be first detected with two separate photodetectors and then electronically correlated. In this approach, the range resolution is limited by both the bandwidth of the chaotic waveform and that of the electronic assembly including the photodetectors, the electronic amplifiers, and the real-time oscilloscope that records and correlates the signals. Figure 5(a) shows the experimental results of ranging a target that is translated for 50 cm in the line of sight. From the separation between the correlation peaks shown in Fig. 5(a), the relative range difference is measured to be 49.5 cm showing a subcentimeter accuracy in ranging. Moreover, with a 0.2-ns FWHM of the cross-correlation peak, a 3-cm range resolution is achieved, which in this case is limited by the bandwidth of the oscilloscope used but not by the bandwidth of the chaotic waveform or that of the photodetectors. To fully utilize the broad bandwidths of

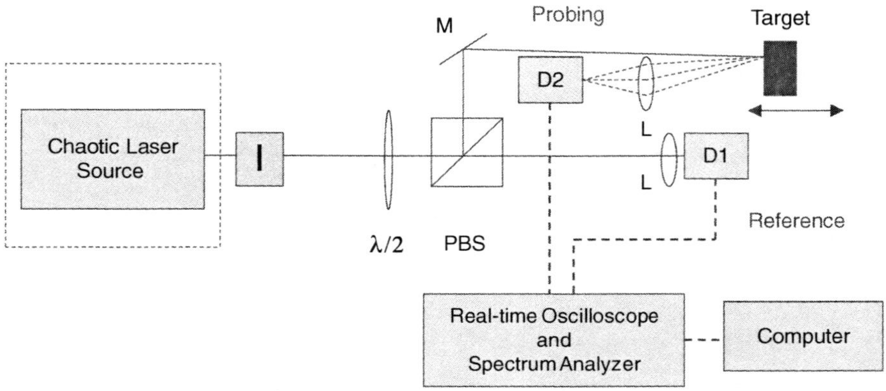

Fig. 4 Schematic setup of a CLIDAR system using semiconductor laser chaos. I: isolator. L: lens. M: mirror. D1, D2: detectors. A: amplifiers. $\lambda/2$: half-wave plate. PBS: polarizing beam splitter

chaotic waveforms and to eliminate the bandwidth limitation from the electronics, a CLIDAR system can optically extract the range information by utilizing a Michelson interferometer that interferes the signal light with the reference light optically. Figure 5(b) shows the result of measuring a target that is located at about 1 m away and is translated for 10.0 mm in the line of sight. As can be seen, a distance of 10.0 mm is measured from the separation of the peaks that has a submillimeter accuracy, while a 5.5-mm range resolution is achieved deriving from the FWHM of the peaks.

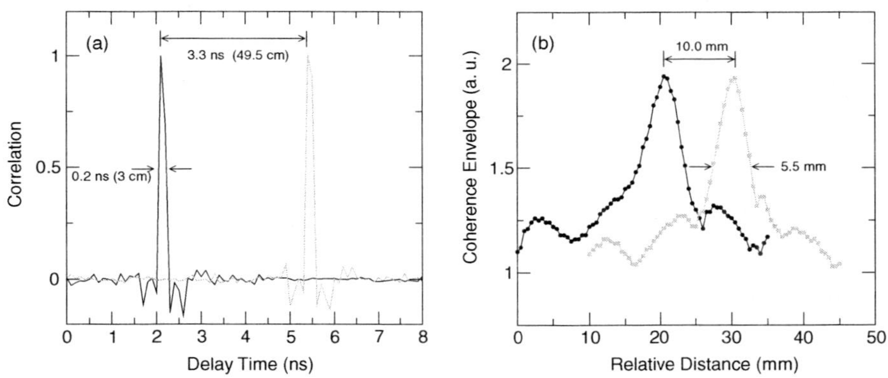

Fig. 5 (**a**) Electronic cross-correlation traces of a target translated for 50.0 cm in the line of sight. (**b**) Optical cross-correlation traces of a target translated for 10.0 mm in the line of sight

3.3 Chaotic Radar

Similar to CLIDAR, chaotic radar (CRADAR) [5, 14–16] also uses the broadband and single-peak correlation characteristics of a chaotic waveform for detection, ranging, and imaging purposes, except that it uses a broadband microwave chaotic waveform that is not carried by an optical carrier. The chaotic waveform can be generated electronically through microwave chaos [14, 15], but microwave chaotic waveforms of much lager bandwidths up to 100 GHz can be generated indirectly through semiconductor laser chaos. The laser-based CRADAR demonstrated in our laboratory uses an optically injected semiconductor laser to generate the chaotic waveform, as shown in Fig. 6. The microwave chaotic waveform is generated through removal of the optical carrier by a broadband photodetector. The microwave chaotic waveform is then amplified and split into two channels, the probe and the reference. The probing signal is transmitted and received by a pair of antennas. Together with the reference waveform, these waveforms are recorded simultaneously with an RF spectrum analyzer and a real-time oscilloscope, where target detection and localization are accomplished through crosscorrelation.

A semiconductor laser-based CRADAR system is capable of delivering a ultra-wide bandwidth up to 100 GHz. Unlike in the case of CLIDAR, the chaotic waveform is radiated as a baseband signal without a high-frequency carrier. At present time, however, the bandwidth of laser-based CRADAR is fundamentally limited by the bandwidths of photodetectors, microwave amplifiers, and most significantly the antennas. Nevertheless, even after significant reduction of the bandwidth after the chaotic waveform is filtered by the bandwidth-limited microwave components, a laser-based CRADAR system still has performance characteristics superior to those of other broadband radar systems, including noise radar. Figure 7 shows (a) the bandwidth-limiting effect imposed by an antenna that has a bandwidth of 1.5–3 GHz and (b) the sharp, single-peak correlation feature of a experimental CRADAR system based on an optically injected semiconductor laser. The chaotic waveform generated by the laser has a bandwidth of 35 GHz, but the reference signal shown in Fig. 7(a) has a bandwidth of only 6 GHz limited by that of the photodetector. The

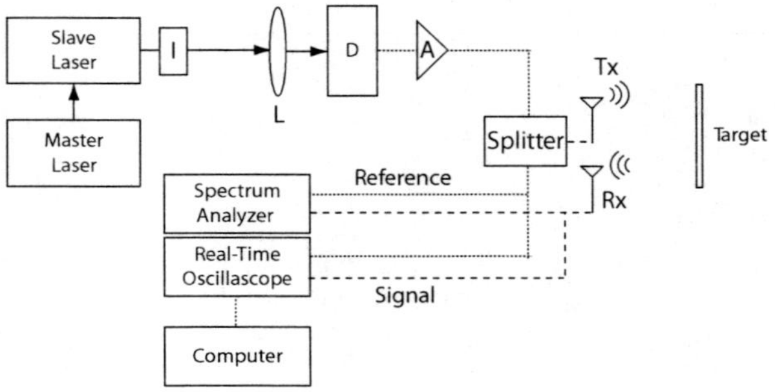

Fig. 6 Schematic setup of a semiconductor-laser-based CRADAR system. I: isolator. L: lens. D: photodetector. A: amplifiers. Tx, Rx: transmitter and receiver antennas

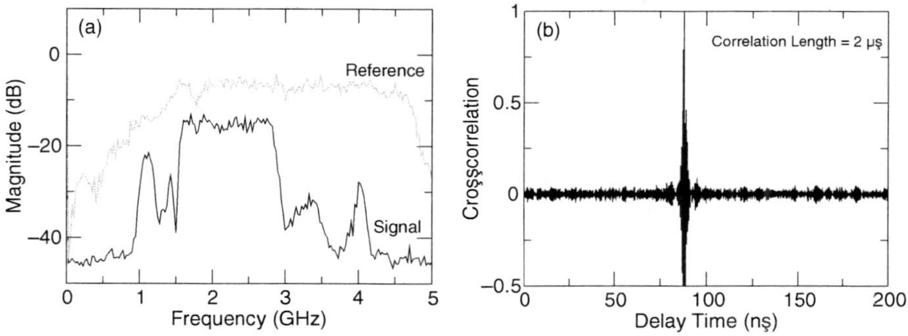

Fig. 7 (**a**) Power spectra and (**b**) the cross-correlation trace of the signal and reference waveforms. The FWHM of the correlation peak is measured to be 0.6 ns, giving a range resolution of 9 cm

radiated and received signal has an even smaller bandwidth limited by the antennas. The correlation trace clearly shows an unambiguous single narrow peak with a range resolution of 9 cm and a low PSL of -27 dB for a correlation time length of 2 μs. Much better performance figures can be accomplished if the bandwidths of the microwave components are improved.

4 Applications of Periodic Dynamics

A semiconductor laser can have many different periodic states. Each type of periodic state, such as the period-one state, can further have many different characteristics, depending on the different perturbation schemes used to excite the periodic oscillation and also on the specific operating conditions under a given perturbation scheme. Here we focus on the period-one state of an optically injected semiconductor laser to demonstrate that even a simple dynamical state like the period-one state can have many exciting new applications [17]. Figure 8 shows the schematics of an optically injected semiconductor laser and its spectral characteristics in the period-one state. In the period-one state, the slave laser is locked by a strong injection signal to oscillate at the injection frequency ν_i with period-one sidebands at frequencies $\nu_\mathrm{i} - f_0$ and $\nu_\mathrm{i} + f_0$. The optical spectrum of the period-one state shown in Fig. 8

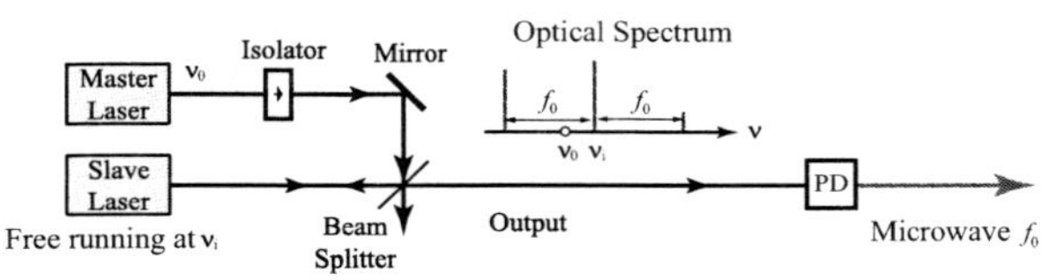

Fig. 8 Schematics of an optically injected semiconductor laser and its spectral characteristics in the period-one state

illustrates a few significant features that are important for the applications discussed below: (1) The microwave frequency f_0 is a large-signal modulation on the optical carrier, and f_0 is broadly tunable. (2) The optical spectrum is highly asymmetric, with the low-frequency sideband at $\nu_i - f_0$ much stronger than the high-frequency sideband at $\nu_i + f_0$. (3) Under a proper operating condition, the high-frequency sideband can be suppressed to a negligibly low level, while the low-frequency sideband can be enhanced to have a magnitude equal to that of the central frequency at ν_i.

4.1 Photonic Microwave Generation

The period-one oscillation generates a laser output that contains a microwave modulation on the optical carrier. Compared to other methods of photonic microwave generation based on semiconductor lasers, such as direct modulation, external modulation, mode locking, self pulsation, heterodyne, and optical phase-lock loop, this technique has several significant advantages [17]. It is an all-optical microwave generation technique that avoids the limitations of microwave electronics and electrical parasitics. Yet, the system is electronically controlled through the DC injection currents to the master and slave lasers. The microwave frequency is broadly tunable by varying the optical injection strength and/or the detuning frequency, $\nu_{rmi} - \nu_0$, between the master and slave lasers. A microwave frequency that is continously tunable from the order of the natural relaxation resonance frequency of the slave laser at its operating point to well beyond six times the natural relaxation resonance frequency can be generated [2, 18]. For an ordinary semiconductor laser, this covers a tunable range from a few GHz to about 30 GHz. For a high-speed semiconductor laser, tunable photonic microwave frequencies up to 100 GHz can be generated [18]. The spectrum seen in Fig. 8 illustrates the fact that this microwave is a single-sideband large-signal modulation on the optical carrier, a characteristic that is very difficult to achieve with any other photonic microwave generation techniques. Indeed, the modulation depth can be close to 100%. Furthermore, the microwave power remains almost constant as its frequency is tuned throughout the entire frequency-tuning range [18]. The linewidth of the photonic microwave generated by the injection scheme shown in Fig. 8 is determined by the optical linewidth of the injection-locked laser, which is typically on the order of 10–200 MHz if a free-running master laser is used. This linewidth can be significantly reduced by three orders of magnitude to the order of 10–200 kHz by adding a passive optoelectronic feedback loop to the slave laser [2]. By slightly modulating the slave laser with an external microwave source, this microwave can be double-locked [19] to have a linewidth as narrow as that of the modulating microwave; a linewidth much less than 1 kHz has been demonstrated [2].

4.2 Dual-Frequency Multifunction Precision Lidar

As discussed above and shown in Fig. 8, the period-one state can have an optical spectrum that consists of basically two optical lines of equal magnitude at $\nu_i - f_0$ and ν_i under proper operating conditions when the high-frequency sideband at $\nu_i + f_0$ is suppressed. As also discussed above, the microwave frequency f_0 can be locked by an external microwave source to a very narrow linewidth less than 1 kHz while the optical linewidth for both optical lines at $\nu_i - f_0$ and ν_i remain large at the order of 10–200 MHz. In this situation, the slave laser has an output of two coherently locked optical lines, which can be used as the optical source for a novel dual-frequency multifunction precision lidar. This dual-frequency lidar has very different characteristics and applications from those of the chaotic lidar discussed above. It can be used as a high-precision Doupler lidar to monitor the movement of a remote target [20]. In this application, the frequencies of both optical lines are proportionally shifted after reflection from a moving target, resulting in a shifted microwave frequency of f_0' between them. The velocity of the moving target can be deduced from the difference $f_D = f_0 - f_0'$. This system can be viewed as a microwave Doppler system, except that the microwave is carried by an optical wave. Doppler measurements depend only on the optical frequency difference, not on the optical frequencies themselves. Because of the narrow linewidth of f_0, the Doppler frequency shift f_D can be measured very accurately even though the linewidths of both optical lines are large. A record low velocity of 26 μm/s has be experimentally measured using this system [20]. The target range for accurate velocity measurement also depends on the microwave stability rather than the optical frequency stability. The narrow microwave linewidth allows long-range measurements of more than 24 km. This system is also capable of measuring the absolute distance to a target [17]. The range information is contained in the microwave phase noise spectrum of the mixer output. It appears as a periodic microwave interference structure separated by the reciprocal of the time of flight. The residual phase noise of the system enables distance measurement. Mixing the reference and target signals essentially measures the time of flight of the phase noise. Furthermore, the system can be used as a differential absorption lidar. The optical frequencies can be tuned "on" and "off" the absorption lines of the molecule to be measured. By detecting the reference and target signals simultaneously, the "on" and "off" backscatter information can be used to deduce the molecular concentration in the probed volume.

4.3 All-Optical AM-to-FM Conversion

The period-one microwave oscillation frequency f_0 of an optically injected semiconductor laser varies linearly with the strength of the optical injection over a wide range of operating parameters [21]. As a result, when a semiconductor laser is injected by an AM baseband optical signal under a condition that make it to oscillate in a period-one state, it generates a corresponding FM microwave signal. The

FM microwave thus obtained has a wide tunability and a large modulation bandwidth [21, 22]. If the optical injection signal is modulated with AM data, such AM data will be directly converted to FM data modulated on the output of the injected laser. Thus, all-optical conversion of an AM baseband signal into a FM microwave signal, both modulated on an optical carrier of the same frequency, is accomplished [23]. This optical AM-to-FM conversion scheme has a few significant features: First, it is all optical, thus circumventing any additional bandwidth limitation, signal distortion, electronic noise, and power consumption. The optical injection system is capable of high frequency operation with a very large bandwidth. Second, it has an enormous bandwidth. By varying the optical injection strength or the detuning frequency, the microwave frequency can be varied widely from 10 GHz to at least 100 GHz. The modulation depth of the output FM increases linearly with that of the input AM; a maximum frequency deviation of more than 10 GHz can be achieved [21]. According to the broad frequency response of the system from DC to more than 1 GHz, the data can contain low-frequency components (or long strings of zeros), and the data rate can be increased. Third, as discussed above, the period-one output of the injected laser can be a single sideband modulation, or at least the high-frequency sideband is always significantly suppressed. Thus, an ordinary double-sideband AM optical signal is converted into a single-sideband FM optical signal. Such a single-sideband FM optical signal has many advantages over a double-sideband AM optical signal in long-distance optical-fiber transmission of the signals [18].

4.4 Single-Sideband Radio-Over-Fiber

In a radio-over-fiber (RoF) system, microwave signals are carried on an optical wave, which is transmitted through optical fibers between the base stations and a central office. The expensive microwave components are centralized at the central office, while the base station is merely responsible for direct optoelectronic conversion. The RoF wireless networks would eventually be connected to the conventional optical communication systems in order to utilize the established optical network. However, the signal frequency bands and the modulation schemes of the two networks are very different. In a RoF system, the optical wave carries a microwave subcarrier that is usually FM for effective wireless broadcasting. In a conventional optical communication system, the optical wave normally carries baseband AM data. The interface of the two systems thus needs an upconversion from the baseband to the microwave band while performing AM-to-FM transformation simultaneously. Furthermore, to reduce the power penalty while transmitting the FM signal through an optical fiber, it is desirable that the signal is carried on a single sideband [18, 22]. All of these desirable characteristics can be accomplished by implementing the optical AM-to-FM conversion discussed above at the central office for an RoF system [23]. Figure 9 illustrates the concept of such an RoF system

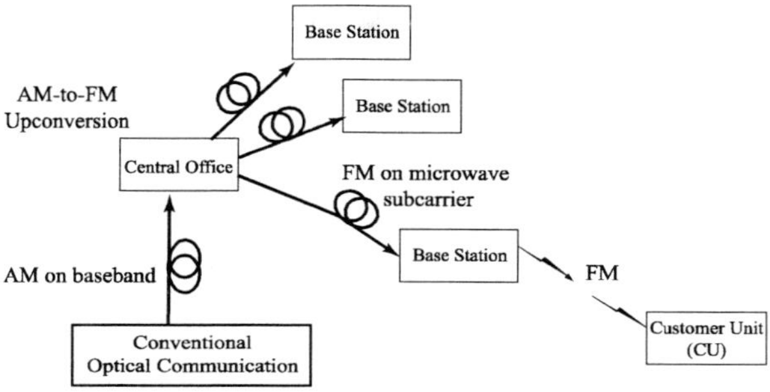

Fig. 9 Schematics of a single-sideband RoF system utilizing all-optical AM-to-FM conversion based on the period-one dynamics of an optically injected semiconductor laser

in which the central office utilizes all-optical AM-to-FM conversion based on the period-one dynamics of an optically injected semiconductor laser.

5 Conclusions

Semiconductor lasers have rich, diverse nonlinear dynamical characteristics ranging from simple period oscillations to chaotic pulsing. After years of detailed studies by many groups in this field, these nonlinear dynamics are well understood and can be precisely controlled in operation. Therefore, it is now possible to take advantage of these dynamical characteristics and their diverse optical and microwave waveforms for many interesting and useful photonic applications. The main purpose of this paper is to illustrate such opportunities through some examples of novel applications on only two states, namely the chaotic and the period-one states, that have been realized. These examples merely demonstrate some possibilities. There are many other possibilities for these two states. And there are many applications for other dynamical states that are not discussed in this paper. The rich nonlinear dynamics of semiconductor lasers clearly offer a wealth of possibilities for novel applications waiting to be explored and developed.

References

1. F.Y. Lin and J.M. Liu, "Diverse waveform generation using semiconductor lasers for radar and microwave applications," IEEE J. Quantum Electron. **40**, 682–689 (2004).
2. S.C. Chan and J.M. Liu,"Tunable narrow-linewidth photonic microwave generation using semiconductor laser dynamics," IEEE J. Sel. Top. Quant. Electron. **10**, 1025–1032 (2004).
3. L. Kocarev, G.M. Maggio, M. Ogorzalek, L. Pecora and K. Yao, *Special Issue on Applications of Chaos in Modern Communication Systems*, IEEE Trans. Circuits Syst. I **48**(12), (Dec. 2001).

4. S. Donati and C. Mirasso, *Feature Section on Optical Chaos and Application to Cryptography*, IEEE J. Quantum Electron. **38**(9), (Sep. 2002).
5. F.Y. Lin and J.M. Liu, "Chaotic radar using nonlinear laser dynamics," IEEE J. Quantum Electron. **40**, 815–820 (2004).
6. F.Y. Lin and J.M. Liu, "Chaotic lidar using laser chaos," IEEE J. Sel. Top. Quantum Electron. **10**, 991–997 (2004).
7. T.B. Simpson and J.M. Liu,"Enhanced modulation bandwidth in injection-locked semiconductor lasers," IEEE Photon. Technol. Lett. **9**, 1322–1324 (1997).
8. Y. Okajima, S.K. Hwang, and J.M. Liu, "Experimental observation of chirp reduction in bandwidth-enhanced semiconductor lasers subject to strong optical injection," Optics Commun. **219**, 357–364 (2003).
9. S.K. Hwang, J.M. Liu, and J.K. White,"35-GHz intrinsic bandwidth for direct modulation in 1.3-μm semiconductor lasers subject to strong injection locking," IEEE Photon. Technol. Lett. **16**, 972–974 (2004).
10. J.M. Liu, H.F. Chen, and S. Tang, "Dynamics and synchronization of semiconductor lasers," in *Digital Communications Using Chaos and Nonlinear Dynamics*, ed. by L.E. Larson, J.M. Liu, and L.S. Tsimring (Springer, New York, 2006), Chapter 10.
11. J.M. Liu, H.F. Chen, and S. Tang, "Synchronized chaotic optical communications at high bit rates," IEEE J. Quantum Electron. **38**, 1184–1196 (2002).
12. S. Tang, H.F. Chen, and J.M. Liu, "Performances of synchronized chaotic optical communication systems," in *Digital Communications Using Chaos and Nonlinear Dynamics*, ed. by L.E. Larson, J.M. Liu, and L.S. Tsimring (Springer, New York, 2006), Chapter 11.
13. S. Tang, and J.M. Liu, "Message encodingdecoding at 2.5 Gbits/s through synchronization of chaotic pulsing semiconductor lasers," Opt. Lett. **26**, 1843–1845 (2001).
14. M.I. Sobhy and A.R. Shehata, "Chaotic radar systems," IEEE MTT-S Int. Microw. Symp. Dig. **3**, 1701–1704 (2000).
15. B.C. Flores, E.A. Solis, and F. Thomas, "Chaotic signals for wideband radar imaging," Proc. SPIE **4727**, 100–111 (2002).
16. F.Y. Lin and J.M. Liu, "Ambiguity functions of laser-based chaotic radar," IEEE J. Quantum Electron. **40**, 1732–1738 (2004).
17. S.C. Chan, R. Diaz, and J.M. Liu, "Novel photonic applications of nonlinear semiconductor laser dynamics," Opt. Qntum Electron. **40**, 83–95 (2008).
18. S.C. Chan, S.K. Hwang, and J.M. Liu, "Period-one oscillation for photonic microwave transmission using an optically injected semiconductor laser," Opt. Express, **15**, 14921–14935 (2007).
19. T.B. Simpson and F. Doft, "Double-locked laser diode for microwave photonics applications," IEEE Photon. Technol. Lett. **11**, 1476–1478 (1999).
20. R. Diaz, S.C. Chan, and J.M. Liu, "Lidar detection using a dual-frequency source," Opt. Lett. **31**, 3600–3602 (2006).
21. S.K. Hwang, J.M. Liu, and J.K. White, "Characteristics of period-one oscillations in semiconductor lasers subject to optical injection," IEEE J. Sel. Top. Quantum Electron. **10**, 974–981 (2004).
22. S.C. Chan and J.M. Liu, "Frequency modulation on single sideband using controlled dynamics of an optically injected semiconductor laser," IEEE J. Quantum Electron. **42**, 699–705 (2006).
23. S.C. Chan, S.K. Hwang, and J.M. Liu, "Radio-over-fiber AM-to-FM upconversion using an optically injected semiconductor laser," Opt. Lett. **31**, 2254–2256 (2006).

Nonlinear Prediction Intervals by the Bootstrap Resampling

Tohru Ikeguchi

Abstract Many nonlinear prediction algorithms have already been proposed to predict complex behavior produced from nonlinear dynamical systems. In this paper, we discuss a novel nonlinear modeling framework, which combines a conventional local linear prediction algorithm and bootstrap resampling scheme. Then, we proposed a new method to evaluate predictability by estimating prediction intervals using a distribution of nonlinear bootstrap predicted points, evaluating the validity of the proposed interval estimation comparing to an ensemble prediction which is one of the conventional interval estimation methods. As a result, we find that the bootstrap prediction interval estimation method is more reasonable to make efficient prediction intervals especially in the case of short term prediction.

1 Background

Many prediction methods have already been proposed to predict complex time series [1–8]. Effective application of these prediction methods makes it possible to predict complex behavior observed in the real world accurately.

Generally, nonlinear prediction methods are classified into two categories: global nonlinear [7, 8] and local linear [1–6, 9]. In this paper, we focused on the second category, or the local linear prediction methods [2–4].

In case of using a local linear prediction method, the most important point is how to extract local information on movement of a predicted point from nearby trajectories as efficiently as possible. In particular, if the observed time series is noisy and the number of the observed time series is short, the local information becomes less reliable. In such cases, the estimated local linear predictors become also unreliable, then it is almost impossible to achieve higher predictability. To solve these issues,

T. Ikeguchi (✉)
Department of Information and Computer Sciences, Saitama University, 255 Shimo-ohkubo, Saitama, 338-8570, Japan, e-mail: tohru@ics.saitama-u.ac.jp

V. In et al. (eds.), *Applications of Nonlinear Dynamics*, Understanding Complex Systems,
DOI 10.1007/978-3-540-85632-0_29,

we have already proposed a new prediction algorithm [10] which combines the local linear prediction algorithm with a bootstrap replication method [11] to effectively use near-neighbor trajectories for constructing local linear predictors.

To evaluate the prediction accuracy, we usually use a normalized root mean square error between true points and predicted points. In such a case, prediction must be described as a point even if its predictability is unreliable due to small size data or noise in the data. We evaluate prediction accuracy from another viewpoint: we estimate prediction intervals in which the future point would fall.

In this paper, we applied the concept of the bootstrap resampling to the estimation problem of nonlinear prediction intervals of a complex time series. We also applied the scheme to the prediction problem of seismic events.

2 Local Linear Prediction

2.1 The Jacobian-Matrix Estimate Prediction

Let us consider a nonlinear dynamical system:

$$x(t+1) = F(x(t)), \tag{1}$$

where F is a d-dimensional nonlinear map, $x(t)$ is a d-dimensional state at time t. At first, to set a small displacement vector of $x(t)$ as $\delta x(t)$, we obtained the following equation,

$$x(t+1) + \delta x(t+1) = F(x(t) + \delta x(t)). \tag{2}$$

Then, to estimate the Jacobian matrix of F, we linearize Eq.(2) as follows:

$$\delta x(t+1) = DF(x(t))\delta x(t), \tag{3}$$

where $DF(x(t))$ is the Jacobian matrix at $x(t)$, and $\delta x(t)$ is a small displacement vector from $x(t)$. If we have the explicit information of the dynamical system $F(x(t))$, we can calculate $DF(x(t))$ directly. However, if we do not know $F(x(t))$, we have to estimate $DF(x(t))$ using information obtained only by the time series. To estimate $DF(x(t))$ only with the information of $x(t)$, first, we extract a near-neighbor set of $x(t)$. Let us denote the ith near neighbor of $x(t)$ as $x(t_{k_i})$, where $i = 0, 1, \ldots, M$. Here, $M+1$ is the total number of near neighbors. After a temporal evolution, $x(t)$ and $x(t_{k_i})$ evolve into $x(t+1)$ and $x(t_{k_i}+1)$, respectively. Then, we denote displacement vectors by $y_i = x(t_{k_i}) - x(t)$ and $z_i = x(t_{k_i}+1) - x(t+1)$. Here, y_i corresponds to $\delta x(t)$, and z_i corresponds to $\delta x(t+1)$ in Eq. (3). If the norms of y_i and z_i and the temporal evolution are small enough, we can approximate the relation between z_i and y_i with the following linear equation:

$$z_i = G(t) y_i,$$

where the matrix $G(t)$ corresponds to the Jacobian matrix $DF(x(t))$ in Eq. (3). Then, to estimate $G(t)$, we use the least-square-error fitting which minimizes the average square error S:

$$S = \frac{1}{M}\sum_{i=1}^{M} |z_i - G(t)y_i|^2.$$

Then, expressing the klth entry of $G(t)$ as g_{kl}, we have to calculate following equation:

$$\frac{\partial S}{\partial g_{kl}} = 0. \tag{4}$$

In other words, we can estimate $G(t)$ using the following equations:

$$G(t)W = C, \tag{5}$$

where W is the variance matrix of y_i, and C is the covariance matrix between y_i and z_i. If W has its inverse matrix, we can obtain $G(t)$ from $G(t) = CW^{-1}$.

2.2 Nonlinear Prediction by the Jacobian Matrix Estimation

Let us introduce the process of a nonlinear prediction of $x(t)$ on an attractor from the dynamical system of Eq. (1) [3].Our purpose is to predict an s step future $x(t+s)$ of $x(t)$. If we have the information of in Eq. (3), we can predict the future state of $x(t)$. Because we do not know a future value $x(t)$, we cannot use the information of z_i, and then cannot have direct information about $G(t)$. To solve the problem, we use the information of the nearest neighbor $x(t_{k_0})$ of $x(t)$. Then, we calculate a displacement vector $y' = x(t) - x(t_{k_0})$. Next, we can estimate the Jacobian matrix $G(t_{k_0})$ at $x(t_{k_0})$ by following the process introduced in Sect. 2.1. If we define $\hat{x}(t+1)$ as a predicted future value of $x(t)$, we can denote the predicted displacement vector $\hat{z}' = \hat{x}(t+1) - x(t_{k_0}+1)$ by $\hat{z}' = G(t_{k_0})y'$. Then, we can predict $\hat{x}(t+1)$ as follows:

$$\hat{x}(t+1) = G(t_{k_0})(x(t) - x(t_{k_0})) + x(t_{k_0}+1). \tag{6}$$

The above prediction algorithm firstly estimates the Jacobian matrices at each point, then predicts a future value by approximating the temporal evolution of the local displacement vector. In other words, the algorithm uses the first-order local information of the first term in Eq. (6) as well as the zeroth order local information of the second term in Eq. (6). Then, repeating the above scheme for s time iteratively, we can predict the s step future of $x(t)$.

3 Bootstrap Prediction

3.1 Bootstrap Replication Method

To statistically analyze an observed data, it is important to consider the characteristics of its population. If we obtain a large size of data sets, we can calculate the statistics of the population. However, if we only obtain a short observed data set X, it is difficult to estimate the statistics of it.

To avoid this issue, we use a bootstrap method which is one of the statistical techniques. The bootstrap method is explained as follows: at first, we get a data set X whose population is defined as U. Next, we perform a sampling with the replacement of X to obtain a new set of data set X_b, where $b = 1, \ldots, B$. We calculate each statistics of X_b, and estimate a confidence interval. Then, we can estimate the population using each statistics of X_b and the confidence interval. That is, the bootstrap method estimates the statistics of the population by changing a combination of data set for many times.

3.2 Point Prediction

If we predict the real time series, unreliable prediction would be led by undesirable situations that the data length is short or the data is disturbed by noise [10]. To solve the issue, we introduced the bootstrap resampling scheme [11] to perform stable estimation of the Jacobian matrices by increasing the variety of near-neighbor sets. First, we selected near neighbors of $x(t)$ in the same way as the conventional method. The nearest neighbor of $x(t)$ is denoted by $x(t_{k_0})$, and the other neighbors are denoted by $D_T = \{x(t_{k_1}), x(t_{k_2}), \ldots, x(t_{k_M})\}$. We performed the resampling with the replacement from D_T to obtain a new set of near neighbors. For example, the first new set D_1^* from the set D_T is $D_1^* = \{x_1^*(t_{k_1}), x_1^*(t_{k_2}), \ldots, x_1^*(t_{k_M})\}$. Next, we estimated the Jacobian matrix $G_1^*(t_{k_0})$ at $x(t_{k_0})$ using D_1^* with the method introduced in Sect. 2.1. Then, we predicted a future point of $x(t)$ by

$$\hat{x}_1^*(t+1) = G_1^*(t_{k_0})(x(t) - x(t_{k_0})) + x(t_{k_0}+1).$$

Here, $\hat{x}_1^*(t+1)$ is the first replica estimated from D_1^*, or the first replica set of the near neighbor set D_T. Then, we repeated such bootstrap estimates for B times. Namely, the b-th bootstrap predicted point is described by

$$\hat{x}_b^*(t+1) = G_b^*(t_{k_0})(x(t) - x(t_{k_0})) + x(t_{k_0}+1),$$

where $G_b^*(t_{k_0})$ is an estimated Jacobian matrix from the bth replica set D_b^* from the data set D_T, and $b = 1,2,\ldots,B$. Finally, we defined the final bootstrap predicted point of $x(t)$ by calculating its mean value as follows:

$$\hat{x}(t+1) = \frac{1}{B}\sum_{b=1}^{B} \hat{x}_b^*(t+1).$$

3.3 Analysis of the Bootstrap Replication for the Jacobian-Matrix Estimation

To estimate the Jacobian matrices we use the following equations:

$$G(t)W = C,$$

where W is the variance matrix of $y_i = x(t_{k_i}) - x(t)$ and C is the covariance matrix of y_i and $z_i = x(t_{k_i} + 1) - x(t + 1)$. However, if we applied the resampling method described above to estimate the Jacobian matrices, it is important not to apply the above algorithm blindly, because it is possible that the matrix W does not always have its inverse. The reason why such a case exists is that resampled near neighbors (the bootstrap samples) can be duplicated or clustered, making the matrix W not a full rank.

In Fig. 1, we show how many times such a case occurs. The horizontal axis shows the amount of near-neighbor data and the vertical axis shows the ratio of when W is not a full rank. In Fig. 1, to estimate the ratio, we use the time series of Japanese vowel /a/, the number of measles patients [6] and chicken pox patients [6], and the Ikeda map [14].

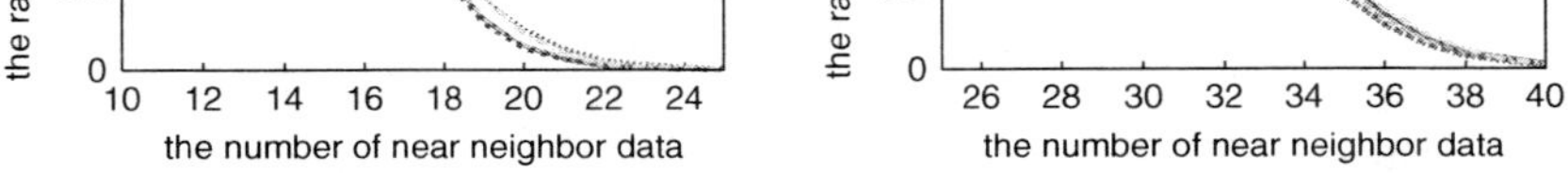

Fig. 1 To estimate the Jacobian matrix $G(t)$, we use the following equations: $G(t)W = C$, where W is the variance matrix and C is the covariance matrix. This figure shows the average ratio when W does not become a fully rank over 10,000 trials. The embedding dimensions [12] are **(a)** 10 and **(b)** 20

In addition, we constructed a mathematical model of this resampling scheme. At first, we prepared N boxes, and indexed them from 1 to N. Next, we sampled U random integers with replacements, and put them into the box of the same number as U. Then, we counted the number of non-empty boxes k. In this model, the parameter N is the amount of near neighbor data. If k exceeds the dimension of the state space d, the matrix W becomes a full rank. Repeating this scheme l times, we calculated the average ratio that $k \leq d$. The results of Fig. 1 show that the ratio does not depend on a characteristic of the time series but on the amount of near neighbor data.

This is one of the most important points we should carefully consider in the case of nonlinear prediction. It is true that the bootstrap resampling scheme could select good near neighbors and improve the local linear predictability. However, it has different aspects from the conventional process of estimating statistics by the bootstrap resampling. To avoid such an undesirable situation in the present case, we calculate W^{-1} as follows: At first, we applied a diagonalization to W, and obtained an orthogonal matrix P and a diagonal matrix $\Lambda = \mathrm{diag}(\lambda_i)(i = 1,\ldots,d)$, where d is the dimension of W and λ_i is the ith eigen value of W. Then, the matrices of W, P, and Λ have the relation $W = P\Lambda P^{-1}$. Next, we calculated the inverse matrix by $W^{-1} = P\Lambda^{-1}P^T$, where Λ^{-1} =diag$(1/\lambda_i)$ [10].

If we encounter an eigen value λ_i that is smaller than 10^{-6}, which corresponds to the case when W is not a full rank, we set the ith diagonal element of Λ as 0, d because W has d no information in the ith eigen direction.

3.4 Bootstrap Interval Estimation

3.4.1 Why Prediction Intervals?

One of the conventional measures to evaluate predictability is the normalized root mean square error:

$$E = \frac{\sqrt{\sum_{t=1}^{N}(x(t+1) - \hat{x}(t+1))^2}}{\sqrt{\sum_{t=1}^{N}(x(t+1) - \bar{x})^2}}, \tag{7}$$

where $x(t+1)$ is a true point and $\hat{x}(t+1)$ is a predicted point. Although the root mean square error of Eq. (7) is basic and essential to evaluate the prediction accuracy, any prediction methods cannot predict future points perfectly. It means that it is important to offer not only a prediction point but also a prediction interval in which the true future point might be included. In this paper, we proposed a new algorithm for evaluating prediction intervals by the nonlinear bootstrap prediction proposed in Sect. 2.1.

3.4.2 Bootstrap Interval Estimation

Because the ensemble prediction [15] needs a large data set to evaluate ensemble properties, estimated prediction intervals might be unreliable if we cannot use large size data. In [10], we have already shown that the bootstrap method [11] is effective for the nonlinear prediction problem even if data size is small [16]. Thus, to evaluate more accurate prediction intervals and to perform stable prediction, it is natural to expect that the bootstrap method may work well. Then, we applied the bootstrap resampling scheme to the local linear prediction estimating the prediction intervals.

The bootstrap resampling scheme [11] is described as follows:

(1) We selected near-neighbor points of $x(t)$ to predict $x(t)$. The data set is denoted by $D = \{x(t_{k_0}), x(t_{k_1}), \ldots, x(t_{k_L})\}$, where L is the number of near-neighbor points to make a predictor. In this paper, we set $L = M$.
(2) We performed a sampling with replacement of D to obtain a new near-neighbor data set $D^{(1)} = \{D^{(1,1)}, \ldots, D^{(1,L)}\}$. Here, $D^{(1,l)}$ means the lth sampling with replacement at the first bootstrap trial.
(3) We estimate a predictor $\tilde{f}^{(1,l)}$ on each $D^{(1,l)}$, and we predict a future point of $x(t)$ by $\tilde{x}^{(1,l)}(t+1) = \tilde{f}^{(1,l)}(D^{(1,l)})$.
(4) We repeated such bootstrap estimates for B times. Namely, the bth bootstrap predicted point is described by $\tilde{x}^{(b,l)}(t+1) = \tilde{f}^{(b,l)}(D^{(b,l)})$, where $b = 1, 2, \ldots, B$.
(5) The equation of an ellipse is defined by

$$E(t+1) = \sum_{i=1}^{d} \frac{\hat{z}_i^2(t+1)}{\sigma_i^2} - 1$$

where $\hat{z}_i^{(t+1)}$ is the ith principle component, obtained by the applying of the principle component analysis to the data set $\hat{x}^{(b)}(t+1), (b = 1, 2, \ldots, B)$, and σ_i is the variance of the ith principle component.
(6) Repeating the above scheme for p times, we can predict the p-step future. However, if we perform a p-step prediction with the bootstrap method directly, the computational load would grow exponentially, because the number of point at the p-step prediction would be $(B \cdot L)^p$. Thus, in this paper, from the second prediction step, we randomly select B elements from $\{\tilde{x}^{(b,l)}(t+2)\}$ to prevent the number of bootstrap samples from exponential explosion.
(7) The prediction interval for p steps future is defined by

$$E(t+p) = \sum_{i=1}^{d} \frac{\hat{z}_i^2(t+p)}{\sigma_i^2} - 1$$

where $\hat{z}_i^{(t+p)}$ is the ith principle component, obtained by the applying of the principle component analysis to the data set $\hat{x}^{(b)}(t+p), (b = 1, 2, \ldots, B^B)$, and σ_i is the variance of the ith principle component.

3.5 Simulations and Results

To estimate prediction accuracy, we proposed a new measure to evaluate the prediction interval: we counted how many times the true point is included in a prescribed prediction interval. The number of the true points included in the prescribed prediction intervals is denoted by V, which is averaged on several trials. We also introduced the size of prediction interval S_r. This measure is defined by

$$S_r(p) = \sqrt{\sum_{d=1}^{k} \tilde{\sigma}_d^2(t+p)}, \tag{8}$$

where p is the prediction step. Then, to evaluate prediction accuracy of the intervals, we use $V(p)/S_r(p)$ which means $V(p)$ is normalized by $S_r(p)$.

Figures 2 and 3 shows the results of the comparisons among four cases to estimate prediction intervals: the bootstrap method with the Jacobian prediction or the Lorenz' method of analogues [1], or the ensemble prediction with the same local prediction methods. In Figs. 2 and 3, the Ikeda maps whose data length N is 200 and 500 with the observational noise are used for simulation.

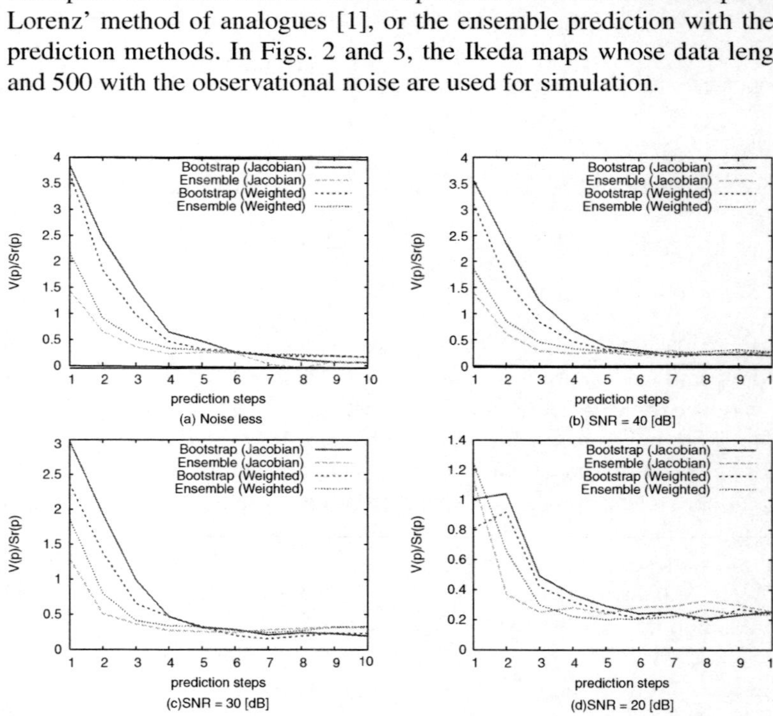

Fig. 2 Comparisons of four combinations to estimate prediction intervals for the Ikeda map whose data length N is fixed to 200. Moreover, the Ikeda map is corrupted by observational noise with different SNRs. The horizontal axis shows the prediction steps p, and the vertical axis shows the estimation accuracy $V(p)/S_r(p)$

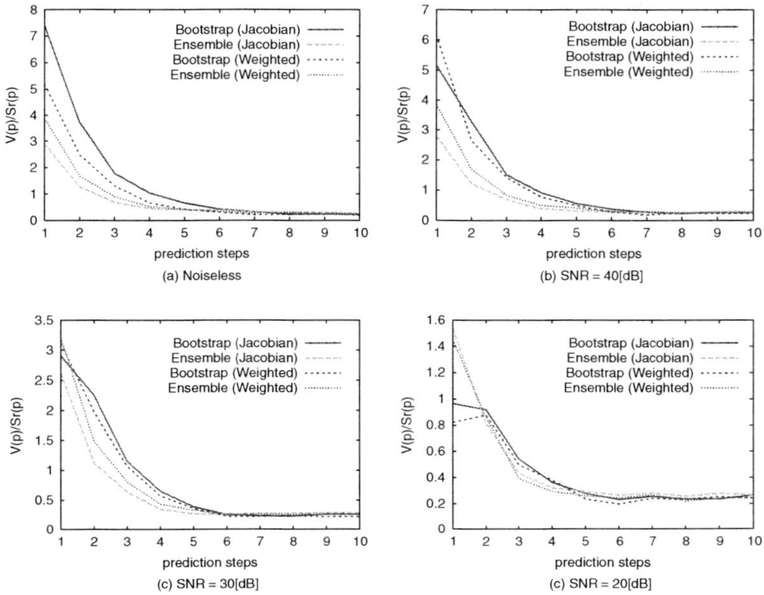

Fig. 3 Comparisons of four combinations to estimate prediction intervals for the Ikeda map whose data length N is fixed to 500. Moreover, the Ikeda map is corrupted by observational noise with different SNRs. The *horizontal axis* shows the prediction steps p, and the vertical axis shows the estimation accuracy $V(p)/S_r(p)$

In the case of the prediction interval estimation, as shown in Fig. 2, we find that the proposed framework improve the prediction accuracy especially in the case of short term prediction. Namely, if we perform short term prediction, the bootstrap prediction method adjusts intervals size more accurately and efficiently than the ensemble prediction method. Although the performance of V/S_r of the bootstrap method and the ensemble method is almost the same as the observational noise level becomes larger, in the case of noiseless data, the bootstrap method with the Jacobian prediction shows the best performance. Figures 2 and 3 shows that the bootstrap method has high prediction accuracy than the ensemble prediction in each basic prediction method. The results indicate that the bootstrap method is also effective in the case of the data is disturbed by the observational noise.

To evaluate applicability of the proposed method to real data, we applied our method to the difference of the number of measles and chicken pox patients [6], and the time series of seismic events observed in Japan. Figure 4 shows the results. For the measles data, we find that the proposed frame work exhibits higher performance for short-term prediction. In. Fig. 5, we show the results for the seismic event time series observed in Japan. The data analyzed and predicted in Fig. 5 are observed from 1979 to 1987, respectively. Although we found that the bootstrap nonlinear

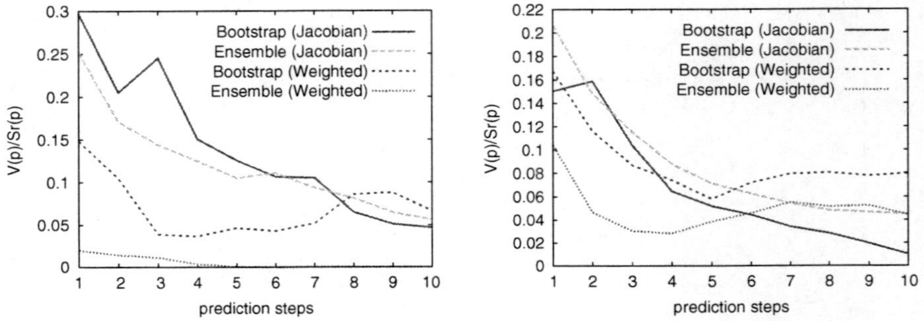

Fig. 4 The same as Figs. 2 and 3, but for the difference of the number of (**a**) measles patients and (**b**) chickenpox patients

prediction with prediction interval is superior to the ensemble prediction, the basic predictability is still low, which indicates that we have to introduce additional information to realize more reliable predictability for seismic events.

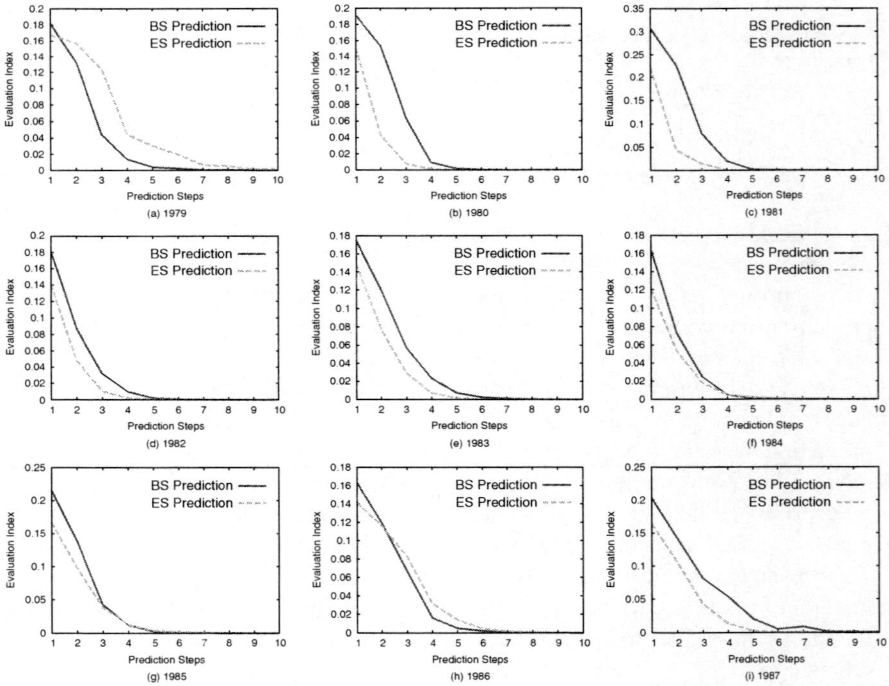

Fig. 5 Comparisons of the bootstrap scheme and the ensemble scheme to estimate prediction intervals for the seismic data. The *horizontal axis* shows the prediction steps p, and the *vertical axis* shows the estimation accuracy $V(p)/S_r(p)$

4 Conclusion

In this paper, to improve the prediction accuracy and to realize reliable predictability in the case of the short or noisy time series, we proposed a new nonlinear prediction interval estimation method which is based on the combination of a basic local linear prediction algorithm with the bootstrap method [10, 16]. To check the validity of the proposed method, we analyzed the proposed method with several real time series.

In particular, the proposed framework improves short-term predictability comparing with an ensemble prediction, one of the conventional frameworks to evaluate the prediction interval. Moreover, the bootstrap method adjusts the size of the prediction intervals effectively according to the difficulty of prediction.

Although we have applied a simple framework of a bootstrap-like resampling procedure to build better predictors, the framework might be a close relation to the nonlinear model selection in statistics [8]. It is also an important future task to discuss the relation not only from a nonlinear dynamical but also from a statistical point of view. In this paper, we improve nonlinear predictability of the local linear prediction algorithms. One of the important next step is to apply the proposed method to local nonlinear prediction algorithm [5]. The application of these frameworks to relatively high dimensional data is one of the most important issues in the nonlinear time-series analysis. Moreover, in this paper, we set the number of resample times as the total number of near neighbors M because we simply follow the original resampling scheme of the bootstrap method [11]. It is also an important future work to evaluate the proposed method with smaller number of the resample time to reduce a computational load of the proposed method.

Acknowledgments The author would like to thank H. Hashiguchi, T. Suzuki, and N. Yabuta for their valuable comments. The research is partially supported by a research grant from the Mazda Foundation.

References

1. E. N. Lorenz, "Atmospheric Predictability as Revealed by Naturally Occurring Analogues," Journal of the Atmospheric Sciences, Vol.26, pp.636–646, 1969.
2. M. Sano and Y. Sawada, "Measurement of the Lyapunov Spectrum from a Chaotic Time Series," Physical Review Letters, Vol.55, No.10, pp.1082–1085, 1985.
3. J. D. Farmer and J. J. Sidorowich, "Predicting Chaotic Time Series," Physical Review Letters, Vol.59, No.8, pp.845–848, 1987.
4. J. P. Eckmann, S. Oliffson Kamphorst, D. Ruelle, and S. Ciliberto, "Lyapunov Exponents from Time Series," Physical Review A, Vol.34, No.6, pp.4971–4979, 1986.
5. K. Briggs, "An Improved Method for Estimating Lyapunov Exponents of Chaotic Time Series," Physics Letters A, Vol.151, Nos.1,2, pp.27–32, 1990.
6. G. Sugihara and R. M. May, "Nonlinear Forecasting as a Way of Distinguishing Chaos from Measurement Error in Time Series," Nature, Vol.344, pp.734–741, 1990.
7. M. Casdagli, "Nonlinear Prediction of Chaotic Time Series," Physica D, Vol.35, pp.335–356, 1989.

8. K. Judd and A. Mees, "On Selecting Models for Nonlinear Time Series," Physica D, Vol.82, No.2, pp.426–444, 1995.
9. P. Bryant, R. Brown and H. D. I. Abarbanel, "Lyapunov Exponents from Observed Time Series," Physical Review Letters, Vol.65, No.13, pp.1523–1526, 1990.
10. D. Haraki, T. Suzuki, and T. Ikeguchi, "Bootstrap Nonlinear Prediction," Physical Review E, Vol.75, 056212, 2007.
11. B. Efron and R. J. Tibshirani, *An Introduction to the Bootstrap*, Chapman and Hall, Boca Raton, FL 1993.
12. F. Takens, "Detecting Strange Attractors in Turbulence," In D. A. Rand and B. S. Young, editors, *Dynamical Systems of Turbulence*, Vol.898 of Lecture Notes in Mathematics, Berlin, Springer-Verlag, pp.366–381, 1981.
13. T. Ikeguchi and K. Aihara, "Estimating Correlation Dimensions of Biological Time Series Using a Reliable Method," Journal of Intelligent and Fuzzy Systems Vol.5, No.1, pp.33–52, 1997.
14. K. Ikeda, "Multiple-Valued Stationary State and Its Instability of the Transmitted Light by a Ring Cavity System," Optics Communications, Vol.30, No.2, pp.257–261, 1979.
15. T. Hurukawa and S. Sakai, *Ensemble Prediction*, Tokyo-doh Press, 2004, in Japanese.
16. D. Haraki, T. Suzuki, and T. Ikeguchi, "Bootstrap Prediction Intervals for Nonlinear Time-Series," Lecture Notes in Computer Science, Vol.4224, pp.155–162, 2006.

Quantum Measurements with Dynamically Bistable Systems

M.I. Dykman

Abstract Periodically modulated nonlinear oscillators often display bistability of forced vibrations. This bistability can be used for new types of quantum measurements. They are based on switching between coexisting vibrational states. Since switching is accompanied by a large change of the amplitude and phase of forced vibrations, the measurements are highly sensitive. Quantum and classical noise plays dual role. It imposes a limitation on sensitivity in the familiar regime of a bifurcation amplifier. On the other hand, it makes it possible to use a bistable modulated oscillator in a new regime of a balanced dynamical bridge. We discuss the switching probabilities and show that they display scaling with control parameters. The critical exponents are found for different types of bifurcations and for different types of noise.

1 Introduction

Bistability of vibrational states in modulated systems and fluctuation-induced switching between these states have attracted much attention recently. Experiments have been done on such diverse systems as electrons [1] and atoms [2, 3] in modulated traps, rf-driven Josephson junctions [4, 5], and nano- and micromechanical resonators [6–9]. These systems have small vibration damping, the quality factor may reach $10^4 - 10^5$. Therefore even a comparatively small resonant field can lead to coexistence of forced vibrations with different phases and amplitudes. This is illustrated in Fig. 1, which refers to a simple model relevant to many of the aforementioned experiments: an underdamped nonlinear classical oscillator driven close to resonance, with equation of motion

M.I. Dykman (✉)
Department of Physics and Astronomy, Michigan State University, East Lansing, MI 48824, USA
e-mail: dykman@pa.msu.edu

V. In et al. (eds.), *Applications of Nonlinear Dynamics*, Understanding Complex Systems, 367
DOI 10.1007/978-3-540-85632-0_30,

$$\ddot{q} + \omega_0^2 q + \gamma q^3 + 2\Gamma\dot{q} = A\cos\omega_F t. \tag{1}$$

Here, q is the oscillator coordinate, ω_0 is its eigenfrequency, Γ is the friction coefficient, $\Gamma \ll \omega_0$, and γ is the nonlinearity parameter. The frequency of the modulating field ω_F is assumed to be close to ω_0. In this case, for comparatively small modulation amplitude A, even where the oscillator becomes bistable its forced vibrations are nearly sinusoidal, $q(t) = a\cos(\omega_F t + \phi)$ [10].

The dynamic bistability of the oscillator is advantageous for measurements. The idea is to make the oscillator switch between the states depending on the value of the parameter to be measured. Switching leads to a strong change in the system that can be easily detected, leading to a high signal-to-noise ratio in a measurement. This has been successfully used for fast and sensitive measurements of the states of different types of Josephson junction based qubits, including quantum non-demolition measurements [5, 11, 12].

So far the experiments were done in the bifurcation amplifier mode, where the control parameter is swept through a bifurcation point (for example, the field amplitude A was swept through A_{B2}, see Fig. 1). The position of the bifurcation point, i.e., the value of A where switching occurs, depends on the state of the measured qubit. However, in contrast to happening for a certain A expected from Fig. 1, switching occurred at random within a certain parameter range of the A near A_{B2}.

The randomness of the switching field is a consequence of fluctuations in the oscillator. They lead to switching even before the control parameter reaches its bifurcational value. This is analogous to activated switching out of a potential well studied by Kramers [13]. However, in the case of an oscillator the stable states are not minima of a potential, and there is no static potential barrier that needs to be overcome. Switching of a modulated oscillator is an example of metastable decay of systems far from thermal equilibrium, the phenomenon of a broad interest.

Theoretical analysis of metastable decay requires developing methods for calculating the decay probability and finding out whether decay displays any universal system-independent features, like scaling dependence on the control parameters. For classical systems, scaling of the decay rate was indeed found for systems close to a bifurcation point, both in the cases of equilibrium [14–16] and nonequilibrium systems [17–19]. In the latter case a scaling crossover may occur as the system goes from the underdamped to overdamped regime while approaching the bifurca-

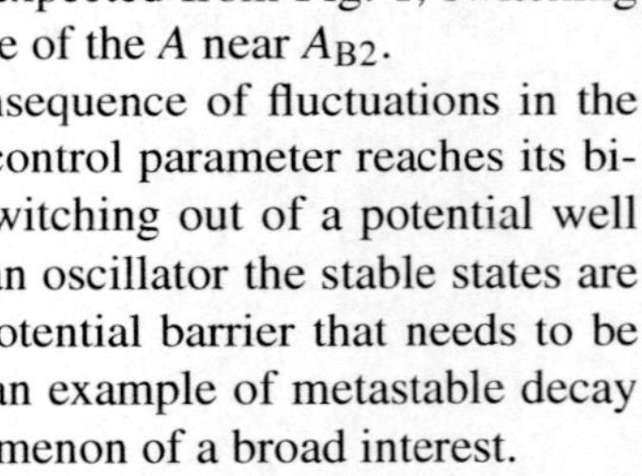

Fig. 1 Bistability of a nonlinear Duffing oscillator (1). The *solid lines* show the squared amplitude of forced vibrations as a function of the squared amplitude of the driving force. At the bifurcational values $A^2_{B1,2}$ one of the stable vibrational states disappears

tion point [20]. Such crossover occurs also for quantum tunneling in equilibrium dissipative systems [21].

In this paper we study decay of metastable vibrational states in quantum dissipative systems close to bifurcation points [22]. This is necessary for understanding the operation of a modulated oscillator in the regime of a bifurcation amplifier. We show that at low temperatures decay occurs via quantum activation. This is a specific process that has no analog in systems in thermal equilibrium [23, 24]. As tunneling, quantum activation is due to quantum fluctuations, but as thermal activation, it involves diffusion over an effective barrier separating the metastable state. As we show, near a bifurcation point quantum activation is more probable than tunneling even for $T \to 0$. We find that the decay rate W scales with the distance to the bifurcation point η as $|\ln W| \propto \eta^{\xi}$. The scaling exponent is $\xi = 3/2$ for resonant driving, cf. (1). We also consider parametric resonance in a nonlinear oscillator and show that in this case $\xi = 2$. In addition, $|\ln W|$ displays a characteristic temperature dependence.

2 Quantum Kinetic Equation for a Resonantly Driven Oscillator

The Hamiltonian of a resonantly driven nonlinear oscillator is

$$H_0(t) = \frac{1}{2}p^2 + \frac{1}{2}\omega_0^2 q^2 + \frac{1}{4}\gamma q^4 - qA\cos(\omega_F t). \tag{2}$$

The notations are the same as in equation of motion (1), p is the oscillator momentum. We assume that the detuning $\delta\omega = \omega_F - \omega_0$ of the modulation frequency ω_F from the oscillator eigenfrequency ω_0 is small and that $\gamma\,\delta\omega > 0$, which is necessary for bistability; for concreteness we set $\gamma > 0$.

It is convenient to switch from q, p to slowly varying operators Q, P, using a transformation

$$q = C_{\rm res}(Q\cos\omega_F t + P\sin\omega_F t), \quad p = -C_{\rm res}\omega_F(Q\sin\omega_F t - P\cos\omega_F t) \tag{3}$$

with $C_{\rm res} = (8\omega_F\delta\omega/3\gamma)^{1/2}$. The variables Q, P are the scaled coordinate and momentum in the rotating frame. They are canonically conjugate,

$$[P, Q] = -i\lambda, \qquad \lambda = 3\hbar\gamma/8\omega_F^2\,\delta\omega. \tag{4}$$

The parameter λ plays the role of the effective Planck constant. We are interested in the semiclassical case; λ is the small parameter of the theory, $\lambda \ll 1$.

In the rotating wave approximation the Hamiltonian (2) becomes $H_0 = (\hbar/\lambda)\delta\omega\,\hat{g}$, with

$$\hat{g} \equiv g(Q, P) = \frac{1}{4}(Q^2 + P^2 - 1)^2 - \beta^{1/2}Q, \qquad \beta = 3\gamma A^2/32\omega_F^3(\delta\omega)^3 \tag{5}$$

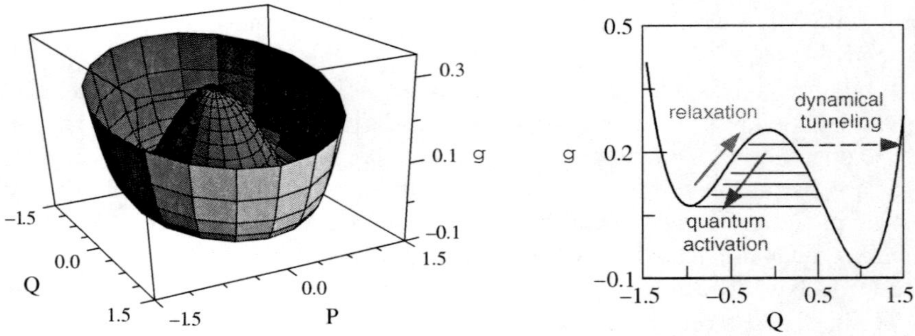

Fig. 2 Quasienergy of a resonantly driven nonlinear oscillator $g(Q,P)$ (*left panel*) and its cross-section by the plane $P = 0$ (*right panel*). The plot refers to $\beta = 1/270$ in (5). *Thin horizontal lines* in the right panel show (schematically) quasienergy levels for quantized motion around the local maximum of $g(Q,P)$. In the presence of dissipation the states at the local maximum and the minimum of $g(Q,P)$ become stable. They correspond, respectively, to forced vibrations with small and large amplitude a in Fig. 1. The *arrows* in the *right panel* show relaxation to the state of small-amplitude vibrations, tunneling from this state with constant quasienergy g, and quantum activation. The latter corresponds to quantum diffusion over quasienergy away from the metastable state, which accompanies relaxation [23, 24]

(in the case $\gamma, \delta\omega < 0$ one should replace $g \to -g, H_0 \to -(\hbar/\lambda)\delta\omega\, g$). The function g is shown in Fig. 2. It plays the role of the oscillator Hamiltonian in dimensionless time $\tau = t|\delta\omega|$. The eigenvalues of $\hat{g}$ give oscillator quasienergies. The parameter β in Eq. (5) is the scaled intensity of the driving field. For weak damping the oscillator is bistable provided $0 < \beta < 4/27$. The Heisenberg equation of motion for an arbitrary operator M is $\dot{M} \equiv dM/d\tau = -i\lambda^{-1}[M,g]$.

We will consider two major relaxation mechanisms of the oscillator: damping due to coupling to a thermal bath and dephasing due to oscillator frequency modulation by an external noise. Usually the most important damping mechanism is transitions between neighboring oscillator energy levels. They result from coupling linear in the oscillator coordinate. Since the energy transfer is $\approx \hbar\omega_0$, in the rotating frame the transitions look instantaneous. Phenomenologically, the resulting relaxation may be described by a friction force proportional to velocity, as in (1). Microscopically, such description applies in the case of Ohmic dissipation, i.e., coupling to Ohmic bath. However, we do not have to assume that dissipation is Ohmic. The only assumption needed for the further analysis is that the density of states of the reservoir weighted with the interaction be smooth in the frequency range, which is centered at ω_0 and has a width that largely exceeds $\Gamma, |\delta\omega|$.

We will assume that the correlation time of the noise that modulates the oscillator frequency is also short compared to $1/|\delta\omega|$, so that the noise is effectively δ-correlated in slow time τ. Then the quantum kinetic equation is Markovian in the rotating frame. It has a familiar form (cf. [25])

$$\dot{\rho} \equiv \partial_\tau \rho = i\lambda^{-1}[\rho, \hat{g}] - \hat{\Gamma}\rho - \hat{\Gamma}^{\mathrm{ph}}\rho, \tag{6}$$

where $\hat{\Gamma}\rho$ describes damping,

$$\hat{\Gamma}\rho = \Gamma|\delta\omega|^{-1}\left[(\bar{n}+1)(\hat{a}^\dagger\hat{a}\rho - 2\hat{a}\rho\hat{a}^\dagger + \rho\hat{a}^\dagger\hat{a}) + \bar{n}(\hat{a}\hat{a}^\dagger\rho - 2\hat{a}^\dagger\rho\hat{a} + \rho\hat{a}\hat{a}^\dagger)\right], \quad (7)$$

and $\hat{\Gamma}^{\mathrm{ph}}\rho$ describes dephasing,

$$\hat{\Gamma}^{\mathrm{ph}}\rho = \Gamma^{\mathrm{ph}}|\delta\omega|^{-1}\left[\hat{a}^\dagger\hat{a},\left[\hat{a}^\dagger\hat{a},\rho\right]\right]. \quad (8)$$

Here, Γ and Γ^{ph} are the damping and dephasing rates, $\hat{a} = (2\lambda)^{-1/2}(Q+iP)$ is the oscillator lowering operator, and $\bar{n} = \left[\exp(\hbar\omega_0/kT) - 1\right]^{-1}$ is the oscillator Planck number. In the classical mean-field limit one can obtain from (6), (7) the same equation of motion as (1) in the rotating wave approximation.

In what follows we use dimensionless parameters

$$\Omega = |\delta\omega|/\Gamma, \qquad \kappa^{\mathrm{ph}} = \Gamma^{\mathrm{ph}}/\lambda\Gamma. \quad (9)$$

We assume that $k^{\mathrm{ph}} \lesssim 1$. This means that the intensity of phase fluctuations may be comparable to the intensity of quantum fluctuations associated with damping, which is $\propto \lambda\Gamma$, see below, but that $\Gamma^{\mathrm{ph}} \ll \Gamma$.

Metastable decay was studied earlier for additively and parametrically driven oscillators at $T = \kappa^{\mathrm{ph}} = 0$ where there is detailed balance [26–28], and the lowest eigenvalue of $\hat{\Gamma}$ was studied numerically [29]. However, the $T = \kappa^{\mathrm{ph}} = 0$ solution is fragile. It can change exponentially strongly already for extremely small T, κ^{ph} [23, 24]. The analysis [23, 24] revealed the mechanism of quantum activation over a quasienergy barrier, but the results referred to the case where the damping-induced broadening of quasienergy levels is small compared to the typical interlevel distance. This condition necessarily breaks sufficiently close to a bifurcation point where the level spacing becomes small as a consequence of the motion slowing down.

3 Wigner Representation

The analysis of metastable decay near a bifurcation point can be conveniently done in the Wigner representation,

$$\rho_W(Q,P) = \int d\xi\, \mathrm{e}^{-i\xi P/\lambda}\rho\left(Q+\frac{1}{2}\xi, Q-\frac{1}{2}\xi\right), \quad (10)$$

where $\rho(Q_1,Q_2) = \langle Q_1|\rho|Q_2\rangle$ is the density matrix in the coordinate representation. Using (4), (5), (6), (7), (8), (9) and (10) one can formally write the equation for ρ_W as a sum of terms proportional to different powers of the effective Planck constant λ,

$$\dot{\rho}_W = -\nabla\cdot(\mathbf{K}\rho_W) + \lambda\hat{L}^{(1)}\rho_W + \lambda^2\hat{L}^{(2)}\rho_W, \quad (11)$$

where $\mathbf{K} = (K_Q, K_P)$ and $\nabla = (\partial_Q, \partial_P)$. Vector $\mathbf{K}$ determines evolution of the density matrix in the absence of quantum and classical fluctuations,

$$K_Q = \partial_P g - \Omega^{-1} Q \qquad K_P = -\partial_Q g - \Omega^{-1} P. \tag{12}$$

This evolution corresponds to classical motion

$$\dot{Q} = K_Q, \qquad \dot{P} = K_P. \tag{13}$$

The condition $\mathbf{K} = \mathbf{0}$ gives the values of Q, P at the stationary states of the oscillator in the rotating frame.

The term $\hat{L}^{(1)}$ in (11) describes classical and quantum fluctuations due to damping and dephasing,

$$\hat{L}^{(1)} = \Omega^{-1} \left[\left(\bar{n} + \frac{1}{2} \right) \nabla^2 + k^{\text{ph}} \left(Q\partial_P - P\partial_Q \right)^2 \right]. \tag{14}$$

These fluctuations lead to diffusion in (Q,P)-space, as seen from the structure of $\hat{L}^{(1)}$: this operator is quadratic in ∂_Q, ∂_P.

The term $\hat{L}^{(2)}$ in (11) describes quantum effects of motion of the isolated oscillator,

$$\hat{L}^{(2)} = -\frac{1}{4} \left(Q\partial_P - P\partial_Q \right) \nabla^2. \tag{15}$$

In contrast to $\hat{L}^{(1)}$, the operator $\hat{L}^{(2)}$ contains third derivatives. Generally the term $\lambda^2 \hat{L}^{(2)} \rho_W$ is not small, because ρ_W varies on distances $\sim \lambda$. However, it becomes small close to bifurcation points, as shown below.

3.1 Vicinity of a Bifurcation Point

From (12) and (13), for given reduced damping Ω^{-1} the oscillator has two stable and one unstable stationary state in the rotating frame (periodic states of forced vibrations) in the range $\beta_B^{(1)}(\Omega) < \beta < \beta_B^{(2)}(\Omega)$ and one stable state outside this range [10], with

$$\beta_B^{(1,2)} = \frac{2}{27} \left[1 + 9\Omega^{-2} \mp \left(1 - 3\Omega^{-2} \right)^{3/2} \right]. \tag{16}$$

At $\beta_B^{(1)}$ and $\beta_B^{(2)}$ the stable states with large and small $Q^2 + P^2$, respectively (large and small vibration amplitudes), merge with the saddle state (saddle-node bifurcation). The values of Q, P at the bifurcation points 1, 2 are

$$Q_B = \beta_B^{-1/2} Y_B (Y_B - 1), \qquad P_B = \beta_B^{-1/2} \Omega^{-1} Y_B \qquad (Y_B = Q_B^2 + P_B^2), \tag{17}$$

with

$$Y_B^{(1,2)} = \frac{1}{3}\left[2 \pm (1-3\Omega^{-2})^{1/2}\right]. \tag{18}$$

In the absence of fluctuations the dynamics of a classical system near a saddle-node bifurcation point is controlled by one slow variable [30]. In our case it can be found by expanding $K_{Q,P}$ in $\delta Q = Q - Q_B$, $\delta P = P - P_B$, and the distance to the bifurcation point $\eta = \beta - \beta_B$. The function K_P does not contain linear terms in $\delta Q, \delta P$. Then, from (13), P slowly varies in time for small $\delta Q, \delta P, \eta$.

On the other hand

$$K_Q \approx -2\Omega^{-1}(\delta Q - a_B\delta P), \quad a_B = \Omega(2Y_B - 1). \tag{19}$$

Therefore the relaxation time of Q is $\Omega/2$, it does not depend on the distance to the bifurcation point. As a consequence, Q follows P adiabatically, i.e., over time $\sim \Omega$ it adjusts to the instantaneous value of P.

4 Metastable Decay near a Bifurcation Point

The adiabatic approximation can be applied also to fluctuating systems, and as we show it allows finding the rate of metastable decay. The approach is well known for classical systems described by the Fokker-Planck equation [31]. It can be extended to the quantum problem by factoring ρ_W into a normalized Gaussian distribution over $\delta\tilde{Q} = \delta Q - a_B\delta P$ and a function $\bar{\rho}_W(\delta P)$ that describes the distribution over δP,

$$\rho_W \approx \text{const} \times \exp\left[-2\frac{(\delta Q - a_B\delta P)^2}{\lambda\,(2\bar{n}+1)\,(1+a_B^2)}\right]\bar{\rho}_W(\delta P).$$

In the spirit of the adiabatic approximation, $\bar{\rho}_W$ can be calculated disregarding small fluctuations of $\delta\tilde{Q}$. Formally, one obtains an equation for $\bar{\rho}_W$ by substituting the factorized distribution into the full kinetic equation (11) and integrating over $\delta\tilde{Q}$. This gives

$$\dot{\bar{\rho}}_W \approx \partial_P\left[\bar{\rho}_W\partial_P U + \lambda\mathscr{D}_B\partial_P\bar{\rho}_W\right], \tag{20}$$

where U and $\mathscr{D}$ have the form

$$U = \frac{1}{3}b(\delta P)^3 - \frac{1}{2}\beta_B^{-1/2}\eta\,\delta P, \qquad \eta = \beta - \beta_B,$$
$$\mathscr{D}_B = \Omega^{-1}\left[\left(\bar{n}+\frac{1}{2}\right) + \frac{1}{2}\kappa^{\text{ph}}(1-Y_B)\right] \tag{21}$$

with

$$b = \frac{1}{2}\beta_B^{1/2}\Omega^2(3Y_B - 2).$$

In (20) and (21) we kept only the lowest order terms in $\delta P, \beta - \beta_B, \lambda$. In particular we dropped the term $-\lambda^2 Q_B \partial_P^3 \bar{\rho}_W/4$ which comes from the operator $\hat{L}^{(2)}$ in (11). One can show that, for typical $|\delta P| \sim |\eta|^{1/2}$, this term leads to corrections $\sim \eta, \lambda$ to $\bar{\rho}_W$.

Equation (20) has a standard form of the equation for classical diffusion in a potential $U(\delta P)$, with diffusion coefficient $\lambda \mathscr{D}_B$. However, in the present case the diffusion is due to quantum processes and the diffusion coefficient is $\propto \hbar$ for $T \to 0$.

4.1 Scaling of the Rate of Metastable Decay

For $\eta b > 0$ the potential U (21) has a minimum and a maximum. They correspond to the stable and saddle states of the oscillator, respectively. The distribution ρ_W has a diffusion-broadened peak at the stable state. Diffusion also leads to escape from the stable state, i.e., to metastable decay. The decay rate W is given by the Kramers theory [13],

$$W = C e^{-R_A/\lambda}, \qquad R_A = \frac{2^{1/2}|\eta|^{3/2}}{3\mathscr{D}_B |b|^{1/2}\beta_B^{3/4}}, \tag{22}$$

with prefactor $C = \pi^{-1}(b\eta/2)^{1/2}\beta_B^{-1/4}|\delta\omega|$ (in unscaled time t).

The rate (22) displays activation dependence on the effective Planck constant λ. The characteristic quantum activation energy R_A scales with the distance to the bifurcation point $\eta = \beta - \beta_B$ as $\eta^{3/2}$. This scaling is independent of temperature. However, the factor $\mathscr{D}_B$ in R_A displays a characteristic T dependence. In the absence of dephasing we have $\mathscr{D}_B = 1/2\Omega$ for $\bar{n} \ll 1$, whereas $\mathscr{D}_B = kT/\hbar\omega_0\Omega$ for $\bar{n} \gg 1$. In the latter case the expression for W coincides with the result [17].

In the limit $\Omega \gg 1$ the activation energy (22) for the small-amplitude state has the same form as in the range of β still close but further away from the bifurcation point, where the distance between quasienergy levels largely exceeds their width [23]. We note that the rate of tunneling decay for this state is exponentially smaller. The tunneling is shown by the dashed line in the right panel of Fig. 2. The tunneling exponent for constant quasienergy scales as $\eta^{5/4}$ [18], which is parametrically larger than $\eta^{3/2}$ for small η [for comparison, for a particle in a cubic potential (21) the tunneling exponent in the strong-damping limit scales as η [21].

For the large-amplitude state the quantum activation energy (22) displays different scaling from that further away from the bifurcation point, where $R_A \propto \beta^{1/2}$ for $\Omega \gg 1$ [23]. For this state we therefore expect a scaling crossover to occur with varying β.

5 Parametrically Modulated Oscillator

The approach to decay of vibrational states can be extended to a parametrically modulated oscillator that displays parametric resonance. The Hamiltonian of such an oscillator is

$$H_0(t) = \frac{1}{2}p^2 + \frac{1}{2}q^2\left[\omega_0^2 + F\cos(\omega_F t)\right] + \frac{1}{4}\gamma q^4 . \tag{23}$$

When the modulation frequency ω_F is close to $2\omega_0$ the oscillator may have two stable states of vibrations at frequency $\omega_F/2$ (period-two states) shifted in phase by π [10]. For $F \ll \omega_0^2$ the oscillator dynamics is characterized by the dimensionless frequency detuning μ, effective Planck constant λ, and relaxation time ζ,

$$\mu = \frac{\omega_F(\omega_F - 2\omega_0)}{F}, \quad \lambda = \frac{3|\gamma|\hbar}{F\omega_F}, \quad \zeta = \frac{F}{2\omega_F\Gamma}. \tag{24}$$

As before, λ will be the small parameter of the theory.

Parametric excitation requires that the modulation be sufficiently strong, $\zeta > 1$. For such ζ the bifurcation values of μ are

$$\mu_B^{(1,2)} = \mp(1 - \zeta^{-2})^{1/2}, \qquad \zeta > 1. \tag{25}$$

If $\gamma > 0$, as we assume, for $\mu < \mu_B^{(1)}$ the oscillator has one stable state; the amplitude of vibrations at $\omega_F/2$ is zero. As μ increases and reaches $\mu_B^{(1)}$ this state becomes unstable and there emerge two stable period two states, which are close in phase space for small $\mu - \mu_B^{(1)}$ (a supercritical pitchfork bifurcation). These states remain stable for larger μ. In addition, when μ reaches $\mu_B^{(2)}$ the zero-amplitude state also becomes stable (a subcritical pitchfork bifurcation). The case $\gamma < 0$ is described by replacing $\mu \to -\mu$.

The classical fluctuation-free dynamics for μ close to μ_B is controlled by one slow variable [30]. The analysis analogous to that for the resonant case shows that, in the Wigner representation, fluctuations are described by one-dimensional diffusion in a potential, which in the present case is quartic in the slow variable. The probability W of switching between the period-two states for small $\mu - \mu_B^{(1)}$ and the decay probability of the zero-amplitude state for small $\mu - \mu_B^{(2)}$ have the form

$$W = C\exp(-R_A/\lambda), \qquad R_A = |\mu_B|\eta^2/2(2\bar{n}+1), \qquad \eta = \mu - \mu_B \tag{26}$$

($\mu_B = \mu_B^{(1,2)}$). The corresponding prefactors are $C_B^{(2)} = 2C_B^{(1)} = 2^{1/2}\pi^{-1}\Gamma\zeta^2|\mu_B||\mu - \mu_B|$. Interestingly, dephasing does not affect the decay rate, to the lowest order in $\mu - \mu_B$. This is a remarkable feature of quantum activation near bifurcation points at parametric resonance.

From (26), at parametric resonance the quantum activation energy R_A scales with the distance to the bifurcation point as η^2. In the limit $\zeta \gg 1$ the same expression as (26) describes switching between period-two states still close but further away from the bifurcation point, where the distance between quasienergy levels largely exceeds their width. In contrast, the exponent for tunneling decay in this case scales as $\eta^{3/2}$ [24].

6 Balanced Dynamical Bridge for Quantum Measurements

Dynamic bistability of a resonantly driven oscillator can be used for quantum measurements in yet another way, which is based on a balanced bridge approach. As a consequence of interstate transitions there is ultimately formed a stationary distribution over coexisting stable states of forced vibrations. The ratio of the state populations w_1, w_2 is

$$w_1/w_2 = W_{21}/W_{12} \propto \exp[-(R_{A2} - R_{A1})/\lambda], \tag{27}$$

where W_{mn} is the switching rate $m \to n$. For most parameter values $|R_{A1} - R_{A2}| \gg \lambda$, and then only one state is predominantly occupied. However, for a certain relation between β and Ω, where $R_1 \approx R_2$, the populations of the two states become equal to each other. This is a kinetic phase transition [32]. A number of unusual effects related to this transition have been observed in recent experiments for the case where fluctuations were dominated by classical noise [9, 33, 34].

In the regime of the kinetic phase transition the oscillator acts as a balanced dynamical bridge: the populations are almost equal with no perturbation, but any perturbation that imbalances the activation energies leads to a dramatic change of $w_{1,2}$, making one of them practically equal to zero and the other to 1. Such a change can be easily detected using, for example, the same detection technique as in the bifurcation amplifier regime. An interesting application of a bridge is measurement of the statistics of white noise in quantum devices. The Gaussian part of the noise does not move a classical oscillator away from the kinetic phase transition. In contrast, the non-Gaussian part, which is extremely important and extremely hard to measure, does, and therefore can be quantitatively characterized.

7 Discussion of Results

It follows from the above results that, both for resonant and parametric modulation, close to bifurcation points decay of metastable vibrational states occurs via quantum activation. The quantum activation energy R_A scales with the distance to the bifurcation point η as $R_A \propto \eta^{\xi}$, with $\xi = 3/2$ for resonant driving and $\xi = 2$ for parametric resonance. The activation energy R_A is smaller than the tunneling exponent. Near bifurcation points these quantities become parametrically different and scale as different powers of η, with the scaling exponent for tunneling (5/4 and 3/2 for resonant driving and parametric resonance, respectively) being always smaller than for quantum activation.

The exponent of the decay rate R_A/λ displays a characteristic dependence on temperature. In the absence of dephasing, for $kT \gg \hbar\omega_0$ we have standard thermal activation, $R_A \propto 1/T$. The low-temperature limit is described by the same expression with kT replaced by $\hbar\omega_0/2$.

Our results show that quantum activation is a characteristic quantum feature of metastable decay of vibrational states. Activated decay may not be eliminated by lowering temperature. It imposes a limit on the sensitivity of bifurcation amplifiers based on modulated Josephson oscillators used for quantum measurements [11, 12]. At the same time, an advantageous feature of dynamically bistable detectors is that they can be conveniently controlled by changing the amplitude and frequency of the modulating signal. Therefore the value of the activation energy may be increased by adjusting these parameters. Another advantageous feature of these detectors, which has made them so attractive, is that they operate at high frequency and have a small response time, which can also be controlled. The results of this work apply to currently studied Josephson junctions, where quantum regime is within reach. They apply also to nano- and micromechanical resonators that can be used for high-sensitivity measurements in the regime of a balanced dynamical bridge.

Acknowledgments I am grateful to M. Devoret for the discussion. This research was supported in part by the NSF through grant No. PHY-0555346 and by the ARO through grant No. W911NF-06-1-0324.

References

1. L.J. Lapidus, D. Enzer, G. Gabrielse, Phys. Rev. Lett. **83**(5), 899 (1999)
2. R. Gommers, P. Douglas, S. Bergamini, M. Goonasekera, P.H. Jones, F. Renzoni, Phys. Rev. Lett. **94**(14), 143001 (2005)
3. K. Kim, M.S. Heo, K.H. Lee, K. Jang, H.R. Noh, D. Kim, W. Jhe, Phys. Rev. Lett. **96**(15), 150601 (2006)
4. I. Siddiqi, R. Vijay, F. Pierre, C.M. Wilson, M. Metcalfe, C. Rigetti, L. Frunzio, M.H. Devoret, Phys. Rev. Lett. **93**(20), 207002 (2004)
5. A. Lupaşcu, E.F.C. Driessen, L. Roschier, C.J.P.M. Harmans, J.E. Mooij, Phys. Rev. Lett. **96**(12), 127003 (2006)
6. J.S. Aldridge, A.N. Cleland, Phys. Rev. Lett. **94**(15), 156403 (2005)
7. R.L. Badzey, G. Zolfagharkhani, A. Gaidarzhy, P. Mohanty, Appl. Phys. Lett. **86**(2), 023106 (2005)
8. C. Stambaugh, H.B. Chan, Phys. Rev. B **73**, 172302 (2006)
9. R. Almog, S. Zaitsev, O. Shtempluck, E. Buks, Appl. Phys. Lett. **90**(1), 013508 (2007)
10. L.D. Landau, E.M. Lifshitz, *Mechanics*, 3rd edn. (Elsevier, Amsterdam, 2004)
11. I. Siddiqi, R. Vijay, M. Metcalfe, E. Boaknin, L. Frunzio, R.J. Schoelkopf, M.H. Devoret, Phys. Rev. B **73**(5), 054510 (2006)
12. A. Lupaşcu, S. Saito, T. Picot, P.C. De Groot, C.J.P.M. Harmans, J.E. Mooij, Nature Physics **3**(2), 119 (2007)
13. H. Kramers, Physica (Utrecht) **7**, 284 (1940)
14. J. Kurkijärvi, Phys. Rev. B **6**, 832 (1972)
15. R. Victora, Phys. Rev. Lett. **63**, 457 (1989)
16. A. Garg, Phys. Rev. B **51**(21), 15592 (1995)

17. M.I. Dykman, M.A. Krivoglaz, Physica A **104**(3), 480 (1980)
18. A.P. Dmitriev, M.I. Dyakonov, Zh. Eksp. Teor. Fiz. **90**(4), 1430 (1986)
19. O.A. Tretiakov, K.A. Matveev, Phys. Rev. B **71**(16), 165326 (2005)
20. M.I. Dykman, I.B. Schwartz, M. Shapiro, Phys. Rev. E **72**(2), 021102 (2005)
21. A.O. Caldeira, A.J. Leggett, Ann. Phys. (N.Y.) **149**(2), 374 (1983)
22. M.I. Dykman, Phys. Rev. E **75**(1), 011101 (2007)
23. M.I. Dykman, V.N. Smelyansky, Zh. Eksp. Teor. Fiz. **94**(9), 61 (1988)
24. M. Marthaler, M.I. Dykman, Phys. Rev. A **73**(4), 042108 (2006)
25. M.I. Dykman, M.A. Krivoglaz, *Soviet Physics Reviews* (Harwood Academic, New York, 1984), vol. 5, pp. 265–441
26. P.D. Drummond, D.F. Walls, J. Phys. A **13**(2), 725 (1980)
27. P.D. Drummond, P. Kinsler, Phys. Rev. A **40**(8), 4813 (1989)
28. G.Y. Kryuchkyan, K.V. Kheruntsyan, Opt. Commun. **127**(4–6), 230 (1996)
29. K. Vogel, H. Risken, Phys. Rev. A **38**(5), 2409 (1988)
30. J. Guckenheimer, P. Holmes, *Nonlinear Oscillators, Dynamical Systems and Bifurcations of Vector Fields* (Springer-Verlag, New York, 1987)
31. H. Haken, *Synergetics: Introduction and Advanced Topics* (Springer-Verlag, Berlin, 2004)
32. M.I. Dykman, M.A. Krivoglaz, Zh. Eksp. Teor. Fiz. **77**(1), 60 (1979)
33. C. Stambaugh, H.B. Chan, Phys. Rev. Lett. **97**(11), 110602 (2006)
34. H.B. Chan, C. Stambaugh, Phys. Rev. B **73**, 224301 (2006)

Poster Session

Dynamics and Noise in dc-SQUID Magnetometer Arrays

John L. Aven, Antonio Palacios, Patrick Longhini, Visarath In and Adi Bulsara

Abstract The natural behavior of most processes is that whenever they are effected by an external phenomenon the outcome of the system is changed. Hence the output of a process can be changed by the input of the system. Many of these processes have the capability of either being in a steady state while at others times they may be induced into a periodic or oscillating state. This type of behavior is common in a system of equations which is used to detect and measure an external phenomena which will change the state of the system. These systems are generally referred to as sensors. Quite often a sensor's effectiveness is increased once it is in a region where it is oscillating. This, however, is not the only way to enhance effectiveness. If one is to connect the sensors in one of many different ways, generally referred to as a coupling, the sensitivity may be increased. In this work we will demonstrate the behavior of a particular type of sensor, known as dc-SQUID (Superconducting Quantum Interference Device) coupled in a nearest-neighboring scheme. We show that the response of the coupled system to an external magnetic field is greatly increased compared to that of a single SQUID element. Applications of this work include: geological equipment, biomedical equipment such as MRI machines, and applications to homeland defense such as mine detection.

1 Introduction

In what follows we will be primarily interested in the behavior of the dc-SQUID magnetometer. This device has the simple 2nd order model [1–3]:

$$\tau_\gamma \dot{\delta}_i = J + \frac{(-1)^i}{\beta}(\delta_1 - \delta_2 - 2\pi x_e) - \sin(\delta_i), \quad i = 1,2, \tag{1}$$

J.L. Aven (✉)
Nonlinear Dynamical System Group, Department of Mathematics, San Diego State University, 5500 Campanile Drive, San Diego, CA 92182, e-mail: chaotic_aven@yahoo.com

V. In et al. (eds.), *Applications of Nonlinear Dynamics*, Understanding Complex Systems, 381
DOI 10.1007/978-3-540-85632-0_31, © Springer-Verlag Berlin Heidelberg 2009

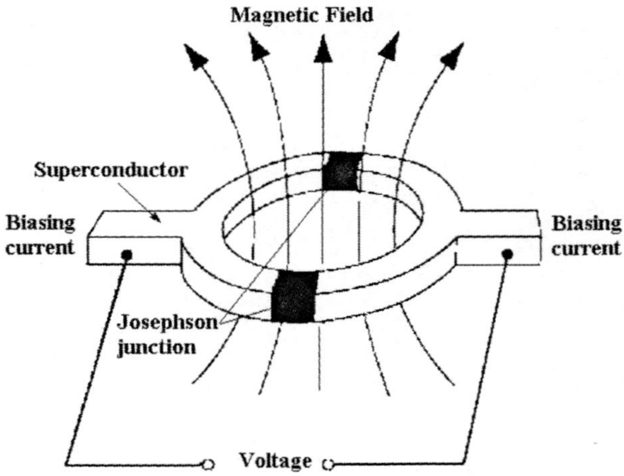

Fig. 1 Schematic diagram of a dc-SQUID magnetometer made from a ring of superconducting material, which is interupted symmetrically by an insulating material, called Josephson Junctions. These junctions short the current flowing through the device while it is driven by a Biasing Current

where τ_γ is the time response rate of the SQUID's internal dynamics (taken to be equal to one from here on), β is the SQUIDs nonlinearity parameter, J is a normalized biasing current, x_e the external magnetic flux normalized by Φ_0, the flux quanta, and $\delta_i (i = 1,2)$ are the phase differences in the current as it crosses the Josephson Junctions (assumed symmetric here) as can be seen in Fig. 1. When we discuss oscillatory behavior we will be referring to the behavior in the variable $\delta = (\delta_1 - \delta_2)/(2\pi)$ which can be easily obtained from solutions of the system

$$\delta = -\frac{1}{\beta}(2\delta - 2\pi x_e)\cos(\Sigma)\sin(\delta), \tag{2}$$

$$\Sigma = J - \cos(\delta)\sin(\Sigma), \tag{3}$$

where $\delta = (\delta_1 - \delta_2)/2$ and $\Sigma = (\delta_1 + \delta_2)/2$. These models have been used to derive a coupled-cell model for an array of N SQUIDs which are coupled either unidirectionally or bidirectionally.

The dynamic behavior of a single, uncoupled, dc-SQUID magnetometer is well understood. The behavior of a coupled system of SQUIDs is, however, more difficult to elucidate. Here we only considered small arrays of dc-SQUIDs coupled in a directed ring manner (unidirectionally and bidirectionally) as is shown in Fig. 2. Using the methods described in [2, 3] we rewrite the governing equations as a Taylor series expansion about the coupling strength. To second order, we arrive at the following set of equations:

$$\dot{\delta}_{kj} = J + \frac{(-1)^j}{\beta}\left(\delta_{k1} - \delta_{k2} - 2\pi x_e + \psi_{k+1} + [\psi_{k-1}]\right) - \sin(\delta_{kj}), \tag{4}$$

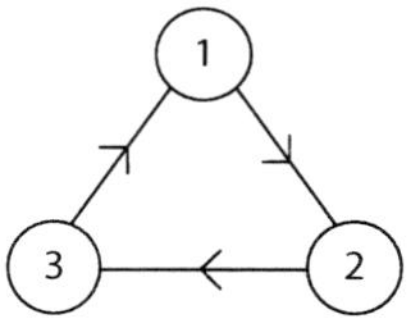

Fig. 2 Schematic diagram of unidirectionally coupled SQUIDs devices represented by circles. Arrows represent the direction of coupling. Bidirectional coupling involves arrows pointing in both directions

where $\psi_k = -2\pi\lambda \frac{I_0}{\beta}(\delta_{k,1} - \delta_{k,2} - 2\pi x_e)$, $k = 1, \ldots, n$, $j = 1, 2$ and I_0 is the critical Josephson Current of the SQUIDs junctions. The term ψ_{k-1} is used only in the bidirectional coupling case, thus it vanishes when we have unidirectional coupling.

2 Bifurcation Analysis

For brevity we examine only the two parameter bifurcations in J and x_e for various values of the coupling parameter λ, as depicted in Fig. 3. For a fixed value of λ, the region where oscillations occur is directly to the right of the cusp point in Fig. 3 with the remaining regions having steady state solutions. We note that as λ increases the region of oscillation decreases, and hence so does the region of detectability. This may be counter constructive yet at the same time it allows the system to be tuned

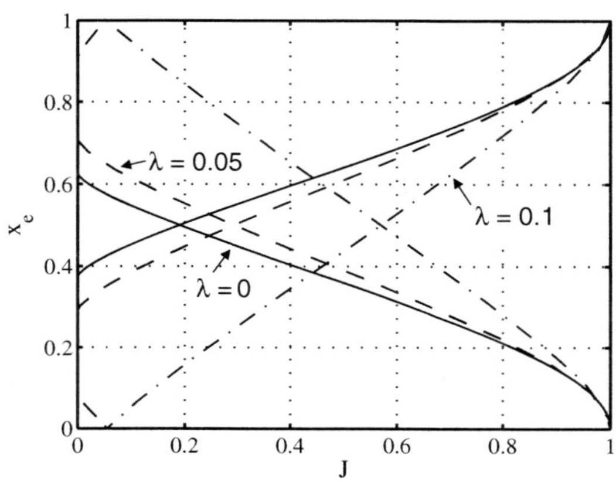

Fig. 3 Two-parameter (J vs. x_e) bifurcation diagram for a system of three SQUIDs unidirectionally coupled. *Solid line* corresponds to the uncoupled system ($\lambda = 0$), *dashed line* is for $\lambda = 0.05$, and *dash-dot line* is for $\lambda = 0.1$. Oscillatory behavior exists for parameter values to the *right* of the cusp, all other behavioral types are steady states [2, 3]

to optimal detection for a given x_e. The bidirectional case yields similar results, for more details the interested reader is referred to [2, 3].

3 Signal Response in the Presence of Noise

In this section we analyze the response of the coupled device in the presence of additive White Gaussian noise. This type of noise is modeled by coupling a standard Wiener Process to each junction with intensity D. Simulating the noise and calculating the Signal-to-Noise Ratio for a fixed intensity of $D = 0.001$ produces Fig. 4, where x_e is set to 0.5, β to 0.8, and J to 0.5. We observe that a maximal signal response occurs for an intermediate coupling strength λ_M. This indicates that the coupled system can yield an optimal response by simply tuning the coupling strength between the SQUIDs.

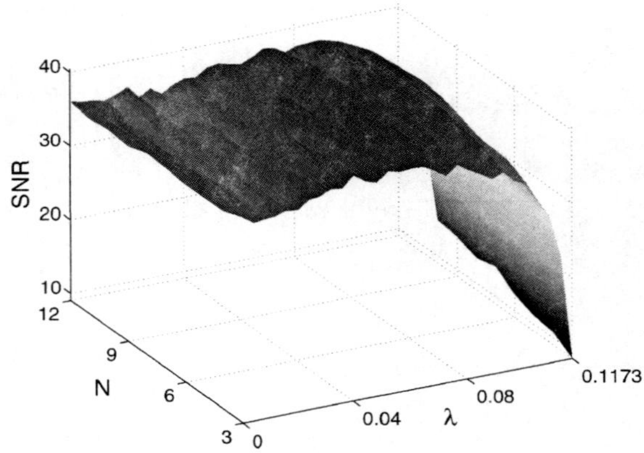

Fig. 4 Signal-to-Noise Ratio of a system of three dc-SQUIDs, coupled unidirectionally (noise intensity $D = .001$) as a function of coupling strength λ and the number N of SQUIDs in the system

Acknowledgments The authors acknowledge support from the Office of Naval Research (Code 331). JA was supported in part by the Office of Naval Research Naval Research Enterprise Intern Program (ONR NREIP). PL was supported in part by the US Space and Naval Warfare Systems Command Grant N66001-06-M-108. AP was supported by National Science Foundation grants CMS-0625427 and CMS-0638814.

References

1. Acebron, J. A., Rappel, W.-J., Bulsara, A. R.: Cooperative Dynamics in a Class of Coupled Two-Dimensional Oscillators. Phys. Rev. E 67, 016210 (2003)
2. Aven, J.: Networks of Coupled SQUID Magnetometers (Master's Thesis). San Diego State University, San Diego CA (2007)
3. Palacios, A., Aven, J., Longhini, P., In, V., Bulsara, A.R.: Cooperative Dynamics in Coupled Noisy Dynamical Systems Near a Critical Point; the dc SQUID as a Case Study. Phys. Rev. E, 74(2), 021122 (2006)

Stochastically Forced Nonlinear Oscillations: Sensitivity, Bifurcations and Control

Irina Bashkirtseva

Abstract Stochastic fluctuations of nonlinear oscillations play an important role for understanding of the corresponding dynamical nonlinear phenomena. The various noise-induced transitions through periodic to more complicated chaotic regimes are a central problem in modern nonlinear dynamics stochastic theory. The sensitivity analysis of random forced oscillations is a basic point for investigation of these transitions. We consider stochastic oscillations for nonlinear randomly forced dynamic systems. For the well-known model of Brusselator, we study the noise-induced transition to chaos. Our approach is based on quasipotential function. The approximation of quasipotential (sensitivity function) gives us the constructive and detailed description of nonlinear system response to the random disturbances. New critical values of Brusselator parameters with the help of sensitivity function have been found. For these values, small background disturbances transfer Brusselator from regular oscillations to chaotic regime. The problem of stochastic attractor synthesis with desired probabilistic characteristics is discussed. Effective algorithms for controlling chaos are suggested.

1 Stochastic Sensitivity of Limit Cycles

For many dynamical processes with regular oscillations, the basic mathematical model is the non-linear deterministic system

$$\dot{x} = f(x) \tag{1}$$

with T-periodic solution $x = \xi(t)$. Here x is $n-$vector, $f(x)$ is $n-$vector function. Let γ be a phase curve (limit cycle) of T-periodic solution $\xi(t)$.

I. Bashkirtseva (✉)
Department of Mathematics, Ural State University, Lenina, 51, 620083,
Ekaterinburg, Russia, e-mail: irina.bashkirtseva@usu.ru

V. In et al. (eds.), *Applications of Nonlinear Dynamics*, Understanding Complex Systems,
DOI 10.1007/978-3-540-85632-0_32,

Consider a system of stochastic differential equations (in Ito's or Stratonovich's sense)

$$\dot{x} = f(x) + \varepsilon\sigma(x)\dot{w}. \tag{2}$$

Here $w(t)$ is a n-dimensional Wiener process, $\sigma(x)$ is $n \times n$-matrix function of disturbances with intensity ε.

The random trajectories of forced system (2) leave the closed curve of deterministic cycle γ and due to cycle stability form some bundle around it.

The detailed description of random distribution dynamics of this bundle is given by Kolmogorov-Fokker-Planck (KFP) equation. If the character of transient is inessential and main interest is connected with regime of steady-state stochastic auto-oscillations then it is possible to restrict the research by analysis of a stationary density function $\rho(x,\varepsilon)$. Analytical research of stationary KFP equation for stochastic limit cycles considered here is a very difficult problem. Under these circumstances asymptotics and approximations based on quasipotential $v(x) = -\lim_{\varepsilon\to 0}\varepsilon^2 \log\rho(x,\varepsilon)$ are actively used.

The probabilistic distribution for the bundle of random trajectories localized near cycle has Gaussian approximation

$$\rho \approx Ke^{-\frac{v(x)}{\varepsilon^2}} \approx K\exp\left(-\frac{(\Delta(x), \Phi^+(\gamma(x))\Delta(x))}{2\varepsilon^2}\right)$$

with covariance matrix $\varepsilon^2\Phi(\gamma)$. This covariance matrix characterizes a dispersion of the points of intersection of random trajectories with hyperplane orthogonal to cycle at the point γ. A function $\Phi(\gamma)$ is a stochastic sensitivity function (SSF) of limit cycle. This function allows to describe non-uniformity of a bundle width along cycle for all directions. It gives the simple way to indicate the most and the least sensitive parts of cycle to external noises.

It is convenient to search for a function $\Phi(\gamma)$ in parametric form. The solution $\xi(t)$ connecting the points of cycle γ with points of an interval $[0,T)$ gives the natural parametrization $\Phi(\xi(t)) = W(t)$. Matrix function $W(t)$ is a solution of Lyapunov equation [3]

$$\dot{W} = F(t)W + WF^\top(t) + P(t)S(t)P(t), \tag{3}$$

with conditions

$$W(0) = W(T) \tag{4}$$

$$W(t)r(t) \equiv 0, \tag{5}$$

Here

$$F(t) = \frac{\partial f}{\partial x}(\xi(t)),\ \ S(t) = \sigma(\xi(t))\sigma^\top(\xi(t)),$$

$$r(t) = f(\xi(t)),\ \ P(t) = P_{r(t)},\ \ P_r = I - rr^\top/r^\top r,$$

where P_r is a projection matrix onto the subspace orthogonal to the vector $r \neq 0$.

1.1 Sensitivity Analysis of 2D-Cycles

For the case $n = 2$ the projection matrix is given by $P(t) = p(t)p^T(t)$, where $p(t)$ is a normalized vector orthogonal to $f(\xi(t))$. As a result the matrix $W(t)$ is written as $W(t) = \mu(t)P(t)$. Here $\mu(t) > 0$ is T-periodic scalar stochastic sensitivity function [1]. The value $M = \max \mu(t),\ t \in [0,T]$ plays an important role in the analysis of stochastic dynamics about a limit cycle. We shall consider M as a *sensitivity factor* of a cycle γ response to random disturbances.

Consider forced system

$$\begin{aligned} \dot{x} &= a - (b+1)x + x^2y + \varepsilon\theta \\ \dot{y} &= bx - x^2y \end{aligned} \tag{6}$$

received by the addition of small disturbances $\varepsilon\theta(t)$ to classical Brusselator.

It is known that for $b > \bar{b} = 1 + a^2$ the unforced system ($\varepsilon = 0$) has a limit cycle ($\bar{b}$ is bifurcation value). We consider the results of the comparative analysis of this system cycles for a fixed $a = 0.2$ and various values of parameter $b > \bar{b} = 1.04$ from an interval [1.06,1.07].

Let disturbances in (6) be stochastic: $\theta(t) = \dot{w}$, where $w(t)$ is an independent Wiener process.

For Lyapunov exponent λ and sensitivity factor M dependence on values b is shown in Fig. 1.

As we can see in Fig. 1a, a parameter λ monotonically decreases with growth b. This means increase of a stability degree of a cycle to disturbances of initial data. One should think it should be accompanied by the appropriate decrease in the sensitivity of a cycle to random disturbances. However, here the converse is observed. The value M behaves absolutely otherwise (see Fig. 1b).

On an examined interval the function $M(b)$ is not monotonic. Its graph has sharp high peak. As a result the function $M(b)$ has an essential overfall of values. By critical value of parameter b here is $b_* = \arg\max_b M(b) = 1.064082$, $M(b_*) = 8.8 \cdot 10^{10}$.

In Fig. 2 (left), the random trajectories found by a direct numerical simulation for b_* and $\varepsilon = 10^{-5}$ are demonstrated. For small stochastic disturbances the burst of response amplitude is observed.

Let disturbances in (6) be periodic: $\theta(t) = \cos(\omega t)$, where ω is a frequency.

For a critical parameter value $b_* = 1.064082$ increase in the periodic force intensity ε results in the period-doubling bifurcations of system (6) attractors: 1-cycle ($\varepsilon = 0.0005$) $\rightarrow$ 2-cycle ($\varepsilon = 0.0007$) $\rightarrow$ 4-cycle ($\varepsilon = 0.000763$) and so on. For $\varepsilon = 0.00085$ the bundle of trajectories (see Fig. 3 (left)) looks chaotic. For a critical parameter value $b_* = 1.064082$ the forced Brusselator is a *generator of chaos* [2]. Thus, the function of sensitivity is the useful analytical tool for the prediction of singular responses of a non-linear system both to stochastic and to periodic disturbances [4, 5].

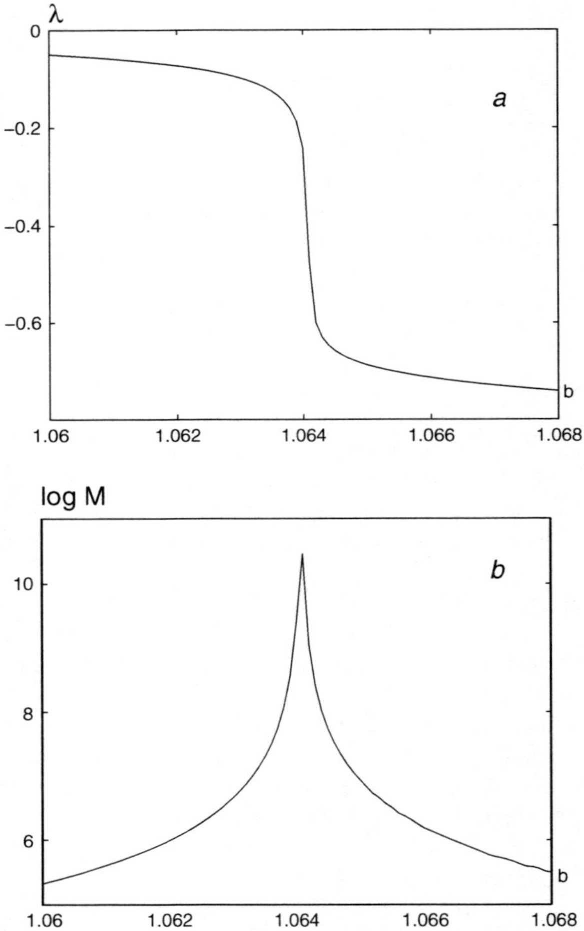

Fig. 1 Deterministic Lyapunov exponent (**a**) and stochastic sensitivity factor (**b**)

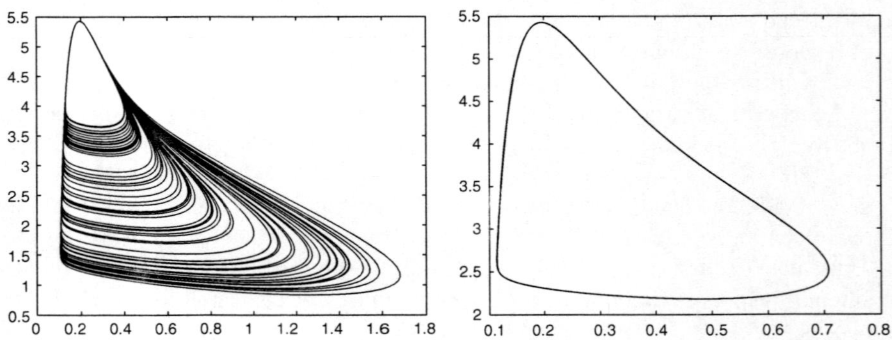

Fig. 2 Stochastic forced Brusselator

2 Sensitivity Control

Consider a stochastic system with a control of the form

$$\dot{x} = f(x,u) + \varepsilon\sigma(x,u)\dot{w}, \tag{7}$$

where x is n-dimensional state variable, u is r-dimensional vector of control functions, $f(x,u)$, $\sigma(x,u)$ are vector functions, $w(t)$ is n-dimensional Wiener process, ε is scalar parameter of disturbances intensity. It is supposed that for $\varepsilon = 0$ and $u = 0$ the system (7) has $T-$periodic solution $x = \xi(t)$ with a phase trajectory γ (cycle).

The stabilizing regulator we shall select from the class U of admissible feedbacks $u = u(x)$ satisfying conditions:

(a) $u(x)$ is sufficiently smooth and $u|_\gamma = 0$;
(b) for the deterministic system

$$\dot{x} = f(x,u)$$

the solution $x = \xi(t)$ is exponentially stable in the neighbourhood Γ of cycle γ.

2.1 Control and SSF

Consider in detail the case $n = 2$. Sensitivity function μ for $u \in U$ is a solution of boundary-value problem

$$\dot{\mu} = a(t)\mu + b(t), \;\; \mu(0) = \mu(T). \tag{8}$$

Here

$$\begin{aligned} a(t) &= p^\top(t)(F^\top(t) + F(t))p(t), \\ b(t) &= p^\top(t)S(t)p(t) \\ F(t) &= \frac{\partial f}{\partial x}(\xi(t),0) + \frac{\partial f}{\partial u}(\xi(t),0)\frac{\partial u}{\partial x}(\xi(t)) \\ S(t) &= \sigma(\xi(t),0)\sigma^\top(\xi(t),0), \end{aligned} \tag{9}$$

$p(t)$ is normalized vector orthogonal to cycle γ at a point $\xi(t)$.

As we see, the variation of control u allows to change the only coefficient $a(t)$ in the eq. (9). Note that outcome of control depends only on values of the derivative $\dfrac{\partial u}{\partial x}$. It gives us possibility to simplify the structure of used regulator. and to restrict our consideration without loss of generality by more simple regulators in the following form

$$u = K(t(x))\Delta(x). \tag{10}$$

Thus the feedback coefficient matrix $K(t)$ completely determines capabilities of the regulator (10) to synthesize SSF $\mu(t)$.

Connect controlled coefficient $a(t)$ in the eq. (8) with feedback matrix $K(t)$ directly. Really, it follows from (9) that

$$\begin{aligned} a(t) &= a_0(t) + a_1(t) \\ a_0(t) &= 2q^\top(t)p(t) \\ a_1(t) &= 2\beta^\top(t)k(t), \end{aligned} \tag{11}$$

where

$$\begin{aligned} q(t) &= A(t)p(t),\ \beta(t) = B(t)p(t) \\ A(t) &= \left[\frac{\partial f}{\partial x}(\xi(t),0)\right]^\top \\ B(t) &= \left[\frac{\partial f}{\partial u}(\xi(t),0)\right]^\top \\ k(t) &= \frac{\partial u}{\partial x}(\xi(t))p(t) \ = \ K(t)p(t) \end{aligned} \tag{12}$$

Note that the vector k is a derivative of a control function u in the direction of normal vector p.

2.2 Control Goal and Choice of Regulator Parameters

The aim of control is the synthesis of desired SSF for cycle γ of stochastic system (7). Let $\bar{\mu}(t) \in M$ be a some assigned SSF. Here

$$M = \{\mu \in C^1_{[0,T]} \mid \mu(t) > 0, \quad \mu(0) = \mu(T)\}.$$

Denote by μ_u a SSF of cycle γ for stochastic system with control $u \in U$.

The function $\bar{\mu}(t)$ is connected with control parameter $k(t)$ by the following equation

$$\beta^\top(t)k(t) = \frac{\bar{a}_1(t)}{2}, \tag{13}$$

where

$$\bar{a_1}(t) = (\dot{\bar{\mu}}(t) - a_0(t)\bar{\mu}(t) - b(t))/\bar{\mu}(t).$$

The eq. (13) has infinite set of the solutions (control is not unique). Consider additional optimal criterion

$$||k(t)||^2 \longrightarrow \min. \tag{14}$$

The problem (13) and (14) has the unique solution

$$\bar{k}(t) = \frac{\bar{a_1}(t)\beta(t)}{2\beta^\top(t)\beta(t)}$$

Here feedback matrix in (10) can be found as follows

$$\bar{K}(t) = \frac{\bar{a_1}(t)}{2\beta^\top(t)\beta(t)}B(t).$$

Consider forced Brusselator (6) with control

$$\dot{x} = a - (b+1)x + x^2y + u_1 + \varepsilon\theta_1$$

$$\dot{y} = bx - x^2y + u_2 + \varepsilon\theta_2, \qquad a = 0.2$$

where ε is an external force intensity, u_1 and u_2 are control functions.

For $b_* = 1.064082$ in absence of control this system for stochastic and periodic disturbances $\theta_i(t)$ is generator of chaos (see Sect. 2.1). An extraordinary sensitivity of Brusselator is connected with huge values of stochastic sensitivity function.

Now we demonstrate the controlling chaos for this model. Let us take $\bar{\mu}(t) \equiv 5$. Such choice of SSF is dictated by the desire to have oscillations with small sensitivity.

The results of a direct numerical simulation of the forced trajectories for controlled Brusselator are presented in Fig. 2 (for stochastic $\theta_i(t) = \dot{w}_i$) and Fig. 3 (for periodic $\theta_1 = cos(\omega t)$, $\theta_2 = 0$. Here left plots demonstrate dynamics of system

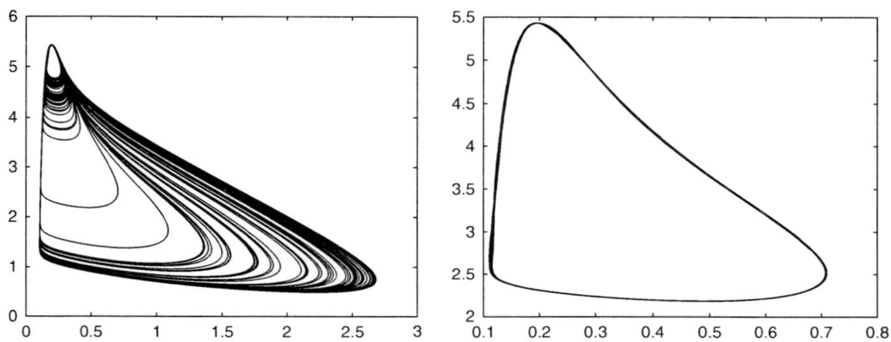

Fig. 3 Periodic forced Brusselator

without control and right plots with control. As we see, constructed regulator gives us the solution of chaos control problem.

3 Conclusion

Suggested stochastic sensitivity function technique is a useful tool for a quantitative description for a system response on the random external disturbances. Using SSF, we can predict some singularities in dynamics of stochastically and periodically forced oscillators. The critical (chaotic) values of Brusselator parameter were found and the thin effects observed in stochastic Lorenz model near chaos in a period-doubling bifurcations zone were investigated. The new method of stochastic cycles control based on SSF allows to form the stochastic attractor with desired features by feedback regulator. This constructed regulator really provides the solution of a controlling chaos problem.

Acknowledgments This work was partially supported by RFBR grants 06-01-00625, 06-08-00396, 07-01-96079ural.

References

1. Bashkirtseva, I. A., Ryashko L. B.: Sensitivity analysis of the stochastically and periodically forced Brusselator. Physica A. **278**, 126–139 (2000)
2. Bashkirtseva, I. A., Ryashko L. B.: Sensitivity analysis of stochastically forced Lorenz model cycles under period-doubling bifurcations. Dynamic Systems and Applications. **11**, 293–309 (2002)
3. Bashkirtseva, I. A., Ryashko L. B.: Stochastic sensitivity of 3D-cycles. Mathematics and Computers in Simulation. **66**, 55–67 (2004)
4. Fradkov, A. L., Pogromsky, A. Yu.: *Introduction to control of oscillations and chaos*. World Scientific Series of Nonlinear Science Singpore (1998)
5. Freidlin, M. I., Wentzell A. D.: *Random perturbations of dynamical systems*. Springer, New York (1984)

Simultaneous, Multi-Frequency, Multi-Beam Antennas Employing Synchronous Oscillator Arrays

J. Cothern, T. Heath, G. Hopkins, R. Kerr, D. Lie, J. Lopez and B. Meadows

The theoretical foundations of coupled, nonlinear oscillator arrays as applied to beam forming have, almost exclusively, presumed the unit cells are well described by a Van der Pol oscillator model. In the past, a "weak" association was made between the spectral output of the differential-pair oscillator and that of an ideal Van der Pol oscillator. By applying the Method of Multiple Scales to the Van der Pol dynamical equations, one finds that only odd order harmonics are present. Moreover, one can determine the key Van der Pol parameters (i.e., the amplitude parameter, p, and nonlinearity parameter, μ) through the power contained in the first and third harmonics. Forgoing the details of the derivation, this relationship is given by:

$$p = \frac{1}{4} \times 10^{H_1/10} \text{and } \mu = \frac{4}{p^{3/2}} \times 10^{H_3/20}$$

where H_1 and H_3 are the powers contained in the first and third harmonics, respectively. The spectral content of an ideal differential-pair oscillator also only possesses odd harmonics of the fundamental frequency. By measuring the power contained in the first and third harmonics, the above relationships can be applied; the resulting parameter values can be used describe the differential-pair oscillator with an "associated" Van der Pol oscillator. Obviously, this approach cannot rigorously prove that the underlying dynamics of the differential-pair oscillator is given by the Van der Pol equations. However, it does provide us with a motivation to use cross-coupled differential oscillators for this purpose. To this end we are developing, in tandem, two oscillator and circuit designs: a discrete board based oscillator and an ASIC design. For the discrete design we have largely used ADS and its' Harmonic Balance tool to simulate the basic oscillator unit and the coupled oscillator array. The discrete design leverages off of earlier non-published work that was somewhat successful in achieving the desired beam-forming results. For the ASIC design we have largely

J. Cothern (✉)
SPAWAR Systems Center, San Diego, CA
Georgia Tech Research Institute, Atlanta, GA
Texas Tech. University, Lubbock, TX

V. In et al. (eds.), *Applications of Nonlinear Dynamics*, Understanding Complex Systems,
DOI 10.1007/978-3-540-85632-0_33,

used CADENCE and it's time based transient analysis tool to simulate the basic oscillator unit and its' arrays. The designs are somewhat independent of each other though the two design teams are coordinated. If successful, the ASIC design could reduce the electronics for a phased array antenna into a single small IC (improved cost, reliability, power optimization) design with multiple Defense and commercial applications. It remains to be seen if either design is Van der pol-like enough to match to the existing beam-forming model. Fundamental mathematical analysis of the circuits has proved non-trivial and we have yet to range test either current design.

The discrete oscillator design uses a cross-coupled differential pair of matched transistors. The differential structure inherently suppresses even-order harmonics of the fundamental oscillation frequency when the output signal is taken differentially across symmetrically located points in the circuit. Due to the resonant length of the inter-oscillator coupling transmission lines, coupling energy spectral components at the fundamental frequency and its odd-order harmonics controls the array inter-oscillator phase-shift stability for a given resonator tuning arrangement, while coupling energy at even-order harmonics tends to decrease phase-shift stability.

Differential symmetry is maintained in the inter-oscillator coupling transmission lines, such that each oscillator is connected to each of its nearest neighbors by a pair of differential transmission lines. (Fig. 1) shows a circuit model, executed in Agilent Advanced Design System (ADS) software, of a single differential oscillator. The model uses microstrip elements to form the inductive part of the oscillator's resonator and for the transmission line elements for inter-oscillator coupling. The coupling lines are connected via DC blocking capacitors to opposite ends of the parallel LC resonator and to the collectors of the two transistors. This oscillator is the central element in an array of three oscillators, and has differential coupling lines connected to its neighbors on either side.

Use of closed commercial software such as ADS for analysis of coupled oscillator arrays is problematic in that oscillator analysis tools provided by the software vendor do not easily find simultaneous solutions for multiple coupled oscillators. Typically the software measures the loop gain and phase of a single oscillator vs. frequency, and uses this information to set up initial circuit conditions for the harmonic balance oscillator plus analysis. These initial conditions are generally not sufficiently accurate for convergence of the harmonic balance analysis of a multiple oscillator array.

Suarez and Quere [1] have shown methods for establishing initial conditions in commercial harmonic balance analysis software through use of auxiliary signal generators. This technique was employed for the present array by installing auxiliary generators in the end oscillator array elements. The initial auxiliary generator voltages and phases are controlled by the initial conditions found by the ADS analyzer for corresponding points in the central oscillator circuit. After an initial harmonic balance solution for the entire array is found with the auxiliary generators in place, a series of optimizations are executed such that at final convergence, the auxiliary signal generators are effectively removed from the circuit because their amplitudes and phases exactly match those of their connection points in the oscillators' circuits at steady state. Figure 1 shows the auxiliary generator connection points in ADS

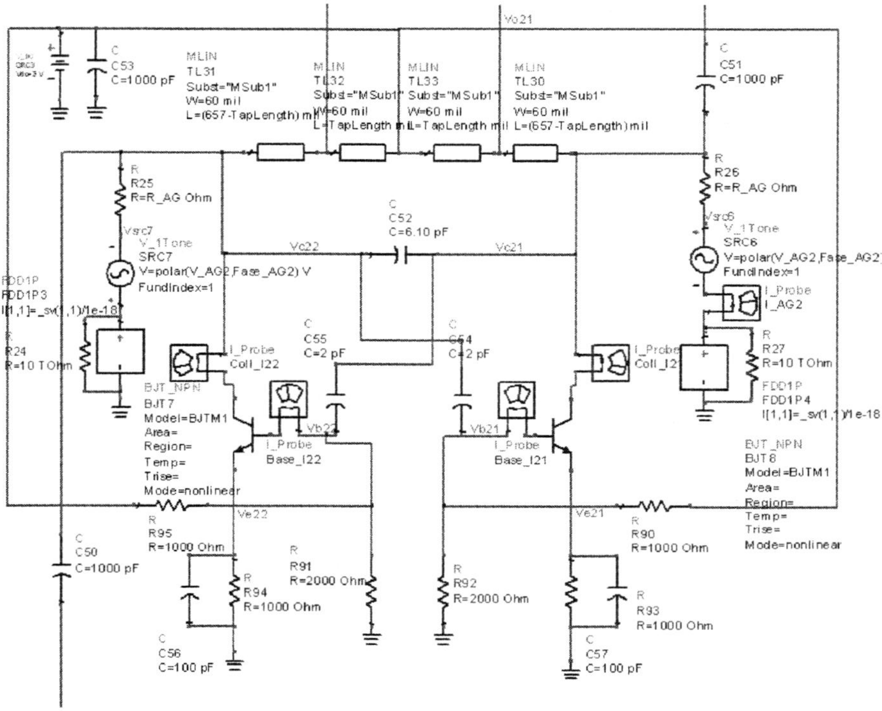

Fig. 1 Discrete Unit Cell Oscillator modified for simulation in array using HB simulator. The auxiliary generators operate only at the fundamental frequency of the harmonic balance analysis. They are connected in series with an ideal model element (*box*) that has very low impedance at the fundamental frequency and very high impedance at all other harmonic frequencies

model for the end elements of the oscillator array. The auxiliary generators operate only at the fundamental frequency of the harmonic balance analysis. They are connected in series with an ideal model element (box) that has very low impedance at the fundamental frequency and very high impedance at all other harmonic frequencies. Simulation Results for broadside (no beam-forming) are shown in Fig. 2. The 3 oscillators are coupled with collector voltage in phase and with suppression of odd order harmonics as required by the use of the oscillator in replacing the ideal Van der pol oscillator. Tuning the oscillators with the voltage control signal creates the phase difference and associated beam-forming.

The ASIC design consists of a fully integrated VCO with various one-dimensional (1-D) arrays already designed and simulated in the IBM 7HP BiCMOS design kit. The VCO topology is a negative-Gm differential L-C oscillator that uses on-chip spiral inductors, on-chip MOS accumulation-mode varactors, and integrated with high-breakdown bipolar SiGe transistors. The VCO exhibits Van der Pol-like behavior as the 2nd order harmonics are suppressed to below -100 dBc, and the 3rd order harmonics fall below ~ -25–45 dBc.

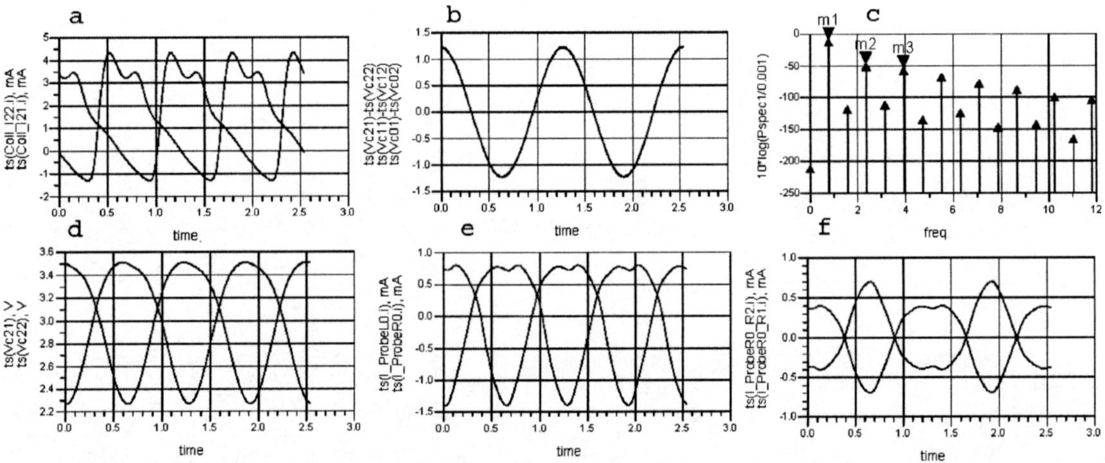

Fig. 2 Coupled Oscillator Simulation Results. (**a**) Differential oscillator transistor collector Currents. (**b**) Differential collector voltages for 3 oscillators – in phase, overlaid (symmetry about y-axis indicates suppression of even order harmonics). (**c**) Oscillator output power spectrum – odd order harmonics dominate. (**d**) Differential oscillator transistor collector voltages. (**e**) Currents feeding differential coupling lines for center oscillator. (**f**) Currents in coupling lines to *left* and *right* of center oscillator – one side of diff. coupling only

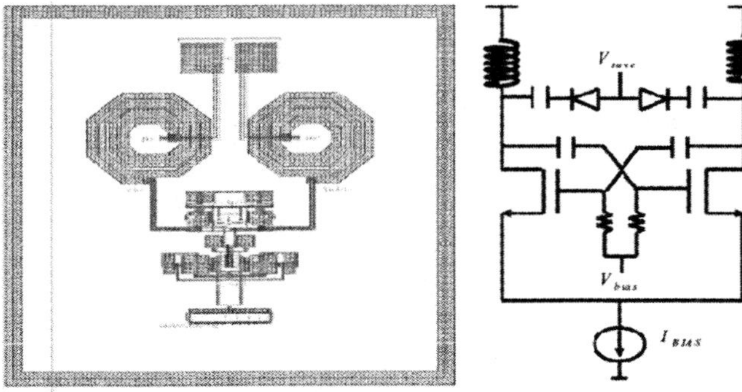

Fig. 3 Simplified schematic and layout for single ASIC VCO element

1-D VCO (simplified schematic and layout shown in Fig. 3) arrays that use both ideal and on-chip resistor elements for passive coupling have been simulated in SPICE. The simulated 1-D arrays include 1×3, 1×5, 1×7 and 1×9 configurations and for several coupling strength (by adjusting values of the ideal coupling resistance) and across several control tuning voltages (i.e., the tuning voltage that changes the oscillation frequencies of each VCO). For most designs, the VCOs at the two edges of these arrays are terminated to 50 ohm for comparing with the theory of nonlinear dynamic VCO-coupled-array previously developed. In some versions of highly-integrated designs, all components of the simulated VCO arrays are integrated on-chip, including an on-chip PTAT (Proportional To Absolute Temperature) bias current cell. In those highly-integrated designs, the only off-chip element for the VCO array is an off-chip precision resistor that can control the bias current within 1% accuracy.

Initial SPICE convergence problems and memory and network issues were overcome to carry out the design for those coupled oscillators in Cadence Spectre. The current consumption for each VCO varies from 4–8 mA at 2.5 V, while in some designs the VCO supply voltage and bias current are adjustable. SPICE simulations show that the tuning range of the 1-D VCO arrays depends on the value of the coupling resistance (therefore the coupling strength). Note these integrated VCO circuits are mainly different from the discrete VCO designs on two accounts: (1) the lower tank quality factor (i.e., lower Q) due to on-chip inductors and varactors, which affect the power consumption, start-up, and probably the coupling behaviors of the VCO array; and (2) the expected lower tuning range due to the availability of on-chip varactors vs. the discrete varactors. Some of the simulated 1-D VCO arrays will be taped out and expected to be back in March 2008 with some preliminary test results available in May 2008. Significant PCB circuit board design will be required to test those integrated ASIC in the Lab.

For the antenna, we needed to design build and test an aperture capable of supporting operation over wide bandwidths for the purpose of demonstrating – at minimal cost – the capabilities of nonlinear oscillator beam control. In particular, at least on octave of instantaneous bandwidth is targeted. To this end, an array of Vivaldi notch radiators (a member of the class of tapered slot radiators) was deemed appropriate. The design was realized using the Finite Element Method program of the High Frequency Structure Simulator (HFSS).

Initially, the design approach was to realize the Vivaldi array in printed circuit form. However, in order to remove the microstrip transitions (and, thereby, increase the operating bandwidth), a double ridged-horn instantiation was chosen. At present, the radiators are designed to support single, linear polarization.

To optimize the single radiator performance, the effect of different tapers on bandwidth and pattern were analyzed. Moreover, the overall size of the model was adjusted to fit the bandwidth specification. Further optimization required adjustment of the ridged waveguide feed line and slot width (to impedance match the 50 ohm coaxial feed).

Due to mutual coupling effects, an optimal single radiator design is likely to be suboptimal in the array environment. To modify the radiator design for optimal performance over the desired bandwidth in an array, the active S-parameters were analyzed. The negative impacts of reflections were minimized through (1) adjusting the distance to the shorting plate and feed and (2) adjusting the distance between the feed and pin. In addition, the ridge tapers were modified in order to achieve good (i.e., non-bifurcated) patterns over the operating bandwidth. However, this result is for an “infinite array” environment; more optimization is required to achieve such results with a finite array. As expected, the patterns broaden significantly at the lower frequencies; this is due to the static inter-element spacing. The Vivaldi radiator separation should be set such that no grating lobes appear at its highest operating frequency.

Initially, modeling predicted reflection coefficient performance that was higher than desired. Additionally, the data was not considered of high value as the model required many more mesh generation iterations and never reached believable convergence.

To correct this problem, a new horn model was developed which had a larger step size and a correspondingly smaller total mesh. The specifications were also relaxed for the new horn as it is believed that the high mutual coupling in the linear array at

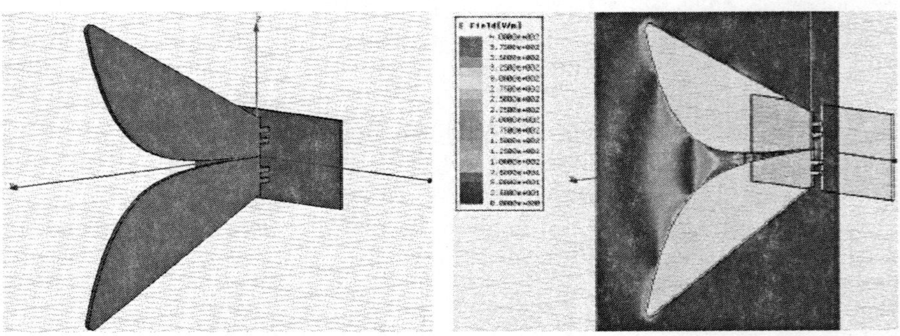

Fig. 4 Single horn radiating element

the lower desired frequencies will not permit a good impedance match. The single element and a view of the magnitude of the electric field for a frequency near the middle of the desired band are shown in Fig. 4.

The single radiator has been replicated in a larger array model comprised of 7 elements and has been simulated. The inter-element spacing of this array has been increased to reduce the mutual coupling and will provide grating-lobe-free operation for the desired frequency band. Simulation difficulties with the array arose when we attempted to simulate a finite array (as opposed to single element or infinite arrays). This was similar to the difficulty in simulating the single element and was solved by setting the correct discretization for the simulation.

Our work over the next year will include range testing for the discrete oscillator circuit and antenna. We will also concentrate on mathematical analysis for the discrete and ASIC circuits such that their compatibility for use in beam forming can be shown in closed form. The ASIC design's first iteration will be tested in the lab and eventually with the 7 element radiator.

References

1. Almadena Suarez and Raymond Quere , "Stability Analysis of Nonlinear Microwave Circuits" Chapter 2, "Nonlinear Analysis Techniques", Artech House, Norwood, MA (2003)

Effects of Nonhomogeneities in Coupled, Overdamped, Bistable Systems

M. Hernandez, V. In, P. Longhini, A. Palacios, A. Bulsara and A. Kho

Abstract Previous work analyzed systems of coupled overdamped bistable elements unidirectionally coupled which exhibited oscillatory behavior due in part to the nature of the coupling scheme. These systems have been assumed to be homogeneous in their thermal- and material-related parameter. In this work, we extend the analysis to study the effects on the network of employing elements that are nonhomogeneous with respect to the thermal- and material-dependent parameter that controls the topology of the potential function and the bistability of each element. In particular, we quantify the effects of nonhomogeneity on the onset of oscillations and the response of the network to external (assumed to be constant) perturbations. Though the potential applications of this work include signal detection and characterization for a large class of sensor systems, we focus on the applicability of this novel coupling scheme and analysis towards fluxgate magnetometers.

1 Introduction

Building on recent theoretical and experimental work [2, 3] which has demonstrated that globally asymptotically stable oscillations in networks of overdamped bistable elements can be induced via well-designed coupling schemes after a control parameter has exceeded a certain threshold value, and with the appropriate use of initial conditions, we extend the analysis here to account for the effects of nonhomogeneities.

Our analysis centers on the effects of nonhomogeneities in the thermal- and material-dependent parameter c that controls the onset of oscillations in a specific prototype system. That system, the "Fluxgate Magnetometer", is a soft-potential

M. Hernandez (✉)
Nonlinear Dynamical System Group, Department of Mathematics, San Diego State University, 5500 Campanile Drive, San Diego, CA 92182, USA,
e-mail: mayra.alina@yahoo.com

V. In et al. (eds.), *Applications of Nonlinear Dynamics*, Understanding Complex Systems,
DOI 10.1007/978-3-540-85632-0_34, © Springer-Verlag Berlin Heidelberg 2009

mean-field hysteritic "singled-domain" ferromagnetic sample read out in the time domain governed by a potential energy function of the form

$$\frac{dx}{dt} = -\Delta U(x), \tag{1}$$

Note that x is the time dependent state-variable and U is a bistable potential function [2, 3].

A Nonhomegeneous Fluxgate network consists of N (odd) unidirectionally coupled elements with cyclic boundary conditions and is described by

$$\dot{x}_i = -x_i + \tanh\left(c_i(x_i + \lambda x_{i+1} + \varepsilon)\right), \qquad i = 1, ..., N \bmod N, \tag{2}$$

where $x_i(t)$ represents the magnetic flux at the output of element i, and $\varepsilon << U_0$ is an external dc target signal, U_0 being the energy barrier height for each element, λ is the coupling strength, and c_i is the thermal- and material dependent parameter for each element i.

2 Onset of Oscillations

Numerical simulations of the network Eq. (2) indicate, see Fig. 1, that a leftward shifting of the threshold values λ_c which marks the onset of oscillations is a consequence of the ring topology's nonhomogeneity in its c_i values.

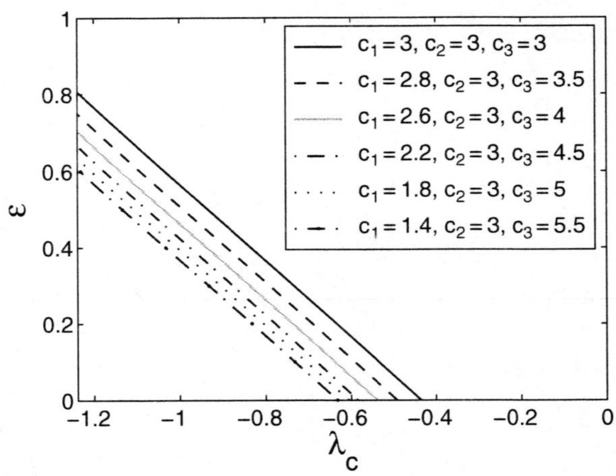

Fig. 1 Two-parameter bifurcation curves outlining the onset of oscillations in the nonhomogenous coupled fluxgate system (2) for different combinations of the parameter c. For any given curve, oscillations occur for values of (ε, λ) that are below that curve. The curve shifts to the left when at least one of the c_i increases [1]

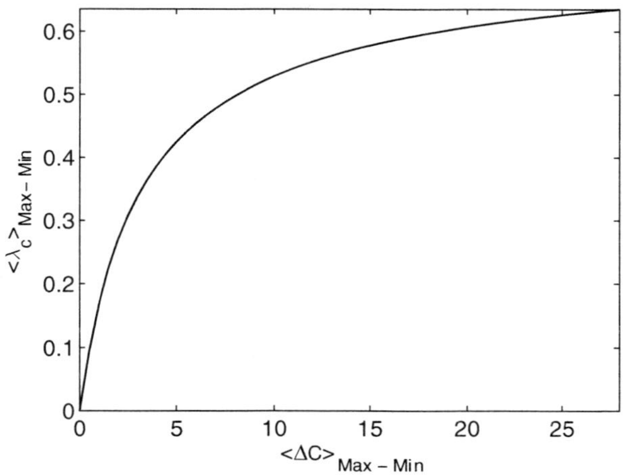

Fig. 2 Critical coupling strength that is required for inducing oscillations, as a function of the separation of the temperature-related parameters $c_{i's}$ between their maximum and minimum values. There is a limit to the minimum coupling strength that will drive self-sustaining oscillations

Hence, we conclude that a nonhomogeneous system behaves like a homogeneous system whose c values equal the largest of the c_i values in the nonhomogeneous system, i.e. $c = max\{c_i\}$ [1]. Interestingly, this leftward movement, associated with increases in c reaches a minimum value of

$$\lambda_c^* = \lim_{c \to \infty} \lambda_c = -\varepsilon - 1 \tag{3}$$

This implies that there is a point at which increasing the value of c, in either the homogeneous or nonhomogeneous systems, will not yield additional engineering benefits to the device [1], see Fig. 2.

3 Frequency Response

From numerical simulations, see Fig. 3, it can be seen that the period of oscillation, T, of individual elements can be obtained by summing the intervals t_i of the time-series simulations. Over any given interval, the partial period t_i can be computed via integration. Thus, after direct calculations we arrive at [1]

$$T = \sum_{i=1}^{3} \frac{\pi}{\sqrt{c_i x_{inf_i}}} \left(\frac{1}{\sqrt{\lambda_{c_i} - \lambda}} + \frac{1}{\sqrt{\lambda_{c_i} - \lambda + 2\varepsilon}} \right). \tag{4}$$

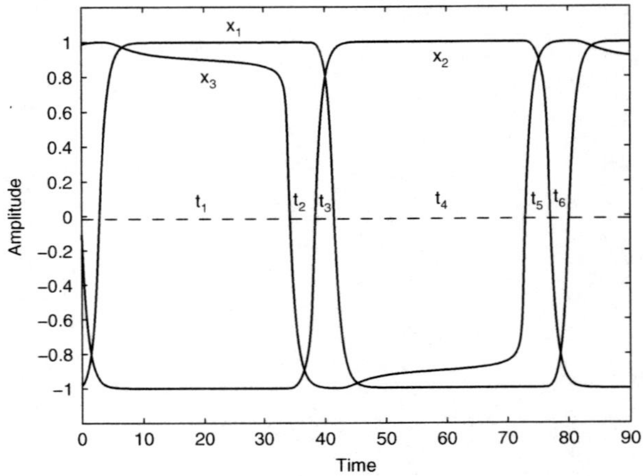

Fig. 3 Time series simulation of a network of three nonhomogeneous bistable elements coupled in a directed ring. Parameters are: $c_1 = 2, c_2 = 3, c_3 = 5$, and $\varepsilon = 0.0$. Time intervals t_i are used (see text) in the calculation of the period of oscillations

4 Sensitivity Response

The network's summed signal response $X(t) = \Sigma x_i(t)$ was used in our previous work [2], since it contained a single frequency. However, when we add nonhomogeneities to the network equations, as Fig. 4 shows, the sum signal response contains at least two distinct frequencies of oscillation. Therefore, we employed the Residence Times Difference (RTD) readout, for signal detection, using individual signals. The difference in zero-crossing times Δt yields [2]:

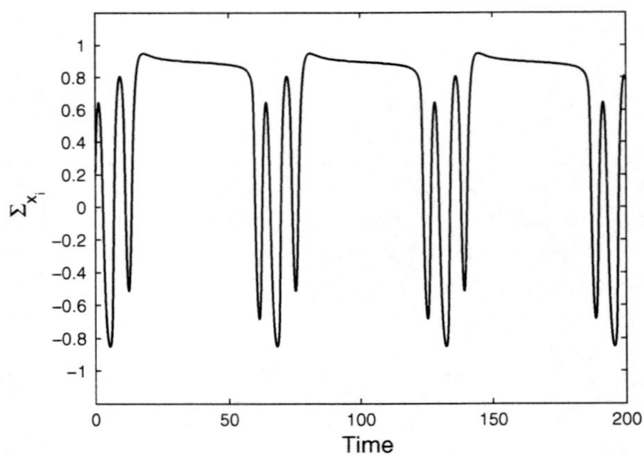

Fig. 4 Sum signal from simulations of a network of three nonhomogeneous bistable elements coupled in a directed ring. Parameters are: $c_1 = 2$, $c_2 = 3$, $c_3 = 5$, and $\varepsilon = 0.0$

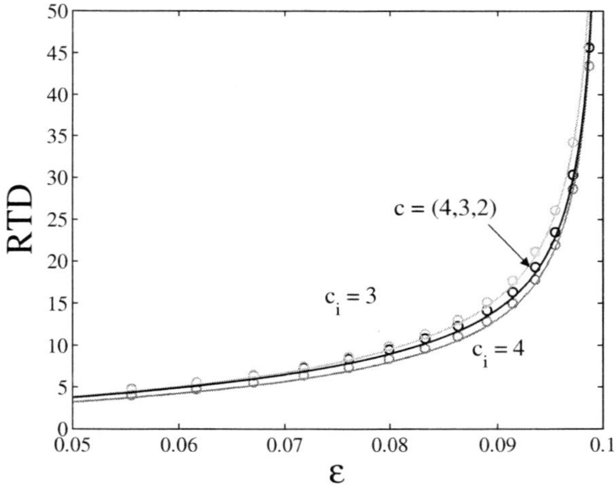

Fig. 5 RTD response to a target signal ε for a non-homogeneous system is compared to that of two closely-related homogeneous systems. Solid lines correspond to analytical calculations while individual markers are associated with numerical computations

$$\Delta t = \sum_{i=2}^{N} \frac{\pi}{\sqrt{c_i x_{inf_i}}} \left(\frac{1}{\sqrt{\lambda_{c_i} - \lambda}} - \frac{1}{\sqrt{\lambda_{c_i} - \lambda + 2\varepsilon}} \right) + \frac{\pi}{\sqrt{c_1 x_{inf_1}}} \left(\frac{1}{\sqrt{\lambda_{c_1} - \lambda + 2\varepsilon}} - \frac{1}{\sqrt{\lambda_{c_1} - \lambda}} \right). \tag{5}$$

Figure 5 shows good agreement between the RTD response calculated analytically, via Eq. (5) and numerically via computer simulations. A nonhomogeneous system is seen to be slightly better performing than a homogeneous system with is c values equal to the highest c value of the nonhomogeneous one [1].

Acknowledgments We gratefully acknowledge support from the Office of Naval Research (Code 331) and the SPAWAR internal research funding (S&T) program. M.H. and A.P. were supported in part by National Science Foundation grants CMS-0625427 and CMS-0638814.

References

1. M. Hernandez, V. In, P. Longhini, A. Palacios, A. Bulsara, A. Kho, Phys. Lett. A **372** (24), 4381–4387 (2008)
2. V. In, A. Bulsara, A. Palacios, P. Longhini, A. Kho, J. Neff, Phys. Rev. **E68**, 045102(R) (2003).
3. V. In, A. Palacios, P. Longhini, J. Neff, B. Meadows, Phys Rev. Lett. **91**, 244101 (2003).

A New Diversification Method to Solve Vehicle Routing Problems Using Chaotic Dynamics

Takashi Hoshino, Takayuki Kimura and Tohru Ikeguchi

Abstract We propose a solving method with two simple local searches separately controlled by chaotic dynamics for vehicle routing problem (VRP). The chaotic neuron memorizes a past searching history with an exponential decay. By the exponential decay, the proposed method can escape from local minima. We confirmed that the proposed method shows good performance for the VRP. Further, we introduce a new diversification method to improve the performance of the proposed method. As a result, the performance of the proposed method is improved.

1 Introduction

Vehicle routing problem (VRP) is an important combinatorial optimization problem. The VRP consists of customers, vehicles which have weight limits, and a depot which is an arrival and departure point of the vehicles. The customers are visited and received services only once by one vehicle. An object of the VRP is to minimize the number of the vehicles and the total travel distance. Because the VRP often occurs in our daily life, it is widely studied.

It is reported that a chaotic search which controls a basic local search using chaotic neurons [1] is very effective for solving combinatorial optimization problems [2], [3]. Thus, we have already proposed the chaotic search which separately controls two simple local searches to solve the VRP [4]. First, we introduce this proposed method. Next, we propose a new diversification method. To find a good solution, it is important to increase the diversification of searched solutions because there is local minima. We confirmed that the performance of the chaotic search is improved by using the diversification method.

T. Hoshino (✉)
Graduate School of Science and Engineering, Saitama University, 255 Shimo-ohkubo, Sakura-ku, Saitama, Saitama, 338-8570 Japan,
e-mail: Hoshino@nls.ics.saitama-u.ac.jp

V. In et al. (eds.), *Applications of Nonlinear Dynamics*, Understanding Complex Systems,
DOI 10.1007/978-3-540-85632-0_35,

2 Chaotic Search

In the chaotic search proposed in [4], two simple local searches are separately controlled by the chaotic neurons [1]. The first local search is an exchange between two customers and the second one is relocation of one customer. This chaotic search needs $2n$ neurons for an n-customer problem. Each neuron corresponds to exchange or relocation of each customer. If the neurons fire, the corresponding local search is executed.

To realize such operations, each neuron has a gain effect ξ and a refractory effect ζ. These effects of the ijth neuron are described as follows:

$$\xi_{ij}(t+1) = \beta \max_{m}\{\Delta_{ijm}\}, \tag{1}$$

$$\zeta_{ij}(t+1) = -\alpha \sum_{d=0}^{t} k_r^d x_{ij}(t-d) + \theta, \tag{2}$$

where $i = 1, \dots, n$ and $j = 1$ or 2. Then, the output of the ijth neuron is defined as follows:

$$x_{ij}(t+1) = f\{\xi_{ij}(t+1) + \zeta_{ij}(t+1)\},$$

where $f(y) = 1/(1+e^{-y/\varepsilon})$ called a sigmoid function and ε is a small positive parameter. If $x_{ij}(t) > 1/2$, the ijth neuron fires at the time t. Then, if the $n1$th neuron fires, the nth customer is exchanged and if the $n2$th neuron fires, the nth customer is relocated.

In Eq. (1), β is a positive scaling parameter of the gain effect, Δ_{ijm} is a gain value of total distance when the ijth neuron fires, and m represents a target of the local searches. $\max_{m}\{\Delta_{ijm}\}$ means that the local searches are chosen so that the gain value is maximum. By the gain effect, the neuron corresponding to a good operation becomes easy to fire. In Eq. (2), α is a positive scaling parameter, $k_r(0 < k_r < 1)$ is a decay factor, and θ is a threshold value. Then, the refractory effect inhibits the firing of a neuron which has just been fired, which realizes a memory effect with an exponential decay. The strength of the refractory effect gradually decays depending on the value of k_r. By the refractory effect, the chaotic search can escape from local minima. In addition, the chaotic search can generate complex firing pattern by using the sigmoid function.

Generally the chaotic neurons [1] are connected by mutual connections. However, in the chaotic search, the mutual connection is not important because it does not directly code information of the optimization, then has little influence on the performance. Thus, the mutual connection is removed in this paper.

In this method, the updating order of the neurons is deterministically defined so that the corresponding customer is placed in a clockwise direction around the depot. A single iteration of the proposed method is finished if all the neurons are updated, and the update is asynchronously conducted. In addition, to reduce the number of vehicles, the ejection chain [5] is executed every iteration. If the number of routes can be reduced, the ejection chain is performed.

3 Diversification Method

3.1 Greedy Firing

In the chaotic search, the refractory effect inhibits the firing of a neuron which has just been fired. As a result, the local searches recently executed are avoided and a diverse search is realized. However, the local searches which can improve the solution may be inhibited by the refractory effect. Thus, we introduce a very simple rule, greedy firing. If the gain value of a neuron is positive, this neuron must fire despite the refractory effect. Using the greedy firing, we expect that good solutions are easily and quickly found.

3.2 Fluctuating a Threshold Value

A threshold value θ included in the refractory effect (Eq. (2)) is also an important factor for the diversification of the chaotic search. By using θ, the neurons which have negative gain values can fire, and then the chaotic search can escape from local minima. However, if we use the greedy firing, the diversification of the chaotic search may decline because the chaotic search approaches to a descent downhill dynamics by using the greedy firing. Thus, we fluctuate the value of θ to increase the diversification of the chaotic search. In this paper, we introduce a very simple way to fluctuate the value of θ. Basically, the value of θ is monotonically decreased. As a result, the solution gradually converges. If all the neurons become impossible to fire, the value of θ is reset to the initial value and start to decrease again. Although the fluctuation is cyclic and linear, the chaotic neurons can generate different rich firing patterns with each cycle by the refractory effect. Thus, we expect that the performance of the chaotic search is improved by combining the greedy firing and the fluctuation of θ.

4 Computational Results

To evaluate the performance of the proposed method, we solved Solomon's benchmark problems of 100 customers [6]. There are six different types called R1, R2, C1, C2, RC1 and RC2. Here, R means a random allocation, C a clustered allocation, and RC their mixture. A total number of the instances is 56. An initial solution

Table 1 Comparison of the performance of the chaotic search with greedy firing or(and) fluctuation of the value of θ

Method	R1	R2	C1	C2	RC1	RC2	CNV/CTD
CS	**12.58**	**3.09**	**10.00**	**3.00**	**12.00**	**3.38**	**422**
	1230.54	961.25	840.19	592.52	1388.11	1195.17	58308
CS with greedy firing	**12.50**	**3.09**	**10.00**	**3.00**	**12.12**	**3.38**	**422**
	1227.87	953.16	885.34	591.80	1380.79	1198.59	58557
CS with fluctuation	**12.42**	**3.09**	**10.00**	**3.00**	**12.12**	**3.38**	**421**
	1224.91	954.24	840.29	590.63	1376.18	1164.92	57812
CS with greedy firing	**12.58**	**3.00**	**10.00**	**3.00**	**12.12**	**3.38**	**422**
and fluctuation	1209.36	959.39	829.29	590.60	1362.63	1183.14	57620

is produced by Bräysy construction heuristic method [5]. The parameters in Eqs. (1) and (2) were set as follows: $\beta = 0.04, \alpha = 0.5,\ k_r = 0.9, \theta = 1.0$, and $\varepsilon = 0.002$. The fluctuating rate of the value of θ was 0.005. When we introduce the fluctuation of the value of θ, the initial value of θ was set to 1.5 to increase the diversification.

Computational results are shown in Table 1. The simulations were cut at 1,000 iterations. Boldfaced numbers indicate average numbers of vehicles, and lightfaced numbers indicate average total distances for each problem type. CNV and CTD mean cumulative numbers of the vehicles and total distance, respectively. We call our proposed method [4] CS.

From the Table 1, we found that the performance of the CS with greedy firing is almost the same as the original CS. On the other hand, the performance of the CS with fluctuation is slightly better than the original CS. However, the CS with greedy firing and fluctuation shows the best performance for the CTD. Thus, it indicates that a combination of the greedy firing and fluctuation of the value of θ generated very good effect. On the instance dependency, the CS with greedy firing and fluctuation is better than the CS with fluctuation at R1, C1, C2 and RC1. That is to say, the CS with greedy firing and fluctuation generally shows the best performance, its performance depends on instance types. Thus, it is an important future work to analyze the effectiveness of the diversification method to clarify the dependency.

The research of TI is partially supported by a research grant from The Mazda Foundation.

References

1. K. Aihara et al., *Phys. Lett. A*, **144**, pp. 333–340, 1990.
2. M. Hasegawa et al., *Neural Networks*, **15**, pp. 271–283, 2002.
3. M. Hasegawa et al., *Euro. J. Oper. Res.*, **139**, pp. 543–556, 2002.
4. T. Hoshino et al., Proc. of the 7th Metaheuristics International Conference, 2007.
5. O. Bräysy, *INFORMS J. Comp.*, **15**(4), pp. 347–368, 2003.
6. M. M. Solomon, Solomon's Benchmark Problems, http://web.cba.neu.edu/~msolomon/problems.htm.

Self-Organized Neural Network Structure Depending on the STDP Learning Rules

Hideyuki Kato, Takayuki Kimura and Tohru Ikeguchi

Abstract Complex systems are widely studied in various fields of science. A neural network is one of the typical examples of the complex systems. Recent studies in neuroscience reported that the neural networks are dynamically self-organized by the spike-timing dependent synaptic plasticity (STDP). Although the neural networks change their structure using the STDP dynamically, neural networks are often analyzed in a static state. Thus, in this paper, we analyze neural network structure from a dynamical point of view. Then, we show that the self-organized neural network to which the STDP learning rule is applied generates the small-world effect and randomness of the inter-spike intervals (ISIs) in the self-organized neural network increases as the small-world effect becomes higher.

1 Introduction

Since Watts and Strogatz proposed the small-world networks in 1998 [1], various kinds of real networks have been analyzed in order to reveal whether the networks are the small-world networks or not. The small-world network has two characteristics: a short characteristic path length and a high clustering coefficient [1]. In [1], they showed that the small-world networks exist everywhere, for example, the collaboration graph of film actors, the power grid network, anatomical structure of a neural network in *C. elegans*, and so on. Neural networks in our brain may also have these characteristics.

On the other hand, in the field of neuroscience, the long-term potentiation (LTP) and the long-term depression (LTD) was found by Markram in 1997 [2], and the spike-timing dependent synaptic plasticity (STDP) was found by Bi and Poo in 1998 [3]. The STDP has been observed in several brain regions, for example, slices of

H. Kato (✉)
Saitama University, 255 Shimo-okubo, Sakura-ku, Saitama-shi, Saitama-ken, 338-8570, e-mail: kato@nls.ics.saitama-u.ac.jp

V. In et al. (eds.), *Applications of Nonlinear Dynamics*, Understanding Complex Systems, 413
DOI 10.1007/978-3-540-85632-0_36, © Springer-Verlag Berlin Heidelberg 2009

hyppocampus and the electrosensory lobe (ELL) [4]. In these regions, the neural networks self-organize their structure dynamically through the STDP.

Recently, it is reported that by redistributing connection weights in a weighted network, the network generates the small-world effect [5]. Neural networks also redistribute their synaptic weights using the STDP. Thus, in this paper, we analyze the neural network structure which self-organizes using the STDP with measures of weighted complex networks, from the viewpoint of the relationship between their network properties and the dynamics of the neurons in the neural network.

2 Model and Measures

In this paper, we use the Izhikevich's simple neuron model [6]. It is defined as follows:

$$\begin{aligned} v' &= 0.04v^2 + 5v + 140 - u + I, \\ u' &= a(bv - u), \end{aligned} \tag{1}$$

where v and u represent a membrane potential and a membrane recovery of the neuron. If v reaches +30[mV], v and u are reset according to Eq. (2).

$$\text{if}\, v \geq +30[\text{mV}],\ \text{then} \begin{cases} v \leftarrow c \\ u \leftarrow u + d. \end{cases} \tag{2}$$

A neural network dynamically modifies their synaptic weight using the STDP learning rule. Let us represent a firing time of a presynaptic neuron by t_{pre}, that of a postsynaptic neuron by t_{post} and an axonal conduction delay by $d_{\text{post,pre}}$. Then, $\Delta t = t_{\text{post}} - d_{\text{post,pre}} - t_{\text{pre}}$ represents a firing time interval between presynaptic and postsynaptic neurons. Thus, the STDP learning rule is modeled as follow:

$$\Delta w = \begin{cases} (1 - \tilde{w})^{\mu} \times e^{-\frac{|\Delta t|}{\tau}} & \text{if}\, \Delta t > 0, \\ -\alpha \tilde{w}^{\mu} \times e^{-\frac{|\Delta t|}{\tau}} & \text{if}\, \Delta t \leq 0, \end{cases} \tag{3}$$

where Δw represents the amount of the synaptic modification, $\tilde{w}$ represents a synaptic weight scaled to [0, 1] and τ represents a time constant. If the parameter μ sets to 0, Eq. (3) corresponds to the additive learning rule. If the parameter μ sets to 1, Eq. (3) corresponds to the multiplicative learning rule.

In this paper, to evaluate the weight effect of the self-organized neural network, we used two characteristics of weighted complex networks: the first one is the clustering coefficient $C = \frac{1}{n} \sum C_i$ where n represents the number of neurons and C_i represents the clustering coefficient of i th neuron. C_i is defined as follow:

$$C_i = \frac{\sum_{j=1,j\neq i}^{n} \sum_{k=j+1,k\neq i}^{n} \tilde{w}_{ij} \tilde{w}_{ik} \tilde{w}_{jk}}{\sum_{j=1,j\neq i}^{n} \sum_{k=j+1,k\neq i}^{n} \tilde{w}_{ij} \tilde{w}_{ik}}, \tag{4}$$

where $\tilde{w}_{ij}$ represents the synaptic weight between the ith and the jth neurons, and is scaled to $[0,1]$.

The second one is the characteristic path length $L = \frac{1}{n(n-1)} \sum\sum d_{ij}$ where d_{ij} represents the shortest distance between the i th and the j th neurons. In addition, we also used the coefficient of variation (CV) [7] to evaluate randomness of the inter-spike intervals (ISIs) in the self-organized neural network.

3 Experimental Results

We used 1,000 neurons as the elements of the neural network. We set 800 neurons to excitatory neurons and the other neurons were set to inhibitory neurons. The STDP learning rule was only applied to the synaptic connections from the excitatory neuron.

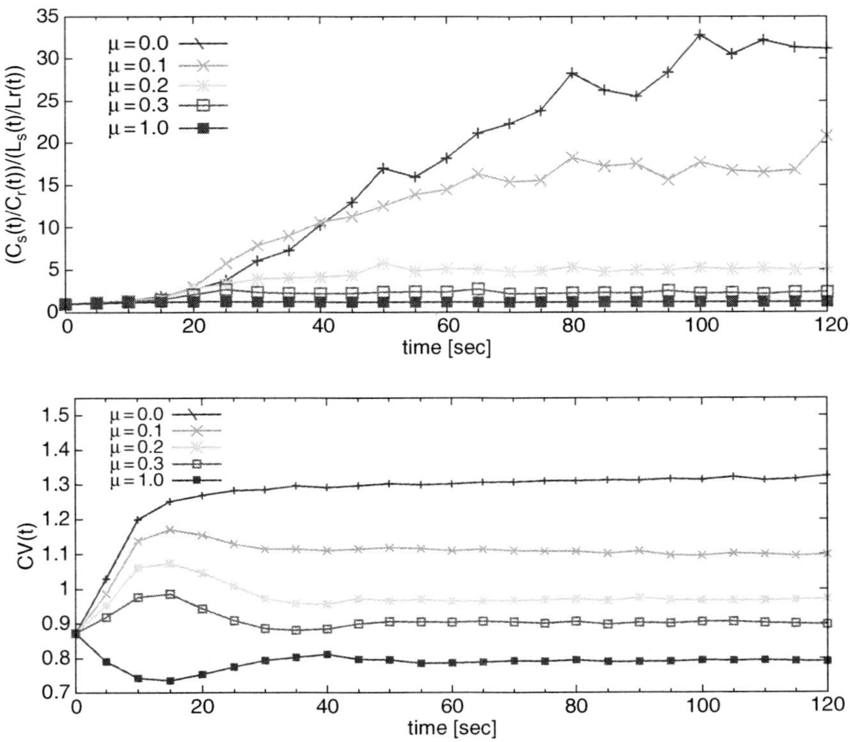

Fig. 1 (**a**) Temporal change of $(C_s(t)/C_r(t))/(L_s(t)/L_r(t))$, and (**b**) $\mathrm{CV}(t)$. The horizontal axes represent time t during which the STDP learning rule is applied to a neural network. The vertical axis of (**a**) represents $(C_s(t)/C_r(t))/(L_s(t)/L_r(t))$. $C_s(t)$ and $L_s(t)$ are the clustering coefficient and the characteristic path length of the STDP network at time t, respectively. $C_r(t)$ and $L_r(t)$ are those of the regular network at time t. The regular network is a weight averaged network. The vertical axis of (**b**) represents $\mathrm{CV}(t)$

The results are shown in Fig. 1(a) is the result of temporal change of $(C_s(t)/C_r(t))/(L_s(t)/L_r(t))$. Here, $C_s(t)$ and $L_s(t)$ are the clustering coefficient and the characteristic path length of a self-organized neural network at time t[sec], respectively. $C_r(t)$ and $L_r(t)$ are those of a corresponding regular network at time t[sec], respectively. In this paper, we defined the weight averaged network as the corresponding regular network of the weighted network. In Fig. 1(a), a larger value of $(C_s(t)/C_r(t))/(L_s(t)/L_r(t))$ means that the self-organized neural network has the higher small-world effect. Thus, we found that smaller parameter μ generates the higher small-world effect in the self-organized neural network. Namely, the additive STDP learning rule generates the highest small-world effect. If the parameter μ is large ($\mu = 0.3, 1.0$), the small-world effect is almost diminished

Figure 1(b) is the result of the temporal change of $\mathrm{CV}(t)$ at the time t[sec]. To evaluate $\mathrm{CV}(t)$, we use spike time series of neurons between t and $t+1$[sec]. In Fig. 1(b), the larger value of $\mathrm{CV}(t)$ means that randomness of the ISIs increases. Then, we found that the smaller μ leads to higher randomness of the ISIs. In the parameters μ except for $\mu = 1.0$, randomness of the ISIs increases with time . These results correspond to the results of $(C_s(t)/C_r(t))/(L_s(t)/L_r(t))$. Namely, the higher small-world effect leads to higher randomness of the ISIs.

4 Conclusion

In this paper, we analyzed the relationship between the weight effect and randomness of the ISIs of the self-organized neural network through the STDP learning. From the experimental results, we found that the self-organized neural network generates the highest small-world effect if we used the additive learning rule. We also showed that the higher small-world effect leads to higher randomness of the ISIs.

In our future works, we analyze the relationship between the small-world effect and randomness of the ISIs more deeply.

References

1. D.J. Watts, S.H. Strogatz, Nature **393**, 440 (1998)
2. H. Markram, M.F. Joachim Lüboke, B. Sakmann, Science **275**, 213 (1997)
3. G. Bi, M. Poo, The Journal of Neuroscience **18**(24), 10464 (1998)
4. L.F. Abbott, S.B. Nelson, Nature neuroscience supplement **3**, 1178 (2000)
5. S.R.R. Gütig *et al.*, The Journal of Neuroscience **23**(9), 3687 (2003)
6. E.M. Izhikevich, IEEE Transactions on Neural Networks **14**(6), 1569 (2003)
7. W.R. Softky, C. Koch, The Journal of Neuroscience **13**(1), 334 (1993)

Communication in the Computer Networks with Chaotic Neurodynamcis

Takayuki Kimura and Tohru Ikeguchi

Abstract Congestion of packets in the Internet is the most undesirable problem to securely communicate between end users. Thus, many approaches have been attempting to resolve such a problem. We have already proposed a routing strategy with chaotic neurodynamics. The chaotic neurodynamics is introduced to alleviate the packet congestion, then, the routing strategy shows higher performance for complex networks than the shortest path approach (the Dijkstra algorithm). In this paper, we extend the routing strategy by combining waiting time at adjacent nodes. We show that the improved routing strategy has high performance for the scale-free networks.

1 Introduction

Packet congestion is the most undesirable issue in the Internet because it leads to loss and delay of packets. The packet congestion is related to a spatial structure of a computer network and a routing strategy. To alleviate the packet congestion of the computer networks, there are two directions of research. The first one is to analyze a topology of a computer network in detail, then, propose a proper routing strategy which works well on the computer network. For example, Arenas et al. focused on Clay-tree networks [1]. The second one is to propose a routing strategy which works well on various topologies of the computer networks. Kimura et al. proposed a routing strategy with a stochastic neural network [2]. In the present work, as the second direction, we proposed an efficient routing strategy with load-balancing using chaotic neurodynamics for the computer networks.

In our computer network model, a node is taken to be both hosts and routers. A link serves as a pathway through which packets are transmitted. When a packet is

T. Kimura (✉)
Graduate School of Science of Engineering, Saitama University, 255 Shimo-ohkubo, Sakura-ku, Saitama, Saitama, 338-8570 Japan, e-mail: Kimura@nls.ics.saitama-u.ac.jp

V. In et al. (eds.), *Applications of Nonlinear Dynamics*, Understanding Complex Systems, DOI 10.1007/978-3-540-85632-0_37,

generated at a node, it is transmitted from a node to another through the links. Then, the packet is stored at the tail of the buffer of the transmitted node.

To alleviate the packet congestion, we have already proposed a routing strategy with chaotic neurodynamics [3]. In the routing strategy [3], each node determines a transmitted node of the packet using the shortest path information and the chaotic selection. Although the routing strategy shows higher performance, if the node obtains additional information, we expect that its performance becomes high. One of the efficient additional information is waiting time at an adjacent node or the number of queueing packets at the adjacent node. In this paper, we improve the previous routing strategy [3] by introducing the additional information. We confirm that the improved routing strategy has high performance for the BA model [4].

2 A Routing Method with Chaotic Neurodynamics

To realize the routing strategy by chaotic neurodynamics, we introduced a chaotic neural network [5]. Our computer network model has N nodes, and the ith node has N_i adjacent nodes ($i = 1,\ldots,N$). In the routing strategy, each node has its own neural network, and N_i adjacent neurons are assigned to each node. If the ijth neuron fires, a packet is transmitted from the ith node to its jth adjacent node. To realize the above function, each neuron has a gain effect, a refractory effect, and a mutual connection effect, which are defined as follows:

$$\xi_{ij}(t+1) = \beta \left\{ H\left(1 - \frac{d_{ij} + d_{jg(p_i(t))}}{d_c}\right) + (1-H)\left(1 - \frac{q_j(t)}{b}\right) \right\}, \tag{1}$$

$$\zeta_{ij}(t+1) = -\alpha \sum_{d=0}^{t} k_r^d x_{ij}(t-d) + \theta, \tag{2}$$

$$\eta_{ij}(t+1) = W - W \sum_{j=1}^{N_i} x_{ij}(t), \tag{3}$$

where d_{ij} is a static distance from the ith node to its jth adjacent node; $p_i(t)$ is a transmitted packet of the ith node at the tth iteration; $g(p_i(t))$ is a destination of $p_i(t)$; $d_{jg(p_i(t))}$ is a dynamic distance from the jth adjacent node to $g(p_i(t))$, that is, $d_{jg(p_i(t))}$ depends on $g(p_i(t))$; d_c is the size of the computer network; β is a normalization parameter; $q_j(t)$ is the number of queueing packets at the jth adjacent node at the tth iteration; b is the buffer size of each node; H decides priority of the first term and the second term; α is a control parameter of the refractoriness; k_r is a decay parameter of the refractoriness; $x_{ij}(t)$ is the output of the ijth neuron at the tth iteration that will be defined in (4); θ is a threshold. $W > 0$ is a parameter and N_i is the number of adjacent nodes at the ith node.

Then, the output of the ijth neuron is defined as follows;

$$x_{ij}(t+1) = f\{\xi_{ij}(t+1) + \zeta_{ij}(t+1) + \eta_{ij}(t+1)\}, \tag{4}$$

where $f(y) = 1/(1+e^{-y/\varepsilon})$.

In the previous proposed routing strategy [3], the first term of (1) is only used to evaluate whether the jth adjacent node is an optimum one or not. The second term of (1) expresses waiting time at the jth adjacent node until $p_i(t)$ will be transmitted from the jth adjacent node to the next transmitted node. Using the waiting time information, each node selects the optimum adjacent node more efficiently and flexibly.

In our proposed routing strategy, the refractory effect is the most important one to route the destinations of the packets efficiently. The refractory effect (2) is used by as a memory effect. Namely, each node can memorize a past routing history using the refractory effect, then, an adjacent node to which many packets have just been transmitted is not selected as a transmitted node of the packets for a while.

3 Computer Simulation

To evaluate the performance of the proposed routing strategy, we compared the proposed routing strategy with two kinds of routing strategies. The first one is the Dijkstra algorithm. The second one is a previous proposed routing strategy with chaotic neurodynamics(the CNN strategy) [3].

The computer simulations are conducted as follows: first, each node calculates the shortest path from the node to the other nodes. Namely, each node always has a static routing table which contains an information list of the shortest distances between any two nodes. In these simulations, at every iteration R packets are generated in the computer network, and sources and destinations of the packets are randomly assigned using uniformly distributed random numbers. Then, at every node, an optimal adjacent node is selected and the packets are simultaneously transmitted to their destinations. If a packet is delivered to its destination, the packet is removed from the computer network. We conducted 10 simulations to average the results.

We repeated the link selection and packet transmission for $10,000$ iterations. We set the parameters in (1), (2), (3) and (4) as follows: $\beta = 1.2$, $H = 0.8$, $\alpha = 0.045$, $k_r = 0.98$, $\theta = 0.5$, $W = 0.05$, and $\varepsilon = 0.05$. We also set d_c as a diameter of the computer network, which is defined as the longest hop between two nodes.

As the topologies of the computer networks, 100 and $1,000$ nodes of the scale-free networks [4] are used in these simulations. To evaluate the performance of the three routing strategies, we measured the number of delivered packets to their destinations, A.

Results for different sizes of the scale-free networks are shown in Fig. 1. In Fig. 1(a), the numbers of delivered packets (A) of the Dijkstra algorithm and the CNN strategy are smaller than the proposed routing strategy when the number of

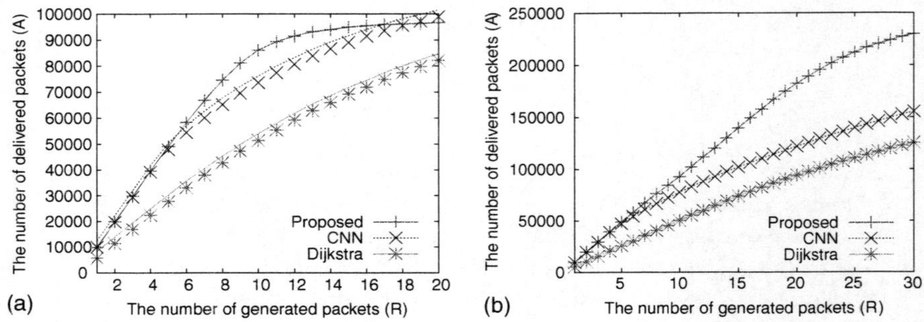

Fig. 1 Relationship between the number of generated packets at each iteration (R) and the number of delivered packets (A) for (**a**) 100 and (**b**) 1000 nodes of the scale-free networks

generated packets (R) becomes large. In addition, in Fig. 1(b), even if the size of the computer networks increases, the proposed routing strategy also transmits more packets to their destinations than the other routing strategies.

Using the waiting time at the adjacent nodes, the proposed routing strategy selects the optimal adjacent nodes more efficiently and flexibly than the conventional routing strategies.

4 Conclusion

In this paper, to improve the performance of the routing strategy with chaotic neurodynamics [3], we introduced the information of the waiting time to transmit the packet at adjacent nodes to each node. Using additional information, the performance is much improved, and the proposed routing strategy has high performance for the scale-free networks comparing with the previous strategy [3]. The research of T.I. is partially supported by a research grant from The Mazda Foundation.

References

1. A. Arenas, A. Diaz-Guimera, R. Guimera, Physical Review Letters **86**(14), 3196 (2001)
2. T. Kimura, H. Nakajima, T. keguchi, Physica A **376**, 658 (2007)
3. T. Kimura, T. Ikeguchi, Neural Computing and Applications (2007) (DOI 10.1007/s00521-007-0099-5), **16**(6), 519–526
4. T.-L. Barábasi, R. Albert, Science, **286**, 509 (1999)
5. K. Aihara, T. Tanabe, M. Toyoda, Physics Letters A **144**, 133 (1990)

Nonlinear DDE Analysis of Repetitive Hand Movements in Parkinson's Disease

Claudia Lainscsek, Luis Schettino, Peter Rowat, Elke van Erp, David Song and Howard Poizner

Abstract Time series analysis with nonlinear delay differential equations (DDEs) is a very powerful tool since it reveals spectral as well as topological properties of the underlying dynamical system and is robust against noise. Here we apply nonlinear DDEs to examine the nature of the spatiotemporal distortions in repetitive finger tapping movements of mild to moderate Parkinson's disease (PD) patients on and off their dopamine replacement therapy and of age-matched controls.
Using DDE analysis, there was a nearly complete separation of the data of all three groups: PD patients were classified separately from control subjects, and PD patients on and off medication were clearly distinguished. The non-linear phase coupling terms were particular important in being able to separate groups. There was an increased degree of multiplicity of frequencies in the temporal patterns when going from control to PD on medication to PD off medication. This analysis was then compared with clinical scores provided by physicians, the UPDRS (United Parkinson's Disease Rating Scale) scores. The values of the nonlinear term of the DDE shows good correlation to this clinical scores.
We conclude that such measures may provide a more objective and precise measure of the spatiotemporal disruption of rhythmic movements in PD, and the reversal of these deficits by pharmacological (or surgical) therapies.

1 Parkinson's Disease

The temporal structure of hand movements in Parkinson's disease (PD) is known to be impaired [4], but the nature of these temporal distortions remains unexplored. Likewise it is unknown how dopamine replacement therapy modulates these temporal abnormalities. To address this issue, we used non-linear time series analysis

C. Lainscsek (✉)
Machine Perception Lab, Institute for Neural Computation, University of California at San Diego, La Jolla, CA, USA, e-mail: clainscsek@ucsd.edu

V. In et al. (eds.), *Applications of Nonlinear Dynamics*, Understanding Complex Systems, DOI 10.1007/978-3-540-85632-0_38,

to examine the nature of the spatiotemporal distortions in repetitive finger tapping movements in PD patients on versus off dopaminergic medications.

6 mild to moderate PD patients were tested on and off their dopaminergic medications and compared to 6 age-matched controls. The clinical state of the PD subjects at the time of testing was rated using UPDRS.

Subjects were asked to tap their right index finger and thumb together making large rapid movements for 10 sec. Three repetitions of the tapping movements were made while subjects had their eyes open. A 12 camera PhaseSpace 3D motion monitoring system (PhaseSpace, Inc.) was used in the newly developed Motion Capture Lab of the Temporal Dynamics of Learning Center at UCSD to record the positions of infra-red markers attached to the subject's index fingertip, thumb, and back of the hand. Data were sampled at 120 Hz.

2 DDE Analysis

A **DDE** is an equation, that relates the velocity of a data point to previous data points of the signal, i.e. $\dot{x} = f(x_{\tau_1}, x_{\tau_2}, ...)$ where $x_\tau = x(t - \tau)$. Here we use DDEs [1–3] to model the temporal evolution in the embedding [5] space. The simplest linear DDE is $\dot{x} = ax_\tau$. If $x(t)$ is a harmonic signal with one frequency, $x(t) = \cos(ft)$, then $\tau = \frac{\pi}{2f}$ and $a = -f$. Therefore $\dot{x} = ax_\tau$ can be used to determine distinguishing frequencies (delays) for different signal classes. The simplest non-linear DDE $\dot{x} = ax_{\tau_1}x_{\tau_2}$ has one non-linear term that is only non-zero for phase couplings.

For three signal classes a DDE $\dot{x} = a_1x_{\tau_1} + a_2x_{\tau_2} + a_3x_{\tau_3} + a_4x_{\tau_1}x_{\tau_2} + a_5x_{\tau_1}x_{\tau_3} + a_6x_{\tau_2}x_{\tau_3}$ with three delays corresponding to three distinguishing frequencies of the three different conditions (control, PD on and off medication) can be used. The non-linear terms in this equation correspond then to the phase couplings between these characteristic frequencies. This equation has six coefficients a_j, $j = 1, 2, \ldots, 6$ and since we want to use this coefficients as features for our further analysis, a reduction to a lower dimensional feature space would be preferable. For this reason we split above mentioned DDE into three DDEs, each of them reading as

$$\dot{x} = a_1x_{\tau_1} + a_2x_{\tau_2} + a_3x_{\tau_1}x_{\tau_2} \,. \tag{1}$$

The feature space for each of these equations is three-dimensional and can be viewed easily. Each equation allows distinguishing between each pair of the three classes.

For our analysis we first have to choose good delays. To do so, we use the simplest linear DDE, $\dot{x} = ax_\tau$ to find characteristic frequencies and therefore estimate the mean coefficient a by using singular value decomposition for each data class as function of the delay. We then choose two delays for two classes to the two most distinguishing coefficients which are related as mentioned above to the most distinguishing frequencies. We get $\tau_1 = 5$ and $\tau_2 = 30$ to separate PD on or off medication from controls and $\tau_1 = 52$ and $\tau_2 = 115$ to separate PD on from PD off medication.

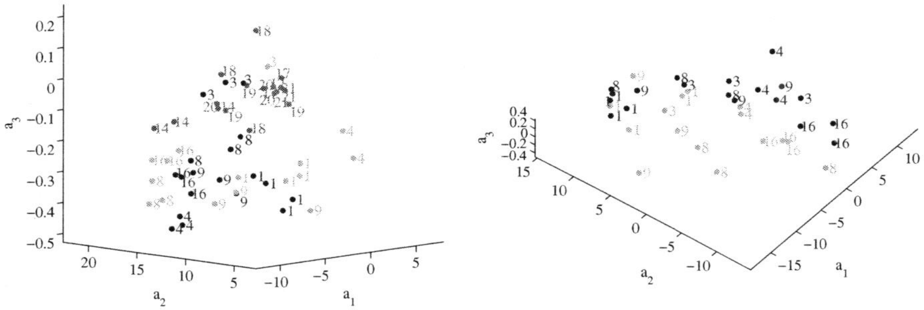

(a) τ1 = 5, τ2 = 30 (b) τ1 = 52, τ2 = 115

Fig. 1 DDE feature space ($a_{1,2,3}$)

Such a choice of the delays in a nonlinear DDE is expected to reveal different types of couplings for the different classes.

For our data analysis we use Eq. (1). The linear terms reflect, as mentioned above the dominant frequencies or dominant time scales and the non-linear terms the phase couplings or feedback. In Fig. 1 the features a_i, $i = 1,2,3$ of the three data sets are plotted in red (control), green (PD off medication), and blue (PD on medication). Subjects 1, 3, 4, 8, 9, and 16 are PD patients and subjects 14, 17, 18, 19, 20, and 21 are controls. For nearly all subjects there is perfect separation of controls vs. PD patients as can be seen in Fig. 1(a). Only subject 3 has features similar to those of the controls. This person suffers only very mild PD and the clinical score for finger movements is only 0.5 (on a scale from 0 to 4). The separation of PD patients on and off medication is also very good as shown in Fig. 1(b). The Table 1 were obtained by training a Support Vector Machine (SVM) on all except one group of either

Table 1 DDE-analysis: area under the ROC curve for leave-one-subject-out cross-validation

(a) $\tau_1 = 5$, $\tau_2 = 30$; PD vs. controls

subject	14	17	18	19	20	21
1	100	100	100	96	100	100
3	0	0	67	0	0	0
4	100	100	100	100	100	100
8	100	100	100	100	100	100
9	100	100	100	100	100	100
16	100	100	100	100	100	100

(b) $\tau_1 = 52$, $\tau_2 = 115$; PD on vs. off med.

subject	1	3	4	8	9	16
1	**144**	100	83	67	50	17
3	25	**100**	100	67	100	0
4	0	100	**100**	67	83	0
8	83	100	100	**100**	100	100
9	67	89	78	67	**67**	56
16	100	100	100	67	100	**100**

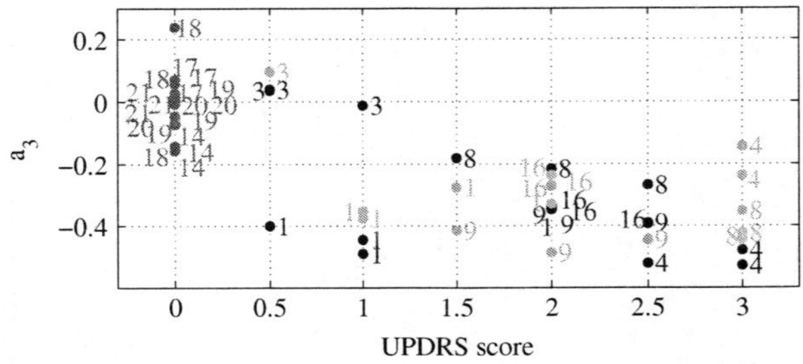

Fig. 2 a_3 vs. UPDRS score; $\tau_1 = 5$, $\tau_2 = 30$; the controls were not clinically rated, but added to the plot with a score of approximately 0 for comparison

condition and testing on the left out groups. The numbers are the areas under the ROC curves on the distances from the hyperplane chosen by the SVM. In Tab. 1(a) the nearly perfect separation of controls from PD patients can be seen. Only subject 3 cannot be separated. In Table 1(b) the diagonal shows the excellent separation of PD on vs. PD off for each subject when the SVM was trained on all other subjects.

We further compared the values of the nonlinear coefficient a_3 to the clinical UPDRS score in Fig. 2. There is a good correlation of the nonlinear coefficient a_3 to the UDPRS score. Each subject shows a value of a_3 closer to the controls when on medication than when off medication.

3 Conclusion

We used nonlinear DDEs to investigate repetitive tapping movement data from Parkinson patients on and off medication and a control group. Our analysis resulted in good separation of PD versus controls as well as PD on from off medication. There was also a good correlation of our analysis to the UDPRS score.

These results are very promising and suggest the investigation of other movement types of PD patients which is planned in future work.

Acknowledgments This work was supported in part by NIH grant # NS036449 and NSF grant # SBE-0542013 to UCSD.

References

1. J. Kadtke. Classification of highly noisy signals using global dynamical models. *Physics Letters A* 203:196, 1995.
2. C. Lainscsek and I. Gorodnitsky. Characterization of various fluids in cylinders from dolphin sonar data in the interval domain. In *Oceans 2003*, volume 2 of *IEEE*, page 629, 2003.

3. C. Lainscsek, C. Letellier, J. Kadtke, G. Gouesbet, and Schürrer F. Equivariance identification using delay differential equations. *Physics Letters A*, 265(4), 2000.
4. R. Agostino, A. Curr, M. Giovannelli, N. Modugno, M. Manfredi, and A. Berardelli Impairment of individual finger movements in parkinson's disease. *Movement Disorders*, 18(5):560, 2003.
5. F. Takens. Detecting strange attractors in turbulence. *Lecture Notes in Mathematics*, 898:366, 1981.

Experimental Results of Coupled E-Field Sensor

Norman Liu

Abstract In this paper, we propose an electric field sensor made up of three unidirectionally coupled bistable Duffing elements. Each element in the sensor will oscillate if the coupling exceeds a threshold value. Then, we present some experimental results on the behavior and sensitivity of the sensor at the onset of bifurcation with and without an external input signal.

1 Introduction

A dynamic system made up of odd number of bistable unidirectional coupled Duffing elements can produce oscillation [1]. This has been the theoretical basis in the continuing development of a sensor for detecting dc or low frequency electric field. The main goal here is to realize and implement the system using electronic circuit components, including the use of a nonlinear ferroelectric capacitor in each element of the system. The ferroelctric capacitor is the main element that provides the bistable properties needed for this system. After constructing the circuit, the oscillatory behavior of the system is tested using different circuit parameters. Then, we look at the sensitivity at the onset of bifurcation of the system with an external signal. That helps us in finding a region of high sensitivity for the operation of the sensor.

2 Background

A dynamic system consisting of N bistable unidirectionally coupled Duffing element can be described using the following differential equation:

N. Liu (✉)
Space and Navel Warfare Systems Center San Diego, San Diego, CA 92152, USA

V. In et al. (eds.), *Applications of Nonlinear Dynamics*, Understanding Complex Systems,
DOI 10.1007/978-3-540-85632-0_39, © Springer-Verlag Berlin Heidelberg 2009

$$\tau\dot{x}_i = ax_i - bx_i^3 + \lambda(x_i - x_{i+1}) + \varepsilon_o \tag{1}$$

where i = 1, 2, ..., N mod N, x_i is the state variable of the ith element, a and b are parameters characterizing each element, λ is the coupling coefficient, τ is the time constant of each element, and ε_o is the external signal.

As derived in [1], when the coupling coefficient, λ, exceeds the critical coupling, λ_c, of the system, or if the external signal is large enough, the system goes into the state of oscillation, and the output of each element is out of phase by 360/N degrees with respect to neighboring elements.

3 Circuit Implementation

In Fig. 1, the circuit schematic of an element of the coupled system is shown. Vdd and Vss are the positive and negative supply of the circuit respectively. FCAP4 is a nonlinear ferroelectric capacitor that is characterized by the parameters a, b, and λ in Eq. (1). The parameter λ in Eq. (1) depends on all the resistors and the feedback capacitor C4 in Fig. 1. The node vout is the output of the element and is connected to node vin of the next element.

Since there isn't any input for an external electric field directly applied to the ferroelectric capacitor, a two-terminal device, a voltage is applied at node vdc of each element. The gain from node vdc of each element to the output of the first amplifier is unity. That means the electric field applied to the ferroelectric capacitor equals to the voltage applied at node vdc divided by the thickness of the capacitor.

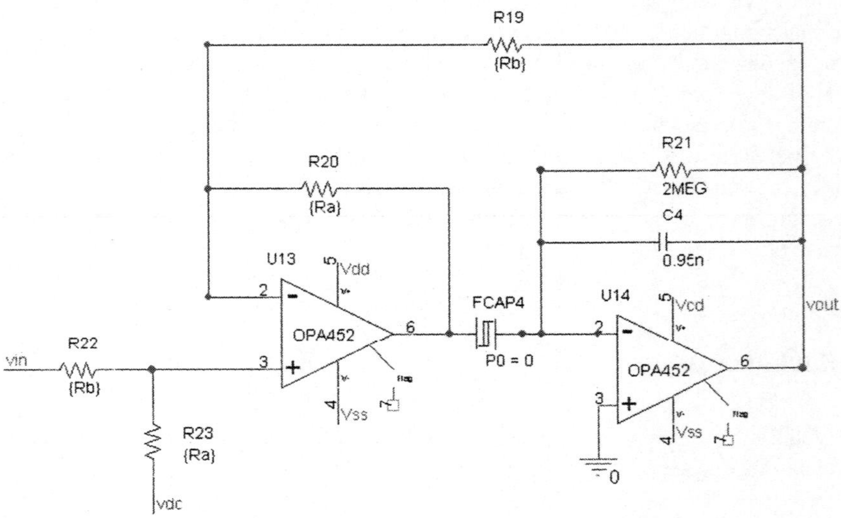

Fig. 1 Circuit schematic of an element

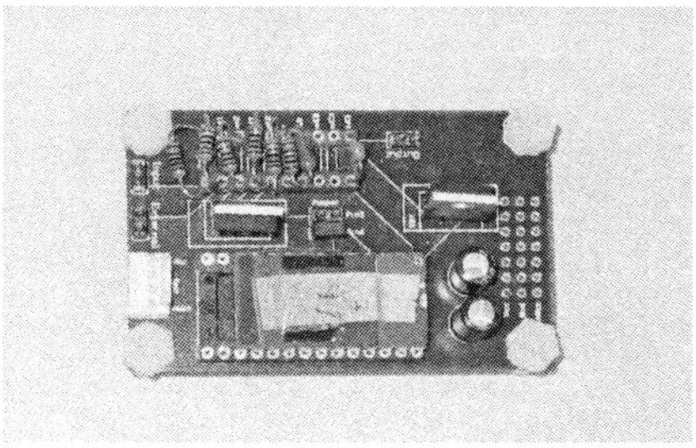

Fig. 2 A single element on PCB

The applied voltage at node vdc and the electric field that exists within the capacitor without an external signal are represented by ε_o in Eq. (1).

Figure 2 shows a picture of a single element implemented in circuit form on a PCB. The package IC on the PCB is actually the ferroelectric capacitor, which is provided by Penn State University. Figure 3 provides a microscopic view of the ferroelectric capacitor. The thickness of the ferroelectric capacitor is about 4μ*m*.

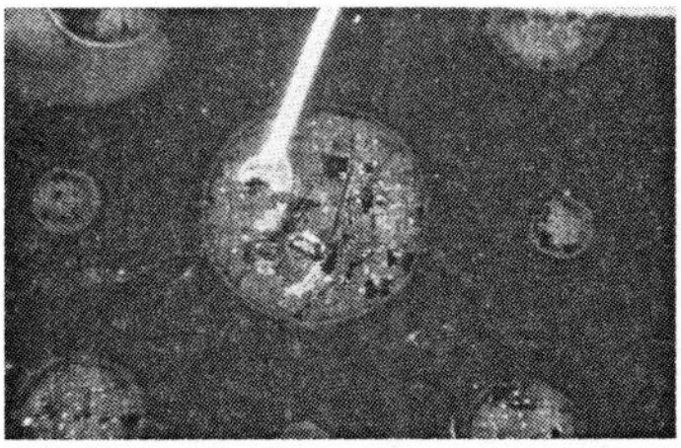

Fig. 3 Microscopic view of the ferroelectric capacitor used

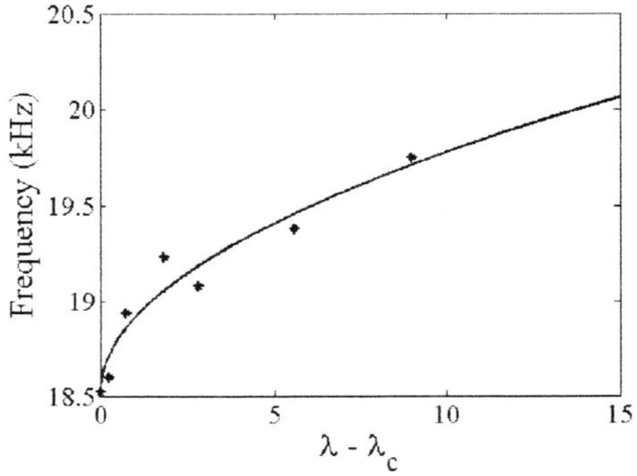

Fig. 4 Plot of $\lambda - \lambda_c$ vs. frequency of output signal

4 Data and Results

Here are some experimental data results of a 3-element coupled E-field sensor. In Fig. 4, the plot of $\lambda - \lambda_c$ versus the frequency of the output of one of the element is shown. The critical coupling of the circuit is estimated to be about 3.5. The solid line in the plot is the theoretical square-root relationship between the coupling and the output frequency as derived in [1], and the dots are data points collected from the actual circuit. As seen from Fig. 4, the data points match up well with the solid line.

In Fig. 5, the transient response of the output of each element is plotted for $\lambda = 6.58$ and without any external voltage input. The output frequency is 19 kHz, and each output is 120 degrees out of phase with its neighboring elements. This result matches with the expected out of phase behavior.

In Fig. 6, the frequency of the output of one of the element is plot against an external dc voltage being applied at node vdc in Fig. 1. When the applied dc voltage is above 15 V, a small change in the applied dc voltage results in a relatively large change in the output frequency. That region of high sensitivity is where the sensor would be operated.

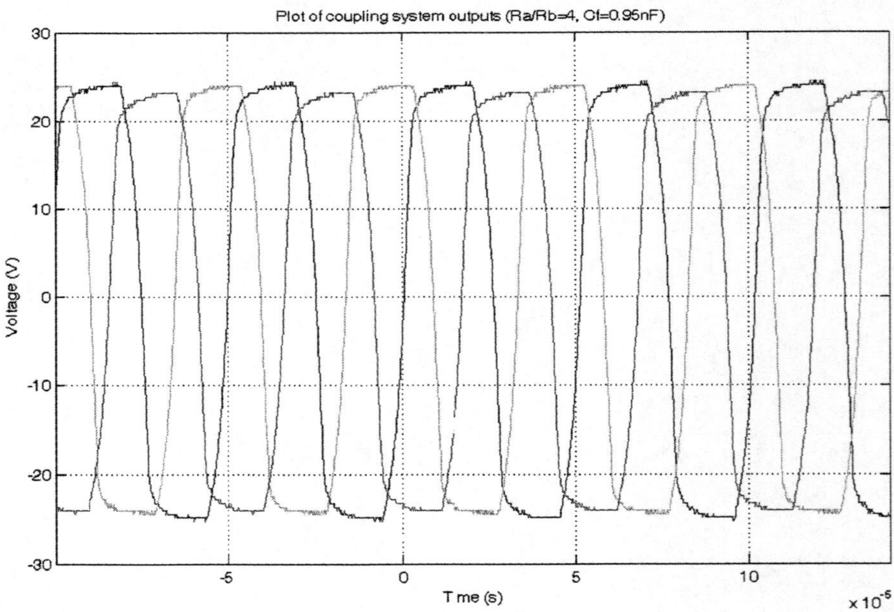

Fig. 5 Plot of the output of each of the 3 elements

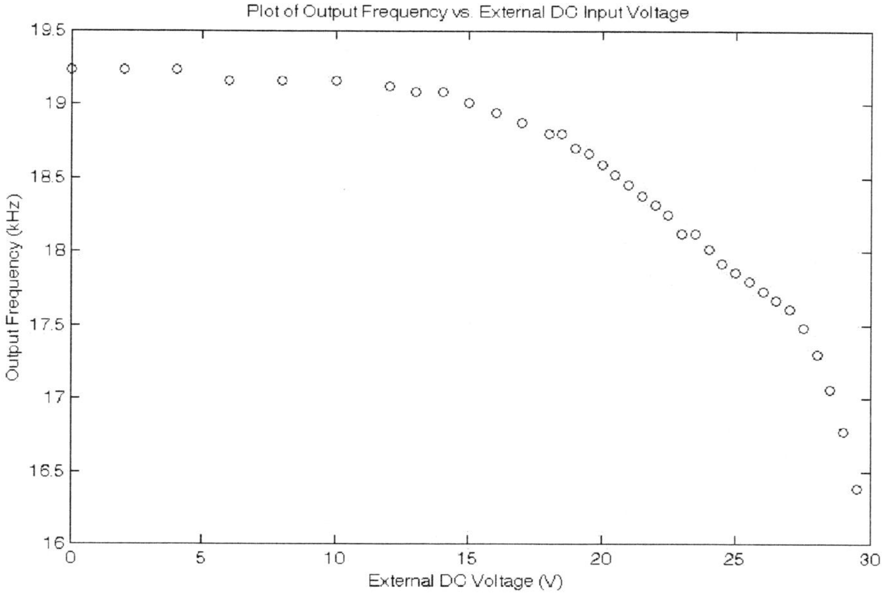

Fig. 6 Plot of external dc voltage vs. output frequency

5 Conclusion

The coupled E-sensor with 3-elements proposed in this paper has been successfully tested, and the behavior of the sensor matches with the theoretical results in [1]. The next step is to obtain a ferroelectric capacitor with an external input to sense the external electric field and to shrink the sensor circuit down to a single chip.

References

1. V. In, A. Palacios, A. R. Bulsara, et al., Phys. Rev. E **73**, 066121 (2006)

Chaos Generators for Noise Radar

K.A. Lukin, V. Kulyk and O.V. Zemlyaniy

Abstract We present results of top-down design of chaos generators that are based on nonlinear electronic systems. Several types of chaotic waveform generators are described such as Weak Resonant BWO, multi-mode resonant systems, time-delayed circuits of ring and reflection feedback types, etc. The Chaos generators presented can be used as sources of noise/random signals in Noise Radars wide frequency range: from P-band up to W-band or as modulating noise sources in radars, operated in microwave band.

1 Introduction

Nowadays there are number of papers devoted to the problem of chaotic waveform generation. Significant interest to this topic is tied with potential usage of chaotic signals instead of traditional application of noise ones. For instance, chaos generators have found practical usage in radars [1–6]. Specific features of chaotic signals give many novel abilities in solving various problems of telecommunication, encryption, information transmission and many others. In this paper we present some results obtained by LNDES research team when developing noise sources based on Dynamical Chaotization Principle [7]. This means that unlike conventional methods for noise generation (amplification and filtration of thermal noise source output) we exploit the dynamical chaos in electronic devices to build up various noise sources. Chaotization of oscillations in electronic circuits has its origin in the fact that nonlinear dynamical systems with 1.5 and more degrees of freedom may demonstrate chaotic behavior. This allows designing chaotic waveform sources for a given frequency band with required parameters, such as power spectrum density, its shape, frequency bandwidth and central frequency. Quite high output power is the very

K.A. Lukin (✉)
Usikov Institute for Radiophysics and Electronics, NAS of Ukraine,
e-mail: LNDES@kharkov.com

V. In et al. (eds.), *Applications of Nonlinear Dynamics*, Understanding Complex Systems,
DOI 10.1007/978-3-540-85632-0_40,

important advantage of chaotic waveform generators, which enables generation of probing signals for noise radar with high power spectrum density. In this paper, we present elaboration of methods for chaotic waveform generation in various frequency bands

2 Dynamical Chaos in Electronic Systems

From mathematical point of view, any dynamical system can be described in terms of differential or difference equations. This can be starting point in exploration of dynamical systems and their classification. A dynamical system can demonstrate chaotic behavior if and only if it has not less than 1.5 degrees of freedom and it is a nonlinear system. There are two general classes of such systems. To the first class may be related systems which are governed by nonlinear/quasi-linear equations. The equations may be either ordinary differential equations (ODE), or partial differential equations (PDE), or functional difference equations (FDE), or differential-difference equations (DDE) which are to be supplied with linear boundary conditions. Chaotic interaction of resonant modes in multi-mode oscillatory system excited by an active element is an example of ODE application. Vacuum electronic system with extended interaction between electromagnetic fields and electron beam is an example where PDE are required, e.g. Weak-Resonant Backward Wave Oscillator (WR-BWO) [7]. To the second class we relate the systems which are governed by linear partial differential equations with nonlinear boundary conditions. This case embraces the electronic systems with delayed feedback: nonlinear amplifier with ring delayed feedback and transmission line loaded by a nonlinear diode. In the following sections of this paper we describe chaotic waveform generators according to this classification, which have been studied in the LNDES of IRE NASU.

3 Chaos in Weak-Resonant BWO

So called Weak Resonant BWO (WR-BWO) was historically the first self-oscillatory electronic system capable of chaotic signals generating in millimeter wave band. The details concerning the tube design and its working chaotic regimes may be found in [8, 9]. In the designed WR-BWO with increase of the working current we observed a series of bifurcations that ends up with excitation of chaotic self-oscillations with wide enough continuous power spectrum density. High power chaotic waveform generators of such type have been developed in selected frequency sub-bands over the frequency range from 30 GHz to 60 GHz having tunable frequency bandwidth from 50 MHz to 700 MHz and CW output power from 20 to 4 W, respectively.

4 Generation of Chaotic Signals in Multi-Resonant Systems

Application of the Dynamical Chaotization Principle enabled us to design various chaotic waveforms generators. Normally, microwave Active Electronic Components (AEC), such as microwave transistors or diodes, are capable of generating within a certain frequency range. If one provides a resonant system having several resonant modes, falling into the same frequency range, a mode interaction will take place in the Nonlinear AEC. This mode interaction may become a chaotic one under certain conditions. Those conditions may be provided via selection/adjusting of the NAEC working regime. As a result, all this enables generation of chaotic self-oscillations due to realization of chaotic interaction of resonant modes [5]. Mathematical model for such generator is a system of quasi-linear ODE that is to be solved numerically with taking into account specific properties of the used NAEC. X-band chaotic waveform generator based upon Gunn-diode placed in rectangular waveguide resonator has been developed and tested. The Gunn-diode is placed inside a rectangular waveguide between two tunable Multi-Resonant Impedance-Transformers (MRIT), which form the main resonant system of the chaotic waveform generator. The Gunn-diode is DC powered through a Low Pass Filter (LPF) that prevents non-desirable leakage of microwave energy and is used as another tunable resonant element. Besides, frequency spectrum of the generated signal also strongly depend on position of Gunn diode itself. Varying parameters of tuning elements one can create conditions for chaotic oscillations generation within rather wide frequency range. Output power of the generated chaotic oscillations is (1–4)dB less than its nominal power for a single frequency regime. Frequency spectrum bandwidth of the output signal and its pattern may be tuned rather efficiently and has Gaussian shape which is most appropriate for noise radar. Generated signal looks like a sequence of spikes with varied period. Such waveform is typical for Gunn-diode chaotic oscillators. It was found that random intervals between spikes get shorter with voltage increase, and signal looks even more random. The working frequency range of the oscillator is determined by the flattened cavity, while its power spectrum is controlled with help of the output impedance transformer. Its output power is comparable with the nominal power of the diode.

Application of microstrip technology is one of the ways to solve the problem of compact noise generators. However, microstrip lines have unavoidable losses because of parasitic radiation. That is why we developed chaotic oscillator where noise signal is generated in a simple circular or rectangular flattened cavity containing Gunn-diode and having electromagnetic coupling to the microstrip oscillatory circuit. This design provides rather high output power and at the same time gives a possibility of making desirable corrections of the output power spectrum density preserving good matching between the diode and the load.

5 Chaos Generators Based on Delayed Feedback

There are two types of chaotic generators with delayed feedback: (1) ring delayed feed back and (2) reflection delayed feedback. Both of them have been studied in LNDES theoretically and experimentally. Theoretical description of both types of systems can be done in terms of initial-boundary problem for linear telegraph equation with nonlinear boundary condition [10–13]. The derivation of the equation is almost the same for both reflection and ring types and consists in reducing of the initial-boundary value problem for the telegraph equation to initial problem for nonlinear FDE or DDE. The details can be found in [10–14]. For instance, for dynamical system compound of a frequency-independent element with nonlinear transfer characteristics and an ideal dispersionless delay line with linear phase response. Nonlinear dynamics in that system can be described by the first order functional-difference equation with continuous time: $u(t) = G(u(t-T))$ where $T = l/v$ is the wave propagation time in the delay line. More complicated system with amplitude dependent time delay is described by equation

$$u(t) = G(u[t - \tau(u(t-T_0))]) \tag{1}$$

where T_0 is the maximal time delay.

The well-known "tent-map" from the class of nonlinear unimodal maps can be chosen as nonlinear function: $G(x) = \gamma x + \theta - |\beta - \alpha x|$, where γ, θ, β and α are the constant coefficients that define position of the map. This map exhibits transition to chaos through period-doubling bifurcation. For certain values of map parameters the sequence of iteration of this map is fully chaotic. We have designed several chaos generators with delayed feedback. One of them is designed for Ka-band and uses delayed feedback of reflection type in a slow-wave waveguide. A periodic structure placed in the waveguide is used for slowing of the electromagnetic wave and therefore shortening of the feedback circuit length. Another chaos generator which uses ring feedback and V-shaped non-linearity was created for generation of chaotic waveforms in P-band with frequency bandwidth of 400 MHz. The output signal parameters of this device.

6 Conclusion

We have presented theoretical and experimental results in design of chaos generators. In these devices we exploit dynamical chaos that enables design of efficient noise waveform sources for a given frequency band with required parameters, such as power spectrum density, its shape, frequency bandwidth and central frequency. This is the most attractive property for rapidly developing Noise Radar Technology. The output power of such generators far exceeds the power attainable in conventional primary thermal noise sources. This is a very important advantage of chaos generators, since it allows generating of probing signals with quite high power spectral density which is required in Noise Radar. It is obvious from our research, that generation of chaos directly in electronic oscillatory system containing nonlinear element is very promising approach in elaboration of efficient noise sources for Noise Radars.

References

1. Lukin K.A., Millimeter Wave Noise Radar Technology, MSMW 98 Symposium Proc., Kharkov, Ukraine, 1998.
2. Lukin K.A., Noise Radar Technology, Telecommunications and Radio Engineering, 55(12), 8–16, 2001.
3. Tarchi D. Leva D. Lukin K.A. et al., Short Range Imaging Applications Noise Radar Technology Proc. of EUSAR 2000, 23–25 May 2000, Munich, Germany, 2000, 361–364.
4. Narayanan R.M., Dawood M., Doppler Estimation Using a Coherent Ultrawide-Band Random Noise Radar IEEE Transactions on Antennas and Propagation, 2000, 48(6), 868–878.
5. Lukin K.A., Noise Radar Technology: The Principles and Short Overview, Applied Radio Electronics, 4(1), 4–13, 2005.
6. Guo-Sui L., Hong-Bo S, Jian-Hui Z. H. G., Wei-Min S., Random Signals Radar – a Winner in Both Military and Civilian Operating Environments, First International Workshop on NRT, Yalta, Crimea, Ukraine, Sept. 18–20, 2002, 43–52.
7. Lukin K.A., Kulyk V. V., Zemlyaniy O.V., Random Waveform Generators for Noise Radar, Applied Radio Electronics, 4(1), 74–79, 2005.
8. Lukin K.A., Rakityansky V.A., Excitation of Chaotic Oscillation in Millimeter BWOs, International Journal of Infrared and Millimeter Waves, 16(6), 1037–1049, 1995.
9. Kulyk V.V., Lukin K.A., Rakityansky V.A., Autodyne Effect in Weak-Resonant BWO with Chaotic Dynamics, International Journal of Microwave and Millimeter Waves, 19(3), 427–440, 1998.
10. Mogel W.S., Lukin K., Zemlyaniy O., Chaotic Wide Band Oscillator with Delay Line, Proc. of NDES95, Dublin, Ireland, 259–262, 1995.
11. Sharkovsky A.N., Ideal Turbulence in an Idealized Time-Delayed Chua's Circuit, International Journal of bifurcation and Chaos, 4(2), 303–309, 1994.
12. Zemlyaniy O.V., Lukin K.A., Correlation and Spectral Properties of Chaos in Delayed Nonlinear Dynamical System with Asymmetric Nonlinear Map, Radiophysics and Electronics, 2(7), 406–414, 2002.
13. Zemlyaniy O.V., Lukin K.A., Fractal Dimension of the Attractor of Nonlinear Dynamical System with Delayed Feedback and Piecewise Linear Map, Radioelektronika i informatika, 3(32), 8–15, 2005.
14. Lukin K.A., Maistrenko Yu.L., Sharkovsky A.N., Shestopalov V.P., Method of Difference Equations in the Resonator Problem with a Nonlinear Reflector, Soviet Physics Doklady. 34, 977, 1988.

Resonance Induced by Repulsive Links

Teresa Vaz Martins and Raúl Toral

Abstract In nonlinear systems, the right amount of noise can amplify the response to a weak periodic signal, by a phenomenon known as stochastic resonance [1]. It was shown that the same constructive role can be played by any source of disorder [2]. We study an Ising model in a network with disorder induced by the presence of both attractive and repulsive links. The system is subjected to a sub-threshold periodic signal, and the goal is to see how the response is enhanced for a given fraction of repulsive links. This can model a network of spinlike neurons with excitatory and inhibitory couplings. By means of numerical simulations, we find that there is an optimal probability of repulsive links, such that the coherent response is maximal, and we propose a mechanism to explain this resonance.

1 Model

We start with a spinlike network, where each node has two possible states: $\mu_i = \pm 1$ These states are assigned randomly at the beginning, and they can represent for instance the firing state of a neuron, or two different opinions

At each time step, a node is selected at random, and then the dynamics evolves with the consecutive application of the following steps:

- the selected node i adopts the majority state of its neighbors. If the weight is zero, it doesn't change

$$\mu_i = sign(\sum \omega_{ij}\mu_j) \tag{1}$$

T.V. Martins (✉)
Centro de Física do Porto (FCUP), 4169-007 Porto, Portugal, FCT, SFRH/BD/23709/2005
e-mail: tvmartins@fc.up.pt

V. In et al. (eds.), *Applications of Nonlinear Dynamics*, Understanding Complex Systems, DOI 10.1007/978-3-540-85632-0_41,

with the relationship matrix

$$\omega_{ij} = \begin{cases} 1 & , prob\ (1-p) \\ -k & , prob\quad p \end{cases} \tag{2}$$

- with probability $|asin(\Omega t)|$, it follows the external signal

$$\mu_i = sign(sin(\Omega t)),\ with\, prob\ |asin(\Omega t)| \tag{3}$$

where Ω is the frequency of the external signal, and a is its amplitude

After that, time increases by $1/N$. Our results concern random networks, the system size is $N = 100$, and the average number of neighbors is 10. Having tried other sizes, regular networks, or topologies such as a replicated random or a square network, we found the same type of phenomenon. The averages are over 1000 runs.

Together with the results for quenched disorder, sometimes we will also show what happens in a temporarily annealed version, where disorder is not correlated in time. The neighborhood is fixed but, for a transient time, the relationship matrix w_{ij} is redefined randomly at each time step. This scenario can model a society where a diffusive animosity finally crystallizes into entrenched positions of friends and enemies, or in another context, the known fact that brain plasticity is higher in initial stages.

2 Results

As a measure of coherent response, we chose the spectral amplification factor R, defined as the ratio of the output power at the corresponding driving frequency to the input power.

The system becomes monostable, as an increasing probability of repulsive links degrades order (Fig. 1). We found that the response to the subthreshold signal is maximal, for an intermediate proportion of repulsive links, within the transition region (Fig. 2).

The amplitude of oscillations is amplified if we add the right proportion of repulsive links. The system oscillates with the signal's frequency (Fig. 3).

When the signal is weak, there is still an optimal probability, but the oscillations amplitude is less magnified (Fig. 4).

In the temporarily annealed disorder system, the signal's weakness is revealed in the coexistence of big and smaller amplitude of oscillations at the optimal probability, and also in a degradation of periodicity (Fig. 5).

In stochastic resonance, with its prototypical bistable potential, we can speak of intra and inter-well oscillations [3]. To see if a similar picture applied, we have looked into how the stable states values were distributed. In the quenched disorder scenario they go from being concentrated at $M = -1$ or $M = 1$, at $p = 0$, to span a continuous wide range, at the optimal probability (Fig. 6).

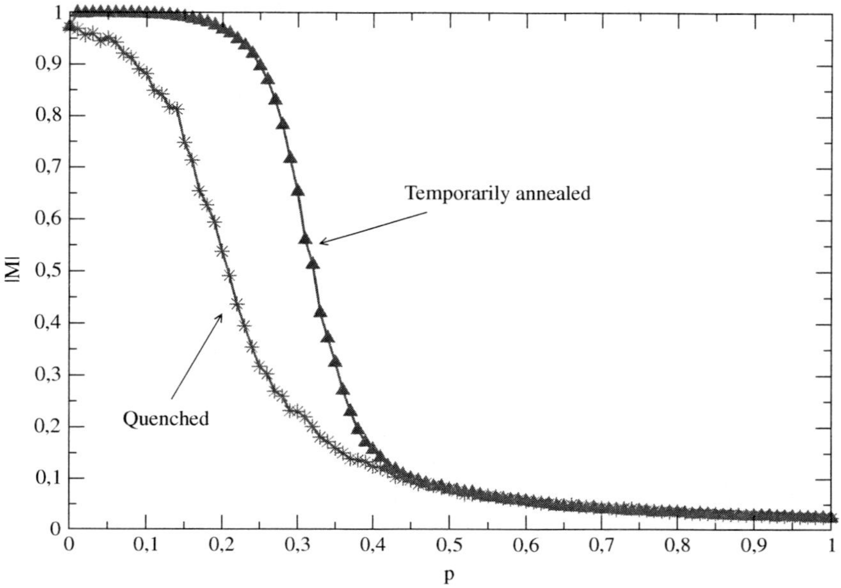

Fig. 1 Total Average Magnetization as a function of the probability of repulsive links

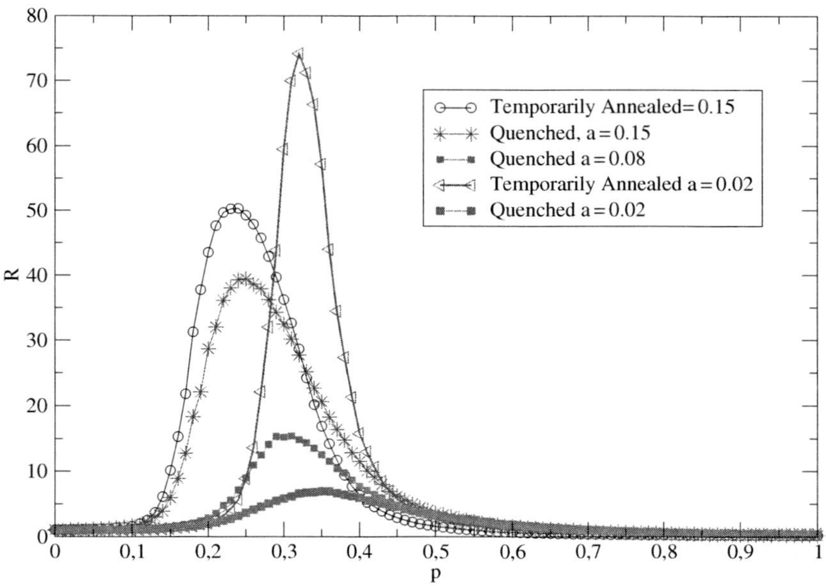

Fig. 2 Spectral amplification factor R as a function of the probability of repulsive links, showing a peak at the right proportion of repulsive links

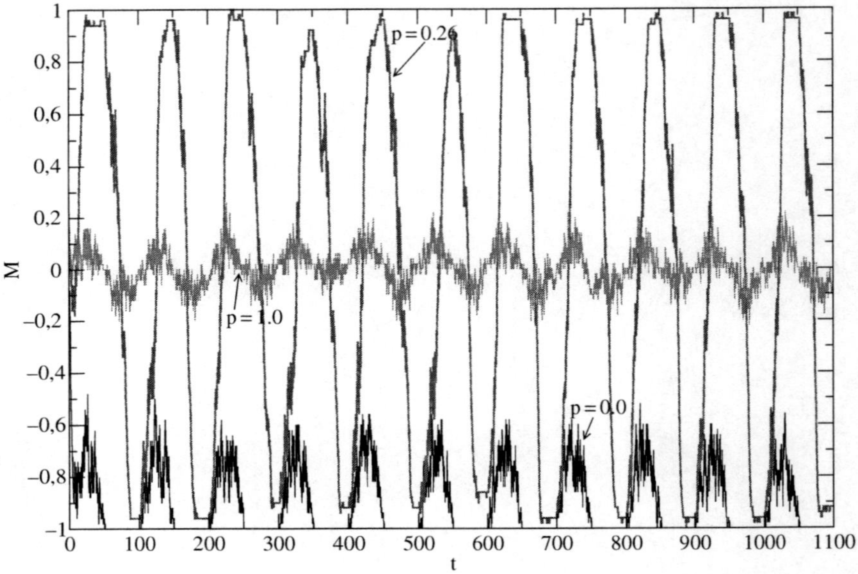

Fig. 3 Evolution of the total average magnetization in time, showing how the amplitude is greater at the optimal probability $p = 0.26$. Here we show a quenched disorder case, but the temporarily annealed looks the same at this signal's intensity

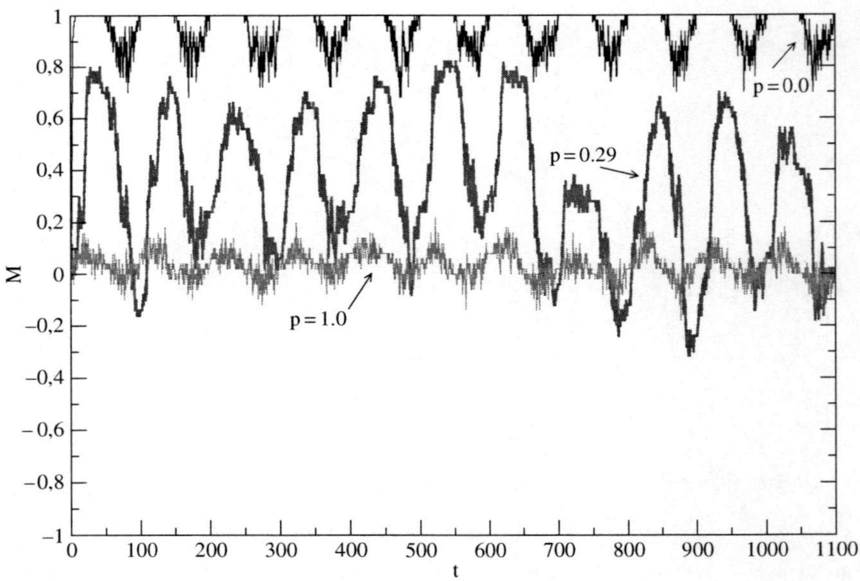

Fig. 4 Evolution of the total average magnetization with time, showing the increase in the amplitude of oscillations at the optimal probability $p = 0.29$. Quenched disorder version. $a = 0.08$

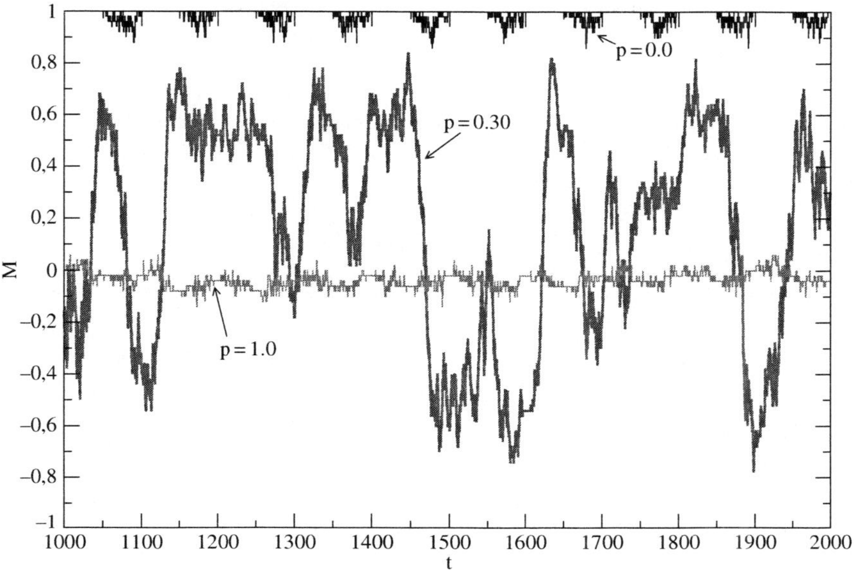

Fig. 5 Evolution of the total average magnetization with time. Temporarily annealed disorder version. $a = 0.02$

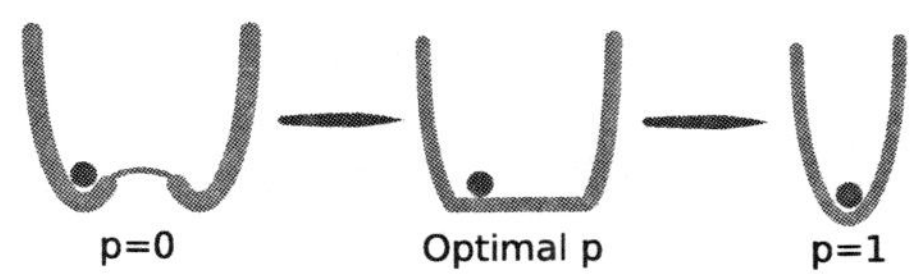

Fig. 6 Quenched disorder version: the optimal region is a region of many stable states, through which the system can progress in multi-steps

In the temporarily annealed version, at the optimal probability they are still concentrated at symmetric values, that get closer as the proportion of repulsive links increases. This system is more prone to respond in switch-like fashion (Fig. 7).

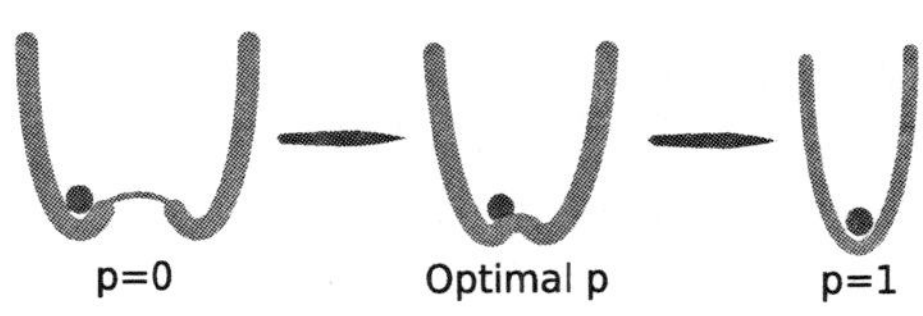

Fig. 7 Temporarily annealed disorder version: Disorder lowers the potential barrier, and moves the wells closer

Meanwhile, the microscopic mechanism is the same. After a node has changed, it has two requirements, that explain where the optimal probability is located: it can not be forced to return to its previous state (the effective local magnetization around a node should be close to zero). And it should spread its perturbation: (it is better to have attractive connections).

3 Conclusions

Competition can enhance a system response to an external signal. The model and rules are generic enough as to predict an application in several systems. Ways to optimize the mechanism: changing the network topology, play with parameters.

References

1. C. Tessone, C.R. Mirasso, R. Toral and J.D. Gunton. Diversity induced resonance, Phys. Rev. Lett. 97, 194101 (2006)
2. L. Gammaitoni, P. Hnggi, P. Jung and F. Marchesoni. Rev. Mod. Phys. 70, 225 (1998)
3. L. Alfonsi et al. Intrawell stochastic resonance versus interwell stochastic resonance in underdamped bistable systems, Phys. Rev. E 62, 299–302 (2000)

Time Scales of Performance Levels During Training of Complex Motor Tasks

Gottfried Mayer-Kress, Yeou-Teh Liu and Karl M. Newell

Abstract Complex motor tasks – such as skydiving or landing an airplane – require extensive training and practice of coordinated movement sequences. But even once the task is learned, experience shows that the proper execution of the task is not guaranteed if the task is to be performed years, months, or even days after its most recent execution. That is why, for example, pilots need to demonstrate a minimum number of flight hours per year. We have developed a dynamical systems model that describes these phenomena in the context of different time scales of adaptation and learning processes. We could show that the performance decrement after interruption of skill practice is exclusively due to the dynamics of adaptive processes, whereas learning continued even without practice. The proper understanding of the separation between these two types of processes can help to predict the type and intensity of warm-up that is necessary to safely and reliably execute a task after a given rest time. Such a situation seems to be especially relevant for complex military missions, which include long pauses between task execution together with short preparation and high stress levels during the relatively infrequent task executions.

1 Introduction

The learning of motor tasks typically takes place over multiple time scales over days, weeks, months and years. Traditionally, theories of learning have been based on a single and universal law of learning that has long been viewed as the power law ([1,2]).

In our dynamical systems theoretical framework of the study of human motor learning we postulate that the persistent and transitory changes in task outcome are products of an evolving attractor landscape [3–5]. The landscape changes include

G. Mayer-Kress (✉)
Department of Kinesiology, PSU, University Park, Pennsylvania PA, USA,
e-mail: gxm21@psu.edu

V. In et al. (eds.), *Applications of Nonlinear Dynamics*, Understanding Complex Systems,
DOI 10.1007/978-3-540-85632-0_42,

those due to bifurcations between attractor organization and the transient phenomena associate with moving toward and away from fixed points. In this theoretical view then, there is not a single universal function for the change in the performance dynamics over time although a small set of dynamical principles can capture the different relative contributions of persistent (learning) and transient (adaptation) changes of performance dynamics.

In the learning of motor tasks that can be represented dynamically as moving to a fixed point we have been able to show that a two time scale model captures the performance dynamics ([6]). The two time scale model has two exponential functions superimposed that represent adaptation and learning processes that are both engaged during and between practice sessions. One characteristic time scale is relatively fast and captures the rapid adaptive change (warm-up) in performance at the beginning of a practice session. The other time scale is relatively slow and captures the persistent change that is more typically associated with learning. The model holds the potential to more fully decompose the performance dynamics and include additional processes, such as fatigue (Fig. 1).

The understanding of the two time scale model of motor learning provides insight into the processes of practice and the distribution of practice and rest in the learning and performance of a task. The model can predict the type and intensity of warm-up that is necessary to safely and reliably execute a task after a given rest interval.

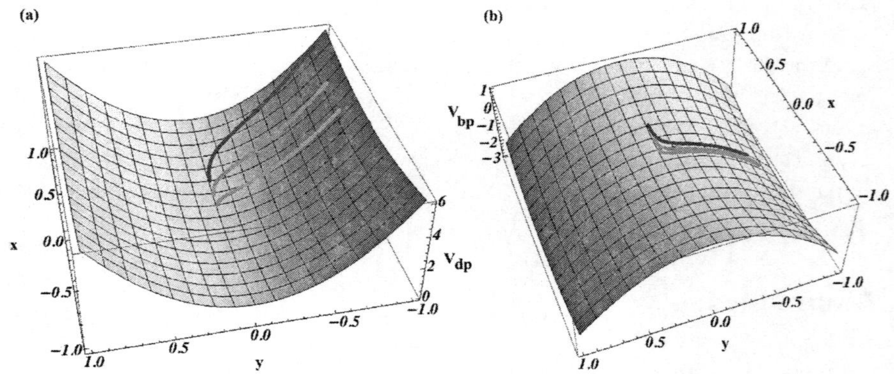

Fig. 1 Two time scale landscape models for learning/adaptation dynamics during practice (**a**) and between practice sessions (**b**). Slow learning dynamics along x-dimension, fast adaptation dynamics evolves along y-dimension. Landscape elevation values determine direction and speed along trajectories in the landscape following the direction of the steepest descent. Note that during the practice sessions (**a**) both the learning dynamics (along x) as well as the adaptation process (along y) evolve towards the learning goal, the attractive fixed point at $(x_g, y_g) = (0,0)$. Between practice sessions (**b**), we assume that for the simplest form of dynamics (i.e. in the absence of interference from other processes) the learning process will still continue towards the learning goal but that the adaptation process relaxes and moves away from the learning goal at a characteristic rate. Depending on the relative size of the two characteristic time scales we can observe either warm-up decrement, reminiscence, or no change in performance during the rest period

Such a situation seems to be especially relevant for complex military missions. Our principles for decomposing performance dynamics also hold relevance for the study of learning in simulators, not only as a precursor to the conduct of performance in the actual system (the standard learning strategy [7–9], but also in the possible use of simulator devices in the actual system to reduce the negative impact of warm-up. The model holds particular significance for tasks where the criterion performance must be executed on the first and perhaps only trial (e.g., take-off and landing an airplane).

2 Two-Time Scale Landscape Model of Adaptation and Learning

In [6] we analyzed a number of motor learning data sets that displayed the phenomenon of warm-up decrement. We could demonstrate that for all of the analyzed data sets we could extract two separate characteristic time scales that differ by an order of magnitude. This was done by fitting the data to the following function:

$$V_j(n) = V_{\mathrm{inf}} + a_s e^{-\gamma_s n} + a_j e^{-\gamma_j (n-n_j)} \tag{1}$$

Here the independent parameter n corresponds to time, measured in units of trial number. V_{inf} corresponds to the target performer that the learner tries to achieve through practice. $V_j(n)$ measures the distance to that goal performance after n practice trials. The characteristic time scales are defined by γ_s (learning) and γ_j (adaptation) and a_s and a_j denote scale parameters for both processes in session j. In [6] we show in detail how the time scales extracted from curve fitting translates into the landscape parameters.

3 Warm-Up Through Mental and Virtual Practice

Based on our theoretical analysis [6] we are able to make predictions about the effect of mental and or virtual practice prior to critical tasks that are often performed under extreme stress. Typical examples are high stakes situations in competitions (final phase of games of golf, tennis etc), racing a formula one car, or landing an airplane on a carrier by night etc. It is known that a mental "walk through" or simulator practice can be highly effective in improving the performance but it is also known that the time between the last rehearsal before a critical task is a relevant factor.

Our multiple time scales model allows to extract the characteristic time scales associated with these preparatory rehearsals (adaptation) and their associated decay during inactivity periods. At the same time it also allows an estimate of the skill level of a participant due to learning, which improves at a much slower time scale. For practical applications we therefore can estimate how many virtual practice runs a participant – at a given skill level – has to perform at a given time interval before a critical task in order to achieve a minimum performance level for the critical task.

4 Conclusions

There are strong, empirical and theoretical arguments that our mathematical model describing motor learning and adaptation through warm-up can be applied to important practical tasks. Of special importance are tasks that are performed at irregular time intervals with or without announcement of the time of the next execution and the value of the achieved score. Our model allows a general strategy to estimate most efficient times at which refresher practice sessions need to be scheduled to achieve a minimum performance level at the next task execution.

Acknowledgments This work was supported by National Science Foundation grant 0518845.

References

1. Anderson, J. R., Fincham, J. M., & Douglass, S. (1999). Practice and retention: A unifying analysis. *Journal of Experimental Psychology: Learning, Memory, and Cognition*, 25, 1120–1136.
2. Newell, A., & Rosenbloom, P. S. (1981). Mechanisms of skill acquisition and the law of practice. In J. R. Anderson (Ed.), *Cognitive skills and their acquisition.* (pp. 1–55). Hillsdale, NJ: Erlbaum.
3. Newell, K. M., Liu, Y-T., & Mayer-Kress, G. (2001). Time scales in motor learning and development. *Journal of Experimental Psychology: Learning, Memory, and Cognition*, 25, 1120–1136.
4. Newell, K. M., Liu, Y-T., & Mayer-Kress, G. (2003). A dynamical systems interpretation of epigenetic landscapes for infant motor development. *Infant Development and Behavior*, 26, 449–472.
5. Newell, K. M., Liu, Y-T., & Mayer-Kress, G. (2005). Learning in the brain-computer interface: Insights about degrees of freedom and degeneracy in a landscape model of motor learning. *Cognitive Processing*, 6, 37–47.
6. Mayer-Kress, G. Characteristic time scales in adaptation and motor learning. Proceedings of the intl workshop on complex dynamics of physiological systems: From Heart of Brain (CDPS 2007) Presidency College, Kolkata 700073, India. February 12–14, 2007.
7. Flexman, R. E., & Stark, E. A. (1987). Training simulators. In G. Salvendy (Ed.), *Teams: Their training and performance.* Norwood, NJ: ABLEX.
8. Tichon J. (2007). Training cognitive skills in virtual reality: Measuring performance. *Cyberpsychology and Behavior*, 10, 286–289.
9. Kosslyn, S. M., & Koenig, O. (1992). *Wet mind: The cognitive neuroscience.* New York: The Free Press.

Analysis of Nonlinear Bistable Circuits

Suketu Naik

Abstract Bifurcations induced in the super-critical regime defined by 1-D parameter space containing two state-space variables offer vital insight into the operation of pertinent systems. For an over-damped system, typical circuit implementation leads to Nonlinear Bi-stable circuits. Some circuits allow direct transformation of the state-space variables into the circuit parameters while others require complex analysis. This paper describes the design of a few of the circuits based on nonlinear bifurcation phenomenon, their simulations and dynamic cooperative behaviors that can be utilized for further signal processing.

1 Bifurcation and the Parameter Space

It has been shown that a bistable system such as a single fluxgate can switch its states in the presence of a periodic forcing function [1]. Furthermore such a periodic forcing function can be obviated by coupling the single elements. Figure 1 shows the effect of coupling and how oscillations are generated once the coupling parameter reaches a critical value [2, 3]. Based on the simluated and measured results of the coupled fluxgates, a new set of systems can be constructed. These systems can be implemented using bistable circuits [4, 5] to achieve desired properties such as downconversion.

Global bifurcation occurs at λ_c, the critical coupling. Figure 1 shows the behavior of a coupled system for oscillation. Here, a potential well function derived from a coupled system of equations vs the coupling parameter, λ_c, is plotted. As shown in Fig. 1, for no coupling (λ=0) the potential function is symmetric, and the solutions (shown as spheres in Fig. 1) remain in steady state or single potential well. Once coupled, the system begins to oscillate for $\lambda > \lambda_c$. For nonzero coupling below

S. Naik (✉)
SSC San Diego, 53560 Hull Street, San Diego, CA 92152 USA,
e-mail: suketu@spawar.navy.mil

V. In et al. (eds.), *Applications of Nonlinear Dynamics*, Understanding Complex Systems, 449
DOI 10.1007/978-3-540-85632-0_43, © Springer-Verlag Berlin Heidelberg 2009

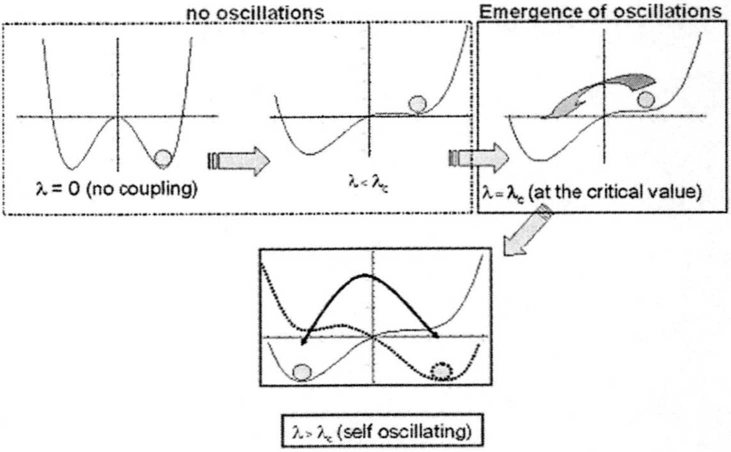

Fig. 1 Oscillations: potential well function vs. the coupling λ [1]

the critical coupling, $\lambda < \lambda_c$, the system becomes asymmetric, but does not switch between the potential wells. At the critical value (global bifurcation) the dynamics of the coupled system change and switching occurs. Above the critical coupling the system is in the self oscillating or supercritical regime hence the oscillation between potential wells begin when $\lambda > \lambda_c$. This behavior is induced without an input signal function [2]. These concepts form the basis for the circuits described in this paper.

1.1 Coupled System with the Input Signal

The behavior of the coupled system can be characterized by varying the peak-to-peak amplitude and the frequency of the input signal. Figure 2 shows such characterization curve that illustrates different oscillating regions. We are interested in operating the system in the middle region where frequency of each individual element oscillates at 1/3rd the frequency of the incoming signal [2]. ($\omega_{Vi} = \omega/3$.). This plot is a compressed version of a full 3D plot where the third axis is represented by coupling parameter λ. Ideally a coupled system would not oscillate in the absence of an input signal. When an input signal is applied the oscillations occur in $\omega/3$ region [2]. In the oscillatory state, each component switches between its two stable states, leading or lagging its neighboring component by $2\pi/N$ radians, where N is an odd number representing the number of oscillators in the array. This is referred to as the out-of-phase pattern. In the supercritical region (I) the oscillator is not frequency locked to the incoming signal, rather it oscillates at its natural frequency. In regions (I) and (II) the components of the oscillator switch in the out-of-phase pattern. In region (II) the oscillator is frequency-locked to the incoming signal so that each component of the oscillator switches at a frequency that is ω/N. In region (III) the dynamics of each component is over-driven by the incoming signal, resulting in an in-phase response with respect to each other and the signal. The significance of regions (II) and (III) is that in those regions the response of system is overwhelmingly driven by those qualities of the incoming signal–in both phase and frequency. In the case of frequency down-conversion, which occurs in region (II), the phase information of the input signal is preserved. This type of synchronization behavior can be used to construct large coupled systems that work as multi-frequency down-converters, e.g. at 1/75th the input frequency.

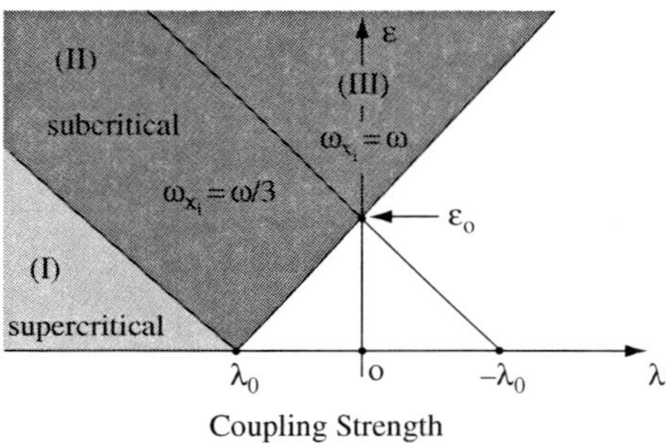

Fig. 2 Parameter space for nonlinear oscillator as a function of signal amplitude and coupling strength [2]. Regions(II) and (III) represent synchronization between the oscillator and the signal. In region (II) each oscillator oscillates at $\omega/3$, where $\omega/3$ is the frequency of the input signal

2 Model

An block diagram of is depicted in Figure 3, which shows an N-element coupled system on the left and an example circuit implementation of a single element on the right. The model that describes nonlinear phenomena, which is the basis for the bistable circuits, is defined by (1). This equation is a representation of a single element in the coupled system that is being utilized in the bistable circuit as shown in Fig. 3.

$$C\frac{\partial V_{out}}{\partial t} \equiv I_o - I_{sc} + I_s \tanh\left[g\left(V_{out} + \lambda V_{in}\right)\right] + V_{sig} \tag{1}$$

$$fori = 1\ldots N \tag{2}$$

where, N = total number of elements; C =load capacitance; I_o = summation of steady state currents of the linear and nonlinear currents; V_{out} = output signal of the element, V_{in} = output signal from the previous element; I_s, g = parameters of the nonlinear portion; $I_{sc} = I_n - I_p$ effective current from the linear portion due to operation of the transistors in saturation; where I_n =drain current thru the nfet, I_p = drain

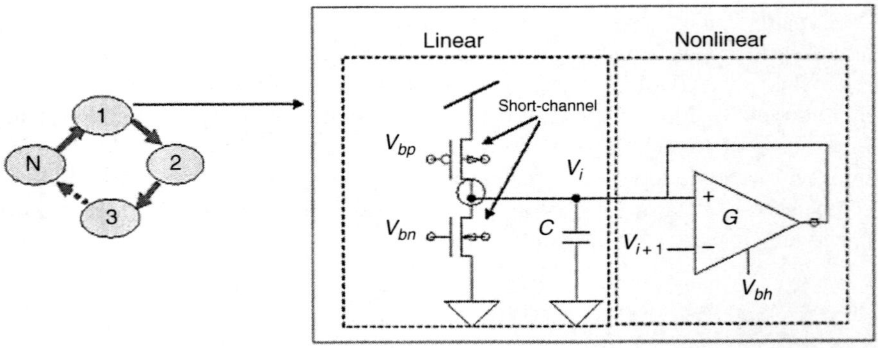

Fig. 3 Basic circuit and the single element of the coupled-system

current thru the pfet; $\lambda =$ coupling parameter between the elements, and $V_{sig} =$ input signal to the coupled system.

3 Circuit Implementation

The model for the single element as described previously is adequate when it's used in the presence of an input signal. However, the coupling parameter, λ is assumed to be fixed in this model. As shown in Fig. 3, linear FETs and the OTA which act as a single element within N-coupled system. The value of λ depends on the ratio of the widths of the transistors in the differential pair (e.g. $W_{\text{element1}}/W_{\text{element2}}$). To improve the functionality of the coupling parameter, a different circuit utilizing two OTAs was used as shown in Fig. 4. This circuit allows a better tuning of the coupling

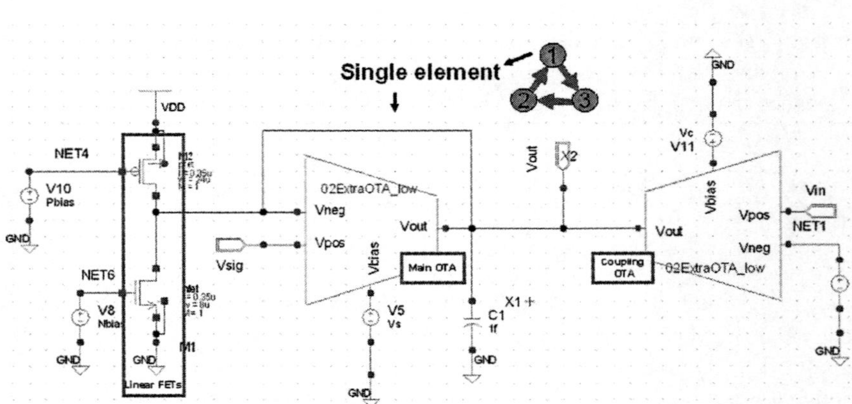

Fig. 4 Improved circuit and the single element of the coupled-system

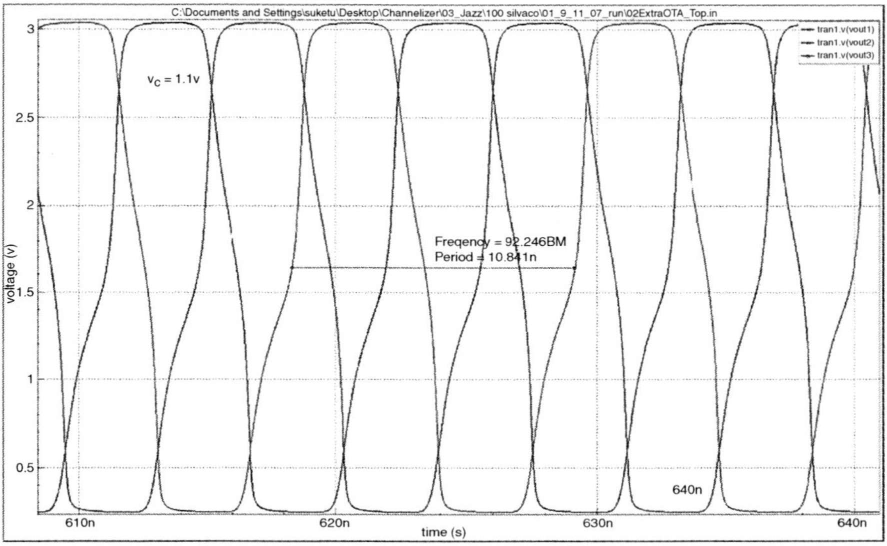

Fig. 5 Simulations results: oscillations at 92.25 MHz with Vc=1.1 V

parameter by simply changing the current, I_c across the transistor at the base of the coupling OTA. Furthermore, the empirical model based on this circuit, as shown in (3) exhibits more sensitivity to the input signal.

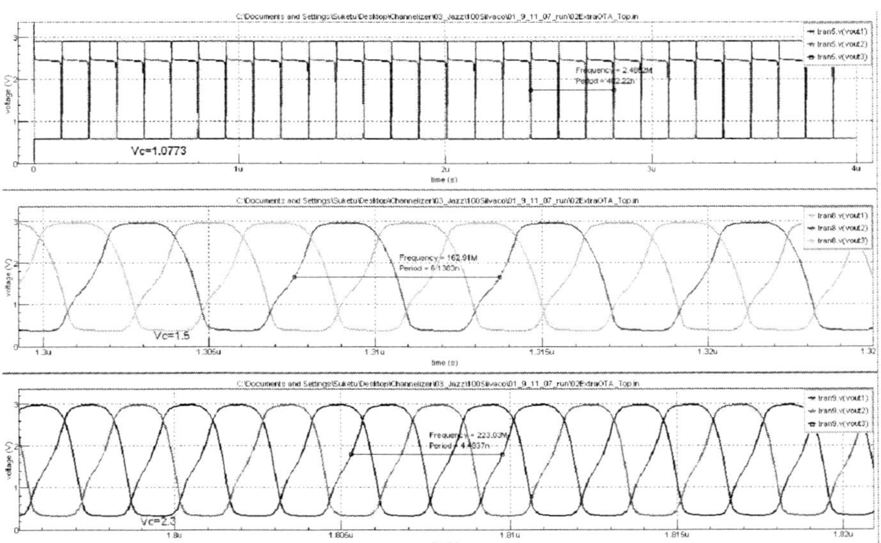

Fig. 6 Simulations results: effects of coupling with varying Vc

$$C\frac{\partial V_{out}}{\partial t} \equiv I_o - I_{sc} + I_s \tanh\left[c_s\left(V_{out} - V_{sig}\right)\right] + I_c tanh\left[c_c\left(V_{dc} - V_{in}\right)\right] \tag{3}$$

$$\text{for } i = 1 \ldots N$$

where, N is the total number of elements; C = load capacitance; I_o = summation of steady state currents of the linear and nonlinear currents; V_{out} = output signal of the element, V_{in} = output signal from the previous element; I_s, g = parameters of the nonlinear portion; I_{sc} = total current out of the linear potion; $c_s = \sqrt{\beta/I_s}$, $c_c = \sqrt{\beta/I_c}$; β = process parameter; I_s = Bias current of the main OTA, I_c = Bias current of the coupling OTA; V_{sig} = input signal to the coupled system.

Figure 5 shows the oscillations generated by tuning the bias voltage, Vc of the coupling OTA. More effects of the coupling are illustrated in Fig. 6 where oscillations start out as being asymmetric at a lower frequency. Increasing the bias voltage of the coupling OTA exhibits symmetric behavior at substantially higher frequency. Once again, this behavior can be very useful when constructing a large coupled system targeted toward the down conversion of an input signal.

References

1. P. Longhini, A. Palacios, V. In, J. Neff, A. Kho, and A. Bulsara. Exploiting dynamical symmetry in coupled nonlinear elements for efficient frequency down-conversion. Physical Review E 76, 026201 (2007).
2. V. In, A. Bulsara, A. Palacios, P. Longhini, and A. Kho. Complex dynamics in unidirectionally coupled overdamped bistable systems subject to a time-periodic external signalt. Physical Review E 72, no. 4, Rapid Communications 045104 (2005).
3. S. Strogatz. Nonlinear Dynamics and Chaos, Addison-Wesley, ReadingMA (1994).
4. B. Razavi. Design of Analog CMOS Integrated Circuits, McGraw-Hill, New York (2001).
5. S. Sedra. Microelectronic Circuits, McGraw-Hill, Saunders College Publishing, Orlando, FL (2002).

Noise-Induced Transitions for Limit Cycles of Nonlinear Systems

Lev Ryashko

Abstract An analysis of noise-induced transitions is challenging and fundamental problem of nonlinear engineering. We consider noise-induced transitions for limit cycles in Hopf bifurcation zone and period-doubling bifurcation zone. Near bifurcation values, even small external disturbances may essentially change behavior of dynamic system. An underlying reason of the unexpected chaotic-looking response of regular oscillations under the small perturbations is a high sensitivity of the nonlinear dynamic system. We present the results of the stochastic analysis for the classical Hopf differential system. The difference in the response of Hopf systems perturbed by additive and multiplicative random noises is demonstrated. A delaying shift of the Hopf bifurcation point induced by multiplicative noise is observed. We show that significant sensitivity of the cycle can cause the noise-induced transitions both from an attractor to another one and between different parts of single attractor. For Lorenz model, the growth of stochastic sensitivity near chaos bifurcations analysis is studied. We demonstrate noise-induced transitions and qualitative transformation of stochastic attractors.

1 Stochastic Hopf Model

Consider the Stratonovich-interpreted stochastic Hopf system

$$\begin{aligned} dx &= \Big(\mu x - y - (x^2+y^2)x\Big)dt + \sigma_1 x \circ dW_1(t) + \sigma_2 dW_2(t) \\ dy &= \Big(x + \mu y - (x^2+y^2)y\Big)dt + \sigma_1 y \circ dW_1(t) + \sigma_2 dW_3(t) \end{aligned},$$

L. Ryashko (✉)
Department of Mathematics, Ural State University, Lenina, 51, 620083, Ekaterinburg, Russia, e-mail: lev.ryashko@usu.ru

V. In et al. (eds.), *Applications of Nonlinear Dynamics*, Understanding Complex Systems, DOI 10.1007/978-3-540-85632-0_44, © Springer-Verlag Berlin Heidelberg 2009

where W_i are standard Wiener processes, $\sigma_1 \geq 0$ is an intensity of parametric (multiplicative) noise perturbing the parameter μ, $\sigma_2 \geq 0$ is an intensity of external (additive) noise.

Fokker-Planck equation for the stationary probabilistic density function $\rho(r)$ ($r = \sqrt{x^2+y^2}$) is following

$$\frac{1}{2}\left[(\sigma_1^2 r^2 + \sigma_2^2)\rho\right]'' = \left[\left(\mu r - r^3 + \frac{\sigma_1^2}{2} r + \frac{\sigma_2^2}{2r}\right)\rho\right]'$$

1.1 Case $\sigma_1 = 0$, $\sigma_2 \neq 0$ (No Parametric Noise)

The stationary probability density of amplitude r can be written in explicit form

$$\rho(r) = Nr\exp\left(\frac{2\mu r^2 - r^4}{2\sigma_2^2}\right),$$

For the probability density $p(x,y)$ of the stationary distribution in terms of Cartesian coordinates x and y, we have

$$p(x,y) = K\exp\left(\frac{2\mu(x^2+y^2) - (x^2+y^2)^2}{2\sigma_2^2}\right)$$

For $\mu \leq 0$, the function p can only have a single maximum at the point $(0,0)$. The graph of p has a single-peak bell-shaped form. The random trajectories of Hopf system under additive noise perturbations are concentrated in the vicinity of point $(0,0)$. It is natural to name this type of stochastic attractor by *stochastic equilibrium point* [1–3].

For $\mu > 0$, the function p possesses a maximum at the points of deterministic cycle characterized by $x^2+y^2 = \mu$ and a single minimum at the point $(0,0)$. The graph of p has a crater-like form. The random trajectories of Hopf system are concentrated in the vicinity of deterministic cycle given by $x^2+y^2 = \mu$. So the randomly forced Hopf model is said to have a *stochastic limit cycle*.

In Fig. 1, the graphs of $p(x,0)$ for two values of parameter $\mu = -1$ (Fig. 1a), $\mu = 1$ (Fig. 1b) and various intensities of additive noise $\sigma_2 = 0.1$ (solid), $\sigma_2 = 0.2$ (dashed), $\sigma_2 = 0.5$ (dotted) are presented in details.

An variation of the parameter σ_2 of additive Gaussian noise does not change the location of extremal points of the related probability density function. An increase of σ_2 just results into a growth of the dispersion of the random Hopf system around the deterministic attractors (i.e. around the stable point $(0,0)$ for $\mu \leq 0$, and around the stable limit cycle $x^2+y^2 = \mu$ for $\mu > 0$). Note that the deterministic bifurcation value $\mu_* = 0$ under additive Gaussian noise perturbations is not changed.

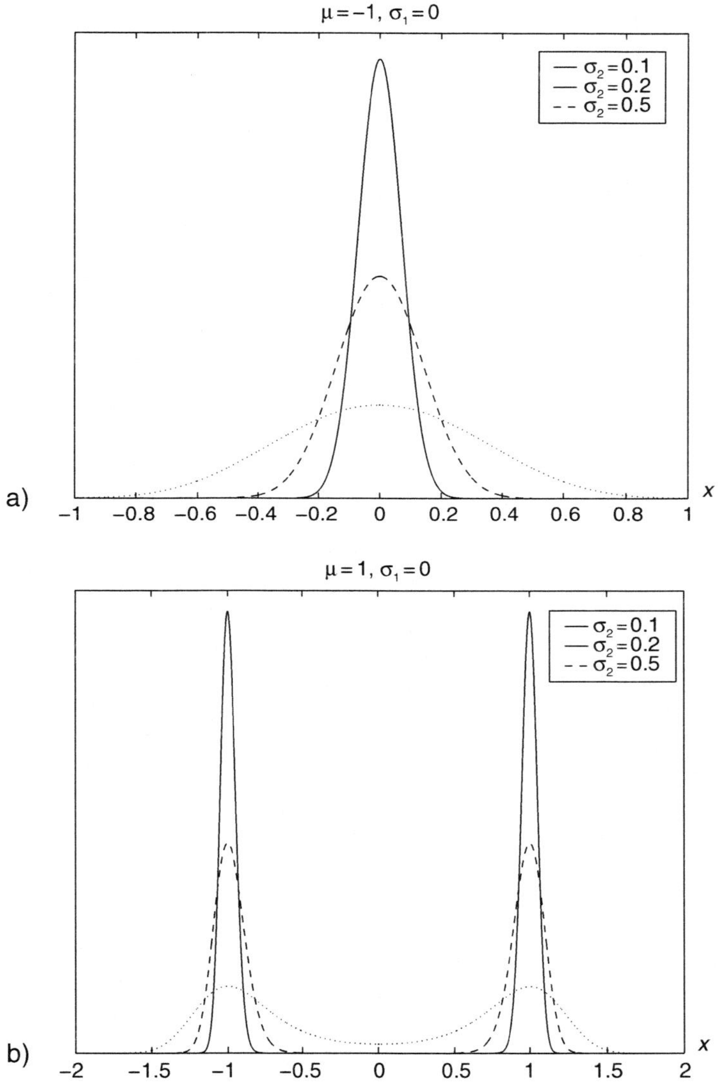

Fig. 1 Stationary probability distribution $p(x,0)$ of Hopf system for $\sigma_1 = 0$, $\sigma_2 = 0.1$ (*solid*), $\sigma_2 = 0.2$ (*dashed*), $\sigma_2 = 0.5$ (*dotted*) for (**a**) $\mu = -1$, (**b**) $\mu = 1$

1.2 Case $\sigma_1 \neq 0$ (with Parametric Noise)

The stationary probability density of amplitude r in this case can be written in explicit form

$$\rho(r) = Nr\exp\left(-\frac{r^2}{\sigma_1^2}\right)(\sigma_1^2 r^2 + \sigma_2^2)^{\alpha}$$

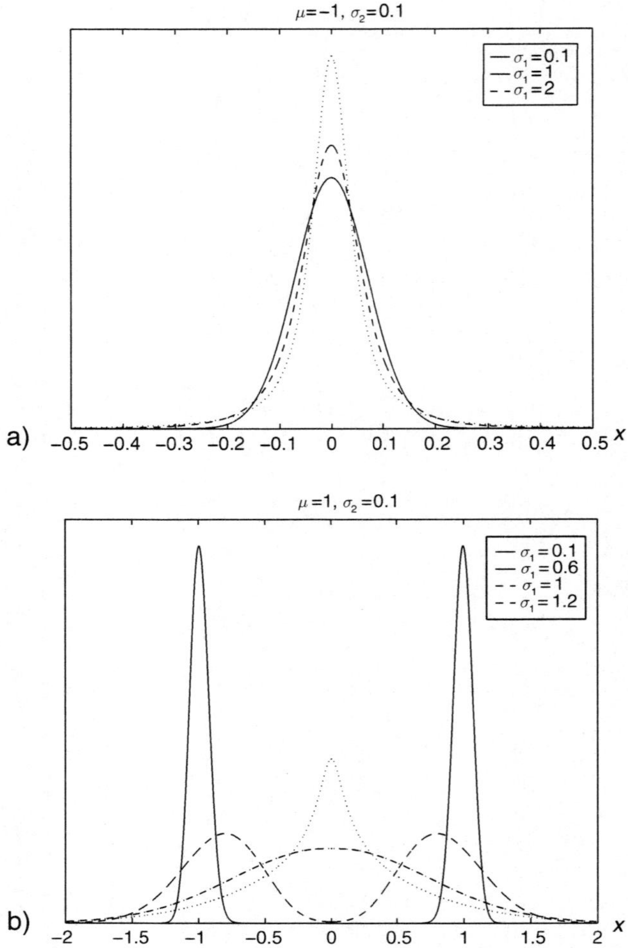

Fig. 2 Stationary probability distribution $p(x,0)$ of Hopf system for $\sigma_2 = 0.1$, (**a**) $\mu = -1$, $\sigma_1 = 0.1$ (*solid*), $\sigma_1 = 1$ (*dashed*), $\sigma_1 = 2$ (*dotted*), (**b**) $\mu = 1$, $\sigma_1 = 0.1$ (*solid*), $\sigma_1 = 0.6$ (*dashed*), $\sigma_1 = 1$ (*dash-dotted*), $\sigma_1 = 1.2$ (*dotted*)

For the probability density $p(x,y)$ of the stationary distribution in terms of Cartesian coordinates x and y for the Hopf system, we obtain

$$p(x,y) = K \exp\left(-\frac{x^2+y^2}{\sigma_1^2}\right)(\sigma_1^2(x^2+y^2)+\sigma_2^2)^{\alpha},$$

$$\alpha = \frac{\mu}{\sigma_1^2} + \frac{\sigma_2^2}{\sigma_1^4} - 1$$

Radius of stochastic cycle

$$r_s = \sqrt{\mu - \sigma_1^2}$$

σ_1^2 is a **shift** value of the bifurcation point

Figure 2 demonstrates a response of the stationary probability density function p subject to changes of intensity σ_1 of multiplicative noise. In Fig. 2a, the graphs of $p(x,0)$ for $\mu = -1$, $\sigma_2 = 0.1$ and three distinct values of intensities $\sigma_1 = 0.1$ (solid), $\sigma_1 = 1$ (dashed), $\sigma_1 = 2$ (dotted) of multiplicative noise are shown. An increase of the intensity of multiplicative noise results into a decrease of the dispersion of random trajectories around equilibrium point $(0,0)$.

In Fig. 2b, the graphs of $p(x,0)$ for $\mu = 1$, $\sigma_2 = 0.1$ and four values of multiplicative noise intensity $\sigma_1 = 0.1$ (solid), $\sigma_1 = 0.6$ (dashed), $\sigma_1 = 1$ (dash-dotted), $\sigma_1 = 1.2$ (dotted) are shown. As we can see in Fig. 2b, the increase of multiplicative noise is accompanied by suppressing auto-oscillations and re-localization of random trajectories in vicinity of zero point.

2 Inverse Stochastic Hopf Bifurcations

As it follows from Fig. 2, a presence of multiplicative noise changes the bifurcation point from $\mu_* = 0$ to $\mu_* = \sigma_1^2$.

In Fig. 3, a bifurcation diagram of the stochastic Hopf system is shown. The curve $\mu = \sigma_1^2$ characterizes the border line between the zones of stochastic equilibrium ($\mu \leq \sigma_1^2$) and stochastic limit cycle ($\mu > \sigma_1^2$).

As we can clearly see, for any fixed $\mu > 0$, an increase of σ_1 results into a transition from the stochastic limit cycle ($0 < \sigma_1 < \sqrt{\mu}$) to the stochastic equilibrium ($\sigma_1 \geq \sqrt{\mu}$) at the point $\sigma_{1*} = \sqrt{\mu}$. Here we observe a so-called *inverse stochastic bifurcation* (see Fig. 4). The underlying reason of this phenomenon rests on the nonlinear response of the Hopf system to the random parametric perturbations.

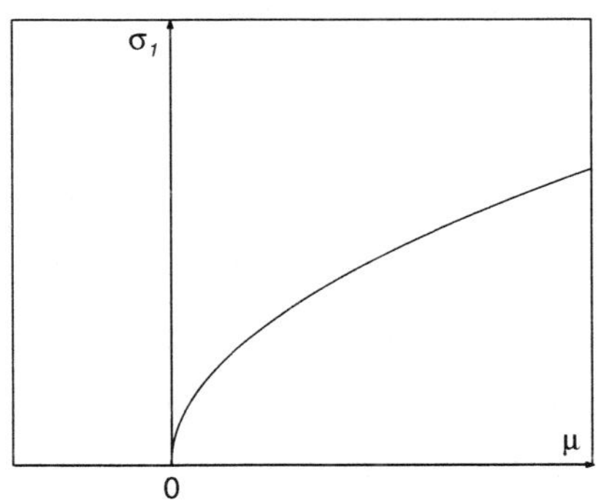

Fig. 3 Bifurcation diagram of stochastic Hopf system: $\mu > \sigma_1^2$ is a stochastic limit cycle zone, $\mu \leq \sigma_1^2$is a stochastic equilibrium zone

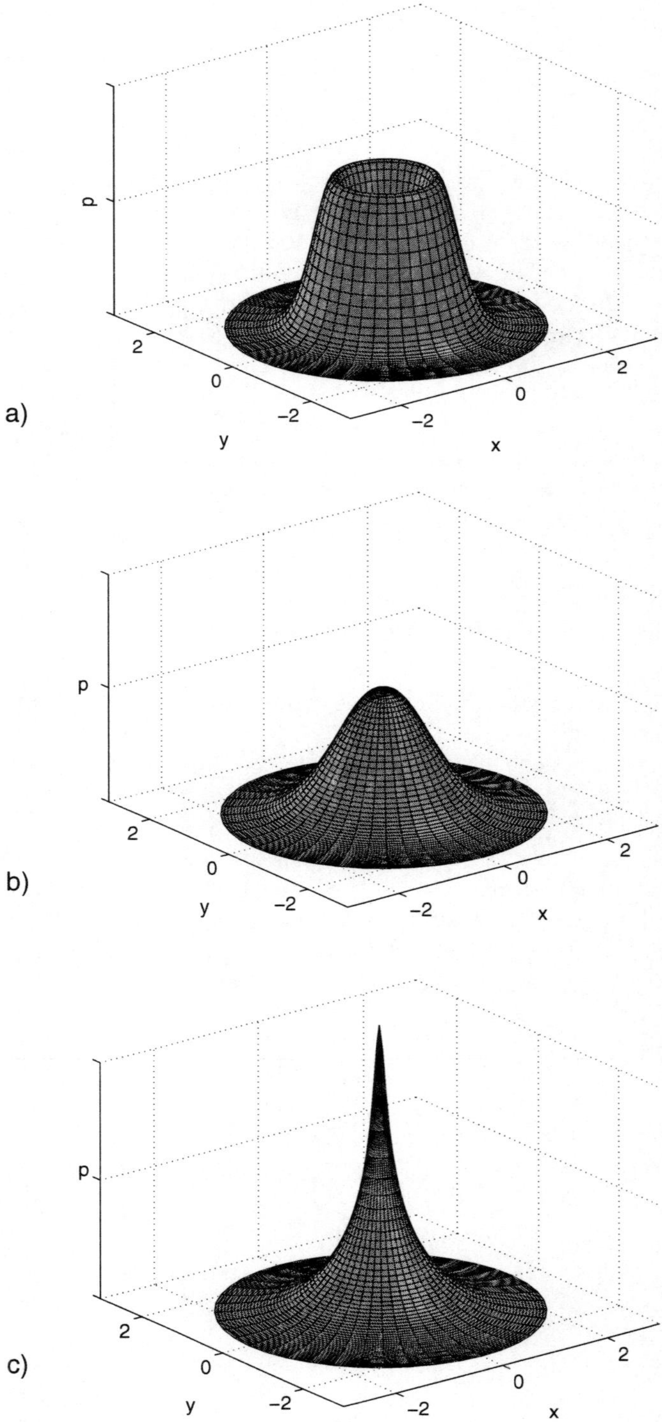

Fig. 4 The stationary probability density of Hopf system for $\mu = 1,\ \ \sigma_2 = 0.1$, **(a)** $\sigma_1 = 0.6$, **(b)** $\sigma_1 = 1$, **(c)** $\sigma_1 = 1.2$

3 Noise-Induced Transitions for Stochastic Lorenz Model

Consider Lorenz system forced by stochastic disturbances

$$\dot{x} = \sigma(-x+y)+\varepsilon\dot{w}_1$$
$$\dot{y} = rx-y-xz+\varepsilon\dot{w}_2$$
$$\dot{z} = -bz+xy+\varepsilon\dot{w}_3$$

with parameters $\sigma = 10, \quad b = 8/3, \quad r \in [190, 330]$.

For $\varepsilon = 0$, Lorenz system demonstrates a well-known scenario of period-doubling bifurcations and transition to chaos. In Fig. 5 one can see 1-cycle (see Fig. 5a) of unforced Lorenz system and a chaotic attractor (see Fig. 5e).

Under the random disturbances, Lorenz system demonstrates noise-induced transition from regular attractor to chaotic one. In Fig. 5, a qualitative transformation of stochastic attractors is shown (see Fig. 5b,c,d).

As we can see, chaotic attractor of unforced Lorenz system (see Fig. 5e) is similar to stochastic attractor (see Fig. 5d).

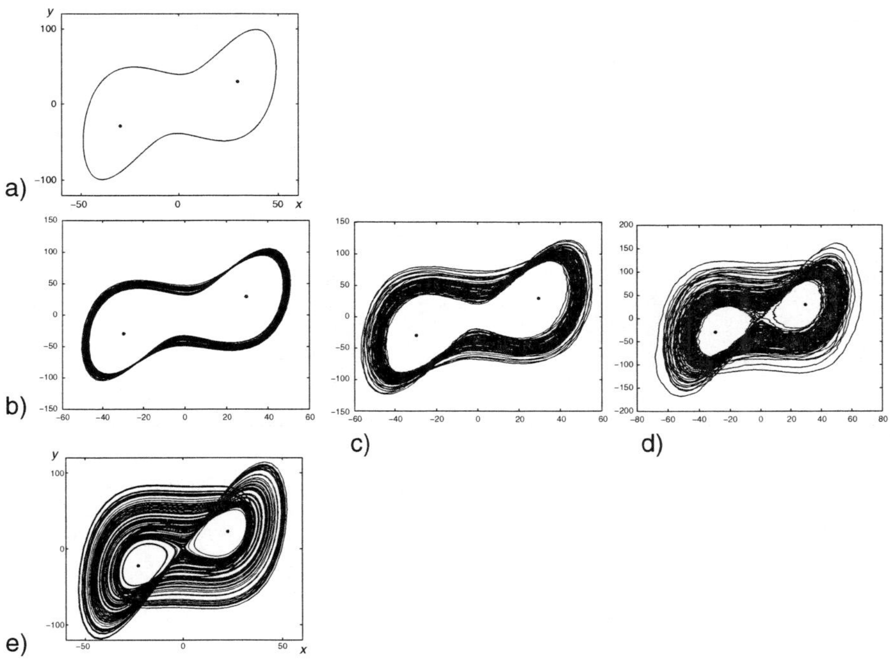

Fig. 5 Attractors of Lorenz system (**a**) $r = 330$, $\varepsilon = 0$, (**b**) $r = 330$, $\varepsilon = 1$, (**c**) $r = 330$, $\varepsilon = 5$, (**d**) $r = 330$, $\varepsilon = 10$, (**e**) $r = 190$, $\varepsilon = 0$

Acknowledgments This work was partially supported by RFBR grants 06-01-00625, 06-08-00396, 07-01-96079ural.

References

1. Bashkirtseva, I. A., Ryashko L. B.: Sensitivity analysis of the stochastically and periodically forced Brusselator. Physica A. **278**, 126–139 (2000)
2. Bashkirtseva, I. A., Ryashko L. B.: Sensitivity analysis of stochastically forced Lorenz model cycles under period-doubling bifurcations. Dynamic Systems and Applications. **11**, 293–309 (2002)
3. Bashkirtseva, I. A., Ryashko L. B.: Stochastic sensitivity of 3D-cycles. Mathematics and Computers in Simulation. **66**, 55–67 (2004)

Torus Bifurcation in Uni-Directional Coupled Gyroscopes

Huy Vu, Antonio Palacios, Visarath In, Adi Bulsara, Joseph Neff and Andy Kho

Abstract Nowadays, the global positioning system (GPS) is popularly used by the U.S. Navy in navigation systems to gain precise position, velocity, and time information. One of the biggest issues for using GPS is its susceptibility to jamming and other inferences. The received GPS signal is approximately 20 dB below the thermal noise level from a distance 11,000 miles away. Because of these weakness and vulnerability, many other alternative navigation methods are needed to improve performance and reduce the dependency on GPS. One of the main alternative methods is the Inertial Guidance System (IGS) that can operate wherever GPS signals are jammed or denied. A prototypical IGS is composed of three accelerometers to measure linear movement and three angular rate sensors (gyroscopes) to gauge the rotational movement. The main benefit of IGS is its low cost relatively to other methods. Current MEMS (Micro-Electro-Mechanical Systems) gyroscopes are compact and inexpensive, but their performance does not meet the requirements for an inertial grade guidance system. In this work, a difference approach was examined on the dynamics of coupled gyroscopes to improve performance through synchronization referred as vibratory coupled gyroscopes with drive amplitudes' coupling. One of the main discoveries from the coupled gyroscopes' mathematical model is a Torus bifurcation, which leads to synchronized behavior in and array of three gyroscopes uni-directionally coupled.

H. Vu (✉)
Nonlinear Dynamical System Group, Department of Computational Sciences,
San Diego State University, 5500 Campanile Drive, San Diego, CA 92182, USA,
e-mail: hvu@sciences.sdsu.edu

V. In et al. (eds.), *Applications of Nonlinear Dynamics*, Understanding Complex Systems,
DOI 10.1007/978-3-540-85632-0_45,

1 Introduction

Although circuit implements have experimentally demonstrated complex behavior in arrarys of coupled simple ring uni-directionally as well as bi-directionally gyroscopes, this research concentrated on theoretical and computational analysis of forward uni-directional coupled gyroscopes with the simplest case of a ring of three gyroscopes. Dynamical behaviors of these three simple uni-directional gyroscopes became plentifully complicated. Surprisingly, the results obtained were different from the initial expectation. Even though it was called "simplest case", each gyroscope constituted a system of four ordinary differential equations and a total of twelve ODE's for three coupled array gyroscopes. This high dimensional system required a substantial time and effort to investigate, especially to obtain high-order numerical accuracy, even though it was assumed there was no stochastic additive noise to the system in this work.

2 The Device

Vibratory gyroscopes are designed to determine the angular rate of a rotating object. As is shown in Fig. (1a), there are a couple of springs connected with a gyroscope. One spring is the drive axis (X-axis) and the other is the sense axis (Y-axis). This model creates two orthogonal vibration modes with restoring coefficients, k_x and k_y. The harmonic motion in the drive axis is caused by an external signal or a forcing oscillator, while the movement in the sense axis depends upon Coriolis force, F_{CY}, through the angular velocity within the Z-axis. Importantly, there is an amplitude coupling within the drive axis to induce phase synchronization [2] among gyroscopes, Fig. (1b), which is a structure of uni-directional forward coupling of three gyroscopes. The operating principle of these vibrating gyroscopes is based upon the transfer of energy from one vibrating mode to another by the Coriolis force. This research focuses upon the modified MEMS gyroscopes with coupling drive displacements in order to achieve the synchronization between the gyroscopes and to maximize the sum of all gyroscopes' sense displacements.

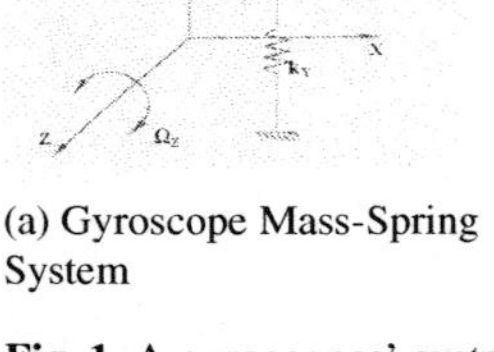

(a) Gyroscope Mass-Spring System

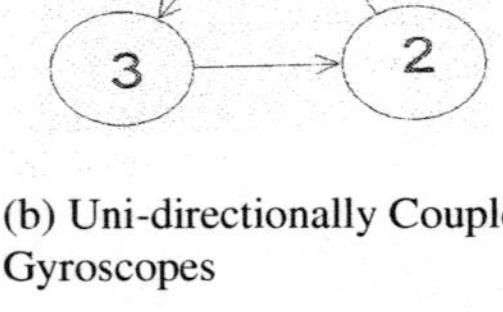

(b) Uni-directionally Coupled Gyroscopes

Fig. 1 A gyroscopes' system

3 Mathematical Model and Background

For a uni-directionally coupled n-gyroscopes system [1], the mathematical model that governs each gyroscope in drive- and sense-coordinate components is given as

$$m_i\ddot{x}_i + C_i\dot{x}_i + k_i x_i + \mu_i x_i^3 - 2m_i\Omega\dot{y}_i + \lambda(x_i - x_{i+1}) = F_{\text{drive}}(t) \tag{1}$$

$$m_i\ddot{y}_i + C_i\dot{y}_i + k_i y_i + \mu_i y_i^3 + 2m_i\Omega\dot{x}_i = N_i(t) \tag{2}$$

where x_i is the displacement of the drive component, y_i is the displacement of the sense component, m_i is the mass of the gyroscope, C_i is the damping coefficient of the gyroscope, k_i is the restoring coefficient of the gyroscope, μ_i is the restoring coefficient with cubic term of the gyroscope, Ω is the angular velocity of the gyroscope, λ is the coupling constant for driving displacement (X-axis) – uni-directional between x_i and x_{i+1}, F_{drive} is the driving force function – assumed to be of the form $[A_d \sin(\omega_d t)]$, A_d is the maximum amplitude of the driving force, ω_d is the frequency of the driving force, t is time, and N_i is the stochastic noise function, assuming to be zero. In this particular research, mass-spring-dampers of the system are identical. The numerical parameters investigated for the coupled gyroscopes' system are specified as λ is varied from -12 to 12, Ω is ranged from -10^4 to 10^4, $A_d = 0.001$, $\omega_d = 51650$, $m_i = 10^{-9}$, $k_i = 5.1472\times10^{-7}$, $C_i = 2.6494$, and $\mu_i = 2.933$.

4 Torus Bifurcation and Analysis of Frequency Response

Torus bifurcation in Fig. 2 is the only type of bifurcation found in the system. These torus bifurcations are symmetric in Ω – between $\Omega > 0$ and $\Omega < 0$ - while it is not necessarily the case for λ – between $\lambda > 0$ and $\lambda < 0$. Generally, the examined

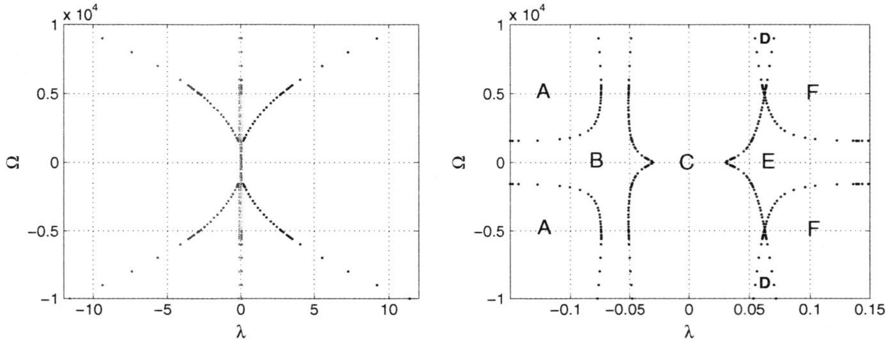

(a) $\lambda \in (-12, 12)$ (b) $\lambda \in (-0.15, 0.15)$

Fig. 2 Two-parameter bifurcation diagram – Ω vs. λ – as $\omega_d = 51650$

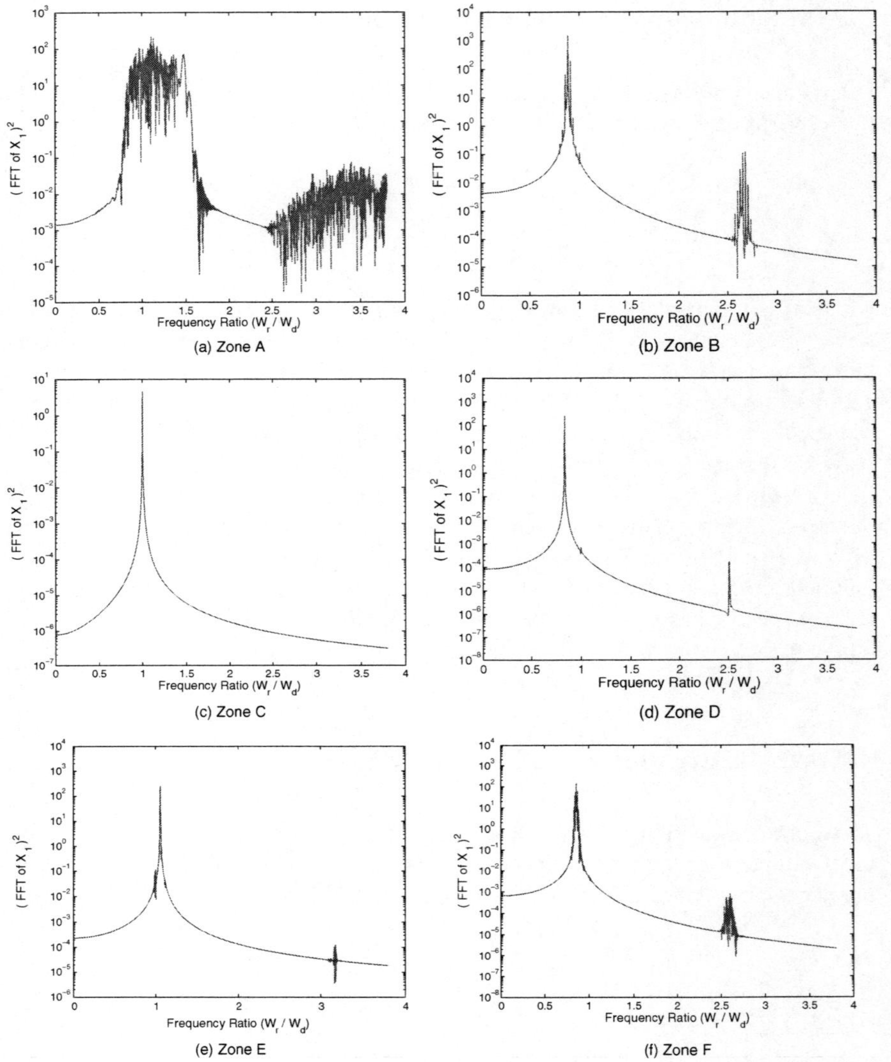

Fig. 3 Plots of $(FFT(X_1))^2$ vs. (ω_r/ω_d)

region was divided into 6 different zones – A, B, C, D, E, and F. Let $R_\omega = [\omega_r/\omega_d]$ be the frequency ratio, the FFT frequency plots and torus-shape plots for the drive-component in Figs. 3 and 4 respectively indicated that: (a) for zone C, one single peak occurs in the FFT plot at $R_\omega = 1$. Thus, the frequency response for gyroscopes in the synchronized region is equal to the driving frequency – (b) for zone B, prominent peaks spread in a small range at ($R_\omega = 1$) and ($R_\omega = 3$) and a ring torus shape is formed. This region is rather quasi-periodic – (c) for zone E, peaks occur at ($R_\omega = 1$)

are more defined than those at ($R_\omega = 3$) and its torus shape has a “loose” ring structure. It indicates this region is in between quasi-periodic and chaotic areas – (d) for zone A, a number of peaks occur in a broad range at ($R_\omega = 1$) and ($R_\omega = 3$) and its corresponding torus shape is well dispersed. The system enters the chaotic region – (e) for zone F, multiple peaks occur in a small range at ($R_\omega = 1$) and ($R_\omega = 3$) and its torus has multiple shapes . This implies the system enters the region between quasi-periodic and chaotic areas – (f) for zone D, a few peaks occur at ($R_\omega = 1$), which more defined than those at ($R_\omega = 3$) and its torus has a ring structure. Thus the region is quasi-periodic – (g) zone D and zone E are different in frequency response and torus shapes. This indicates the presence of the intersection between two branches of torus bifurcations distinguishing those two zones.

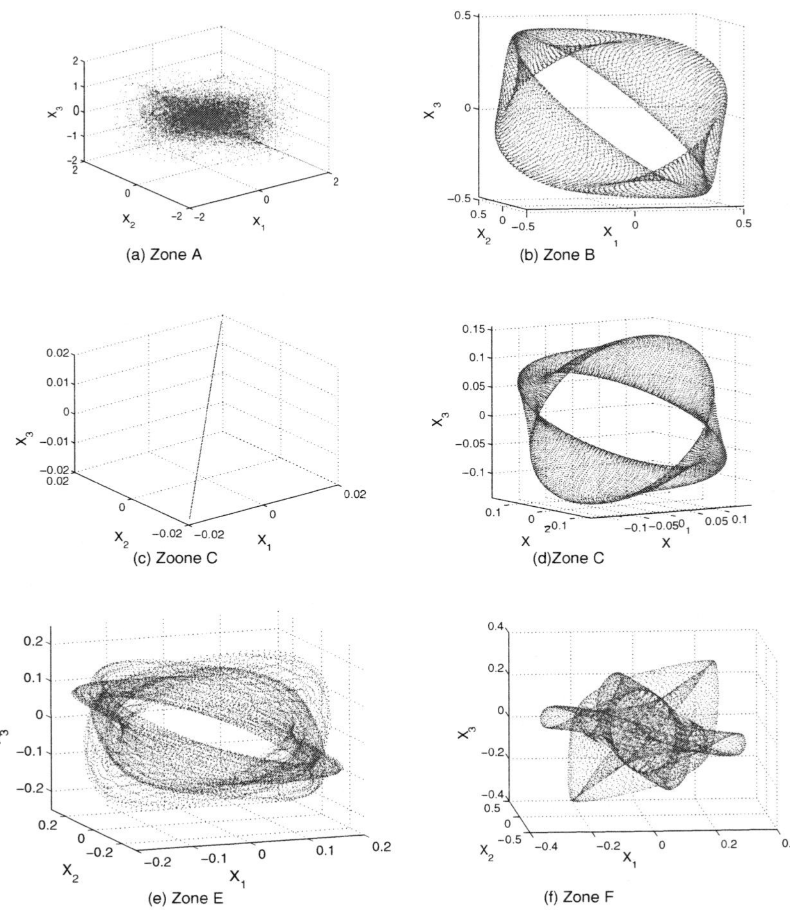

Fig. 4 Phase space plots of [X_1 vs. X_2 vs. X_3]

Acknowledgments We gratefully acknowledge support from the Office of Naval Research (Code 331) and the SPAWAR internal research funding (S&T) program. This work was supported in part by National Science Foundation grants CMS-0625427 and CMS-0638814.

References

1. Brian Meadows, Joseph Neff, Visarath In, and Adi Bulsara (2005), United States Patent No. US 6,880,400 B1
2. Arkady Pikovsky, Michael Rosenblum, and Jurgen Kurths (2001), Synchronization – A Universal Concept in Nonlinear Sciences, Cambridge University Press, Cambridge

Index